Aerobiology
The Toxicology of Airborne Pathogens and Toxins

Issues in Toxicology

Series Editors:
Professor Diana Anderson, *University of Bradford, UK*
Dr Michael D. Waters, *Michael Waters Consulting, N. Carolina, USA*
Dr Timothy C. Marrs, *Edentox Associates, Kent, UK*

Advisor to the Board:
Dr Alok Dhawan, *Ahmedabad University, India*

Titles in the Series:
1: Hair in Toxicology: An Important Bio-Monitor
2: Male-mediated Developmental Toxicity
3: Cytochrome P450: Role in the Metabolism and Toxicity of Drugs and other Xenobiotics
4: Bile Acids: Toxicology and Bioactivity
5: The Comet Assay in Toxicology
6: Silver in Healthcare
7: *In Silico* Toxicology: Principles and Applications
8: Environmental Cardiology
9: Biomarkers and Human Biomonitoring, Volume 1: Ongoing Programs and Exposures
10: Biomarkers and Human Biomonitoring, Volume 2: Selected Biomarkers of Current Interest
11: Hormone-Disruptive Chemical Contaminants in Food
12: Mammalian Toxicology of Insecticides
13: The Cellular Response to the Genotoxic Insult: The Question of Threshold for Genotoxic Carcinogens
14: Toxicological Effects of Veterinary Medicinal Products in Humans: Volume 1
15: Toxicological Effects of Veterinary Medicinal Products in Humans: Volume 2
16: Aging and Vulnerability to Environmental Chemicals: Age-related Disorders and their Origins in Environmental Exposures
17: Chemical Toxicity Prediction: Category Formation and Read-Across
18: The Carcinogenicity of Metals: Human risk through occupational and environmental exposure
19: Reducing, Refining and Replacing the Use of Animals in Toxicity Testing
20: Advances in Dermatological Sciences
21: Metabolic Profiling: Disease and Xenobiotics

22: Manganese in Health and Disease
23: Toxicology, Survival and Health Hazards of Combustion Products
24: Masked Mycotoxins in Food: Formation, Occurrence and Toxicological Relevance
25: Aerobiology: The Toxicology of Airborne Pathogens and Toxins

How to obtain future titles on publication:
A standing order plan is available for this series. A standing order will bring delivery of each new volume immediately on publication.

For further information please contact:
Book Sales Department, Royal Society of Chemistry, Thomas Graham House, Science Park, Milton Road, Cambridge CB4 0WF, UK
Telephone: +44 (0)1223 420066, Fax: +44 (0)1223 420247
Email: booksales@rsc.org
Visit our website at www.rsc.org/books

Aerobiology
The Toxicology of Airborne Pathogens and Toxins

Edited by

Harry Salem
US Army Edgewood Chemical Biological Center, Edgewood, MD, USA
Email: harry.salem@us.army.mil

Sidney Katz
Rutgers University, Cherry Hill, NJ, USA
Email: skatz@camden.rutgers.edu

ROYAL SOCIETY OF CHEMISTRY

THE QUEEN'S AWARDS FOR ENTERPRISE: INTERNATIONAL TRADE 2013

Issues in Toxicology No. 25

Print ISBN: 978-1-84973-594-0
PDF eISBN: 978-1-84973-791-3
EPUB eISBN: 978-1-78262-783-8
ISSN: 1757-7179

A catalogue record for this book is available from the British Library

Published by The Royal Society of Chemistry,
Thomas Graham House, Science Park, Milton Road,
Cambridge CB4 0WF, UK

Registered Charity Number 207890

For further information see our web site at www.rsc.org

Printed in the United Kingdom by CPI Group (UK) Ltd, Croydon, CR0 4YY, UK

Preface

With a surface area of 100 m^2 and a capacity of some 6 L, approximately 10% of which is exchanged during each of the15 breaths taken every minute, the lungs are the most likely portal for systemic intoxication by airborne pathogens and toxins. Pathogens are disease-producing microorganisms. Bacteria, mycoplasma, rickettsia, fungi and viruses are among the naturally occurring pathogens. Toxins are poisons produced through the metabolic activities of living organisms. They are organic chemical compounds such as proteins, polypeptides and alkaloids that come from a variety of biological sources.

In Chapter 1, Henderson and Salem discuss the relationships between the structure and historical development the atmosphere of and the presence of airborne microbial life. They address the questions of contributions from the atmosphere to the origin and evolution of microbial life and whether the atmosphere can be considered as the true habitat for airborne microorganisms. These are indeed relevant questions considering the diverse array of airborne microbes identified in collaborative research by the University of Colorado Denver and the North Carolina State University. The study reported more than 110 000 bacterial species and more than 55 000 fungal species in dust samples collected from 1200 homes representing locations in all 50 states. The average dust sample contained 4700 bacterial species and 1400 fungal species.[1] Some of these microbes could be carried on airborne dusts of terrestrial origin and/or in airborne mists of aquatic origin. At present, it is undecided if the atmosphere is a habitat for microbes or merely a conduit for their dispersal.

Airborne bacteria are responsible for diseases such as anthrax (*Bacillus anthracis*), diphtheria (*Corynebacterium*), legionellosis (*Legionella pneumophila*), meningitis (*Neisseria* species), pneumonia (*Mycoplasma pneumoniae*, *Streptococcus* species), tuberculosis (*Mycobacterium tuberculosis*) and

Issues in Toxicology No. 25
Aerobiology: The Toxicology of Airborne Pathogens and Toxins
Edited by Harry Salem and Sidney Katz

Published by the Royal Society of Chemistry, www.rsc.org

whooping cough (*Bordetella pertussis*). Chickenpox (varicella zoster virus) and smallpox (variola major virus), influenza (influenza virus), measles (morbillivirus) and German measles (rubellavirus) and mumps (rubulavirus) are among the diseases of viral origin communicable by airborne transport. Psittacosis (*Chlamydia psittaci*), aspergillosis (*Aspergillus fumigatus*, *A. flavus*, *A. niger*), histoplasmosis (*Histoplasma capsulatum*) and coccidioidomycosis (*Coccidioides immitis*) are examples of infections in humans initiated by the inhalation of fungal spores and their deposition in the alveoli. In addition to infections with microbial pathogens, some of their airborne metabolic by products are toxic. Some examples of such toxins are aflatoxin, a hepatotoxin of fungal origin (*Aspergillus flavus*, *A. parasiticus*), botulinum toxin, a neurotoxin of bacterial origin (*Clostridium botulinum*) and ricin, a cellular toxin extracted from the caster oil bean (*Ricinus communis*).

The recent outbreak of measles demonstrates the importance of immunization in providing effective protection against this viral disease.[2] Measles is only one of the two dozen or so vaccine-preventable diseases. During the 225 years since Jenner "vaccinated" 8-year-old James Phipps with exudate taken from a cowpox lesion on the hand of dairy maid Sarah Nelms,[3] routine vaccination against smallpox has virtually eliminated this disease from human infection. However, Vora *et al.*[4] reported the infection of three unvaccinated Georgian dairy men with an orthopoxvirus as recently as 2013. In Chapter 2, Ibrahim and Meyer describe animal models, pathogenesis, vaccine and drug studies for smallpox and the other orthopoxviruses. Smallpox is a human disease. Each of the animal models was able to mimic some features of the human disease and collectively the mousepox, rabbitpox and monkeypox models contributed significantly to understanding the pathogenesis of the disease and to developing new generations of vaccines.

Recent media reports[5–7] on the unintentional shipment of live anthrax spores from the Dugway Proving Ground to as many a 68 external institutions on 22 May 2015 are reminders of the continuing need for planning, developing and implementing emergency response strategies. Hamilton *et al.*[8] have presented an analysis of post-attack strategies for mitigating risks associated with reoccupying areas contaminated with *Bacillus anthracis*. In Chapter 3, Falk and Eisenkraft evaluate the inhalation hazard and the dose–response relationships for anthrax, which is another vaccine-preventable disease. They focus on their relevance to risk analysis and response planning and on their relevance to the mitigation of biological terrorism and biological warfare attacks. The review by Shah *et al.* in Chapter 7 considers the aerodynamics of anthrax particles, their mechanisms of infection at the molecular level and the manifestations of infection at the clinical level, in addition to diagnosis and treatment of the inhalational, cutaneous, gastrointestinal and injectional forms of the disease. Timely and relevant considerations of anthrax protection, detection and decontamination are included in this chapter.

Shannon Guess Richardson[9] and Nicholas Helman[10] attempted to use "ricin letters" as tools for assassinating a sitting president and a rival suitor, respectively. Both were captured, tried, found guilty and incarcerated. The Bulgarian defector Gregori Markov is thought to have been assassinated with a ricin-injecting umbrella during the cold war.[11] Pincus *et al.*[12] have focused on the potential use of aerosolized ricin as a bioweapon for use against civilian and military personnel and they have reported on the clinical aspects of inhalation exposure to ricin. Henderson *et al.* present a detailed description of ricin toxicity at the molecular level in Chapter 5.

The Working Group on Civilian Biodefense[13] developed consensus-based recommendations for measures to be taken by medical personnel and public health officers in the event that botulinum neurotoxin was used as a biological weapon against a human population. The 23 members of the Working Group, representing academic, government and private institutions, were experts in public health, emergency management and clinical medicine. They concluded that an aerosol- or food-borne botulinum neurotoxin weapon would cause acute systemic, descending flaccid paralysis with bulbar palsies such as diplopia, dysarthria, dysphonia and dysphagia that would typically present 12–72 h after exposure. Effective response to a deliberate release of botulinum toxin would depend on timely clinical diagnosis, case reporting and epidemiological investigation. Persons potentially exposed to botulinum neurotoxin should be closely observed and those with signs of botulism would require prompt treatment with antitoxin and supportive care that would include assisted ventilation for weeks of months. The treatment with antitoxin should not be delayed by microbiological evaluation. Park and Simpson[14] reported that rats vaccinated with the heavy chain component of the botulinum neurotoxin were completely protected when exposed to doses up to several thousand times the LD_{50}. In Chapter 4, Adler and Franz review the consequences of exposure to botulinum neurotoxin by both the ingestion and inhalation routes. They cite the effectiveness of treatment with equine antitoxins in non-human primates exposed to botulinum neurotoxin by the inhalation route. Like Park and Simpson,[14] Adler and Franz stress the need for initiating treatment promptly.

The threat of bioterrorism and pandemics has highlighted the urgency for rapid and reliable bioaerosol detection.[15,16] Early detection of airborne pathogens and toxins is essential for reducing contagion and initiating protective measures. In Chapter 6, Santarpia identifies the compositions of ambient biological aerosols, discusses the roles of bacteria, viruses and fungi and reviews methods for their measurement. In Chapter 8, Trebše *et al.* review the chemical, physiological, biochemical and immunochemical principles serving as the basis of methods for the detection of airborne pathogens and toxins and the applications of these principles to their detection and measurement.

Several aspects of particle physics were considered by Polymenakou[17] in his review on bioaerosols. In Chapter 9, Corriveau discusses these aspects as they relate to the weaponization of biological agents. In Chapter 10, Kesavan *et al.* discuss the impacts of particle size, shape, density, surface area, mass and concentration on predicting the movements of bioaerosols.

In Chapter 11, Bona and Katz include these parameters in their discussion of the models for air filtration, and also identify some of the airborne pathogens and toxins, compare air filtration devices and describe some applications of respiratory protection devices.

Aerial dispersion of pathogens is a potential route for the spread of infection. King *et al.*[18] have combined computational fluid dynamics simulations of bioaerosol deposition with a probabilistic healthcare workers surface contact model to estimate pathogen accrual. In Chapter 12, McClellan *et al.* describe the Deposition And Response in the Respiratory Tract (DARRT) model, which accounts for variations in human response caused by differences in particle size, in particular, for coarse particles that may be present near an aerosol dissemination source and that may remain suspended long enough in an urban environment to expose large numbers of people. In Chapter 13, Reed *et al.* describe and discuss several reproducible exposure methods for producing standardized infections by inhaled viral and bacterial agents. In Chapter 14, Ingersoll and Williams present computer code for agent-based disease models in the mathematical programming language R allowing characterizations of the dynamics of infectious diseases in host communities.

The contents of these chapters are intended to provide information for the protection of human health from the harmful effects of airborne pathogens and toxins. Continued research and development will undoubtedly change some of these perspectives. This is especially true in the areas of detection and protection. Consequently, the work presented here is a starting point for further enhancement of the quality of life in all parts of Planet Earth.

Sidney A. Katz
Cherry Hill, NJ
Harry Salem
Edgewood, MD

References

1. ResearchersProduceFirstAtlasofAirborneMicrobesAcrosstheUnitedStates, http://colorado.edu/news/releases/2015/04/20/researchers-produce-first-atlas-of-airborne-microbes-across-the-united-states#st-hash.4UCO8Ud5.dpuf.
2. N. S. Clemmons, P. A. Gastanaduy, A. P. Fiebelkorn, S. B. Redd and G. S. Wallace, Measles – United States, January 4 – April 5, 2015, Morbidity and Mortality Weekly Report (MMWR), Center for Disease Control and Prevention, www.cdc.gov/mmwr/preview/mmwrhtm/mm6414a.htm.
3. S. Riedel, Edward Jenner and the History of Smallpox and Vaccination, *Baylor University Medical Center Proceedings*, 2005, **18**, 21–25.
4. M. L. Vora, Y. Li, M. Geleishvil, G. L. Emerson, E. Khmaladze, G. Maghlakelidze, A. Navdarashvili, K. Zakhasvili, M. Endeladze, G. Mokverashvili, P. S. Satheshkumar, N. Gallardo-Romero, C. S. Goldsmith, M. G.

Metcalfe, I. Damon, E. F. Maes, M. G. Reynolds, J. Morgan and D. S. Carroll, Human Infection with a Zoonotic Orthopoxvirus in the Country of Georgia, *N. Engl. J. Med.*, 2015, **372**(**13**), 1223–1230.

5. CBS News, 27 May 2015, http://cbsnews.com, accessed 10 June 2015.
6. ABC News, 03 June, http://abcnews.com, accessed 10 June 2015.
7. Military Times, 09 June 2015, http://militarytimes.com, accessed 10 June 2015.
8. M. A. Hamilton, T. Hong, E. Casman and P. L. Gurian, Risk-Based Decision Making for Reoccupation of Contaminated Areas Following a Wide-Area Anthrax Release, *Risk Anal.*, 2015, **35**(7), 1348–1363.
9. Daily Mail, 15 June 2015, http://dailymail.co.uk, acessed 20 June 2015.
10. CNN News, 16 July 2014, http://cnnnews.com, accessed 20 June 2015.
11. The Telegraph, 19 June 2008, http://telegraph.co.uk, accessed 20 June 2015.
12. S. H. Pincus, M. Bhaskaran, R. N. Brey, P. J. Didier, L. A. Doyle-Meyers, C. J. Roy, Clinical and Pathological Findings Associated with Aerosol Exposure of Macaques to Ricin Toxin, *Toxins*, 2015, 7(6), 2121–2133.
13. S. S. Arnon, R. Schechter, T. V. Inglesby, D. A. Henderson, J. G. Bartlett, M. S. Ascher, E. Eitzen, A. D. Fine, J. Hauer, M. Layton, S. Lillibridge, M. T. Osterholm, T. O'Toole, G. Parker, T. M. Perl, P. K. Russell, D. L. Swerdlow and K. Tonat, Botulinum Toxin as a Weapon: Medical and Public Health Management, *J. Am. Med. Assoc.*, 2001, **285**(**8**), 1059–1070, DOI: 10.1001/jama.285.8.1059.
14. J.-B. Park and L. L. Simpson, Inhalation Poisoning by Botulinum Toxin and Inhalation Vaccination with Its Heavy-Chain Component, *Infect. Immun.*, 2003, **71**(**3**), 1147–1154, DOI: 10.1128/1AI.71,3, 1147-1154.2003.
15. A. Sharma, E. Clark, J. D. McGlothlin and S. K. Mittal, Efficiency of Airborne Sample Analysis Platform (ASAP) Bioaerosol Sampler for Pathogen Detection, *Front. Microbiol.*, 2015, **6**, 1–7.
16. C. F. Fronczek and J. Y. Yoon, Biosensors for Monitoring Airborne Pathogens, *J. Lab. Autom.*, 2015, **20**(4), 309–410.
17. P. N. Polymenakou, Atmosphere: A Source of Pathogenic or beneficial Microbes?, *Atmosphere*, 2012, **8**, 87–102, DOI: 10.3390/atmos3010087, accessed 20 June 2015.
18. M. F. King, C. J. Noakes and P. A. Sleigh, Monitoring Environmental Contamination in Hospital Single- and Four-Bed Rooms, *Indoor Air*, 2015, **25**(6), 694–707.

Acknowledgements

The editors gratefully acknowledge the authors of the chapters for their contributions to this book. Clearly, there would be no book without their individual and collective efforts in researching and preparing the material for their individual manuscripts. It is equally clear to the editors there would be no book without the active involvement of Donna Hoffman in skillfully maintaining communications between the editors and the authors and the editors and the publisher, and meticulously organizing the voluminous correspondence associated with queries about chapter format, copyright releases, manuscript submission deadlines and extensions, and other details too numerous to mention. The editors also gratefully acknowledge Rowan Frame and the staff at the Royal Society of Chemistry for their patience and forbearing in dealing with the delays in delivering the manuscripts, their guidance and assistance in obtaining copyright releases and their willingness to assist individual authors in matters of manuscript preparation. In addition to acknowledging those directly involved in the preparation of *Aerobiology: The Toxicology of Airborne Pathogens and Toxins*, the editors gratefully acknowledge the continued support and encouragement of their respective families for tolerating the time spent away from the domestic scene and for the occasional reversal of family priorities.

Issues in Toxicology No. 25
Aerobiology: The Toxicology of Airborne Pathogens and Toxins
Edited by Harry Salem and Sidney Katz

Published by the Royal Society of Chemistry, www.rsc.org

Contents

Issues in Toxicology No. 25
Aerobiology: The Toxicology of Airborne Pathogens and Toxins
Edited by Harry Salem and Sidney Katz

Published by the Royal Society of Chemistry, www.rsc.org

CHAPTER 1

The Atmosphere: Its Developmental History and Contributions to Microbial Evolution and Habitat†

TERRY J. HENDERSON*[a] AND HARRY SALEM[a]

[a]Research and Technology Directorate, US Army Edgewood Chemical Biological Center, Aberdeen Proving Ground, MD 21010–5424, USA
*E-mail: terry.j.henderson.civ@mail.mil, harry.salem.civ@mail.mil

1.1 Introduction

Research efforts focused on the Earth's biota have been limited almost exclusively to land, soil and aquatic habitats; however, this focus is now beginning to change. There is a renewed and building interest in the ecology of living microorganisms found in the atmosphere. Hundreds of thousands of individual microorganisms can exist in a cubic meter of air,[1] which can represent hundreds of different taxa.[2,3] The atmosphere is one of our Planet's most intriguing habitats to investigate because extreme cold temperatures,

†The opinions expressed in this chapter are the private views of the authors and are not to be construed as an official Department of the Army position unless so designated by other authorizing documents. This chapter has been approved for public release.

Issues in Toxicology No. 25
Aerobiology: The Toxicology of Airborne Pathogens and Toxins
Edited by Harry Salem and Sidney Katz

Published by the Royal Society of Chemistry, www.rsc.org

hypobaria (low pressures), desiccation and ultraviolet (UV) irradiation make it more indicative of the surface conditions on Mars rather than anywhere else on Earth.[4] In a recent report,[5] the atmosphere was described as "one of the last frontiers of biological exploration on Earth."

The Planet's atmospheric biota, however, remains one of the most challenging to investigate, which has been reflected by the many shortcomings encountered throughout the history of aerobiology. In the past, aerobiological studies relied almost exclusively on culture-based analyses that neglected the vast majority of microbes present in air samples.[1,6,7] This is because only 0.1–10% of the total airborne microbial flora are able to grow in culture.[8–10] Airborne microorganisms can also become damaged or killed by desiccation, UV irradiation or the extreme low temperatures that occur in the atmosphere. Acquiring culture-independent microbiological data, on the other hand, can be difficult because the density of airborne microorganisms decreases with increasing altitude.[1,11] Large, or sometimes enormous, volumes of air must be collected and processed for detecting microorganisms with molecular detection assays and air-sampling systems must also be designed to prevent cell trauma and damage and sample contamination. To complicate matters, a modern, standard method for reading and scoring microorganisms in aerobiological samples does not exist, making the interpretation and intercomparison of results difficult.[12] Finally, studies of atmospheric residence times (the lifetime of particles aloft in the atmosphere) for microorganisms and their dispersal patterns rely on computer simulations based on theoretical mathematical models that are difficult to correlate with experimental data.[4,6,13] Such experimental design and engineering challenges explain why the upper atmosphere is one of the least explored biological environments on Earth, rivaling deep oceanic and subsurface environments.

Given the renewed interest in aerobiology and the inherent difficulties in aerobiological research, it should be of no surprise that the field is characterized by a remarkable lack of knowledge and a great deal of speculation. Some of the most important and fundamental questions regarding microorganisms in the atmosphere cannot be answered definitively and are active topics of debate. Undoubtedly, one of the more fundamental questions concerns the contributions of the atmosphere to the origin and evolution of microbial life. A complete account of this topic should include the evolutionary history of airborne microorganisms, the details of which have never been addressed to date. Another question central to aerobiology is whether the atmosphere can be considered a true habitat for airborne microorganisms. Several recent reports allude to the atmosphere as a habitat or ecosystem,[14] especially in the case of microorganisms,[6,15] but others argue that the debate is far from settled[4,16] and subscribe to the more traditional point of view that the atmosphere is merely a conduit for dispersing microbes to distant locations. Other important problems in aerobiology have somewhat vague and tenuous answers simply because they are difficult to answer by experimentation. For microorganisms in particular, the most fundamental of these concern deriving realistic values for atmospheric residence times[17] and understanding cosmopolitan dispersal by atmospheric transport (see, for example, Fröhlich-Nowoisky *et al.*,[18] Wilkinson *et al.*[13] or Smith *et al.*[19]). Atmospheric air is the primary medium for dispersing microorganisms across the globe and connecting all microbial habitats on the Earth's surface, but it is not known whether the atmosphere contains biogeographic regions similar to those found on the Planet's surface.[6,18] We discuss the most recent research focused on establishing more concrete answers to these questions and, when appropriate, present our own speculation. Because the developmental history of the atmosphere is intimately related to microbial evolution and because some understanding of atmospheric structure is necessary to discuss aerobiology in detail, we begin by outlining this history and the Earth's present-day atmosphere.

1.2 The Origin and Evolution of the Earth's Atmosphere

The processes by which the current atmosphere arose from earlier conditions are exceedingly complex; however, evidence related to these processes, although indirect, is abundant. Ancient sediments and rocks record the past changes in the atmospheric composition from chemical reactions within the crust, and also the evolutionary history of living systems. Several processes, including plate tectonics, weathering and photosynthesis, were internal to the Planet. However, extra-planetary processes such as the slowly and ever-increasing luminosity of the Sun over billions of years, gradual changes in the Earth's orbit over many tens of thousands of years and the rare but catastrophic impacts of giant meteorites and comets, have also played important roles. Collectively, these factors have forged three distinct atmospheres for the Earth: an initial, tremendously hot atmosphere composed principally of H_2 (hydrogen) and He (helium) gases, a second, rich in gaseous N_2 (nitrogen) and CO_2 (carbon dioxide), and today's atmosphere, rich in gaseous N_2 and O_2 (oxygen).

1.2.1 The Primary Atmosphere of the Earth

The oldest materials ever found in the Solar System occur in meteorites ~4.57 Ga (gigaannum or billion years ago) of age,[20] marking the starting point for the condensation of the first solids in our Solar System. At this time, Earth was tremendously hot and inhospitable due to collisions and compressions of matter during accretion (the growth of a massive object by gravitationally attracting more and more matter, typically gaseous matter), heat released from the formation of an early planetary core and the ubiquitous, constant decay of radioactive elements. The Planet's earliest surface was molten or a thin and unstable basaltic crust with constant volcanism. This was the birth of planet Earth and, as shown in Table 1.1, marked the start of geological time. Earth's atmosphere likely consisted of gases captured from the solar nebula (the gaseous cloud from which the Sun and planets are believed to have formed by condensation)[21–23] with H_2, by far the most abundant element in the Universe, as its principal component. Other atmospheric gases

Table 1.1 Geological time scale.

Eon	Era	Period	Millions of years before the present
Precambrian			4500–570
	Archean		4500–2500
		Hadean	4500–3900
		Early Archean	3900–2900
		Late Archean	2900–2500
	Proterozoic		2500–570
		Early Proterozoic	2500–1600
		Middle Proterozoic	1600–900
		Late Proterozoic	900–540
Phanerozoic			570–present
	Paleozoic		540–225
		Cambrian	540–500
		Ordovician	500–430
		Silurian	430–395
		Devonian	395–45
		Carboniferous	345–280
		Permian	280–225
	Mesozoic		225–65
		Triassic	225–190
		Jurassic	190–136
		Cretaceous	136–65
	Cenozoic		65–present
		Tertiary	65–1
		Quaternary	1–present

most likely would have included He and simple hydrides such as those now found on Jupiter, Saturn, Uranus and Neptune, with CH_4 (methane), NH_3 (ammonia) and water vapor being the most notable. Details of the Earth's primary atmosphere are very difficult to determine simply because there is very little evidence of the primary atmosphere left to investigate.[24]

As the solar nebula began to dissipate, so too did the Earth's primary atmosphere. Atmospheric gases escaped because the early Planet's gravity was not strong enough to hold lighter gases. Gases were also driven off by the solar wind, a stream of plasma released from the upper atmosphere of the Sun containing high-energy electrons and protons. This was a consequence of the young, premature Planet not yet having a differentiated core (a solid inner and liquid outer core) to create a planetary magnetic field capable of deflecting the solar wind.[24] Other events soon followed that profoundly changed the Planet and its atmosphere.

1.2.2 The Secondary Atmosphere of the Earth

The loss of gases from the primary atmosphere was accompanied by a loss of the Planet's primordial heat into space, the condensation of water as rain and its accumulation on the surface of the cooling Planet to form lakes, seas and oceans. The interaction of water, heat and rock set the stage for the origin of life. Once the Earth's core differentiated, heavier gases were finally retained in the atmosphere, ultimately giving rise to the secondary atmosphere.

Earth's secondary atmosphere first appeared at ~4.5 Ga, soon after the Earth and Moon completed their formational phase, and was produced by out-gassing from volcanism together with gases produced during the late heavy bombardment of the Earth by huge asteroids.[24] The gases constituting this atmosphere were probably similar to those created by modern volcanoes, likely including H_2, water, CO (carbon monoxide), CO_2, N_2, S_2 (sulfur), SO_2 (sulfur dioxide) and Cl_2 (chlorine). The secondary atmosphere was probably several times denser than the present atmosphere and almost certainly was dominated by CO_2, a major greenhouse gas. A tremendous greenhouse effect must have accompanied this atmosphere, especially since the Sun was ~30% dimmer at this time and supplied less solar radiation to warm the Planet.[25,26] Such a warming effect would have been necessary to maintain water in a liquid state and ensure that Planet would not become a frozen wasteland without any hope for the earliest forms of life to take hold.[27] With the exception of one cold glacial period phase at ~2.4 Ga, the geological record reveals a warm surface during the Archean Era (the geological era 4.5–2.5 Ga and part of the Precambrian Supereon; see Table 1.1) suitable for sustaining life. Free O_2 began to appear in the atmosphere late in the Archean, apparently produced by photosynthesizing cyanobacteria (referred to as the Great Oxygenation Event).

1.2.3 Evolution of an O_2-Rich Atmosphere

The evolution of an O_2-rich atmosphere was intimately coupled to the evolutionary history of life. As the Planet's biosphere and atmosphere co-evolved over the following billions of years, free O_2 created from photosynthesis began to dominate the chemistry of the atmosphere. Some O_2 was transformed into O_3 (ozone) in a process using UV radiation from the Sun, and a slow but progressive accumulation of O_3 began in the upper atmosphere. During this same period, the atmospheric concentration of free CO_2 progressively decreased and stabilized at its present level. The decreasing CO_2 levels appear to have been a response of the environment to the Sun's steady increase in luminosity in order gradually to attenuate greenhouse warming at the same time. The carbon cycle, which contains both biological and geological processes, began early in the evolutionary history of life as living systems took over the production of organic matter and O_2 began to regulate the balance of carbon between the atmosphere and the oceans. The cycle helped to regulate the Earth's surface temperature by balancing the CO_2 output from volcanoes and weathering and the burial of organic matter in sediments. Clearly, the evolution of life was central for creating the unique conditions for habitability on Earth; life regulates the global environment.[28] As shown in Table 1.2, comparisons of the Earth's atmosphere today with those of its nearest neighbors, Mars and Venus, illustrate this concept. Our nearest

Table 1.2 Atmospheric pressure and composition of the earth in comparison with its nearest planetary neighbors. Adapted from Mojzsis.[27]

Planet	Pressure/ bar	CO_2/% v/v	N_2/% v/v	^{36}Ar/% v/v	H_2O/% v/v	O_2/% v/v
Venus	92	96.5	3.5	0.00007	<0.00003	Trace
Earth	1.013	0.033	78	0.01	<3	21
Mars	0.006	95.3	2.7	0.016	<0.0001	Trace

neighbor planets have negligible amounts of O_2 in their atmospheres, both of which are dominated by CO_2. This would have been the fate of the Earth's atmosphere also if were not for life appearing and changing the atmosphere to keep the Planet habitable. Lastly, the constant rearrangement of continents by plate tectonics influenced the long-term evolution of the atmosphere by transferring CO_2 to and from continental carbonate stores. The theoretical decrease in atmospheric CO_2 levels over geological time, based on the model of Kasting,[29] is illustrated in Figure 1.1.

Evidence from paleosols (ancient soils) and also a variety of other geologic evidence indicates that beginning ~2.5 Ga ago, sedimentary rocks began to become increasingly affected by the rising concentrations of free atmospheric O_2 and appeared red in color as if "rusted." Resulting from the oxidation of iron in the rock, the "rusted" appearance is possible only in the presence of O_2. Atmospheric O_2 concentrations are believed to have increased rapidly at about this time as a consequence of more O_2 being produced through photosynthesis than was consumed by chemical reactions within the atmosphere, oceans and rocks. A threshold O_2 concentration was finally reached that transformed the earlier reducing atmosphere to an oxidizing atmosphere. At about the same time, more complex microbial life evolved in response to the higher O_2 concentrations, including eukaryotes, and a powerful O_3 screen began to form in the upper atmosphere when the atmospheric O_2 level reached ~1% of today's level. The theoretical increase in atmospheric O_2 levels over geological time according to the model of Kasting[29] is shown in Figure 1.2.

1.2.4 Present-Day Composition and Structure of the Atmosphere

The chemical composition of the present-day atmosphere is shown in Table 1.2. Water vapor concentrations vary significantly from ~10 parts per million by volume (ppmv) in the coldest regions of the atmosphere to as much as 50 000 ppmv (5% by volume) in hot, humid air masses (the value in the table is an average value). Many substances of natural origin may be present as aerosols, including dusts of different mineral and organic compositions, pollens and spores, sea spray and volcanic ash. Various industrial pollutants may also be present as gases or aerosols, including chlorine (in both elemental Cl_2 and molecular forms), fluorine compounds and elemental mercury

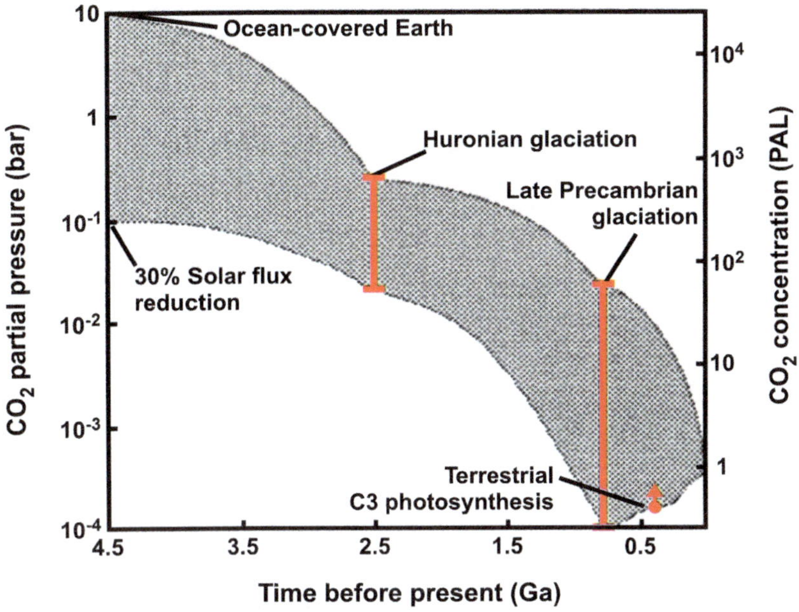

Figure 1.1 Estimated CO_2 concentrations in Earth's atmosphere over geological time in response to the steady increase in solar luminosity. The shaded areas represent the range in CO_2 concentrations required to keep the Earth's surface warm enough for liquid water to exist in the presence of the lower solar output. The figure starts ~4.5 Ga with an ocean-covered Earth and a solar flux ~30% lower than that today and extends to present-day conditions. Major glaciation periods are indicated, as is the rise in terrestrial plants and C_3 photosynthesis (the most common type of photosynthesis where CO_2 is first incorporated into 3-phosphoglycerate, a three-carbon or C_3 compound). PAL is the present atmospheric level. Adapted from Kasting.[29]

vapor. Sulfur compounds such as H_2S (hydrogen sulfide) and SO_2 may be derived from natural sources or industrial air pollution.

In general, both air pressure and density decrease with increasing altitude in the atmosphere. Temperature, however, has a more complex relationship with altitude and may remain constant or even increase with altitude in some regions. However, because the relationship of temperature with altitude is generally constant and recognizable through measurements such as balloon soundings, temperature behavior provides a useful means to distinguish between atmospheric layers. Based exclusively on temperature, the atmosphere can be divided into five principal layers or atmospheric stratifications: the troposphere, stratosphere, mesosphere, thermosphere and exosphere. The order of these stratifications is illustrated schematically in Figure 1.3 along with the atmospheric temperature profile. Of these, only the troposphere, stratosphere and mesosphere are considered part of the biosphere[30] and are relevant to aerobiology.

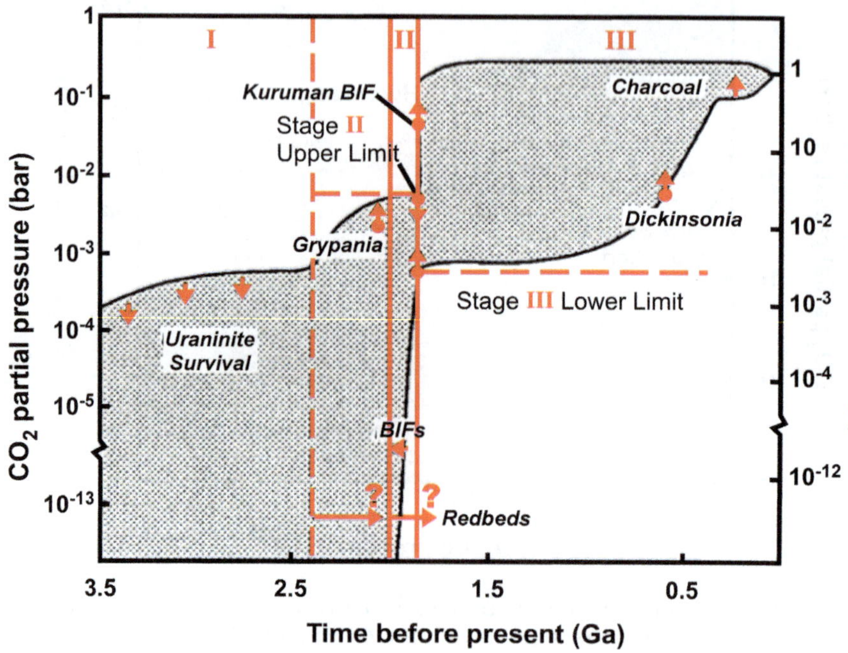

Figure 1.2 Estimated free O_2 concentrations in Earth's atmosphere over geological time. The shaded areas represent the range of possible O_2 concentrations from model calculations and investigations of paleosols (ancient soils), fossil organisms and marine sediments that only form in the absence of O_2. The sediments are preserved as banded iron formations (BIFs) and only appear in the geological record up to ~1.8 Ga. The point labeled Kruman BIF represents a lower limit for the partial pressure of O_2 (pO_2) based on a paleoweathering rate analysis (Hollard and Beukes, 1990). Solid vertical lines represent the most probable dates for transitions between three unique stages in the development of the atmosphere and ocean as O_2 reservoirs:[137] (I) the atmosphere and entire ocean were anoxic, with localized O_2 oases in highly productive regions of the surface ocean, (II) the atmosphere and surface ocean were O_2-rich and the deep ocean remained anoxic and (III) both the atmosphere and ocean were O_2-rich. The dashed vertical line indicates an earlier date for the beginning of stage II consistent with redbed (sedimentary rocks appearing red from the presence of ferric oxides) data[138] and the appearance of *Grypania* (a corkscrew-shaped organism from the Proterozoic era likely to have been a photosynthetic alga). The dashed horizontal lines represent the theoretical pO_2 limits derived from the three-stage model. Additional constraints can be imposed on the upper pO_2 limit during stage I when considering the survival of uraninite and other detrital grains[139] that rapidly weather in the present-day atmosphere. The lower pO_2 limits, labeled Dickinsonia and Charcoal, are from Runnegar[140] and Jones and Chaloner,[141] respectively. PAL is the present atmospheric level. Adapted from Kasting.[29]

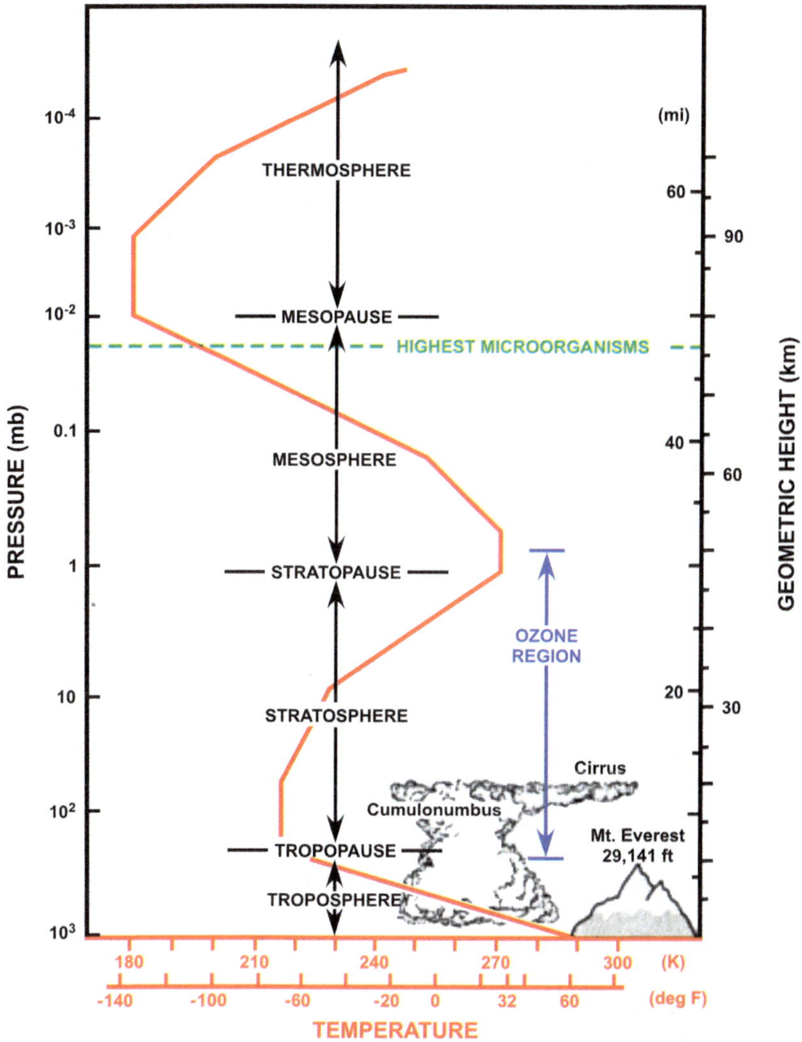

Figure 1.3 Schematic structure and temperature profile of the Earth's atmosphere. The geometric height above sea level is shown in kilometers (km) and miles (mi). Temperature is shown in kelvin (K) and degrees Fahrenheit (deg F), while pressure is shown in millibars (mb). Illustrations of cumulonimbus and cirrus clouds, and also Mount Everest, are included as vertical scale references. The horizontal dashed line at 77 km represents the highest altitude at which living microorganisms have been found.[30]

The troposphere extends from the Earth's surface to an average height of ~12 km (~40 000 ft) ASL (above sea level), although it actually varies from ~8 km (~26 000 ft) ASL at the Poles to ~17 km (~56 000 ft) ASL at the Equator, with some local variations due to weather. This atmospheric stratification contains the Planet's boundary layer (0–2 km or ~0–6600 ft ASL), where air is heavily mixed with the surface by wind and weather systems to create vertical mixing. Temperatures typically decrease with increasing altitude in the troposphere (see Figure 1.3) because energy transfer from the surface provides most of the heating in this stratification, creating turbulence. The troposphere contains ~80% of the entire mass of the Earth's atmosphere and is denser than all overlying atmospheric stratifications. Nearly all atmospheric water vapor is found in the troposphere and, therefore, is where most all of the Planet's weather occurs. It has virtually all weather-associated cloud genus types generated by active wind circulation, although very tall cumulonimbus thunderclouds can penetrate the tropopause, the troposphere–stratosphere boundary, from below and rise into the stratosphere. The troposphere is the most important stratification of the atmosphere in the study of aerobiology by an exceptionally considerable margin. Most conventional aviation occurs in the troposphere and it is the only atmospheric stratification that propeller-driven aircraft can access.

The stratosphere extends from the tropopause to an altitude of 50–55 km (~160 000–180 000 ft) ASL and contains the O_3 layer. Temperatures rise with increasing altitude in the stratosphere (see Figure 1.3) because the absorption of UV radiation by O_3 releases heat. Although the temperature may be −60 °C (−76 °F) at the tropopause, the top of the stratosphere is much warmer, reaching up to ~0 °C (32 °F) with a pressure ~0.1% that at sea level. The stratospheric temperature profile creates very stable atmospheric conditions, so the atmospheric layer lacks the weather-producing air turbulence that is so prevalent in the troposphere. Consequently, the stratosphere is almost completely free of clouds and weather. Polar stratospheric or nacreous clouds, however, are occasionally seen in the lower extremities of this stratification where the air is coldest. This is the highest atmospheric stratification that can be accessed by jet-powered aircraft.

Finally, the mesosphere extends from the stratopause, the stratosphere–mesosphere boundary, to ~80–85 km (~260 000–280 000 ft) ASL. Within the mesosphere, temperatures decrease with increasing altitude right up to the mesopause, the boundary between the mesosphere and thermosphere (see Figure 1.3). The mesopause is the coldest place on Earth, with an average temperature ~−85 °C (−120 °F),[31,32] and is considered to be the upper boundary of the biosphere. The mesosphere is usually accessed by rocket-powered aircraft and unmanned sounding rockets.

1.2.5 The Atmosphere As Habitat

As with any habitat, the atmosphere must provide conditions amenable to the living systems that it hosts. Many of the physical and chemical features of the atmosphere that most directly impact living organisms vary tremendously with latitude, longitude or altitude and both seasonal and weather effects can also dramatically change local conditions. The atmosphere, however, can be hospitable to microbial life, especially within the troposphere; this is illustrated in the case of temperature. As revealed in the atmospheric temperature profile in Figure 1.3 (and discussed above), temperatures decrease with increasing altitude in the troposphere, ranging from an average of 15 °C (59 °F) at sea level to ≥−60 °C at the tropopause, where they increase steadily with increasing altitude right up to the stratopause at ~0 °C. Many microorganisms grow readily at these freezing temperatures,[33] with some found to be metabolically active down to 18 °C (~0 °F).[5] Both pressure and air density decrease steadily with increasing altitude, with pressure ranging between ~1 atm (0.987 bar) at sea level to >0.1 atm (0.00987 bar) in the lower mesosphere.

UV radiation (wavelengths of 100–400 nm) is one of the only physical characteristics of the atmosphere that increases directly with increasing altitude.[34] With every 1000 m increase in altitude, UV levels generally increase by 10–20%, depending upon latitude, weather and both the time of day and year.[35–38] Increased UV radiation at higher altitudes, however, does not necessarily represent greater UV exposures for airborne microorganisms in comparison with those for terrestrial organisms. In addition to the suite of DNA repair mechanisms found in all microorganisms,[39] airborne microbes can adopt one or more additional measures for affording UV protection. For example, it has been suggested that airborne microorganisms may become embedded within larger particles having UV-reflecting or UV-protecting properties,[11,40] and some UV-resistant bacterial strains readily form cell clumps or agglomerates.[41] A correlation of pigmented microorganisms in the atmosphere with high levels of solar radiation[42] indicates that microbes may also use pigments for UV protection.[30,43] Protective mechanisms such as these are especially important in the upper stratosphere and mesosphere, where DNA-damaging UV-B (280–315 nm) and UV-C (100–280 nm) radiation is not attenuated by the O_3 layer.[44] The amount of cosmic radiation also increases directly with increasing altitude,[45] but is much less effective in producing biological damage than UV radiation.[46]

Finally, the hydrogen ion (H^+) concentrations found in clouds and rainwater fall in an acidic range between pH 3 and 7,[47] which is a narrower range than that found in other microbial habitats. The acidic conditions are the result of gases and compounds from aerosols dissolving into cloud water, which varies with their underlying terrestrial sources.[48] Microorganisms in general have adapted to survive a significantly broader range of H^+ concentrations, from extremely acidic conditions of pH ~0 [49] to highly alkaline conditions of pH ≥ 11.[50]

As a habitat, the atmosphere must also be able to provide nutrients for its heterotrophic life forms. Most of these nutrients are found in the troposphere, carried by water in the form of clouds and rain. For example, concentrations of sulfate and nitrate nutrients in cloud water and rainwater reach levels typically found in oligotrophic lakes.[40] Several potential carbon sources are found in clouds and tropospheric air, including carboxylic acids and alcohols at concentrations ≤1 mg L^{-1}, (see Section 1.3.2) and a variety of hydrocarbons at concentrations ≤4 ng L^{-1}.[47]

1.3 Evolutionary History of the Atmosphere

Despite considerable deficiencies in our current understanding of atmospheric ecology, the wealth of information provided by geological records, fossil records and paleobiology in general can be used as a foundation for speculation about the evolutionary history of atmospheric life. At least provisionally, and in the broadest of terms, we outline this history below, framing it around the Precambrian Supereon between 4.6 Ga and 541 Ma (megaannum or million years ago; see Table 1.1), as it accounts for 88% of all geological time and encompasses all of the early evolutionary events leading to prokaryotic and eukaryotic life forms, the colonization of land by living systems and the evolution of an O_2-rich atmosphere, among other important events. The Supereon literally was the age of microbes, macroscopically expressed in colonial structures referred to as stromatolites. A photograph of an ancient fossil stromatolite, and also examples of living extant stromatolites, are shown in Figure 1.4. The structures are ubiquitous throughout the fossil record of the Precambrian, but are rare today. Except for phylogenetic determinations from conserved DNA sequences and some subtle molecular fossils, stromatolite fossils remain the only visual portal into the very early history of Precambrian life.

1.3.1 Aquatic Microbial Life of the Early Precambrian

Life began sometime during the Early Precambrian, the broad expanse of time starting with the formation of the Earth ~4.6 Ga and extending over 2 billion years to 2.5 Ga. The oldest reported fossils, those discovered in the Strelley Pool Formation of Western Australia, date back 3.5–3.4 Ga and provide some idea of Early Precambrian living systems. Sedimentary rocks from the Strelley Pool Formation contain what appear to be fossil stromatolites.[51–53] Microscopic forms reminiscent of bacteria can be observed within the stromatolites, indicating that if the primitive bacteria-like cells are direct ancestors of extant photosynthetic bacteria, life was already well developed at that time[54] and free O_2 was being added to the atmosphere on a continuous basis (see Figure 1.2). Other evidence indicates that the divergence between bacteria and archaea also occurred at about this time,[55] marking the end of the last universal common ancestor.[56] In general terms, it appears that the evolutionary process was well under way for microbial life during the Early Precambrian. The prevailing biota at this time consisted of primitive, unicellular prokaryotic microorganisms of limited biological diversity and relatively little morphological complexity.[57] Bacteria and blue-green algae appear to have constituted a significant part of this biota, occurring earliest as isolated cells and somewhat later as components of stromatological communities. Chemical analyses of shale formations from Western Australia suggest that bacterial life migrated from the oceans onto land towards the close of the Early Precambrian at ~2.75 Ga.[58] Other evidence of terrestrial life at roughly the same time has been discovered in organic matter-rich

a

b

Figure 1.4 Photographs of (a) an ancient, fossilized stromatolite from the Strelley Pool chert in Western Australia formed ≥3.4 Ga and (b) modern living stromatolites from Shark Bay, Western Australia. Stromatolites are layered biochemical accretionary structures formed in shallow water by the trapping, binding and cementation of sedimentary grains by biofilms (microbial mats) of microorganisms. The stromatolite fossil in (a) dates back to the Precambrian age. Photograph (a) from Hallmann[142] and (b) from Caldicott.[143]

fossilized soil[59] and small, isolated wetlands that periodically became dry on a regular basis.[60] This latter environment and other, shallow, aquatic environments that periodically exposed their microbial inhabitants to desiccated conditions due to changing tides, currents or evaporating water are believed to be the ancient venues where aquatic microbes first developed adaptations necessary for colonizing land.[61] The periodic desiccation allowed microbial dispersion by wind and weather from these sub-aerial habitats, affording a practical mechanism for land colonization. Because fossil records place the colonization of land towards the close of the Early Precambrian, it is likely that some of the unsophisticated microorganisms at this time were just beginning to adapt for surviving sub-aerial habitats. Such adaptations probably did not become common in microbial populations until roughly the start of the Middle Precambrian.

Careful examination of early fossils lends support to the idea that microbial dispersal through the airspace, in all likelihood, occurred even during the Early Precambrian. Microfossils observed in sedimentary rocks formed during this period often reveal bacteria-like cells of sizes similar to those of extant microorganisms believed to have atmospheric residence times of ~1 week (see Section 1.4.2). Electron micrographs of such rocks from South Africa dating to ~3.1 Ga, for instance, contain what appear to be ancient bacterial cells that are spherical and $\ll 0.5$ μm in diameter, rod-shaped cells ~0.2 μm in diameter and ~0.6 μm long, and other, rod-shaped cells ~0.3 μm in diameter and ~7.0 μm long.[57,62] Optical photomicrographs of thin sections prepared from other sedimentary rocks from this same formation show algal-like microfossils 10–20 μm in diameter.[63] Also, spiny, spherical acritarchs (small, non-acid-soluble organic structures that cannot be positively categorized) appearing 1–2 μm in diameter have been identified in electron micrographs of >2.7 Ga sedimentary rocks from Zimbabwe.[57] It is not difficult to conclude that these small organisms became lofted into the Early Precambrian atmosphere, not unlike extant microorganisms that become lofted into the atmosphere today. This is especially the case when considering the 300 m (~1000 ft) tides believed to have occurred during this time, a consequence of the Moon being much closer to the Earth than today. On the other hand, it is difficult to imagine that these organisms could have survived in the atmosphere for any significant length of time, at least for most of the Early Precambrian. Living systems were entirely aquatic and unsophisticated for essentially all of the Early Precambrian and most likely not vigorous or robust enough to survive anywhere but in their native aquatic environments. Moreover, the absence of a well-formed O_3 layer would only add to the difficulties of survival in the atmosphere due to severe UV exposure. Additional complications for airborne survival during the Early Precambrian include (1) the increasing poisonous effect from free O_2 on many bacteria as a result of its increasing atmospheric concentrations starting ~3.5 Ga (see Figure 1.2) and (2) the ravages of the atmosphere's recurring desiccant conditions on the aquatic-derived living systems. The Early Precambrian atmosphere must have been a very inhospitable environment for the primitive living systems of the time.

1.3.2 Middle Precambrian Migration of Microbial Life from Aquatic Habitats

The best-known and most significant microfossil assemblage of the Middle Precambrian has been organically preserved in sedimentary rocks formed ~2 Ga in the Gunflint Iron Formation[64,65] of Ontario, Canada. Gunflint microfossils are diverse and abundant and typically resemble extant bacteria.[66–68]

These microfossils, and several others, indicate that during the Middle Precambrian between 2.5 and 1.7 Ga, stromatological algal communities flourished in shallow seas, estuaries and other littoral environments (areas of a lake, river or sea close to shore) on a global scale and planktonic algae were abundant locally. For most of the Middle Precambrian, the Planet's biota was composed entirely of prokaryotic organisms, including members of several extant cyanophycean families, including Chroococcaceae, Nostocaceae, Oscillatoriaceae and a variety of chemosynthetic bacteria. Other taxa of less certain biological relationships were also represented; however, it is not known whether these organisms were prokaryotic, eukaryotic or somehow intermediate in cellular organization.[57] Many microfossils indicate that eukaryotic cells first appeared at the close of the Middle Precambrian ~1.85 Ga,[69] derived from prokaryotic cells engulfing each other by phagocytosis to form new cells with membrane-bound organelles of diverse functions existing in symbiotic relationships. The Middle Precambrian environment was oxygenic[70] on a global scale (see Figure 1.2), favoring the spread of the new eukaryotic life forms.

Evidence for the bacterial colonization of land ~2.75 Ga[58–60] carries profound implications for living systems around the turn of the Middle Precambrian. In particular, the primitive prokaryotic life forms so prevalent during the Early Precambrian must have developed reasonably high levels of vigor and robustness in order to migrate from their natural aquatic habitats on to dry land and not just survive after landfall, but also thrive. As mentioned previously, some of the most essential features for such robustness, especially tolerances against (1) an O_2-containing atmosphere, (2) strong UV radiation and (3) recurring desiccant conditions, are also central for survival in the atmosphere. Bearing this in mind, it becomes straightforward to conclude that the bacterial life forms at the time must have been able to endure at least short excursions into the atmosphere. Although it is likely that most bacterial cells lofted into the atmosphere did not return to the surface alive days or even hours later, the return of living cells to the surface ultimately became possible, providing a completely new mechanism for introducing novel genes into distant gene pools. Microbial life in the atmosphere may well have had its origins in the Middle Precambrian, involving mostly prokaryotic organisms, if not entirely. However, it is difficult to understand how long their newly developed vigor and robustness could keep the prokaryotic life forms alivc in the atmosphere. Water and nutrients could have been found within clouds, but the stratosphere still did not have a well-developed O_3 layer to absorb harmful UV radiation. The one obvious conclusion that can be drawn from the colonization of land is that it would have allowed microorganisms to enter the atmosphere directly from land for the first time, supplementing their already established lofting into the atmosphere from sea spray.[1] Because land-to-air lofting of microbial particles is more efficient than ocean-to-air lofting,[1,18] the new lofting mechanism had the potential to increase appreciably the numbers and density of microorganisms in the Middle Precambrian atmosphere. However, terrestrial life is not believed to have been very pervasive at this time, leaving the Middle Precambrian landmasses barren and sterile.[71–75] This point of view is not universally accepted, however, as some believe that terrestrial microbes could have existed in substantial numbers at this time.[61]

1.3.3 Diversification of Microbial Life During the Late Precambrian

The Late Precambrian, between 1.7 Ga and 541 Ma, appears to be a time of increasing complexity and diversity for microbial life. The fossil record demonstrates that stromatological communities reached their greatest numbers ~1.25 Ga[76] and subsequently declined in both abundance and diversity,[77] so that by the end of the Precambrian age, they had fallen to just ~20% of these numbers. The Late Precambrian featured diverse assemblages of prokaryotic and eukaryotic microorganisms, some of which are morphologically similar to extant taxa.[57] Filamentous and coccoid blue-green algae were the dominant organisms in widespread biohermal (a mound of material laid down by marine organisms) communities and stromatolites also contained nucleated algae. Sometime ~1.2 Ga, mitotic algae began to supplant the prokaryotic microorganisms that dominated the Early and Middle Precambrian. Eukaryotic algae gave rise to unicellular life forms during the Late Precambrian and the transition to multicellular organization occurred sometime between 1.2 Ga and 700 Ma. The constant, long accumulation of atmospheric O_2 finally culminated into a well-established O_3 layer ~600 Ma,[78] greatly facilitating the colonization of land by living systems. There is some fossil evidence that fungi existed during this period, but identification of these fossils is far from certain. On the other hand, DNA analysis indicates that the earliest fungi appeared just before the end of the Precambrian period ~570 Ma.[79]

It is evident that the developing complexity of microorganisms during the Late Precambrian included a further evolution of the robustness that permitted their survival on dry land and in the atmosphere. Relative to earlier living systems, Late Precambrian microorganisms almost certainly experienced improved survivability and diversity in the atmosphere, ultimately leading to larger numbers of living microorganisms aloft. For the first time in evolutionary history, prokaryotic and eukaryotic cells could coexist in the atmosphere. The rise of eukaryotes introduced sexual reproduction, enabling both genetic diversity and an enhanced ability to adapt to and survive in new or changing environments. This is particularly important because the increased atmospheric survivability would have increased the introduction of new genes into distant gene pools by aerial dispersal. The biocidal effects from UV radiation attenuated steadily with the development of the O_3 layer throughout the Late Precambrian and the lofting of land-based microorganisms into the atmosphere became increasingly more common. The exact numbers of airborne microorganisms originating from landmasses in comparison with those from aquatic environments is difficult to determine, especially because evidence of early terrestrial microorganisms is rare.[61] Although the overwhelming opinion is that Late Precambrian landmasses were still relatively barren and sterile,[71–75] life must have become more established on land and more prevalent in the atmosphere as the dawn of the Cambrian grew increasingly closer.

1.3.4 Evolutionary History of Airborne Microorganisms from the Cambrian Explosion

Starting from ~540 Ma, the Cambrian period (see Table 1.1) marked an extraordinary change for life on Earth. Although complex, multicellular organisms gradually became more common in the millions of years immediately preceding the Cambrian,[80] a rapid and explosive diversification of life forms during this period produced the first representatives of all modern animal phyla. The age of microbes had come to an end, but microbial evolution was still an ongoing process. In terms of atmospheric life and survivability, both prokaryotic and eukaryotic organisms just might have come into their own during this time. Since the start of the Cambrian, both organisms became increasingly more diversified and established, with some becoming well suited for terrestrial life. Ultimately, these terrestrial life forms grew to be pervasive on land and flourished. There is an analogous story to tell about the evolutionary history of bacteria and fungi in the atmosphere. However, the timeline of this story is not clear because, unlike the case of terrestrial life, it cannot be extrapolated from the fossil record. However, when considering the tremendous diversity and pervasiveness of bacteria and fungi in the atmosphere today,[81] it is apparent that their robustness and vigor for surviving excursions into the atmosphere and the diversity of microbial life in the airspace continued to evolve and become more sophisticated in the ~540 Ma since the dawn of the Cambrian.

1.3.5 Atmospheric Contributions to Microbial Evolution

There has been some speculation regarding the extreme environmental conditions in the atmosphere and their influences on the origin and evolutionary history of life. Dobson *et al.*[82] hypothesized that organic aerosols may have acted as pre-biotic chemical reactors during the start of the Precambrian that ultimately gave rise to primitive living cells; the general concept of their hypothesis is illustrated in Figure 1.5. Smith[4] suggested that when microorganisms were first able to survive extended excursions into the atmosphere, time aloft could have accelerated the processes of natural selection and speciation by imposing unique combinations of stresses on microorganisms that cannot be found anywhere else. Longer residence times increase the impact of atmospheric stress on airborne species, especially cellular damage from desiccation and UV radiation. The suggestion has led to two

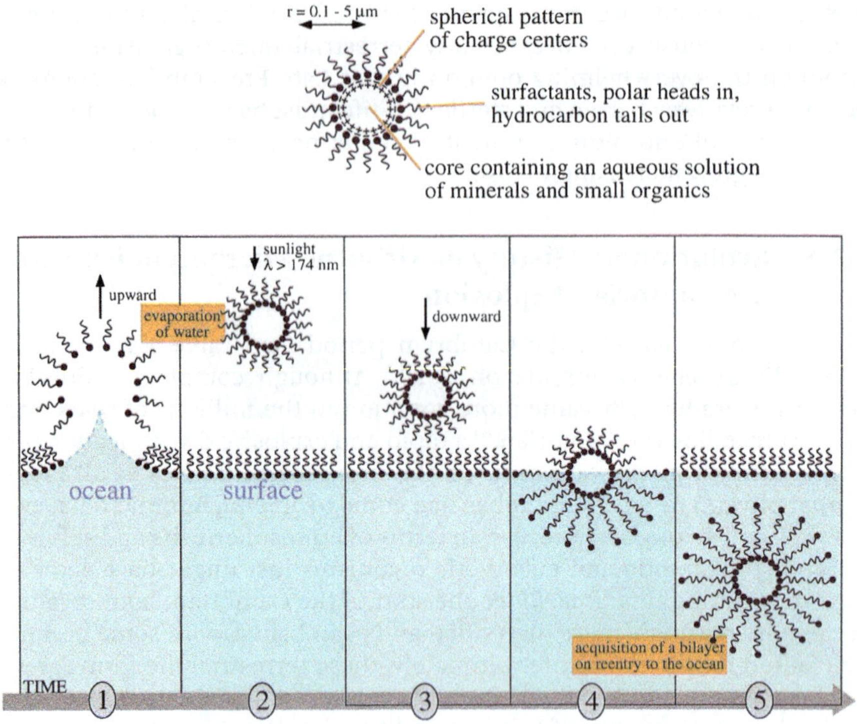

Figure 1.5 Schematic representation of an organic aerosol particle (top) and a sequence of five events leading to the formation of stable vesicle in the Precambrian Ocean (bottom). The five events illustrated are the following: (1) the formation of an aerosol particle with a surfactant partial monolayer at the ocean surface, (2) its exposure to sunlight while airborne, which drives the evaporation of water from its interior to produce a complete monolayer, (3) its inevitable fallout from the atmosphere, (4) re-entry at the ocean surface and (5) acquisition of a second layer of surfactant to permit survivability in the ocean as a stable vesicle. The ocean surface acts as a global scale concentrator of surfactants, which have their polar groups in the water and their hydrocarbon chains in the air. Very large numbers of this repetitive process would have to have occurred for life to originate from the vesicles. Adapted from Dobson *et al.*[82]

opposing hypotheses concerning microbial evolution and survival in the atmosphere.[4] One is that microorganisms have developed adaptations for quickly returning to the surface after billions of years of drifting perilously through the atmosphere. Examples of such adaptations include membranes that promote cloud and ice condensation and cell aggregation mechanisms to reduce residence times. On the other hand, some microorganisms could have evolved cloud condensation adaptations to collect water and nutrients to metabolize while airborne. The latter hypothesis is consistent with large number of reports on the metabolic activities of airborne bacterial and fungal strains (discussed immediately below). There have also been reports addressing potential stratospheric contributions to evolution.[19,41,83,84] Wainwright *et al.*[84] proposed that the lofting of microorganisms into the stratosphere particularly accelerates the process of evolution. Microorganisms that are mutated and not killed by exposure to stratospheric UV radiation will ultimately return to Earth, many with one or more novel mutations. When introduced into soil-derived gene pools, these mutations will increase the numbers and diversities of mutations available for natural selection. As a result, the rate of microbial evolution will have been greater than that without stratospheric influence.

1.4 Microbial Life in the Atmosphere

There is a long-standing preconception in ecology that both landmasses and bodies of water constitute habitats for living systems whereas the atmosphere does not. Early definitions of habitat focused sharply on species occupancy[85,86], but later definitions also included the survival and reproduction of the species.[87] Consistent with the latter definition, Green and co-workers[6] argued that the true "residents" of the atmosphere are airborne microorganisms that are active metabolically and reproductively. While this can include bacteria, archaea and fungi (including yeasts and molds), it explicitly eliminates viruses, spores and other inactive propagules that are not metabolically active, and also organisms such as birds and insects that are metabolically active, but not reproductively active, when airborne.

Similarities can be drawn between the atmosphere and other environments, especially ocean environments. Physically, both are fluid environments with currents (horizontal fluid movement) and eddies (swirling of fluid) that are influenced by locations, topographies and seasons. Environmentally, both contain microorganisms that are transported by the currents and eddies. Such transport through the atmosphere is the simplest means of accounting for the global distribution of many microbial species,[88] and some have suggested that the currents and eddies associated with this transport can give rise to biogeographic regions in the atmosphere analogous to those found on the Earth's surface.

1.4.1 Metabolic Activity of Microorganisms in the Atmosphere

The atmosphere is literally teeming with living microorganisms; however, direct *in situ* measurements of their metabolic activities while airborne are tremendously difficult to carry out. This has restricted research on the subject primarily to experimental, culture-based approaches conducted in laboratory settings. The first such investigation was reported in 1975, with Dimmick *et al.*[89] revealing that aerosolized bacteria are capable of glucose metabolism. Over the next 4 years, the same group demonstrated that aerosolized bacteria incorporate thymidine into their DNA[90] and are also capable of cell division.[91,92]

Most culture-based studies addressing the metabolic activities of microorganisms in the atmosphere involve isolating bacteria or yeast from cloud water, rainwater, snow or ambient air specifically for laboratory testing. In 1987, a report by Herlihy *et al.*[93] presented evidence for the metabolic degradation of both formic and acetic acid by bacteria isolated from rainwater, and others have reported similar findings.[94] Fifteen years later, Amyot and co-workers[95] presented strong evidence that airborne bacteria and fungi existing in the tropospheric boundary layer can efficiently degrade dicarboxylic acids and use them as nutrients. The same group also reported a wide range of variation in degradation lifetimes for dicarboxylic acids as a result of the microbial taxon under investigation, and also other factors such as H^+ concentration and temperature.[96] Other investigators[97,98] revealed that bacterial activity could occur in rainwater held at sub-freezing temperatures.

Delort's group have reported numerous investigations of bacteria and yeast strains isolated from cloud water samples collected at the Puy de Dôme station (a reference site of the European Aerosols, Clouds and Trace Gases Research Infrastructure Network at 1465 m or 4086 ft ASL). Their first investigation[99] found that of a total of 17 bacterial strains isolated from Puy de Dôme cloud events, most were able to degrade various organic compounds, such as formate, acetate, lactate, methanol and formaldehyde, the major organic compounds commonly found in cloud water. This study was extended to include 60 total strains of bacteria and yeasts isolated from the cloud water samples and included succinate as an additional test compound.[100] The extended study revealed that (1) all the compounds could be biodegraded when used as a single carbonaceous substrate, with activities dependent upon both the microbial strain and test compound, and (2) formaldehyde could be biodegraded through many different metabolic pathways, leading to the production of formate and/or methanol. In a somewhat related study,[101] ATP concentrations measured in cloud water samples from 14 independent cloud events indicated that a significant fraction of the microorganisms present in clouds are metabolically active. Subsequently, Vaïtilingom *et al.*[102] found that four bacterial strains and a single yeast strain isolated from Puy de Dôme clouds degraded formate, acetate and succinate when cultured in simulated cloud water at 5 and 17 °C (41.0 and 62.6 °F), temperatures representative of the clouds in winter and summer, respectively. The bacteria and yeast strains in the study were also found to have lifetimes ranging between 0.6 and 69.1 days at the two temperatures. These results suggest that biological activity alone drives 90–99% of the oxidation of carbonaceous compounds at night, and during the day, when competing with photochemistry,[103,104] biological activity contributes only 2–37% of the total reactivity. A year later, Vaïtilingom *et al.*[105] examined the contributions of photochemistry and microbial activity to the degradation of atmospheric carboxylic acids, again using bacterial strains from Puy de Dôme cloud water. Their major conclusions were that (1) some test compounds could only be photodegraded, others only biodegraded and the remainder could be degraded by both mechanisms, (2) the biodegradation rates did not decrease drastically when incubation temperatures decreased from 17 to 5 °C and (3) all the observed biodegradation rates, even those at 5 °C, are competitive with measured and calculated photodegradation rates. Around the same time, Husárová *et al.*[106] reported that bacterial strains isolated from Puy de Dôme clouds degraded methanol and formaldehyde at 5 and 17 °C and the observed biodegradation rates were within the same range of magnitude as those from free radical chemistry. The most recent report from Vaïtilingom *et al.*[107] investigating microorganisms from Puy de Dôme cloud water revealed that (1) cloud microorganisms remain metabolically active in the presence of ˙OH radicals photo-produced from H_2O_2 (reactive oxygenated species such as H_2O_2 and free radicals are found in clouds) and (2) cloud microorganisms can metabolize H_2O_2 to O_2 and H_2O using catalases. Lastly, Durand *et al.*[108] reported the first pure bacterial strain capable of rapidly degrading the herbicide mesotrione, which was also isolated from Puy de Dôme cloud samples.

Considered collectively, the culture-based investigations present overwhelming evidence that microorganisms from atmospheric habitats are remarkably capable of metabolic activity. This conclusion is substantiated by the results of Sattler *et al.*[109] and Hill *et al.*,[110] from direct observation of pure, unadulterated cloud water. Sattler *et al.*[109] showed that the microorganisms present in cloud water held at 0 °C (32 °F) incorporated thymidine into their DNA at $(1.1–4.8) \times 10^{-15}$ mol L^{-1} h^{-1} and are also capable of reproduction. The results of Hill *et al.*[110] demonstrated the processing of atmospheric nitrogen by cloud water and also found that 76% of the bacterial cells in their cloud water samples absorbed and reduced the dye 5-cyano-2,3-ditolyltetrazolium chloride, indicating that most of the cells in the samples were living and capable of metabolic activity. There now appears to be a general consensus taking shape that although microbial metabolic activity has never been directly measured in the atmosphere *in situ*, at least some microorganisms in the troposphere must be metabolically active. Vaïtilingom *et al.*[107] estimated that microorganisms in clouds metabolize the release of $(51–215) \times 10^6$ tons of CO_2 each year on a global scale.

1.4.2 Microorganism Residence Times and Reproduction in the Atmosphere

In striking contrast to the case of metabolic activity, studies of microbial reproduction in the atmosphere are severely limited, with only three reported over the last 35 years.[91,92,109] To complicate matters, determining whether microorganisms reproduce while airborne has relied on considering their atmospheric residence times, as they might be the principal factor limiting microorganism reproduction while airborne.[17,109] In a theoretical sense, and to a first approximation, residence times for microorganisms are inversely proportional to the sizes of the particles they form when airborne, but other factors, including air temperature, relative humidity, wind and weather conditions, can contribute significantly to residence time values.[17,40,111]

Residence times, however, cannot be measured *in situ* with contemporary technologies and, therefore, are computed from virtual simulations of particle transport. Table 1.3 summarizes data from three very different simulations modeling particles as rigid spheres between 0.7 and 60.0 µm in diameter.[13,17,111] Considerable variation can be seen in the table, revealed by the broad ranges of residence times for any single sphere diameter, and also the very different ranges between similar sphere diameters reported from different references such as those for the 1.0 and 1.1 µm sphere diameters. The broad nature of any one reported range is a consequence of the different conditions used for any single simulation protocol. Burrows *et al.*,[17] for example, conducted simulations for 10 different ecosystems (costal, desert, grassland, *etc.*) under conditions that explicitly included or eliminated cloud condensation nucleation events to generate a total of 20 unique residence times. The variations between similar sphere diameters from different references are largely a result of their different simulation protocols. In a more general sense, the variations in the table illustrate the complexities of deriving accurate and realistic residence time values from computational approaches.

Notwithstanding these shortcomings, Table 1.3 can be used to arrive at meaningful conclusions about airborne microorganisms by relating their values to those of naturally occurring microbes in the atmosphere. Airborne bacteria can occur as individual cells but are more likely to be attached to other particles, including soil or plant fragments, or agglomerated with other bacterial cells.[11,41] Whereas individual bacterial cells are ≤1 µm in diameter,[112] the median diameter of airborne particles containing bacteria has been reported to be somewhere between 2 and 4 µm.[17,113–116] In Table 1.3, therefore, sphere diameters between 0.7 and 3.0 µm can be considered to represent airborne bacteria and the 3.0 µm diameter can also be used to represent fungal spores in the airspace.[116] The diameters in Table 1.3 between 20 and 60 µm were specifically chosen to model eukaryotic microbes,[13] such as testate amoebae.[117,118]

Table 1.3 Summary atmospheric residence times for 0.7–9.0 µm diameter spheres derived from virtual simulations.

Sphere diameter/µm	Residence time/days: Range	Average	No. of values	Reference
0.7	0.7–6.0	3.9	3	Williams *et al.*[111]
0.9	0.7–6.0	2.9	3	Williams *et al.*[111]
1.0	3.8–14.4	7.6	20	Burrows *et al.*[17]
1.1	0.6–5.0	2.9	3	Williams *et al.*[111]
1.4	0.6–5.0	2.8	2	Williams *et al.*[111]
1.8	0.5–3.0	1.2	2	Williams *et al.*[111]
3.0	2.2–10.8	5.8	20	Burrows *et al.*[17]
9.0	2.2–57.0	13.9	12	Wilkinson *et al.*[13]
20.0	0.6–12.8	4.5	12	Wilkinson *et al.*[13]
40.0	0.1–9.8	2.3	12	Wilkinson *et al.*[13]
60.0	0.03–9.0	2.0	12	Wilkinson *et al.*[13]

Using generation times measured for bacteria, Table 1.3 can be used to evaluate the likelihood of bacterial reproduction in the atmosphere. The most relevant data for this type of evaluation are probably those of Sattler *et al.*[109] using cloud water collected at 3106 m (10 190 ft) and temperatures of −4 and −9 °C (24.8 and 15.8 °F, respectively). Bacterial generation times were measured by culturing the cloud water at 0 °C and counting the numbers of bacterial cells over time. Ten different generation times ranging between 3.6 and 19.5 d were reported, with an average value of 10.9 d. In Table 1.3, residence times for sphere diameters between 1.0 and 3.0 μm range between 0.5 and 14.4 d with average values between 1.2 and 7.6 d. Extrapolating from the overlap between the lower range in generation times (about one new generation every 4–5 d) and the upper range of residence times (~10–14 d), it is conceivable that at least some microorganisms could divide while aloft in the atmosphere. More specifically, the extrapolation indicates that bacterial excursions into the atmosphere could possibly include one or two cell divisions and occurrences of more divisions for a single bacterial excursion are rare. The fact that the residence and generation times do not completely overlap implies that significant numbers of bacterial excursions into the atmosphere, if not most of them, do not involve bacterial reproduction. Others[1,119] using the generation time values have arrived at similar conclusions, but based their extrapolations on data suggesting that bacterial cells spend an average of only 5–6% of their time aloft inside cloud droplets[120] where nutrients are available. Without more investigations, these conclusions should be considered as speculative and preliminary.

1.4.3 The Movement of Microorganisms Through the Atmosphere

As with all particles in the atmosphere, microorganisms are consistently moving through its three-dimensional airspace. The movement is the resultant sum of two different types of motions, vertical and horizontal motion, each driven by very different mechanisms. Vertical motion includes the lofting of particles into the atmosphere and their inevitable fallout back to the surface, whereas horizontal motion, that along the east–west and north–south dimensions in atmospheric space, gives rise to particle dispersal. In a general sense, these motions do not occur separately as isolated events, but tend to be inter-related to varying extents depending largely on local conditions. Particles must first be lofted high into the atmosphere, for example, for their dispersal over appreciable distances such as across oceans or continents,[4,83,121,122] and dispersal distances can be considerably enhanced by local storms or hurricanes. Particle lofting occurs mostly within the troposphere and particularly in the boundary layer, where wind, weather and solar heating of the Earth's surface cause substantial mixing between the surface and the airspace. Surface heating causes warm air masses to form near the surface, which cool as they rise, and then fall back to the surface to be warmed again, leading to vertical mixing or convection, of atmospheric gases.

The collective result of convection is the lofting of aerosol particles such as dust, smog, smoke, insects, pollen and microorganisms, with most of the microorganisms deriving from arid topsoil and marine sea spray.[1,6,18,123] In addition to naturally occurring environments, bacteria may become lofted into the atmosphere from man-made point sources such as landfills, waste-water treatment facilities, farms and also construction sites, the last from the disturbance of settled dust. Airborne microorganisms and other particles fall out of the atmosphere by gravitational settling and both ice nucleation[119] and cloud condensation nucleation.[124]

The tropopause is thought to act as a barrier to the vertical motion of particles, but microorganisms routinely reach into the stratosphere[4,41,83,84] and even the mesosphere.[30] Observations such as these have driven much speculation about mechanisms that can rationalize the presence of microbes above the tropopause. Some mechanisms proposed in recent decades include strong upward winds,[83,84] thunderstorms[125] and blue lightning strikes,[126] hurricanes[127] and monsoons,[128] stratosphere–troposphere exchange by deep convection in the tropics,[129] electrostatic levitation,[130] volcanic eruptions[131] and commercial aviation. The US Bureau of Transportation Statistics[132] estimates that the US airline industry alone sends ~800 000 domestic and commercial international flights into the upper atmosphere every month, with each aircraft shedding countless microorganisms into the upper troposphere and lower stratosphere.

The horizontal movement of particles through the airspace is a consequence of atmospheric circulation occurring on a global scale. Patterns for atmospheric circulation vary somewhat from year to year due to mid-latitude depressions, tropical convective cells and other individual weather systems; however, the forces driving the circulation are always acting on the airspace and their patterns are stable over longer periods of time. The patterns are a fundamental property of the Earth's size, rotation rate, atmospheric depth and the solar heating of its surface. In the absence of thermal energy from the Sun, the Earth would be a wind-less planet. Solar energy heats the Planet's surface, but the heating is never even across the surface and convection cell systems form in the atmosphere. The Earth's latitudinal convection cell systems are shown schematically in Figure 1.6, which reveals three unique cell system types: Hadley cells (named after George Hadley, who first proposed the convection cell to explain the trade winds), mid-latitude cells or Ferrel cells (named after the meteorologist William Ferrel, who first proposed their existence) and Polar cells, with one of each type found in both hemispheres.

The atmospheric circulation patterns first described by George Hadley for the tropics (latitudes within ~30° north and south of the Equator) agree well with the trade winds observed in the region. The Hadley cell systems are closed-loop circulations driven largely by surface heating near the Equator in the ITCZ (intertropical convergence zone), where solar radiation is most intense. This causes the heated air to rise vertically towards the tropopause, creating an area of low pressure at the ITCZ. Because the environmental lapse rate changes from positive to negative (the variation in air

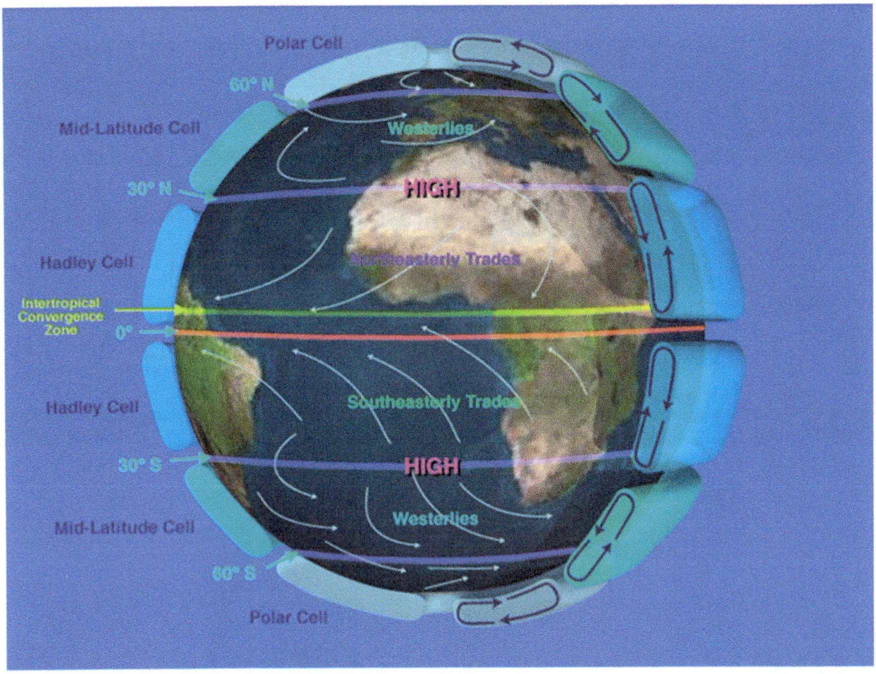

Figure 1.6 Schematic representation of the Earth's global circulation. The parallels at 30 and 60° north (N) and south (S) of the Equator (0°) are shown. Violet lines labeled "HIGH" are areas of high atmospheric pressure. Adapted from NASA.[144]

temperature changes from decreasing to increasing with height; see Figure 1.3) at the tropopause, it acts as a ceiling and diverts much of the rising air horizontally in a either northerly or southerly direction. The air moving in a northerly direction just below the tropopause becomes part of the Hadley circulation in the Northern Hemisphere, while its southerly-moving counterpart becomes part of the Hadley circulation in the Southern Hemisphere. For both hemispheres, the air circulating towards the Poles constantly decreases in temperature and concomitantly increases in density, causing the tropopause to decrease in altitude at the same time; this is illustrated schematically in Figure 1.7a. The tropopause is ~15 km (~9 mi) ASL at the ITCZ but reaches only ~10 km (~6 mi) ASL at about the 30th parallel N/S (latitudes of 30° north and south), creating a "downhill" effect that accelerates the air moving towards the Poles. At the same time, the circulating air experiences a Coriolis force that deflects its northerly or southerly progression eastwards, a consequence of the nearly spherical Earth rotating about its axis. The Earth's latitudinal parallels encompass increasingly less distance around the Planet with increasing latitude (progressing from the Equator towards the Poles), so that for every complete rotation about its axis, less distance is covered over the Earth's surface as latitude increases. The velocity of rotation, therefore,

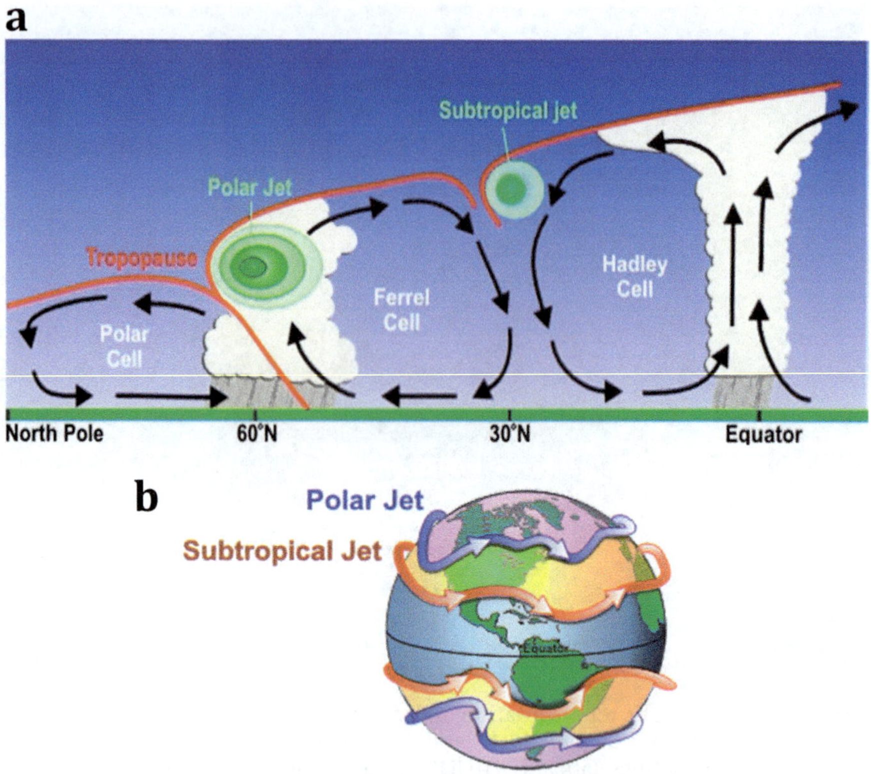

Figure 1.7 Schematic representations of atmospheric circulation systems and jet streams. (a) Cross-sectional view of the Hadley circulation, Ferrel or mid-latitude circulation and Polar circulation cells of the Northern Hemisphere, with circulation directions shown with arrows and locations for the tropopause and both the subtropical and polar jet streams indicated. The tropopause is shown to decrease in altitude moving from the Equator to the pole. (b) The general paths of the four jet streams across the Earth are shown with their directions indicated by arrows. The jet streams take a meandering path around the Earth, dipping and rising in both altitude and latitude, splitting at times and forming eddies and even disappearing altogether to appear elsewhere in the troposphere. Adapted from NOAA.[145]

is fastest at the Equator and decreases with increasing latitude, which acts to deflect or "turn" the path of air circulating in Hadley cells towards the east. Somewhere around the 30th parallel north or south, this air converges into one of the two subtropical jet streams that travel in easterly directions. Jet streams form just below the tropopause by the pressure differences resulting from two air masses of very different temperatures and densities converging around the 30th parallel N/S or 60th parallel N/S. The subtropical jets streams vary in height between 10 and 16 km (33 000 and 52 000 ft) ASL and often exceed speeds of 50 knots (57 mi h^{-1}), although speeds of >240 kts

(>275 mi h^{-1}) have been measured along its meandering path around the Earth; see Figure 1.7b. Air circulating in both Hadley cells experiences the greatest descent towards the surface at about these same latitudes, creating areas of high atmospheric pressure. These high-pressure areas and the low-pressure area at the ITCZ form a pressure gradient that drives air near the surface towards the ITCZ, completing the Hadley circulation loop. The moving air also experiences a Coriolis force that turns its path towards the ITCZ in a westerly direction, creating the trade winds (see Figure 1.6).

Polar circulation cells occur in each hemisphere roughly between their respective 60th parallel and the Polar region and are driven by the same mechanisms that drive the Hadley cells. Both the Polar and Hadley cells are thermally direct cells because they convert thermal energy directly into kinetic energy, resulting in rising branches occurring over warm temperature zones and descending branches over cold temperature zones. The vertical rise in Polar cells, however, is limited by a lower tropopause ceiling only ~8 km (~5 mi) ASL and contributes to the cells having substantially less vertical movement than the Hadley cells; this is represented schematically in Figure 1.7a. The rising branches in each Polar cell are found somewhere around their corresponding 60th parallel, where the rising air creates low-pressure areas. Upon reaching the tropopause, the rising air turns and begins its circulation towards the Poles, progressively becoming cooler and denser along the way and ultimately descending towards the surface at or near the Poles. The descending air creates a high-pressure area and, together with the low-pressure area at the corresponding 60th parallel, forms a pressure gradient that drives the surface circulation towards the low-pressure area, completing the Polar cell circulation loop. The circulating surface air experiences a Coriolis force on its path towards its corresponding 60th parallel that deflects it eastwards (see Figure 1.6).

Mid-latitude circulation cells are found in both hemispheres between their respective Hadley and Polar cells. In contrast to the other cell types, the mid-latitude cells are thermally indirect cells, as they are not driven by thermal forcing but by eddy forcing (weather systems). The rising branches of mid-latitude cells occur over cold temperature zones with descending branches over warm temperature zones, driving cell circulations in opposite directions as those of the thermally direct systems. Mid-latitude surface air moves towards the Poles rather than the Equator and experiences a Coriolis force that turns them in an easterly direction (see Figure 1.6). The upper branches of the mid-latitude cells are not well defined, partly because the cells occur between the Hadley and Polar cells with neither a strong heat source nor strong cold sink to drive their convection. Polar jet streams form in each hemisphere where air masses from their mid-latitude and Polar circulation cells converge. The Polar jets travel at higher speeds than the subtropical jets, but occur only 7–12 km (23 000–39 000 ft) ASL.

The Hadley, mid-latitude and Polar circulation cells do not act alone as major contributors to the global heat transport driving the horizontal movement of atmospheric air, as disparities in temperature also drive a set of longitudinal circulation cells. Longitudinal circulations arise because water has a higher specific heat capacity than landmasses. Large bodies of water such as lakes and oceans, therefore, absorb and release more heat than landmasses, but upon doing so, their temperatures change less than those of landmasses. We experience this phenomenon in coastal areas as sea breezes (air cooled by large bodies of water) during the day and land breezes (air cooled by landmasses) at night, but on a much larger scale, the effects are seasonal. Warm air rises over the equatorial, continental and western Pacific Ocean regions, moves eastwards or westwards (depending on its location) upon reaching the tropopause and descends onto the Atlantic and Indian Oceans and the eastern Pacific Ocean. The Pacific Ocean circulation plays an especially important part in influencing the Earth's weather. This ocean-based circulation cell results from the marked difference in surface temperatures across the Pacific Ocean, which contains warm western waters and cool eastern waters. Strong convective activity over equatorial East Asia coupled with descending cool air masses offshore of South America's west coast creates strong winds that force Pacific water westwards into the western Pacific Ocean. Changes in the Pacific Ocean circulation every few years generate El Niño or La Niña effects that can bring unusually warm or cold winters and changes in the frequency of hurricanes for indeterminate periods of time.

1.4.4 Biogeography and the Atmosphere

Evolutionary biologists have divided the Earth's surface into six biogeographic regions, each of which contains unique plants and animals.[133,134] It is widely accepted that these unique biotas exist because of vicariance, the evolutionary isolation of species due to historic barriers to dispersal. For microorganisms in particular, including microscopic eukaryotes, such a biogeographic distribution is currently a subject of ongoing debate. This is especially the case for biogeographic regions in the atmosphere, as only a very limited number of studies have been conducted to determine whether large-scale patterns exist in the distribution of airborne microorganisms. Womack *et al.*[6] argued that analogous patterns are possible for the atmospheric habitat and, like others,[135,136] suggested that the Hadley, mid-latitude and Polar circulation cells might be responsible in large part for creating such patterns. Proponents of this idea point out that the mixing of air is more frequent within circulation cells than between them, resulting in barriers to air movement and the potential for vicariance. Although such patterns have yet to be discovered, there has been at least one report of large-scale distribution patterns for microorganisms in the atmosphere. Fröhlich-Nowoisky *et al.*[18] found that the ratio of species richness between Basidiomycota and Ascomycota, the two major phyla of higher fungi, is much higher in the airspace over continents than that over oceans. This is a significant discovery because it suggests that there might be biogeographic regions in the atmosphere and global atmospheric circulation is important for understanding the biodiversity of microorganisms.

Particle transport through the troposphere is especially relevant to microbial biogeography, as it is central to the cosmopolitan distribution of many microbial species on the Planet's surface.[88] This transport provides rapid, long-distance dispersal of microorganisms, allowing some species to overcome geographic barriers. The recent computer simulations of Wilkinson *et al.*[13] illustrate the potential for the cosmopolitan dispersal of airborne microorganisms by tropospheric air circulation. Their simulations used virtual microbes with diameters of 9, 20, 40 and 60 μm released from the surface and at an altitude of 6.5 km (~21 300 ft) ASL from one site in each hemisphere. Results of their 16 individual simulations over a virtual 1 year time frame are summarized graphically in Figure 1.8. The most striking result is the extensive within-hemisphere dispersal of virtual microorganisms 9 and 20 μm in diameter, representative of smaller eukaryotic microorganisms.[117,118] Because bacteria are significantly smaller in size, their dispersal can be anticipated to be at least as extensive than that for the virtual microbes with 9 μm diameters. The well-known inverse correlation between organism size and dispersal area is readily apparent by comparing the results for the 9 and 20 μm diameter virtual microbes with those for the 40 and 60 μm diameter virtual microbes, which are seen to disperse over substantially less area over the virtual 1 year time frame. Another striking result is the lack of dispersal between the Northern and Southern Hemispheres, which is a direct consequence of the atmospheric circulation patterns around the ITCZ. Other mechanisms may be important for dispersal between hemispheres, including bird migration and ship and airline travel. The results also illustrate the importance of prevailing wind directions, with long-distance dispersal from west to east more likely than from east to west.

1.5 Conclusion

The Earth's atmosphere is unique to the Solar System because of its intimate relationship with living systems and their evolutionary history, a relationship that has existed over geological time. The oldest known fossils suggest that this relationship was well established even 3.4 Ga, with primitive bacteria-like cells already adding free O_2 into the atmosphere as a product of photosynthesis.[51–54] Over the next billion years, photosynthesis completely transformed the Earth's CO_2-dominated atmosphere to one with significant concentrations of O_2,[29] favoring the proliferation of the new eukaryotic life forms at the time. Some hundred million years later, free O_2 began to dominate the chemistry of the atmosphere and O_3 formed in the upper atmosphere from the recombination of O_2 by solar radiation. Finally, the formation of a powerful O_3 screen ~600 Ma[78] in the stratosphere protected living systems on the Planet's surface from the Sun's harmful UV radiation.

Life originated in the oceans and, roughly 1 billion years later, began to colonize land. This terrestrial migration implies that the primitive, anaerobic prokaryotes so prevalent at the time must have developed reasonably high levels of vigor and robustness in order to migrate from their natural aquatic habitats

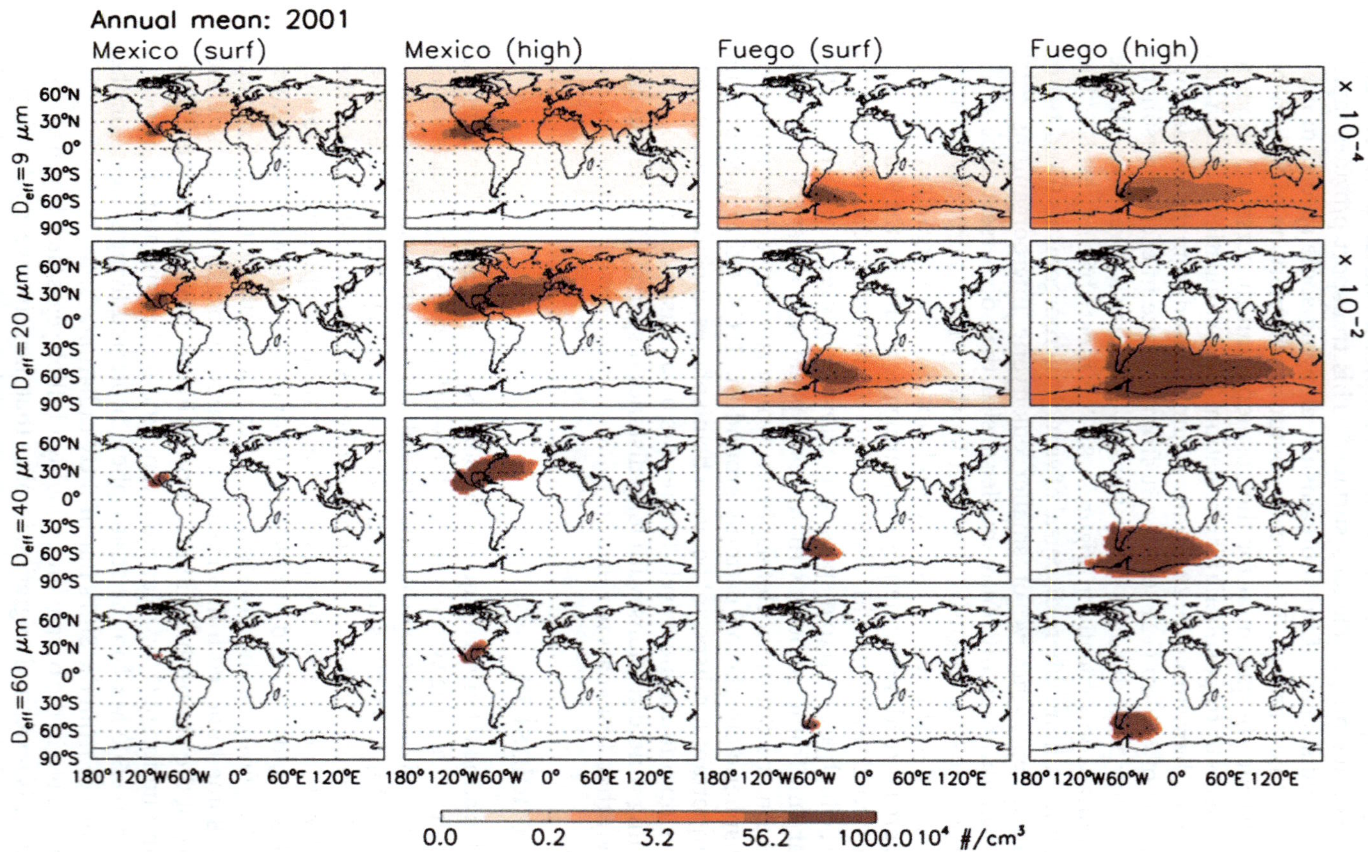

Figure 1.8 Results from computer simulations over a 1 year period for the atmospheric dispersal of particles released from locations in both hemispheres. The simulations were conducted with the GEOS-Chem model[146,147] for the year 2001, with releases from sites in Mexico and Tierra del Fuego. Both sites included releases of virtual microorganisms with 9, 20, 40 and 60 μm diameters from the surface (surf) and an altitude of 6.5 km (~21 300 ft) ASL (high). Shading illustrates the extent of dispersal for virtual microbes in each simulation on a logarithmic scale. Adapted from Wilkinson *et al.*[13]

onto dry land and ultimately thrive. Some of the most essential features for this robustness, especially tolerances against an O_2-containing atmosphere, strong UV radiation and recurring desiccant conditions, are also central for airborne survival in the atmosphere. It is almost certain that at about this time in evolutionary history, or perhaps some time later, these primitive prokaryotic life forms were able to survive short excursions into the airspace and return to the surface alive. Microfossils reveal that living systems existing ~3.1 Ga[57,62,63] were of sizes similar to those of extant microorganisms believed to have residence times of ~1 week,[17,111] revealing that the lofting of the early prokaryotes into the atmosphere at the time was very likely. Over the following billions of years, prokaryotic life forms must have evolved to survive excursions into the atmosphere better, and eukaryotic cells, single-celled microorganisms at first followed by multicellular life forms, appeared and also became lofted into the atmosphere. The atmosphere is believed to contribute to microbial evolution in two important ways: it imposes unique combinations of stresses on microorganisms to drive mutation[4] and it provides a means to disperse microorganisms over great distances and introduce novel genes into distant gene pools.[84] The evolutionary history of living microorganisms surviving atmospheric excursions is exceedingly difficult to understand with certainty. Direct evidence of early living systems aloft in the airspace cannot be found in the fossil record and facts must be extrapolated from fossils of terrestrial and aquatic living systems from the distant past. Details on the subject, therefore, are largely speculative.

There is now abundant evidence that the atmosphere is indeed a true habitat for microorganisms. The present-day airspace is literally teeming with living microorganisms and a wealth of culture-based investigations present overwhelming evidence that microorganisms collected from atmospheric environments are remarkably capable of metabolic activity. These investigations, and others involving direct observations of cloud water,[109,110] strongly suggest that at least some microorganisms must be metabolically active while airborne. Investigations of microbial reproduction in the atmosphere, on the other hand, have been severely limited. Generation times for microorganisms in the atmosphere are usually extrapolated from atmospheric residence times of microorganisms, as they may be the principal factor limiting microorganism reproduction while airborne.[17] Residence times derived from particle transport simulations[13,17,111] together with generation times measured for bacteria in cloud water held at typical tropospheric temperatures[109] suggest that bacterial excursions into the atmosphere could potentially include one or two cell divisions and that significant numbers of bacterial excursions into the atmosphere, if not most of them, are likely not to involve bacterial reproduction at all. Like many investigative efforts conducted to answer intriguing questions in the field of aerobiology, the largest obstacle to studying microbial reproduction in the airspace is the difficulty of designing direct *in situ* methods to investigate live microorganisms aloft in the atmosphere.

Like oceans, the atmosphere has currents and eddies that are influenced by solar heating, convection, Coriolis forces and topographies.

In the troposphere, the currents and eddies constitute the atmospheric circulation, which is largely responsible for the worldwide dispersal of microorganisms and connecting all microbial habitats across the Planet's surface to one another. Microbial dispersal through the airspace is the most straightforward means of explaining the cosmopolitan distribution of many microbial species,[88] as it provides a mechanism for microorganisms to overcome historic barriers to dispersal. Some have suggested[6,135,136] that analogous barriers to microbial dispersal might exist in the atmosphere, possibly giving rise to its own biogeographic regions. Characterizing the microbiological variability in the atmosphere and also the forces responsible for generating them holds real promise for advancing our understanding of microbial biogeography in general, currently a subject of ongoing debate among many microbiologists.

Acknowledgements

We wish to thank Ms Joy L. Henderson and Ms Donna M. Hoffman for their support and expert assistance in preparing this chapter.

References

1. S. M. Burrows, W. Elbert, M. G. Lawrence and U. Pöschl, Bacteria in the global atmosphere – Part 1: review and synthesis of literature data for different ecosystems, *Atmos. Chem. Phys.*, 2009, **9**, 9263–9280.
2. N. Fierer, Z. Liu, M. Rodríguez-Hernández, R. Knight, M. Henn and M. T. Hernandez, Short-term temporal variability in airborne bacterial and fungal populations, *Appl. Environ. Microbiol.*, 2008, **74**, 200–207.
3. R. Bowers, C. Lauber, C. Wiedinmyer, M. Hamady, A. Hallar, R. Fall, R. Knight and N. Fierer, Characterization of airborne microbial communities at a high-elevation site and their potential to act as atmospheric ice nuclei, *Appl. Environ. Microbiol.*, 2009, **75**, 5121–5130.
4. D. J. Smith, Microbes in the upper atmosphere and unique opportunities for astrobiology research, *Astrobiology*, 2013, **13**, 981–990.
5. L. J. Rothschild and R. L. Mancinelli, Life in extreme environments, *Nature*, 2001, **409**, 1092–1101.
6. A. M. Womack, B. J. Bohannan and J. L. Green, Biodiversity and biogeography of the atmosphere, *Philos. Trans. R. Soc., B*, 2010, **365**, 3645–3653.
7. I. Gandolfi, V. Bertolini, G. Bestetti and A. Franzetti, Unraveling the bacterial diversity in the atmosphere, *Appl. Microbiol. Biotechnol.*, 2013, **97**, 4727–4736.
8. J. Heidelberg, M. Shahamat, M. Levin, I. Rahman, G. Stelma, C. Grim and R. Coldwell, Effect of aerosolization on culturability and variability of gram-negative bacteria, *Appl. Environ. Microbiol.*, 1997, **63**, 3585–3588.
9. B. Lighthart, Mini-review of the concentration variations found in the alfresco atmospheric bacterial populations, *Aerobiologia*, 2000, **16**, 7–16.

10. M. Chi, Fluorochrome in monitoring atmospheric bioaerosols and correlations with meteorological factors and air pollutants, *Aerosol Sci. Technol.*, 2007, **41**, 672–678.
11. B. Lighthart, The ecology of bacteria in the alfresco atmosphere, *FEMS Microbiol. Ecol.*, 1997, **23**, 263–274.
12. G. Frenguelli, New trends in aerobiology, *Rev. Allergy Clin. Immunol.*, 2013, **20**, 3–9.
13. D. M. Wilkinson, S. Koumoutsaris, E. A. Mitchell and I. Bey, Modelling the effect of size in the aerial dispersal of microorganisms, *J. Biogeogr.*, 2012, **39**, 89–97.
14. R. H. Diehl, The airspace is habitat, *Trends Ecol. Evol.*, 2013, **28**, 377–379.
15. P. Polymenakou, Atmosphere: a source of pathogenic or beneficial microbes?, *Atmosphere*, 2012, **3**, 87–102.
16. G. Horneck, D. M. Klaus and R. L. Mancinelli, Space microbiology, *Microbiol. Mol. Biol. Rev.*, 2010, **74**, 121–156.
17. S. M. Burrows, T. Butler, P. Jöckel, H. Tost, A. Kerkweg, U. Pöschl and M. G. Lawrence, Bacteria in the global atmosphere – Part 2: modeling of emissions and transport between different ecosystems, *Atmos. Chem. Phys.*, 2009, **9**, 9281–9297.
18. J. Fröhlich-Nowoisky, S. M. Burrows, Z. Xie, G. Engling, P. A. Solomon, M. P. Fraser, O. L. Mayol-Bracero, P. Artaxo, D. Begerow, R. Conrad, M. O. Andreae, V. R. Després and U. Pöschl, Biogeography in the air: fungal diversity over land and oceans, *Biogeosciences*, 2012, **9**, 1125–1136.
19. D. J. Smith, H. J. Timonen, D. A. Jaffe, D. W. Griffin, M. N. Birmele, K. D. Perry, P. D. Ward and M. S. Roberts, Intercontinental dispersal of bacteria and archaea by transpacific winds, *Appl. Environ. Microbiol.*, 2013, **79**, 1134–1139.
20. A. Bouvier and M. Wadhwa, The age of the Solar System redefined by the oldest Pb-Pb age of meteoric inclusion, *Nat. Geosci.*, 2010, **3**, 673–674.
21. C. Hayashi, K. Nakazawa and H. Mizuno, Earth's melting due to the blanketing effect of the primordial dense atmosphere, *Earth Planet. Sci. Lett.*, 1979, **43**, 22–28.
22. J. S. Lewis and R. G. Prinn, *Planets and their atmospheres – Origin and evolution*, Academic Press, Orlando, 1984, p. 470.
23. M. Ikoma and H. Genda, Constraints on the mass of a habitable planet with water of nebular origin, *Astrophys. J.*, 2006, **648**, 696–706.
24. K. Zahnle, L. Schaefer and B. Fegley, Earth's earliest atmospheres, *Cold Spring Harbor Perspect. Biol.*, 2010, **2**(**10**), a004895, DOI: 10.1101/cshperspect.a004895, PMC 2944365. PMID 20573713.
25. C. Sagan and G. Mullen, Earth and Mars: evolution of atmospheres and surface temperatures, *Science*, 1972, **177**, 52–56.
26. D. O. Gough, Solar interior structure and luminosity variations, *Sol. Phys.*, 1981, **74**, 21–34.
27. S. J. Mojzsis, *Life and the evolution of Earth's atmosphere*, 2014, http://www.lopdf.net/preview/JAZrqd53nvaJYfRV113Nehvbax1W-9eIC7wx4A-h3osk,/Life-and-the-Evolution-of-Earth-s-Atmosphere.html?query=What-is-the-Definition-of-Earth-s-Crust, retrieved August 31.

28. J. Lovelock, *Gaia: a new look at life on Earth*, Oxford University Press, New York, 1979, p. 148.
29. J. F. Kasting, Earth's early atmosphere, *Science*, 1993, **259**, 920–926.
30. A. A. Imshenetsky, S. V. Lysenko and G. A. Kazakov, Upper boundary of the biosphere, *Appl. Environ. Microbiol.*, 1978, **35**, 1–5.
31. R. J. States and C. S. Gardner, Thermal structure of the mesopause region (80–105 km) at 40° N latitude. Part I: seasonal variations, *J. Atmos. Sci.*, 2000, **57**, 66–77.
32. J. Buchdahl, *Atmosphere, Climate & Environment Information Program*, 2012, http://www.ace.mmu.ac.uk/eae/Atmosphere/Older/Mesosphere.html, retrieved April 18.
33. R. Morita, Psychrophillic bacteria, *Microbiol. Mol. Biol. Rev.*, 1975, **39**, 144–167.
34. M. Blumthaler, W. Ambach and W. Rehwald, Solar UV-A and UV-B radiation fluxes at two alpine stations at different altitudes, *Theor. Appl. Climatol.*, 1992, **46**, 39–44.
35. M. Blumthaler, W. Ambach and R. Ellinger, Increase in solar UV radiation with altitude, *J. Photochem. Photobiol.*, 1997, **39**, 130–134.
36. G. Seckmeyer, B. Mayer, G. Bernhard, R. Erb, A. Albold, H. Jäger and W. R. Stockwell, New maximum UV irradiance levels observed in Central Europe, *Atmos. Environ.*, 1997, **31**, 2971–2976.
37. D. A. Schmucki and R. Philipona, Ultraviolet radiation in the Alps: the altitude effect, *Opt. Eng.*, 2002, **41**, 3093–3095.
38. WHO, World Health Organization, *Global solar UV index: a practical guide*, A joint recommendation of the World Health Organization, United Nations Environment Programme and International Commission on Non-Ionizing Radiation Protection, 2002, p. 28.
39. E. M. Witkin, Ultraviolet mutagenesis and inducible DNA repair in Escherichia coli, *Microbiol. Mol. Biol. Rev.*, 1976, **40**, 869–907.
40. D. A. Pearce, P. D. Bridge, K. A. Hughes, B. Sattler, R. Psenner and N. J. Russell, Microorganisms in the atmosphere over Antarctica, *FEMS Microbiol. Ecol.*, 2009, **69**, 143–157.
41. Y. Yang, S. Itahashi, S.-i. Yokobori and A. Yamagishi, UV-resistant bacteria isolated from upper troposphere and lower stratosphere, *Biol. Sci. Space*, 2008, **22**, 18–25.
42. Y. Tong and B. Lighthart, Solar radiation is shown to select for pigmented bacteria in the ambient outdoor atmosphere, *Photochem. Photobiol.*, 1997, **65**, 103–106.
43. A. A. Imshenetsky, S. V. Lysenko and S. P. Lach, Microorganisms of the upper layer of the atmosphere and the protective role of their cell pigments, *Life Sci. Space Res.*, 1979, **17**, 105–110.
44. R. C. Smith, B. B. Prézelin, K. S. Baker, R. R. Bidigare, N. P. Boucher, T. Coley, D. Karentz, S. MacIntyre, H. A. Matlick, D. Menzies, M. Ondrusek, Z. Wan and K. J. Walters, Ozone depletion: ultraviolet radiation and phytoplankton biology in Antarctic waters, *Science*, 1992, **255**, 239–242.

45. G. M. Kendall, Factors affecting cosmic ray exposures in civil aviation, *Int. Congr. Ser.*, 2005, **1276**, 129–132.
46. D. Galanet and J. E. Horvath, Biological effects of gamma-ray bursts: distances for severe damage on the biota, *Int. J. Astrobiol.*, 2007, **6**, 19–26.
47. P. Warneck, *Chemistry of the natural atmosphere*, Academic Press, New York, 1988, p. 757.
48. P. Amato, *Clouds provide atmospheric oases for microbes*, 2014, http://www.microbemagazine.org/index.php?option=com_content&view=article&id=4547:clouds-provide-atmospheric-oases-for-microbes&catid=939:07-2011-features&Itemid=1210, Microbe Magazine, American Society of Microbiology, retrieved October 21.
49. C. Schleper, G. Puehler, I. Holz, A. Gambacorta, D. Janekovic, U. Santarius, H. Klenk and W. Zillig, *Picrophilus* gen. nov., fam. nov.: a novel aerobic, heterotrophic, thermoacidophilic genus and family comprising Archaea capable of growth around pH 0, *J. Bacteriol.*, 1995, **177**, 7050–7059.
50. B. Jones, W. Grant, A. Duckworth and G. Owenson, Microbial diversity of soda lakes, *Extremophiles*, 1998, **2**, 191–200.
51. D. R. Lowe, Stromatolites 3,400 myr old from the Archean of Western Australia, *Nature*, 1980, **284**, 441–443.
52. D. R. Lowe, Restricted shallow water sedimentation of Early Archean stromatolitic and evaporitic strata of the Strelley Pool Chert, Pilbara Block, Western Australia, *Precambrian Res.*, 1983, **19**, 239–283.
53. J. W. Schopf, *Earth's Earliest Biosphere: Its Origin and Evolution*, Princeton University Press, Princeton, NJ, 1983, p. 610.
54. C. Woese and J. P. Gogarten, *When did eukaryotic cells (cells with nuclei and other internal organelles) first evolve? What do we know about how they evolved from earlier life-forms?*, 2014, http://www.scientificamerican.com/article/when-did-eukaryotic-cells/, retrieved September 6.
55. J. Hahn and P. Haug, Traces of Archaebacteria in ancient sediments, *Syst. Appl. Microbiol.*, 1986, **7**, 178–183.
56. N. Glansdorff, Y. Xu and B. Labedan, The last universal common ancestor: emergence, constitution and genetic legacy of an elusive forerunner, *Biol. Direct*, 2008, **3**, 29, DOI: 10.1186/1745–6150–3-29.
57. J. W. Schopf, Recent advances in precambrian paleobiology, *Grana Palynologica*, 1969, **9**, 147–168.
58. V. Stricherz, *Large bacterial population colonized land 2.75 billion years ago*, 2014, http://www.washington.edu/news/2012/09/23/large-bacterial-population-colonized-land-2–75-billion-years-ago/, retrieved October 11.
59. Y. Watanabe, J. E. L. Martini and H. Ohmoto, Geochemical evidence for terrestrial ecosystems 2.6 billion years ago, *Nature*, 2000, **408**, 574–578.
60. R. Rye and H. D. Holland, Life associated with a 2.76 Ga ephermal pond? Evidence from Mount Roe #2 paleosol, *Geology*, 2000, **28**, 483–486.
61. H. Beraldi-Campesi, Early life on land and the first terrestrial ecosystems, *Ecol. Processes*, 2013, **2**, 1, DOI: 10.1186/2192–1790–2-1.

62. E. S. Barghoorn and J. W. Schopf, Microorganisms three billion years old from the Precambrian of South Africa, *Science*, 1966, **152**, 758–763.
63. J. W. Schopf and E. S. Barghoorn, Algal-like fossils from the Early Precambrian of South Africa, *Science*, 1967, **156**, 508–512.
64. P. M. Hurley, H. W. Fairbain and W. H. Pinson, Jr., Unmetamorphosed minerals in the Gunflint Formation used to test the age of the Animikie, *J. Geol.*, 1962, **70**, 489–492.
65. C. F. Davidson, Geochemical aspects of atmospheric evolution, *Proc. Natl. Acad. Sci. U. S. A.*, 1965, **53**, 1194–1205.
66. P. E. Cloud, The significance of the Gunflint (Precambrian) microflora, *Science*, 1965, **148**, 27–35.
67. P. E. Cloud and H. Hagen, Electron Microscopy of the Gunflint microflora: preliminary results, *Proc. Natl. Acad. Sci. U. S. A.*, 1965, **54**, 1–8.
68. A. H. Knoll, E. S. Barghorn and S. M. Awramik, New organisms from the Aphebian Gunflint Iron Formation, *J. Paleontol.*, 1978, **52**, 1074–1082.
69. A. H. Knoll, E. J. Javaux, D. Hewitt and P. Cohen, Eukaryotic organisms in Proterozoic oceans, *Philos. Trans. R. Soc., B*, 2006, **365**, 1023–1038.
70. M. Bjornerud, *Reading the rocks: The autobiography of the Earth*, Westview Press, Cambridge, MA, 2006, p. 237.
71. R. K. Bambach, Energetics in the global marine fauna: a connection between terrestrial diversification and change in the marine biosphere, *Geobios*, 1999, **32**, 131–144.
72. M. Blackwell, Terrestrial life: fungal from the start?, *Science*, 2000, **289**, 1884–1885.
73. P. L. Corcoran and W. U. Mueller, Aggressive Archean weathering, in *The Precambrian Earth: tempos and events. Developments in Precambrian geology*, ed. P. G. Eriksson, W. Altermann, D. R. Nelson, W. U. Mueller and O. Catuneanu, Elsevier Science, Amsterdam, 2004, vol. 12, pp. 494–504.
74. H. W. Nesbitt and G. M. Young, Aggressive Archean weathering, in *The Precambrian Earth: tempos and events. Developments in Precambrian geology*, ed. P. G. Eriksson, W. Altermann, D. R. Nelson, W. U. Mueller and O. Catuneanu, Elsevier Science, Amsterdam, 2004, vol. 12, pp. 482–493.
75. P. G. Gensel, The earliest land plants, *Annu. Rev. Ecol. Evol. Systemat.*, 2008, **39**, 459–477.
76. A. C. Allwood, J. P. Grotzinger, A. H. Knoll, I. W. Burch, M. S. Anderson, M. L. Coleman and I. Kanik, Controls on development and diversity of Early Archean stromatolites, *Proc. Natl. Acad. Sci. U. S. A.*, 2009, **106**, 9548–9555.
77. M. A. S. McMenamin, Precambrian conical stromatolites from California and Sonora, *Bull. South. Calif. Palenontol. Soc.*, 1982, **14**, 103–105.
78. NASA, National Aeronautics and Space Administration, *Formation of the ozone layer*, 2014, http://disc.sci.gsfc.nasa.gov/ozone/additional/science-focus/about-ozone/ozone_formation.shtml, Goddard Earth Sciences and Information Services, NASA, retrieved October 11.
79. M. C. Brundrett, Coevolution of roots and mycorrhizas of land plants, *New Phytol.*, 2002, **154**, 275–304.

80. S. G. Staupe, *Concepts of Biology: digestion/nutrition*, 2014, http://employees.csbsju.edu/SSAUPE/biol116/Zoology/digestion.htm, St. Benedict/St. John's University, retrieved October 21.
81. P. Amato, M. Parazols, M. Sancelme, P. Laj, G. Mailhot and A.-M. Delort, Microorganisms isolated from the water phase of tropospheric clouds at the Puy de Dôme: major groups and growth abilities at low temperatures, *FEMS Microbiol. Ecol.*, 2007, **59**, 242–254.
82. C. M. Dobson, G. B. Ellison, A. F. Tuck and V. Viada, Atmospheric aerosols as prebiotic chemical reactors, *Proc. Natl. Acad. Sci. U. S. A.*, 2000, **97**, 11864–11868.
83. D. J. Smith, D. W. Griffin, R. D. McPeters, P. D. Ward and A. C. Schuerger, Microbial survival in the stratosphere and implications for global dispersal, *Aerobiologia*, 2011, **27**, 319–322.
84. M. Wainwright, S. Alharbi and N. C. Wickramsainghe, How do microorganisms reach the stratosphere, *Int. J. Astrobiol.*, 2006, **5**, 13–15.
85. E. P. Odum, *Fundamentals of Ecology*, W. B. Saunders, Philadelphia, 3rd edn, 1971, p. 574.
86. R. H. Whittaker, S. A. Levin and R. B. Root, Niche, habitat and ecotope, *Am. Nat.*, 1973, **107**, 321–338.
87. M. L. Morrison, B. Marcot and W. Mannan, *Wildlife-habitat Relationships: Concepts and Applications*, Island Press, Washington, DC, 3rd edn, 2006, p. 520.
88. M. A. O'Malley, 'Everything is everywhere: but the environment selects': ubiquitous distribution and ecological determinism in microbial biogeography, *Stud. Hist. Philos. Biol. Biomed. Sci.*, 2008, **39**, 314–325.
89. R. L. Dimmick, P. Straat, H. Wolochow, G. Levin, M. Chatigny and J. Schrot, Evidence for metabolic activity of airborne bacteria, *J. Aerosol Sci.*, 1975, **6**, 387–393.
90. P. A. Straat, H. Wolochow, R. L. Dimmick and M. A. Chatigny, Evidence for incorporation of thymidine into deoxyribonucleic acid in airborne bacterial cells, *Appl. Environ. Microbiol.*, 1977, **34**, 292–296.
91. R. L. Dimmick, H. Wolochow and M. A. Chatigny, Evidence that bacteria can form new cells in airborne particles, *Appl. Environ. Microbiol.*, 1979, **37**, 924–927.
92. R. L. Dimmick, H. Wolochow and M. A. Chatigny, Evidence for more than one division of bacteria within airborne particles, *Appl. Environ. Microbiol.*, 1979, **38**, 642–643.
93. L. J. Herlihy, J. N. Galloway and A. L. Mills, Bacterial utilization of formic and acetic acid in rainwater, *Atmos. Environ.*, 1987, **21**, 2397–2402.
94. K. Kawamura and I. R. Kaplan, Stabilities of carboxylic acids and phenols in Los Angeles rainwaters during storage, *Water Res.*, 1990, **24**, 1419–1423.
95. P. A. Ariya, O. Nepotchatykh, O. Ignatova and M. Amyot, Microbial degradation of atmospheric organic compounds, *Geophys. Res. Lett.*, 2002, **29**, 2077–2080.
96. P. A. Ariya and M. Amyot, New directions: the role of bioaerosols in atmospheric chemistry and physics, *Atmos. Environ.*, 2004, **38**, 1231–1232.

97. B. C. Christner, R. Cai, C. E. Morris, K. S. McCarter, C. M. Foreman, M. L. Skidmore, S. N. Montross and D. C. Sands, Geographic, seasonal and precipitation chemistry influence on the abundance and activity of biological ice nucleators in rain and snow, *Proc. Natl. Acad. Sci. U. S. A.*, 2008, **105**, 18854–18859.
98. B. C. Christner, C. E. Morris, C. M. Foreman, R. Cai and D. C. Sands, Ubiquity of biological ice nucleators in snowfall, *Science*, 2008, **319**, 1214.
99. P. Amato, M. Ménager, M. Sancelme, P. Laj, G. Mailhot and A.-M. Delort, Microbial population in cloud water at the Puy de Dôme: implications for the chemistry of clouds, *Atmos. Environ.*, 2005, **39**, 4143–4153.
100. P. Amato, F. Demeer, A. Melaouhi, S. Fontanella, A.-S. Martin-Biesse, M. Sancelme, P. Laj and A.-M. Delort, A fate for organic acids, formaldehyde and methanol in cloud water: their biotransformation by microorganisms, *Atmos. Chem. Phys.*, 2007, **7**, 4159–4169.
101. P. Amato, M. Parazolis, M. Sancelme, G. Mailhot, P. Laj and A.-M. Delort, An important oceanic source of micro-organisms for cloud water at the Puy de Dôme (France), *Atmos. Environ.*, 2007, **41**, 8253–8263.
102. M. Vaïtilingom, P. Amato, M. Sancelme, P. Laj, M. Leriche and A.-M. Delort, Contribution of microbial activity to carbon chemistry in clouds, *Appl. Environ. Microbiol.*, 2010, **76**, 23–29.
103. M. C. Barth, The importance of cloud drop representation on cloud photochemistry, *Atmos. Res.*, 2006, **82**, 294–309.
104. J. L. Neu, M. J. Prather and J. E. Penner, Global atmospheric chemistry: integrating over fractional cloud cover, *J. Geophys. Res.*, 2007, **112**, D11306.
105. M. Vaïtilingom, T. Charbouillot, L. Deguillaume, R. Maisonobe, M. Parazols, P. Amato, M. Sancelme and A.-M. Delort, Atmospheric chemistry of carboxylic acids: microbial implication versus photochemistry, *Atmos. Chem. Phys.*, 2011, **11**, 8721–8733.
106. S. Husárová, M. Vaïtilingom, L. Deguillaume, M. Traikia, V. Vinatier, M. Sancelme, P. Amato, M. Matulová and A.-M. Delort, Biotransformation of methanol and formaldehyde by bacteria isolated from clouds. Comparison with radical chemistry, *Atmos. Environ.*, 2011, **45**, 6093–6102.
107. M. Vaïtilingom, L. Deguillaume, V. Vianter, M. Sancelme, P. Amato, N. Chaumerliac and A.-M. Delort, Potential impact on microbial activity on the oxidant capacity and organic carbon budget in clouds, *Proc. Natl. Acad. Sci. U. S. A.*, 2013, **110**, 559–564.
108. S. Durand, P. Amato, M. Sancelme, A.-M. Delort, B. Combourieu and P. Besse-Hoggan, First isolation and characterization of a bacterial strain that transforms the herbicide mesotrione, *Lett. Appl. Microbiol.*, 2006, **43**, 222–228.
109. B. Sattler, H. Puxbaum and R. Pessner, Bacterial growth in supercooled cloud droplets, *Geophys. Res. Lett.*, 2001, **28**, 239–242.
110. K. A. Hill, P. B. Shepson, E. S. Galbavy, C. Anastasio, P. S. Kourtev, A. Konopka and B. H. Strim, Processing of atmospheric nitrogen by clouds above a forest environment, *J. Geophys. Res.*, 2007, **112**, 1–16.

111. J. Williams, M. De Reus, R. Krejci, H. Fischer and J. Ström, Application of the variability-size relationship to atmospheric aerosol studies: estimating aerosol lifetimes and ages, *Atmos. Chem. Phys.*, 2002, **2**, 133–145.
112. H. E. Kubitschek, Cell volume increase in *Escherichia coli* after shifts to richer media, *J. Bacteriol.*, 1990, **172**, 94–101.
113. Y. Tong and B. Lighthart, Diurnal distribution of total and culturable atmospheric bacteria at a rural site, *Aerosol Sci. Technol.*, 1999, **30**, 246–254.
114. B. T. Shaffer and B. Lighthart, Survey of culturable airborne bacteria at four diverse locations in Oregon: urban, rural, forest and costal, *Microb. Ecol.*, 1997, **34**, 167–177.
115. C. C. Wang, G. C. Fang and L. Lee, Bioaerosols study in central Taiwan during summer season, *Toxicol. Ind. Health*, 2007, **23**, 133–139.
116. J. A. Huffman, B. Treutlein and U. Pöschl, Fluorescent biological aerosol particle concentrations and size distributions measured with an ultraviolet aerodynamic particle sizer (UV-APS) in central Europe, *Atmos. Chem. Phys.*, 2010, **10**, 3215–3233.
117. R. Meisterfeld, Order *Arcellinida Kent*, 1880, in *The Illustrated guide to Protozoa*, ed. J. J. Lee, G. F. Leedale and P. Bradbury, Society of Protozoologists, Lawrence, KS, 2002, pp. 827–860.
118. R. Meisterfeld, Testate amoebae with filopodia, in *The Illustrated guide to Protozoa*, ed. J. J. Lee, G. F. Leedale and P. Bradbury, Society of Protozoologists, Lawrence, KS, 2002, pp. 1054–1084.
119. V. R. Després, J. A. Huffman, S. M. Burrows, C. Hoose, A. S. Safatov, G. Buryak, J. Fröhlich-Nowoisky, W. Elbert, M. O. Andreae, U. Pöschl and R. Jaenicke, Primary biological aerosol particles in the atmosphere: a review, *Tellus, Ser. B*, 2012, **64**, 15598, DOI: 10.3402/tellusb.v64i0.15598.
120. J. Lelieveld and J. Heintzenberg, Sulfate cooling effect on climate through in-cloud oxidation of anthropogenic SO_2, *Science*, 1992, **258**, 117–120.
121. D. J. Smith, D. A. Jaffe, M. N. Birmele, D. W. Griffin, A. C. Schuerger, J. Hee and M. S. Roberts, Free tropospheric transport of microorganisms from Asia to North America, *Microb. Ecol.*, 2012, **64**, 973–985.
122. I. Toepfer, J. Favet, A. Schulte, M. Schmölling, W. Butte, E. W. Triplett, W. J. Broughton and A. A. Gorbushina, Pathogens as potential hitchhikers on intercontinental dust, *Aerobiogloia*, 2011, **28**, 221–231.
123. C. A. Kellogg and D. W. Griffin, Aerobiology and the global transport of desert dust, *Trends Ecol. Evol.*, 2006, **21**, 638–644.
124. H. Bauer, H. Giebl, R. Hitzenberger, A. Kasper-Giebl, G. Reischl, F. Zibuschka and P. Pauxbaum, Airborne bacteria as cloud condensation nuclei, *J. Biophys. Res.*, 2003, **108(d21)**, 4658, DOI: 10.1029/2003jd003545.
125. T. Dehel, F. Lorge and M. Dickinson, Uplift of microorganisms by electric fields above thunderstorms, *J. Electrost.*, 2008, **66**, 463–466.
126. V. P. Pasko, M. A. Stanley, J. D. Mathews, U. S. Inan and T. G. Wood, Electrical discharge from thundercloud tops to the lower ionosphere, *Nature*, 2002, **416**, 152–154.

127. N. Deleon-Rodriguez, T. L. Lathem, R. L. Rodriguez, J. M. Barazesh, B. E. Anderson, A. J. Beyersdorf, L. D. Ziemba, M. Bergin, A. Nenes and K. T. Konstantinidis, Microbiome of the upper troposphere: species composition and prevalence, effects of tropical storms and atmospheric implication, *Proc. Natl. Acad. Sci. U. S. A.*, 2013, **110**, 2575–2580.
128. W. J. Randel, M. Park, L. Emmons, D. Kinnison, P. Bernath, K. A. Walker, C. Boone and H. Pumphrey, Asian monsoon transport of pollen to the stratosphere, *Science*, 2010, **328**, 611–613.
129. J. R. Holton, P. H. Haynes, M. E. McIntyre, A. R. Douglass, R. W. Rood and L. Pfister, Stratosphere-troposphere exchange, *Rev. Geophys.*, 1995, **33**, 403–440.
130. H. Rohatschek, Levitation of stratospheric and mesospheric aerosols by gravito-photophoresis, *J. Aerosol Sci.*, 1966, **27**, 467–475.
131. A. R. Van Eaton, M. A. Harper and C. J. N. Wilson, High flying diatoms: widespread dispersal of microorganisms in an explosive volcanic explosion, *Geology*, 2013, **41**, 1187–1190.
132. BTS, Bureau of Transportation Statistics, *March 2013 U.S. airline system passengers up 1.0% from March 2012*, BTS, U.S. Department of Transportation, 2013.
133. A. R. Wallace, *The geographical distribution of animals, with a study of the relations of living and extinct faunas as elucidating the past changes of the Earth's surface*, Harper Brothers, New York, 1876, vol. 1, p. 503.
134. M. V. Lomolino, B. R. Riddle, R. J. Whittaker and J. H. Brown, *Biogeography*, Sinauer Associates, Sunderland, MA, 4th edn, 2010, p. 560.
135. J. K. Brown and M. S. Hovmøler, Aerial dispersal of pathogens on the global and continental scales and its impact on plant disease, *Science*, 2002, **297**, 537–541.
136. D. W. Griffin, C. A. Kellogg, V. H. Garrison and E. A. Shinn, The transport of dust: an intercontinental river of dust, microorganisms and toxic chemicals flows through Earth's atmosphere, *Am. Sci.*, 2002, **90**, 228–235.
137. J. F. Kasting, Theoretical constraints on oxygen and carbon dioxide concentrations in the Precambrian atmosphere, *Precambrian Res.*, 1987, **34**, 205–229.
138. J. C. G. Walker, C. Klein, M. Schidlowski, J. W. Schopf, D. J. Stevenson and M. R. Walter, Environmental evolution of the Archean-early Proterozoic Earth, in *Earth's Earliest Biosphere: Its Origin and Evolution*, ed. J. W. Schopf, Princeton University Press, Princeton, NJ, 1983, pp. 260–290.
139. J. E. Johnson, A. Gerpheide, M. P. Lamb and W. W. Fischer, O_2 constraints from paleoproterozoic detrital pyrite and uraninite, *Geol. Soc. Am. Bull.*, 2014, **126**, 813–830.
140. B. Runnegar, Precambrian oxygen levels estimated from the biochemistry and physiology of early eukaryotes, *Global Planet Change*, 1991, **5**, 97–111.
141. T. P. Jones and W. G. Chaloner, Fossil charcoal, its recognition and paleoatmospheric significance, *Global Planet Change*, 1991, **5**, 39–50.

142. C. Hallmann, *Max Planck Research Group: Organic Paleobiogeochemistry*, Max Plank Institute for Biogeochemistry, Bremen, Germany, 2014, https://www.bgc-jena.mpg.de/index.php/Groups/OrganicPaleobiogeochemistry, retrieved December 3.
143. A. Caldicott, *Alistair Caldicott: Making hard places easy reading*, Alistair Caldicott, 2014, http://www.alitravelstheworld.com/australia/shark_bay/shark_bay_stromatolites, retrieved December 3.
144. NASA, National Aeronautics and Space Administration, *Global circulation of Earth's atmosphere*, Jet Propulsion Laboratory, NASA, 2005, http://sealevel.jpl.nasa.gov/overview/climate-climatic.html sealevel.jpl.nasa.gov, retrieved June 22.
145. NOAA, National Oceanic and Atmospheric Administration, *National Weather Service: Jetstream – online school for weather*, National Weather Service, NOAA, 2014, http://srh.noaa.gov/jetstream/global/jet.htm, retrieved December 3.
146. I. Bey, D. J. Jacob, R. M. Yantosoa, J. A. Logon, B. D. Field, A. M. Fiore, Q. B. Li, H. G. Y. Liu, L. J. Mickley and M. G. Schultz, Global modeling of tropospheric chemistry with assimilated meterology: model description and evaluation, *J. Geophys. Res.*, 2001, **106**, 23073–23906.
147. M. Park, W. J. Randel, D. E. Kinnison, R. R. Garcia and W. Choi, Seasonal variation of methane, water vapor and nitrogen oxides near the tropopause: satellite observations and model simulations, *J. Geophys. Res.*, 2004, **109**, D03302, DOI: 10.1029/2003JD003706.

CHAPTER 2

Orthopoxviruses *and Animal Models for Pathogenesis, Vaccine and Drug Studies*[†]

M. SOFI IBRAHIM*[a,b] AND HERMANN MEYER[a]

[a]Bundeswehr Institute of Microbiology, Munich, Germany; [b]US Army Research Development and Engineering Command, Edgewood Chemical Biological Center, Aberdeen Proving Ground, MD 21010, USA
*E-mail: sofi.ibrahim@us.army.mil

2.1 Introduction

Airborne transmissibility of the variola virus was a key factor for its pandemicity. It is also a factor for the increased concerns about its potential use as a biological weapon by rogue states and clandestine or terrorist organizations.[1] Although smallpox is now eradicated,[2] the variola virus still exists in high-security WHO repositories in the United States and the Russian Federation and may also exist in unaccounted for locations throughout the world. If the virus is released, deliberately or accidentally, there would be global

[†]The views and opinions expressed in this chapter are those of the author (M.S.I.) and should not be construed as an official Department of the Army position, policy, or decision, unless so designated by other official documentation. Citation of trade names in this chapter does not constitute an official Department of the Army endorsement or approval of the use of such commercial items.

Issues in Toxicology No. 25
Aerobiology: The Toxicology of Airborne Pathogens and Toxins
Edited by Harry Salem and Sidney Katz

Published by the Royal Society of Chemistry, www.rsc.org

catastrophic consequences, because most of the world population today is immunologically naïve against smallpox after the cessation of mandatory vaccination in the early 1970s in the United States and Europe and 1980s in the rest of the world and the prevalence of immunocompromised individuals has grown. Since the pace of worldwide travel continues to increase and the incubation period of the infection is relatively long (up to 2 weeks), it would be difficult to trace all contacts to contain an outbreak efficiently.[1] Considerable efforts have been made by the United States and other countries to develop new generations of antiviral drugs and vaccines. However, owing to justifiable safety, security and ethical considerations, the use of variola virus in research and development is highly restricted. Development and testing of therapeutics and vaccines for smallpox require US Food and Drug Administration (FDA) approval under the "Animal Rule" (21 CFR 314.600 for drugs and 21 CFR 601.90 for biological products). Therefore, a number of animal models using different orthopoxviruses have been established[3] and, although no single model mimics all aspects of the human smallpox, these models have collectively contributed to significant understanding of pathogenicity and host response and for evaluating the safety and efficacy of drugs and vaccines. Since the respiratory system has been well documented to be the primary route of entry of the variola virus,[4–8] aerosol biological studies using closely related poxviruses have been used in rodent, rabbit and non-human primate animal models. However, because of differences in virus–host specificity, susceptibility and disease outcome, research on poxvirus pathogenicity, drugs and vaccines has encompassed various poxvirus species and strains, and other routes of infections in different laboratory animals to investigate certain aspects of smallpox that could be useful in developing prevention and therapeutic strategies. This chapter focuses on the utility of some of the animal models in understanding the systemic nature of the disease, its pathogenesis, host defense mechanisms and for the development of intervention strategies.

2.2 The Poxviruses

2.2.1 Taxonomy and Morphology

Poxviruses are large, enveloped, double-stranded DNA viruses that include a number of medically and veterinary important viruses, the most notable of which is the variola virus, the causative agent of smallpox. The Poxviridae family is divided into two subfamilies Chordopoxvirinae and Entomopoxvirinae, which infect vertebrates and insects, respectively.[9] There are 10 recognized genera with 37 species (plus one unassigned) within Chordopoxvirinae (Table 2.1) and three genera with 26 species (plus two unassigned) within Entomopoxvirinae. Although most poxviruses are zoonotic, several viruses in the genera *Molluscipoxvirus*, *Orthopoxvirus*, *Parapoxvirus* and *Yatapoxvirus* of subfamily Chrodopoxvirinae can also infect humans (Table 2.1). These viruses, collectively, have a broad host range, but vary with respect to host

Table 2.1 Chordopoxviruses.

Subfamily	Genus	Virus
Chordopoxvirinae	*Avipoxvirus*	Canarypox, fowlpox[a], Juncopox, mynahpox, pigeonpox, psittacinepox, quailpox, sparrowpox, starlingpox, turkeypox
	Capripoxvirus	Goatpox, lumpy skin disease, sheeppox[a]
	Cervidpoxvirus	Mule deerpox[a]
	Crocodylidpoxvirus	Nile crocodilepox[a]
	Leporipoxvirus	Hare fibroma, myxoma[a], rabbit fibroma, squirrel fibroma
	Molluscipoxvirus	Molluscum contagiosum[a,b]
	Orthopoxvirus	Camelpox, cowpox[b], ectromelia, monkeypox[b], raccoonpox, skunkpox, taterapox, vaccinia[a,b], variola[b], volepox virus
	Parapoxvirus	Bovine papular stomatitis, orf[a,b], parapoxvirus of red deer in New Zealand, pseudocowpox[b]
	Suipoxvirus	Swinepox[a]
	Yatapoxvirus	Tanapox[b], Yaba monkey tumor[a]
	Unassigned	Squirrelpox

[a]Prototype virus within the genus.
[b]The virus causes human infection.

specificity. Whereas vaccinia, cowpox and monkeypox viruses have multiple hosts (cattle, rodents and domestic and zoo animals), camelpox, ectromelia, taterapox and variola viruses have a highly restricted host range (camels, mice, gerbils and humans, respectively). Such host range adaptation seems to be linked to a set of viral genes that have been selected over time during virus association with particular hosts.[10]

Poxviruses share common morphological features. The viral particles are oval or brick shaped, ranging in size from approximately 100–300 nm by 200–500 nm. The outer envelope is composed of a lipoprotein layer, which, like other enveloped viruses, is derived from host cells during virus assembly and exit. The viral core (or nucleoprotein), usually described as biconcave, is filled with two masses of lateral bodies, one on each side of the concavity. It contains the viral genome, enzymes, proteins, fibrillar and granular structures, which give poxviruses their characteristic textured appearance (Figure 2.1).

2.2.2 Genomes and Genetic Relationships

Poxviruses have double-stranded DNA genomes with covalently closed ends, ranging in size from about 130 to 380 kbp. The genomes encode from about 130 to over 300 genes. Unlike other DNA viruses, which replicate in the nucleus, poxviruses replicate in the cytoplasm, as they possess viral-encoded enzymes, other proteins and transcription factors that are necessary for replication.

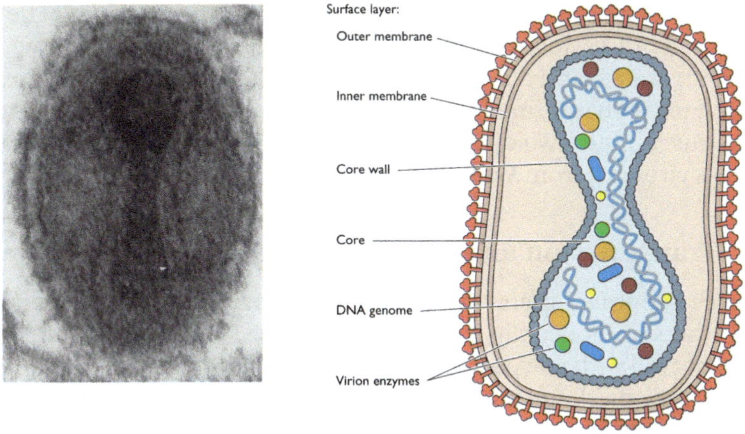

Figure 2.1 Structure of the vaccinia virus. Principles of Virology, S. J. Flint, V. R. Racaniello, L. W. Enquist and A. M. Skalka, 2009, ASM Press, Washington, DC, 2009.[156]

There are 49 genes that are conserved in all members of the Poxviridae family and 90 genes that are conserved only within members of the Chordopoxvirinae subfamily.[11] All conserved genes lie within the central region of the genome in both subfamilies.[12,13] This conserved region, which comprises about two-thirds of the genome, is flanked by terminal regions that contain hairpins and inverted terminal repeats (ITRs), which vary considerably in length between different species and strains. These terminal regions contain the host-range and virulence genes and the ITRs play an important role in viral genome replication.

Complete genome sequencing of many poxviruses over the past two decades has elucidated their phylogenetic relationships. Entomopoxvirinae and Chordopoxvirinae are quite divergent.[13,14] Within the Chordopoxvirinae subfamily, genus *Avipoxvirus* is the most divergent, followed by *Molluscipoxvirus* and *Parapoxvirus*.[12,15] The *Capripoxvirus*, *Cervidpoxvirus*, *Leporipoxvirus*, *Suipoxvirus* and *Yatapoxvirus* genera cluster together and are distinct from the genus *Orthopoxvirus*.[15] Within the *Orthopoxvirus* genus, monkeypox, ectromelia and cowpox Brighton viruses do not group closely with any other orthopoxvirus species; variola, camelpox and taterapox viruses form a subgroup;[10] and vaccinia virus is most closely related to cowpox virus GRI-90.[12]

The evolution of poxviruses has been subjected to a number of studies. The prevailing hypothesis is that horizontal gene transfer, *i.e.* acquisition of host genes by viruses, plays an important role in poxvirus evolution and host adaptation.[16,17] A substantial number of these host-acquired genes are involved in evasion or subversion of host defense mechanisms and pathology. Recombination has played a significant role in the evolution of at least some poxviruses.[10,12,18,19]

Strains of variola virus show two major clades. One clade contains isolates of variola major from Asia that are associated with high mortality and isolates from Africa that are associated with varying mortalities. The second clade consists of variola minor (alastrim) from South America, which has lower fatality rates, and also closely related isolates from West Africa, which are associated with intermediate levels of disease severity.[20,21] Strains of monkeypox virus fall into two distinct groups, one from the Congo basin and the other (less virulent) from West Africa.[22]

2.2.3 Transmission and Routes of Infection

Poxviruses can infect their host *via* skin, oral and respiratory routes. Entry through the skin is usually through abrasions. This route of entry is commonly observed with cowpox, vaccinia and buffalopox virus infections in cattle with lesions on their teats and on the hands of humans who milk them.[23] It is notable that this route has been used to deliver the smallpox vaccine, by scarification. Oral entry occurs through small abrasions of the lips, oral mucous membranes or epithelial cells of the small intestine.[23]

Virus entry through the respiratory tract by aerosols or microdroplets is the primary route of infection in smallpox. This has been documented from epidemiological data, where the most important source of infection was entry of the variola virus into the respiratory tract *via* airborne droplets from the mouth and nose of enanthemic patients. This usually occurred when persons came into face-to-face contact with patients during the first week of the rash, prior to release of virus from skin lesions.[23] Nosocomial airborne transmission of variola virus was reported in Meschede Hospital in Germany in 1970, where 19 patients had contracted smallpox from a patient with whom they had no direct contact. Airborne transmission was confirmed by tests of air movements from the room occupied by the index case to all rooms in which secondary cases occurred.[24] The last documented fatal case of smallpox, which occurred in 1978 at the University of Birmingham, UK, is believed to have been acquired by the 40-year-old woman patient through an air duct that connected a smallpox virology laboratory to the "telephone room."[25]

Experimental infections using cowpox, ectromelia, monkeypox, rabbitpox, vaccinia and variola viruses have been carried out in different laboratory animals, mostly mice, rabbits, squirrels, prairie dogs and monkeys.

2.3 Clinical Aspects of Poxvirus Infections

2.3.1 Smallpox

The most striking clinical features of poxviruses infections are manifested by rash leading to the characteristic cutaneous lesions known as pocks. In smallpox, the lesions appear after an incubation period of approximately 8–14 days, starting with rashes in different parts of the body and ending with scabs (and, in some cases, permanent facial pockmarks), passing through stages of macules, papules and vesicles. All lesions, in smallpox infections, in a given area progress through these stages synchronously.[26] Ulcerating lesions in the mucous membranes of the nose and mouth produce large amounts of infectious virus into the mouth and throat, constituting the main source of respiratory transmission. The prodromal phase, which occurs a few days prior to the rash, is usually accompanied by fever, headache, backache and sore throat. The profusion and severity of the rash and skin lesions vary depending on the virus strain and the individual's susceptibility and immune system status.

Smallpox infections have historically been recognized in two different forms. The severe form was caused by variola major virus with case fatality rate up to 10–30% in patients with discrete pustules, 60% in patients with confluent pustules and 100% in patients with hemorrhagic and flat-type rash.[26,27] The less severe form was caused by variola minor (alastrim) with a case fatality rate of about 2%. In severe and fatal variola major cases, the skin lesions were likened to severe burns and, together with other systemic (vascular, alveolar and nephritic) lesions, they caused shock due to volume depletion, hypotension and renal failure.[7,28–32]

2.3.2 Monkeypox

Monkeypox was first described as a rash-causing illness in zoo monkeys by von Magnus *et al.*[33] Subsequently, the disease was recognized in other captive animals in zoos and animal holding facilities elsewhere. Human monkeypox was first reported in 1970 in the Democratic Republic of Congo (DRC).[34,35] Recent epidemiological investigations suggested a 20-fold increase in human monkeypox in rural DRC since the 1980s.[36] Monkeypox virus infections are acquired through the respiratory system or skin abrasions. The clinical appearance of human monkeypox is much like that of smallpox, with fever, headache, muscle pain, vesiculo-pustular rash and respiratory distress. During the prodrome phase, lymphadenopathy in the submandibular, cervical, axillary and inguinal nodes occurs in most cases.[37] Individual skin lesions develop through stages of macule, papule, vesicle, pustule and scabs; however, progression of skin lesions through the stages is not synchronous, as is the case with smallpox lesions. Severe cases involving pulmonary disease and respiratory distress are associated with fatalities,[38] which can be as high as 10% in children under 10 years old.

Strains of monkeypox virus fall into two distinct groups, one from the Congo basin and the other from West Africa.[22] The West African strains have been shown to be significantly less virulent than the Central African strains in aerosol studies with cynomolgus monkeys and at least five genes were attributed to differences in virulence.

2.3.3 Other Poxvirus Infections

Members of the species cowpox virus have significant genomic variability and have been isolated from humans and a variety of animals in Europe and western Asia.[39] Cowpox virus infections are usually acquired by direct contact with infected cows and other domestic and zoo animals. Serosurveys of wild animals in Great Britain and Germany showed orthopoxvirus antibodies in bank and field voles, wood mice and foxes.[40,41] The clinical features of cowpox infections are characterized by single pock-like lesions on the hands, legs or face with reddening and swelling, although multiple lesions are seen in some cases.[39,42–45] The lesions progress over 2–3 weeks from macules to papules to vesicles to pustules to scabs. Complete healing occurs within 6–8 weeks. A few severe generalized cowpox infections have been reported.[46]

Vaccinia virus is the prototype species of genus *Orthopoxvirus* and the most thoroughly studied poxvirus, although its origin is uncertain. Vaccinia virus preparations have been used as vaccines against smallpox, which ultimately led to its eradication. However, severe condition known as eczema vaccinatum, sometimes fatal, can develop in individuals with atopic dermatitis (eczema) as a result of vaccination.[47] Other serious adverse reactions including progressive vaccinia and post-vaccinial encephalitis have been reported.[7] Vaccinia virus infections are not generally regarded as occurring naturally; however, multiple bovine vaccinia outbreaks in cattle and humans have been reported in Brazil since 1999.[48–50] Inadvertent exposure to vaccinia-vectored recombinant rabies virus vaccine dispersed to control rabies in wildlife has resulted in at least one instance of human infection.[51] Sporadic outbreaks of infection caused by the vaccinia virus subspecies buffalopox virus in buffalo milkers have been reported, mainly in India.[52]

Several other poxviruses, including bovine popular stomatitis, deerpox, molluscum contagiosum, orf, sealpox and tanapox viruses, can infect humans, causing a few cutaneous lesions with low morbidity.[53]

2.4 Animal Models and Aerosol Studies

The challenge in aerosol studies is to set up reproducible experimental conditions that mimic natural smallpox infection. Since the variola virus is solely a human pathogen, it is not possible to reproduce all aspects of disease characteristics in one laboratory model. Therefore, various models, different routes of infections and different orthopoxviruses are used to capture certain disease characteristics. Different laboratory animals have been investigated using the intradermal, intravenous, intraperitoneal, intranasal, intratracheal or aerosol routes of infection with several orthopoxviruses. The most widely used routes of infection to induce smallpox-like disease in laboratory animals are the intradermal route for mousepox in mice, cowpox in mice and rabbitpox in rabbits. The need to develop aerosol models is driven by the fact that the respiratory system is the main route of variola virus entry in natural smallpox infection and it is the most likely route by which the virus could be used in a biological warfare scenario. The respiratory route of infection, by aerosol, intranasal or intratracheal exposure, has been used with mousepox, cowpox, vaccinia and rabbitpox in mice and monkeypox in monkeys and other animals.

In aerosol studies, determination of the proper dose for each model is a critical factor, which in turn relies on a number of factors, including the viability, count and size of aerosolized viral particles, air flow rate, size and weight of the animal, anatomy of its respiratory system and the animal's respiration volume and rate.[54] Since poxviruses are highly stable in the ambient environment,[55] it is assumed that the virus attains near 100% viability over the duration of the study, which usually lasts from minutes to hours. Assessment of virus viability is usually performed by plaque assays on samples collected during the study.

In early aerosol studies, animals were exposed to poxviruses (and other microorganisms) in the form of a cloud of dried particles in a Henderson aerosol apparatus.[6,56] The apparatus is essentially a closed chamber that produces a particulate cloud from a liquid suspension. It allows animals to be exposed at varying periods of time and samples of the cloud to be collected for estimating the concentration of viable organisms. The virus (or other microorganisms) aerosol is generated by a collision atomizer operated by compressed air at one end of the chamber. Constant, unidirectional airflow is maintained in the chamber using an air suction pump with a pressure regulator at the other end. The air is kept dry by means of silica gel driers inside the chamber. Using this device, particles measuring about 1 μm or less in diameter produced systemic infections in mice, rabbits and monkeys with ectromelia, rabbitpox and variola viruses, respectively.[6] Modifications of this apparatus have been made over the past six decades to improve particle size homogeneity, distribution, dosage, sampling and safety and accommodate for simultaneous exposure of multiple animals and different animal species.[57–60]

Infectious dose has traditionally been expressed in terms of plaque-forming units (PFU) for poxviruses. Much of the informative laboratory work with regard to infectious doses of poxviruses has been carried out with vaccinia virus.[61] Although there are differences in virulence between vaccinia and variola viruses, studies have shown similar ratios between the concentration of viral particles and the concentration of infectious units. Much of the informative laboratory work with regard to infectious doses of poxviruses has been carried out with vaccinia virus, which raises concerns about extrapolating experimental results from vaccinia to variola viruses infections.[61] However, despite the differences in virulence between vaccinia and variola viruses, studies have shown similar ratios between the concentration of viral particles and the concentration of infectious units, suggesting that infectivity trait is conserved across *Orthopoxvirus* species.[61,62,63]

2.4.1 Pathogenesis

The characteristic cutaneous lesions of poxviruses infections are just the tip of the iceberg of the virus–host interactions and pathogenesis processes. The classic work of Fenner[64,65] on ectromelia virus infection in mice elucidated for the first time how the virus is disseminated from local sites of inoculation through the lymphatic and circulatory systems into internal organs, involving the host defense system and resulting in rash (Figure 2.2). These findings were subsequently confirmed in rabbits infected with rabbitpox[66] and monkeys infected with monkeypox.[67] Further knowledge about poxvirus pathogenesis over the past four decades has been unraveled by many studies using different species and strains of poxviruses in a variety of laboratory animal models.[68–71]

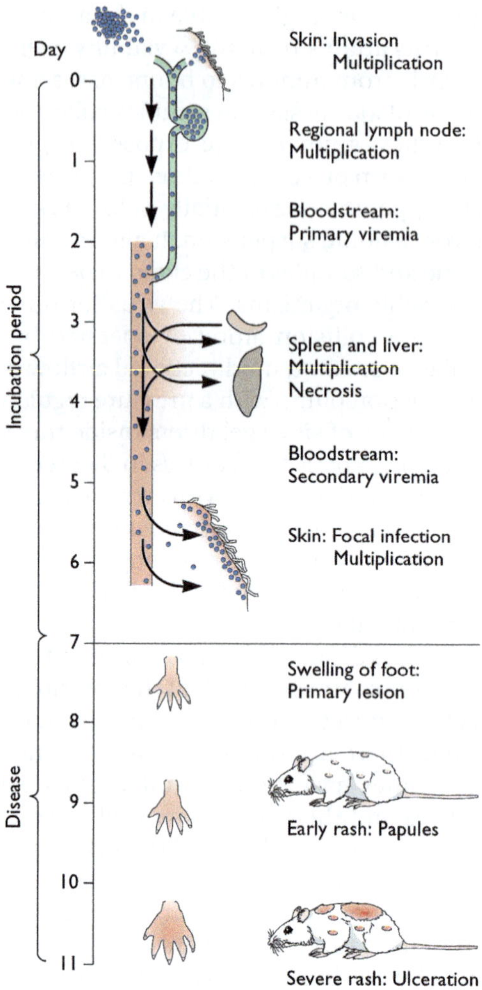

Figure 2.2 Pathogenesis of mousepox. Inoculation of ectromelia virus in the mouse footpad results in disseminated infection. After initial virus replication in the site of infection, the virus is transported to internal organs *via* the lymphatic and circulatory systems. After viremia, the host response to virus replication results in skin rash. Reprinted with permission from, Principles of Virology, S. J. Flint, V. R. Racaniello, L. W. Enquist and A. M. Skalka, 2009, ASM Press, Washington, DC, 2009. (Adapted with permission from, The Biology of Animal Viruses, F. Fenner *et al.*, 1974, Academic Press, New York).

In both natural and experimental infections, poxvirus pathogenesis is the outcome of cytopathic effects that cause broad-ranging dysregulation of cellular functions and host defense mechanisms. An examination of smallpox pathology reports published over a period of 200 years suggested that the cause of death was cytopathic effects,[32] which begin at the cellular level during virus replication in the cell cytoplasm in areas known as "viral factories" or "inclusion bodies." Replication starts with attachment of viral particles to the cell *via* a host of viral proteins and cell receptors.[72] Entry of the virion core into the cell occurs within the first hour. The entire replication cycle (attachment, entry, transcription and translation, genome replication, assembly and exit) may take up to 12 h. Virus replication occurs in three main stages: early, intermediate and late. After virion fusion with cell membrane, the virion core is internalized into the cytoplasm by receptor-mediated phagocytosis or macropinocytosis, and early mRNAs are synthesized and translated by cell machinery. The early proteins induce the release of viral nucleoprotein from the core coat, proliferation of neighboring cells, modulation of host defense mechanisms and replication of genomic DNA. Newly synthesized DNAs serve as templates for new cycles of genome replication and transcription of intermediate genes. Proteins translated from the intermediate mRNAs induce transcription of late genes, which produce structural proteins in addition to nonstructural proteins. These proteins, along with the genomic DNA, are assembled and packaged into membrane-bound intracellular immature virions, which then evolve into brick-shaped intracellular mature virions (IMVs). These IMVs are infectious when released during cell lysis. Some IMVs acquire a second membrane to form intracellular enveloped virions (IEVs), migrate to the cell surface *via* microtubules and exit the cell *via* actin filaments. These virions are known as extracellular mature virions (EEVs) and can directly infect neighboring cells.[73]

During the replication cycles, cell toxicity is manifested by morphological and molecular aberrations, with far-reaching pathological consequences on the infected host as a whole. The signs of gross toxic effects include rounding, granulations and clumping of infected cells.[23] These effects are seen within about 1 h after infection. As virus replication progresses, the cell architecture continues to degenerate with ultimate cell lysis and death, which may occur as early as 24 h post-infection.[68]

As mentioned above, typical poxvirus infection, as illustrated by Fenner,[64,65] involves hematogenous spread of the virus *via* the lymphatic and circulatory systems. The first round of replication occurs in epithelial cells at the site of inoculation, followed by virus transport through the lymphatic system into blood vessels, where primary viremia occurs. More rounds of viral replication occur at multiple organs, including the lungs, spleen and kidney. From these organs, the virus is transported *via* capillaries, where a second viremia occurs, to the skin, oropharyngeal and other visceral tissues and organs immediately before the prodromal phase. In animal studies, evidence of replicating virus was found in the lymphoid tissues, skin, oral mucosa, gastrointestinal tract, respiratory system, reproductive system, kidneys and other organs and correlated with pathology in these tissues and organs.[37]

Comprehensive descriptions of the pathology of smallpox have been presented by Councilman *et al.*,[29] Lillie,[30] MacCallum and Moody,[74] Michelson and Ikeda,[75] Bras[31,76] and Fenner *et al.*[7]

In ordinary-type smallpox, viremia was usually restricted to the pre-eruptive and early eruptive stages of the disease; in contrast, hemorrhagic-type smallpox appears to have been associated with overwhelming infection and continued release of virus into the bloodstream.[7] Oropharyngeal lesions ulcerate very soon after their formation, releasing large amounts of highly infectious virus into the saliva. Following virus dissemination throughout the body and onset of rash, toxemia occurs, particularly in patients with flat-type or hemorrhagic-type smallpox. Even in ordinary variola major infections, patients appeared very sick compared with patients with variola minor infections who had a similar number of skin lesions. It is believed, from experimental rabbitpox studies, that the cause of toxemia was hypotension leading to shock-like syndrome, hyperuricemia and hyperkalemia.[77]

Anatomical and histological studies from autopsy and biopsy specimens from smallpox victims showed that the skin and mucous membranes had the most striking pathological changes. Histological studies of the skin lesions showed that initial pathological changes occur in the dermal blood vessels, followed by degeneration of surrounding cells with appearance of the characteristic B-type inclusion (Guarnieri) bodies and vesiculation. Pustulation occurs by the migration of the polymorphonuclear cells. Umbilication was mainly due to the swelling of the cells around the vesicle and the proliferation of basal cells surrounding the lesion, so that the periphery of the vesicle was raised above the level of its center, and also above the surrounding unaffected skin.[7] Scabbing occurs as a result of effective immune response in patients recovering from the infections. As healing begins, the contents of pustules desiccate and become crusts or scabs. Scars, remaining mostly in the face of fully recovered patients, result from the filling of dermal defects with granulation tissues, which subsequently shrink, leaving localized facial pockmarks. The major difference between ordinary and hemorrhagic-type lesions is that hemorrhage occurs beneath the pustules and patients usually die before vesiculation.

Lesions of the mucous membranes of the respiratory and digestive tracts (pharynx, uvula, tongue and trachea) evolve more rapidly than those in the skin. Pathological changes in the lungs were complicated by secondary, streptococcal infections; however, the case fatality rate was not reduced with antibiotic treatment, indicating that the death was due to the variola virus, not bacterial infections.[31] Microscopic examinations of fatal cases revealed involvement of the reticuloendothelial system. The endothelial lining of the sinusoids of the liver was often swollen and occasionally necrotic and parenchymal cells usually showed intense cloudy swelling. The spleen was usually enlarged and contained numerous lymphoid cells. Small necrotic foci were observed in various parts of the testis.

2.4.1.1 Smallpox Model

An animal model in which variola virus produces a disease similar to smallpox would be ideal to demonstrate the protective efficacy of vaccines and antiviral drugs. Early attempts to infect and produce disease with variola virus in small laboratory animals such as rodents and rabbits were unsuccessful.[78] Monkeys have been used for many decades to study variola virus pathogenesis since they are the only experimental animals to develop generalized infection with rash. Studies using airborne exposure of different species of monkeys to different variola virus strains resulted in mild exanthema.[4,5] Noble and Rich[79] demonstrated cross-infections of cynomolgus monkeys with variola virus when placed in contact with other infected monkeys during the period of rash, suggesting natural aerosol infection. Higher doses of variola virus were found to be necessary to establish a model for lethal smallpox by aerosol and other routes of infection. Aerosol exposure of cynomolgus monkeys to 10^9 PFU of Yamada or Lee variola virus strains resulted in infection but no serious disease;[80] however, when monkeys were exposed to Harper or India 7124 strains at the same or higher doses, acute lethality resulted.[71] It is generally acknowledged that intravenous infection of cynomolgus monkeys leads to a sequence of disease manifestations that is similar to the disease in humans,[27] although it is accelerated due to the elimination of a prodromal period.

2.4.1.2 Monkeypox Model

Monkeypox virus has a broad host range. It produces many of the smallpox signs and symptoms in several vertebrate species, including rodents, monkeys and humans.[23,81–85] Despite its name "monkeypox," the virus is a significant human disease and is maintained in nature in rodent reservoirs, including squirrels.[86–88] There is evidence that monkeypox virus strains from West Africa are less virulent than those from the Congo basin.[22,89]

Experimental infection of common and ground squirrels with the West African or Congo basin strains at 10^6 PFU *via* the respiratory (intranasal) route is 100% fatal within 6–9 days.[90,91] High intranasal doses of monkeypox virus produce systemic infections leading to liver and splenic necrosis and interstitial inflammation in the lungs, suggesting that monkeypox in squirrels might be developed further into a useful animal model for testing countermeasures for monkeypox and smallpox.

Prairie dogs have been considered as a model after the occurrence of the first cluster of human monkeypox cases in the United States in 2003.[92,93] Most human cases presenting with febrile vesicular rash illness were believed to have acquired the infection from infected prairie dogs that were housed with other rodents that were imported from Africa. Intranasal infection of prairie dogs with West African strains was shown to produce generalized rash, oropharyngeal discharges and respiratory symptoms with 60% mortality.[94] These and other studies suggest that the monkeypox prairie dog model may be useful for evaluating countermeasures since the clinical time course of disease in these animals is close to what has been described with human systemic smallpox or monkeypox diseases.[93,95]

STAT1-deficient C57BL/6 mice are susceptible to low doses of monkeypox virus *via* the intranasal route and these mice were useful in demonstrating the efficacy of two antiviral drugs, CMX001 and ST-246.[96] Non-human primates have been extensively investigated using aerosol, intramuscular, intratracheal, intrabronchial and intravenous infections.[6,70,97–99] The clinical features in monkeys exposed to aerosolized monkeypox virus reportedly occurred in a sequence similar to those in humans, but at a greater rate.[27] Aerosol infections with 3×10^6 PFU killed five of six animals between days 9 and 12 post-infection. The monkeys exhibited significant fever but mild exanthema. Respiratory symptoms were profound, as necropsied animals showed high viral loads in the lungs, suggesting that the cause of death was severe fibrinonecrotic bronchopneumonia.[70,98] The skin lesions were resolved within 2 weeks in animals that survived the infections. Intravenous infection of cynomolgus monkeys with monkeypox virus, although an unnatural route of infection, has also been used in support of pathogenesis, vaccines and antiviral drug research. Infected monkeys developed skin lesions within 4–5 days, but the severity and disease outcome depended on the inoculation dose. Animals that were inoculated 10^5, 10^6 and 10^7 PFU developed mild, moderate and severe disease, respectively. Most of the animals (92%) died at the higher dose within 8–12 days.[88] There has been some reluctance to accept the intravenous challenge model, on the grounds that the challenge should be *via* the "natural" route; however, this model proved to be useful in testing the efficacy of a candidate vaccine for smallpox, the highly attenuated modified vaccinia virus Ankara (MVA)[100] and the antiviral drugs cidofavir[101] and ST-246.[102]

Another method for infecting cynomolgus monkeys with monkeypox virus using a liquid micro sprayer has recently been developed.[103] Although in this method a larger particle size than with traditional collision nebulizers is delivered, it was shown to produce a systemic disease reminiscent of smallpox at lower doses than with intravenous inoculation.

2.4.1.3 Rabbitpox Model

Aerosol in addition to intradermal infection of rabbits with rabbitpox virus produces a syndrome similar to that in humans with smallpox.[6,104–106] After an incubation period of about 4–6 days, infected rabbits show signs of fever, nasal and ocular discharges, respiratory complications and skin lesions. The skin lesions progress through the papular, vesicular and pustular stages in animals that survive the infection. Natural airborne transmission of rabbitpox in rabbits has been observed and reported in several studies and reviews.[6,104,106–110] In aerosol exposure, the dose and particle size are important in determining the outcome of infection. A dose of about 15 PFU at a particle size of about 1 μm or less induces severe disease with high mortality. In an aerosol study by Westwood *et al.*,[6] after a 4–6 day incubation period, signs of severe illness appeared, manifested by fever, weakness and weight loss, followed by generalized rash, pock-like edematous lesions and purulent discharges from the eyes and nose. The extent of the rash varied from a few scattered lesions to confluency. In surviving animals, the lesions progressed from macules to papule to a pseudo-pustule with "caseous contents" before scarring. Pronounced respiratory involvement, including hyperventilation, dyspnea and respiratory distress, has been reported.[111] Necropsy examination of aerosol-infected rabbits showed enlarged, heavy lungs with patches of congestion, hemorrhage and edema.[109] Exposure of rabbits to high doses (>200 PFU) usually produces lethal infection.[112] Intradermal infection of rabbits produces similar symptoms to aerosol infections, but with relatively less severe respiratory involvement. As with aerosol infections, intradermal infections produce mild to severe diseases depending on the inoculation dose. Animals that are infected intradermally can transmit the infection to other animals by aerosol and droplets.[106,113]

2.4.1.4 Mousepox Model

Ectromelia virus infection in mice, known as mousepox disease, has been extensively studied as a model for smallpox pathogenesis. One advantage of using mousepox as a model for smallpox pathogenesis is the capability to evaluate an *Orthopoxvirus* in its natural host species. Other advantages are that the mousepox and smallpox diseases have some common features, the mouse is a convenient laboratory animal model and a great deal of knowledge is available about mouse genetics and immunology, which makes it possible to investigate the genetic basis of susceptibility or resistance and host defense mechanisms in response to infection. The disadvantages of this model, however, are that unlike smallpox, the skin is the natural route of infection and liver disease is the major cause of death. Nevertheless, the model has been valuable in demonstrating the hematogenous spread of poxviruses, which was a major milestone in understanding the pathogenesis of poxviruses and the role of the innate and adaptive immune responses in disease pathogenesis. The mortality of susceptible mice strains such as A, CBA, C3H, BALB/c and DBA/2 may be up to 90%. Mice that survive the infection develop smallpox-like clinical signs, including rash, cutaneous pustules and ulcerations on the face, limbs and tail. Early signs of the disease after footpad, subcutaneous or intradermal inoculation are edematous skin lesion at the site of inoculation. These are followed by generalized rash as the virus disseminates through the lymphatic system to the bloodstream. Hepatocellular necrosis and necrosis of the spleen, thymus and lymph nodes are the hallmarks of mousepox disease. Resistant mice strains such as C57BL/6, C57BL/10 and AKR show no clinical signs of infection, but can transmit the infection to other susceptible mice. At least four genetic loci have been attributed to mousepox resistance. The resistance to mousepox-1 (Rmp-1) locus maps to the NK gene complex, the Rmp-2 locus maps near the complement component C5 gene, the Rmp-3 locus is linked to the MHC and the Rmp-4 locus maps near the selection gene complex.[114–117]

It is noteworthy that the ectromelia mouse model played a very important role in raising concerns over the possibility of a new generation of biological weapons. Insertion of interleukin-4 (IL-4) gene into the genome of ectromelia virus rendered the virus lethal to naturally resistant as well as vaccinated mice.[118]

2.4.1.5 Other Models

Cowpox virus has been used in a number of studies. In BALB/c mice infected by aerosol at a dose of 5×10^6 PFU or by intranasal inoculation at a dose of 10^4–10^6 PFU, viral antigens were detected in bronchiolar epithelial and smooth muscle cells of airways and vessels, macrophages and alveolar septal lining cells within 2 days of infection.[119] Lesions in the lungs, airways, trachea, nasal passages, sinuses and occasionally the skin become evident within 10 days of infection, indicating that both aerosol and intranasal infections induce systemic smallpox-like disease. However, there are differences in frequency, distribution and severity between the two routes of infection. Aerosol exposure tends to target the lower respiratory tract and usually causes more sever pulmonary disease than intranasal exposure, which targets the upper respiratory tract due the deposition of the larger particle sizes.

Vaccinia virus has been used in several aerosol studies in mice and rabbits. The outcome of infections of mice with vaccinia virus depends on the mouse strain, virus strain, dose and method of infection. Mice inoculated with vaccinia virus *via* the respiratory route generally require high viral doses (10^4–10^5 PFU) to generate lethal disease.[120,121] Higher doses of intranasal vaccinia Western Reserve (WR) virus infections are required to produce lethal infections in BALB/c mice than C57BL/6 mice.[88] The vaccinia virus New York City Board of Health strain was more virulent than the vaccinia WR in BALB/c mice.[122] Hairless mice (SKH-1) have been used to establish dermal infections with vaccinia virus, where the severity of infection can be quantified by counting skin lesions.[121]

In a study by Westwood *et al.*,[6] aerosol infection of rabbits with vaccinia virus resulted in mild disease with scanty papular rash and occasional nasal discharge. Intradermal inoculation, the preferred route in rabbits, of vaccinia virus WR produced lethal, systemic disease and exhibited aspects desired of a small-animal model for variola virus in humans.[106]

2.4.2 Vaccines

Smallpox is the only human infectious disease that was successfully eradicated by the global immunization campaign that was started by the WHO in 1958 and lasted over 20 years. However, the first-generation vaccinia virus-derived vaccines, including Dryvax and Lister, that were used in the new and old worlds, respectively, during the smallpox eradication campaign were associated with rare but serious adverse effects, particularly in immunocompromised individuals.[123–125] Additionally, these vaccines do not meet today's safety and good manufacturing practice (GMP) standards. Many animal models using various orthopoxviruses have been used to develop new generations of vaccine with improved safety and immunogenic properties under current GMP rules. One of the second-generation vaccines, ACAM2000, was produced in Vero cells from a plaque-purified clone derived from the vaccinia virus Dryvax vaccine and was licensed in the United States in 2007. The vaccine has been demonstrated to be effective in an aerosol monkeypox model. Cynomolgus monkeys infected with aerosolized monkeypox virus at an average dose of 2.6×10^5 PFU were all protected, although a few skin lesions and low-level viremia were present.[126] Although preclinical and clinical trials showed that ACAM2000 closely matches the efficacy of Dryvax, there were concern that it might be associated with similar health risks to its predecessor, the Dryvax vaccines. A third-generation vaccine, IMVAMUNE, is derived from the highly attenuated Modified Vaccinia Ankara (MVA) vaccine. This vaccine contains deletions and mutations in the host-range terminal regions of the genome and, therefore, lost its capacity to replicate efficiently in human cells, but retains most of the protective antigenic properties.[100,123] It induces an antibody profile similar to that induced by Dryvax and a single dose was found to be protective in nonhuman primates.[100,126] Studies in rabbits vaccinated with a single high dose of this vaccine showed that animals exhibited moderate clinical signs of the disease, but all survived aerosolized rabbitpox virus challenge.[112] The post-exposure protection ability of MVA vaccine was demonstrated in the mousepox model.[127] This vaccine was licensed in Germany and was used to vaccinate more than 100 000 individuals with no serious side effects.[128,129] Recently it has been licensed in Canada under the trade name IMVAMUNE and in the European Union under trade name IMVANEX. However, because of the high attenuation and low replicative efficiency of this vaccine, two doses or a single high-titer dose are required to induce protective antibody response in human subjects. Another third-generation vaccine, LC16m8, which has been licensed in Japan since 1975, is derived from Vaccinia Lister strain by serial passages in primary rabbit kidney cells at 30 °C. The low-temperature passage resulted in mutation of the B5R open reading frame, which encodes a protein similar to the superfamily of complement control proteins. Vaccination of BALB/c and nude mice with LC16m8 was shown to protect 80% of the animals against intranasal challenge with 100 or 250 LD_{50} of vaccinia virus WR.[130] In the same study, all Dryvax-vaccinated mice survived the 100 LD_{50} and 60% survived the 250 LD_{50} challenge. This vaccine has been shown to protect cynomolgus monkeys against both intranasal and lethal subcutaneous challenge with monkeypox virus.[131] Despite the outstanding efficacy and safety profiles of LC16m8, the occurrence of large plaque-forming virulent revertants following passage in cell culture is a major drawback.[132]

All previous vaccines were generated by propagation in animal hosts (Dryvax, Lister), propagation in chick embryos followed by attenuation by serial passages in cell cultures (IMVAIMUNE/IMVAVEX) or attenuation by serial passages in cell cultures at low temperature (LCM16M8). The discovery that LC16m8 contained a deletion in a virulence gene promoted investigations into the development of fourth-generation vaccines that are attenuated by using genetic engineering techniques. One of the promising fourth-generation vaccines, LC16m8Δ, was produced by deleting the entire B5R gene from the LC16m8 genome by homologous recombination.[133] Kidokoro and Shida[132] showed that this vaccine was as effective as Dryvax and superior to MVA and LC16m8 vaccines against lethal intranasal challenge of vaccinia WR in BALB/c mice. They also showed that the vaccine was stable (non-revertant) in primary rabbit kidney cells and safe in SCID mice.

2.4.3 Therapeutics

Currently, there is no licensed drug for treating poxvirus infections. Three candidate drugs, cidofovir, CMX001 and ST-246, have extensively been investigated *in vitro* and in rodents, rabbits and nonhuman primates and gained investigational new drug (IND) status by the US FDA.

Cidofovir is a nucleoside cytosine analog that selectively inhibits viral DNA polymerase. It has broad-spectrum activity against DNA viruses, including herpes-, adeno-, polyoma-, papilloma- and poxviruses. Its anti-poxvirus activity has been demonstrated *in vitro* against vaccinia, variola, cowpox, monkeypox, camelpox, ectromelia, molluscum contagiosum and orf viruses.[134] Its anti-poxvirus activities were evaluated mostly in mice and rabbits and a few studies were conducted in nonhuman primates using the intravenous infection route.[135–137] Several studies have been conducted to investigate the efficacy of cidofovir against vaccinia virus respiratory, cutaneous, intraperitoneal and intravenous infections. In a study by Smee *et al.*[138] of BALB/c mice infected intranasally with vaccinia WR virus, 60–100% of the animals survived the infection, whereas all placebo controls died within 8 days of exposure.

Bray *et al.*[139] demonstrated the efficacy of cidofovir in BALB/c mice infected by the aerosol or intranasal routes with cowpox Brighton virus. Intramuscular doses of 100 mg kg^{-1} given on days 0, 2 or 4 after exposure resulted in 90–100% protection.[139] Aerosol treatment with cidofovir at doses of 0.5–5 mg kg^{-1} was effective in protecting BALC/c mice infected with aerosolized or intranasal cowpox Brighton virus.[140,141] Intranasal treatment with a single codofovir dose (10–40 mg kg^{-1}) protected 90–100% of BALB/c mice against lethal intranasal cowpox Brighton virus infection when the drug was given within 24 h of exposure.[142] An aerosol rabbitpox model was used to evaluate cidofovir as a post-exposure treatment.[143] All rabbits that were given intravenous injection of cidofovir for 3 days survived the infection, whereas all untreated animal died. Monkeypox, ectromelia and variola viruses have been used to a lesser extent than the other viruses. Cynomolgus monkeys infected intravenously with monkeypox or variola viruses survived the infection and showed highly significant reductions in lesions when treated with cidofovir.[137]

However, cidofovir is poorly bioavailable and therefore it is administered in large doses or intravenously. It also causes nephrotoxicity. To improve its bioavailability so that it can be administered orally, cidofovir prodrugs were made with lipid conjugates and early studies on several derivatives showed promising results in a lethal respiratory mousepox model.[144] CMX001 (hexadecyloxypropylcidofovir) is one of several lipid ester derivatives of cidofovir that proved effective when delivered orally and has no nephrotoxic side effects. Its activity against orthopoxviruses has been studied *in vitro* and in mice, rabbits and nonhuman primates. *In vitro* studies showed its potency against ectromelia, vaccinia, rabbitpox, cowpox and variola viruses.[145] In mice, CMX001 was found to be effective in preventing mortality after lethal intranasal infection with ectromelia, cowpox, vaccinia and monkeypox viruses. CMX001 given at 10 mg kg^{-1} followed by a dose of 2.5 mg kg^{-1} every other day for 14 days protected A/Ncr mice against a lethal intranasal challenge (100–200 × LD_{50}) of ectromelia virus.[146] Against intranasal cowpox infection, in BALB/c mice, a single oral dose of 12.5 mg kg^{-1} was 100% effective when given 1 day post-exposure and 87% protective when given 3 days post-exposure.[121] In the same study, 86% protection was reported against lethal intranasal vaccinia virus infection when the drug was given 24 h post-infection. CMX001 administered prophylactically or soon after infection prior to the onset of symptoms was also found to be very effective in preventing death and lessening clinical symptoms in rabbits infected intradermally with rabbitpox virus and rabbits naturally infected by aerosol.[147,148]

ST-246 (tecovirimat) is a small-molecule agent active against several orthopoxviruses including vaccinia, cowpox, camelpox, ectromelia and variola viruses in cell culture. The compound has been shown to inhibit virion release by targeting the poxvirus p37 protein, which is required for the production of extracellular forms of the virus.[149] This drug is currently in the late stage of development under the US FDA Animal Rule, and Phase I and II clinical trials to evaluate its safety, tolerability and pharmacokinetics have been conducted.

Numerous studies have been conducted in animals to evaluate the efficacy and safety of ST-246 in mice infected with vaccinia, cowpox or ectromelia viruses,[150–153] prairie dogs and golden ground squirrels infected with monkeypox virus,[154] rabbits infected with rabbitpox virus[111] and cynomolgus monkeys infected with monkeypox and variola viruses.[102,155] In each of these models, an efficacious dose was determined that prevented mortality and significantly reduced clinical symptoms of disease. A dose of 100 mg kg^{-1} protected 93–100% of mice challenged with vaccinia or cowpox viruses and 100% of squirrels infected with monkeypox virus. A dose of 40 mg kg^{-1} protected all rabbit infected with rabbitpox. Doses of 3 and 10 mg kg^{-1} protected all monkeys infected with monkeypox and variola viruses, respectively. Further, the drug was found to be efficacious against lethal challenge with vaccinia virus in immunocompromised mice, except in situations of severe combined immunodeficiency.[152]

2.5 Conclusion

It has been over 30 years since smallpox was declared eradicated but the prospects of its use as a potential biological agent raise substantial concerns. Although known variola virus stocks are contained in internationally sanctioned high-security repositories in the United States and the Russian Federation, other stocks may exist in unknown locations around the world. Deliberate release of the variola virus as a biological warfare or bioterrorism weapon in any country would have grave consequences throughout the world. The virus can spread in asymptomatic travelers days or weeks before suspicions are aroused. Significant investments have been by made the United States and EU over the last decade to develop anti-smallpox countermeasures. Since the variola virus is strictly a human virus and its use in research is highly regulated, only animal models can be used to develop such countermeasures. Under the US FDA's Animal Rule, drugs and vaccines can be approved on the basis of adequate and well-controlled animal efficacy studies if human challenge studies would not be feasible or ethical. Numerous studies have been made to develop appropriate animal models using different animal species, different orthopoxvirus species and different routes of exposure. Because the variola virus is transmitted mainly through the respiratory system and the aerosol route of exposure would be the most likely vehicle in a bioterrorism event, a respiratory, nonhuman primate model that mimics smallpox infection would be ideal. However, to induce smallpox-like disease in nonhuman primates using the aerosol route of exposure, a very high infectious dose is required. Therefore, in addition to the aerosol exposure, research using intranasal, intratracheal and even intradermal and intravenous routes of exposures have been used with different animals to induce certain aspects of the smallpox disease. Each model has advantages and disadvantage and there is no one model that can mimic all features of the human diseases. Nevertheless, collectively, the mousepox, rabbitpox and monkeypox models, using mice, rabbits and nonhuman primates, respectively, have contributed significantly not only to understanding disease pathogenesis, but also to successfully developing new generations of effective vaccines such as ACAM2000 and drugs such as CMX001 and ST-246.

References

1. D. A. Henderson, T. V. Inglesby, J. G. Bartlett, M. S. Ascher, E. Eitzen, P. B. Jahrling, J. Hauner, M. Layton, J. McDade, M. T. Osterholm, T. O'Toole, G. Parker, T. Perl, P. K. Russell and K. Tonat, Smallpox as a Biological weapon: medical and public health management, *JAMA*, 1999, **281**, 2127–2137.
2. J. G. Breman and I. Arita, The confirmation and maintenance of smallpox eradication, *N. Engl. J. Med.*, 1980, **303**(**22**), 1263–1273.
3. J. L. Chapman, D. K. Nichols, M. J. Martinez and J. W. Raymond, Animals models of orthopoxvirus infection, *Vet. Pathol.*, 2010, **47**(**5**), 852–870.
4. N. Hahon and B. J. Wilson, Pathogenesis of variola in *Macaca irus* monkeys, *Am. J. Hyg.*, 1960, **71**, 69–80.
5. N. Hahon, Smallpox and related poxvirus infections in the simian host, *Bacteriol. Rev.*, 1961, **25**, 459–476.

6. J. C. Westwood, E. A. Boulter, E. T. Bowen and H. B. Maber, Experimental respiratory infection with poxviruses. I. Clinical virological and epidemiological studies, *Br. J. Exp. Pathol.*, 1966, **47**(**5**), 453–465.
7. F. Fenner, D. A. Henderson, I. Arita, Z. Jezek and I. D. Ladnyi, in The pathogenesis, pathology and immunology of smallpox and vaccinia, *Smallpox and Its Eradication*, ed. F. Fenner, D. A. Henderson, I. Arita, Z. Jezek and I. D. Ladnyi, WHO, Geneva, Switzerland, 1988, ch. 3, pp. 121–168.
8. D. K. Milton, What was the primary mode of smallpox transmission? Implications for biodefense, *Front. Cell. Infect. Microbiol.*, 2012, **2**, 150.
9. ICTV, *Virus taxonomy*, 2013, http://www.ictvonline.org/virusTaxonomy.asp.
10. C. Smithson, A. Purdy, A. J. Verster and C. Upton, Prediction of steps in the evolution of variola virus host range, *PLoS One*, 2014, **9**(**3**), e91520.
11. C. Upton, S. Slack, A. L. Hunter, A. Ehlers and R. L. Roper, Poxvirus orthologous clusters: toward defining the minimum essential poxvirus genome, *J. Virol.*, 2003, 77(**13**), 7590–7600.
12. C. Gubser, S. Hue, P. Kellam and J. Smith, Poxvirus genomes: a phylogenetic analysis, *J. Gen. Virol.*, 2004, **85**, 105–117.
13. J. Thézé, J. Takatsuka, Z. Li, J. Gallais, D. Doucet, B. Arif, M. Nakai and E. Herniou, New insights into the evolution of *Entomopoxvirinae* from the complete genome sequences of four entomopoxviruses infecting *Adoxophyes honmai, Choristoneura biennis, Choristoneura rosaceana*, and Mythimna separate, *J. Virol.*, 2013, **87**(**14**), 7992–8003.
14. E. Lefkowitz, C. Wang and C. Upton, Poxviruses: past, present and future, *Virus Res.*, 2006, **117**(**1**), 105–118.
15. P. P. Afonso, P. M. Silva, L. C. Schnellrath, D. M. Jesus, J. Hu, Y. Yang, R. Renne, M. Attias, R. C. Condit, N. Moussatché and C. R. Damaso, Biological characterization and next-generation genome sequencing of the unclassified Cotia virus SPAn232 (*Poxviridae*), *J. Virol.*, 2012, **86**(**9**), 5039–5054.
16. A. McLysaght, P. F. Baldi and B. S. Gaut, Extensive gene gain associated with adaptive evolution of poxviruses, *Proc. Natl. Acad. Sci. U. S. A.*, 2003, **100**(**26**), 15655–15660.
17. A. Hughes and R. Friedman, Poxvirus genome evolution by gene gain and loss, *Mol. Phylogenet. Evol.*, 2005, **35**(**1**), 186–195.
18. D. Coulson and C. Upton, Characterization of indels in poxvirus genomes, *Virus Genes*, 2011, **42**(**2**), 171–177.
19. L. Qin, C. Upton, B. Hazes and D. H. Evans, Genomic analysis of the vaccinia virus strain variants found in Dryvax vaccine, *J. Virol.*, 2011, **85**(**24**), 13049–13060.
20. J. J. Esposito, S. A. Sammons, A. M. Frace, J. D. Osborne, M. Olsen-Rasmussen, M. Zhang, D. Govril, I. K. Damon, R. Kline, M. Lakes, Y. Li, G. L. Smith, H. Meyer, J. W. LeDuc and R. M. Wohlhueter, Genome sequence diversity and clues to the evolution of variola (smallpox) virus, *Science*, 2006, **313**, 807–812.

21. Y. Li, D. S. Carroll, S. N. Gardner, M. C. Walsh, E. A. Vitalis and I. K. Damon, On the origin of smallpox: correlating variola phylogenics with historical smallpox records, *Proc. Natl. Acad. Sci. U. S. A.*, 2007, **104**(**40**), 15787–15792.
22. N. Chen, G. Li, M. K. Liszewski, J. P. Atkinson, P. B. Jahrling, Z. Feng, J. Schriewer, C. Buck, C. Wang, E. J. Lefkowitz, J. J. Esposito, T. Harms, I. K. Damon, R. L. Roper, C. Upton and R. M. Buller, Virulence differences between monkeypox virus isolates from West Africa and the Congo basin, *Virology*, 2005, **340**(**1**), 46–63.
23. F. Fenner, R. Wittek and K. R. Dumbell, The pathogenesis, pathology, and immunology of orthopoxvirus infections, in *The Orthopoxviruses*, ed. F. Fenner, R. Wittek and K. R. Dumbell, Academic Press, Inc., San Diego, CA, USA, 1989, ch. 4, pp. 85–141.
24. P. E. Wehrel, J. Posch, K. H. Richeter and D. A. Henderson, An airborne outbreak of smallpox in a German hospital and its significance with respect to other recent outbreaks in Europe, *Bull. World Health Organ.*, 1970, **43**(**5**), 669–679.
25. R. A. Shooter, C. C. Booth, D. Evans, J. R. McDonald, D. A. J. Tyrrell and R. Williams, in Conclusions on the source of Mrs Parker's infection, *Report of the Investigation into the Case of the 1978 Birmingham Smallpox Occurrence*, House of Commons. Her Majesty's Stationary Office, London, UK, 1980, ch. 8, p. 29.
26. D. R. Franz, P. B. Jahrling, D. J. McClain, D. L. Hoover, W. R. Byrne, J. A. Pavlin, G. W. Christopher, T. J. Cieslak, A. M. Friedlander and E. M. Eitzen, Clinical recognition and management of patients exposed to biological warfare agents, *Clin. Lab. Med.*, 2001, **21**(**3**), 435–473.
27. J. G. Breman and D. A. Henderson, Diagnosis and management of smallpox, *N. Engl. J. Med.*, 2002, **346**(**17**), 1300–1308.
28. N. D. Brayton, Pathology and treatment of smallpox: an analysis of over two thousand cases and fifty autopsies, *J. Am. Med. Assoc.*, 1903, **41**(**4**), 233–239.
29. W. T. Councilman, G. B. Magrath and W. R. Brinckerhoff, The pathological anatomy and histology of variola, *J. Med. Res.*, 1904, **11**(**1**), 12–135.
30. R. D. Lillie, Smallpox and vaccinia: the pathologic histology, *Arch. Pathol.*, 1930, **10**, 241–291.
31. G. Bras, The morbid anatomy of smallpox, *Doc. Med. Geogr. Trop.*, 1952, **4**(**4**), 303–351.
32. D. B. Martin, The cause of death in smallpox: an examination of the pathology record, *Mil. Med.*, 2002, **167**(7), 546–551.
33. P. von Magnus, E. Andersen, K. Petersen and A. Birch-Andersen, A pox-like disease in cynomolgus monkeys, *Acta Pathol. Microbiol. Scand.*, 1959, **46**(**2**), 156–176.
34. I. D. Ladnyj, P. Ziegler and E. Kima, A human infection caused by monkeypox virus in Basankusu Territory, Democratic Republic of the Congo, *Bull. World Health Organ.*, 1972, **46**(**5**), 593–597.
35. Z. Jezek, S. S. Marennikova, M. Mutumbo, J. H. Nakano, K. M. Paluku and M. Szczeniowski, Human monkeypox: a study of 2,510 contacts of 214 patients, *J. Infect. Dis.*, 1986, **154**(**4**), 551–555.

36. A. W. Rimoin, P. M. Mulembakani, S. C. Johnston, J. O. Lloyd Smith, N. K. Kisalu, T. L. Kinkela, S. Blumberg, H. A. Thomassen, B. L. Pike, J. N. Fair, N. D. Wolfe, R. L. Shongo, B. S. Graham, P. Formenty, E. Okitolonda, L. E. Hensley, H. Meyer, L. L. Wright and J. J. Muyembe, Major increase in human monkeypox incidence 30 years after smallpox vaccination campaigns cease in the Democratic Republic of Congo, *Proc. Natl. Acad. Sci. U. S. A.*, 2010, **107**(**37**), 16262–16267.
37. P. B. Jahrling, J. W. Huggins, M. S. Ibrahim, J. V. Lawler and J. W. Martin, Smallpox and related orthopoxviruses, in *Textbook of Military Medicine – Medical Aspects of Biological Warfare*, ed. M. K. Lenhart, D. E. Lounsbury and J. W. Martin, Borden Institute, Washington, DC, 2007, pp. 215–240.
38. Z. Jezek, J. Szczeniowski and K. M. Paluku, Human monkeypox: clinical features of 282 patients, *J. Infect. Dis.*, 1987, **156**(**2**), 293–298.
39. D. Baxby and M. Bennett, Cowpox: a re-evaluation of the risks of human cowpox based on new epidemiological information, *Arch. Virol., Suppl.*, 1997, **13**, 1–12.
40. M. Bennett, A. J. Crouch, M. Begon, B. Duffy, S. Feore, R. M. Gaskell, D. F. Kelly, C. M. McCracken, L. Vicary and D. Baxby, Cowpox in British voles and mice, *J. Comp. Pathol.*, 1997, **116**(**1**), 35–44.
41. M. Pfeffer, S. Pfleghaar, D. von Bomhard, O. R. Kaaden and H. Meyer, Retrospective investigation of feline cowpox in Germany, *Vet. Rec.*, 2002, **150**, 50–51.
42. D. Baxby, M. Bennett and B. Getty, Human cowpox 1969–93: a review based on 54 cases, *Br. J. Dermatol.*, 1994, **131**(**5**), 598–607.
43. M. Bennett and D. Baxby, Cowpox, *J. Med. Microbiol.*, 1996, **45**, 157–158.
44. A. T. Silva-Fernandesa, C. E. P. F. Travassosa, J. M. S. Ferreirab, J. S. Abrahãob, E. S. D. Rochab, F. Viana-Ferreirab, J. R. dos Santosb, C. A. Bonjardimb, P. C. P. Ferreirab and E. G. Erna Geessien Kroonb, Natural human infections with Vaccinia virus during bovine vaccinia outbreaks, *J. Clin. Virol.*, 2009, **44**(**4**), 308–313.
45. E. G. Kroon, B. E. F. Mota, J. S. Abrahão, F. G. da Fonseca and G. D. Trindade, Zoonotic Brazilian Vaccinia virus: from field to therapy, *Antiviral Res.*, 2011, **92**(**2**), 150–163.
46. P. M. Pelkonen, K. Tarvainen, A. Hynninen, E. R. Kallio, K. Henttonen, A. Palva, A. Vaheri and O. Vapalahti, Cowpox with severe generalized eruption, Finland, *Emerging Infect. Dis.*, 2003, **9**(**11**), 1458–1461.
47. J. L. Reed, D. E. Scott and M. Bray, Eczema Vaccinatum, *Clin. Infect. Dis.*, 2012, **54**(**6**), 832–840.
48. G. S. Trindade, Z. I. Lobato, B. P. Drumond, J. A. Leite, R. C. Trigueiro, M. I. Guedes, F. G. da Fonseca, J. R. dos Santos, C. A. Bonjardim, P. C. Ferreira and E. G. Kroon, Isolation of two vaccinia virus strains from a single bovine vaccinia outbreak in rural area from Brazil: implications on the emergence of zoonotic orthopoxviruses, *Am. J. Trop. Med. Hyg.*, 2006, **75**(**3**), 486–490.
49. G. S. Trindade, G. L. Emerson, D. S. Carroll, E. G. Kroon and I. K. Damon, Brazilian vaccinia viruses and their origins, *Emerging Infect. Dis.*, 2007, **13**(**7**), 965–972.

50. G. P. Oliveira, A. T. S. Fernandes, F. L. de Assis, P. A. Alves, A. P. M. F. Luiz, L. B. Figueiredo, C. M. C. Almeida, C. E. P. F. Travassos, G. D. Trindade, J. S. Abrahao and E. G. Kroon, Intrafamilial transmission of vaccinia virus during a bovine vaccinia outbreak in Brazil: a new insight in viral transmission chain, *Am. J. Trop. Med. Hyg.*, 2014, **90**(**6**), 1021–1023.
51. C. E. Rupprecht, L. Blass, K. Smith, L. A. Orciari, M. Niezgoda, S. G. Whitfield, R. V. Gibbons, M. Guerra and C. A. Hanlon, Human infection due to recombinant vaccinia-rabies glycoprotein virus, *N. Engl. J. Med.*, 2001, **345**(**8**), 582–586.
52. R. K. Singh, M. Hosamani, V. Balamurugan, V. Bhanuprakash, T. J. Rasool and M. P. Yadav, Buffalopox: an emerging and re-emerging zoonosis, *Anim. Health Res. Rev.*, 2007, **8**(**1**), 105–114.
53. S. Lewis-Jones, Zoonotic poxvirus infections in humans, *Curr. Opin. Infect. Dis.*, 2004, **17**(**2**), 81–89.
54. R. F. Phalen, M. J. Oldham and R. K. Wolff, The Relevance of Animal Models for Aerosol Studies, *J. Aerosol Med. Pulm. Drug Delivery*, 2008, **21**(**1**), 113–124.
55. F. V. Rheinbaben, J. Gebel, M. Exner and A. Schmidt, Environmental resistance, disinfection, and sterilization of poxviruses, in *Poxviruses*, ed. A. A. Mercer, A. Schmidt and O. Weber, Burkhauser Verlag, Basel, Switzerland, 2007, pp. 397–405.
56. D. W. Henderson, An apparatus for the study of airborne infection, *J. Hyg.*, 1952, **50**(**1**), 53–68.
57. W. C. Cannon, E. F. Blanton and K. E. McDonald, The flow-past chamber: an improved nose-only exposure system for rodents, *Am. Ind. Hyg. Assoc. J.*, 1983, **44**(**12**), 923–928.
58. J. M. Hartings and C. J. Roy, The automated bioaerosol exposure system: preclinical platform development and a respiratory dosimetry application with nonhuman primates, *J. Pharmacol. Toxicol. Methods*, 2004, **49**(**1**), 39–55.
59. C. J. Roy, D. S. Reed and J. A. Hutt, Aerobiology and inhalation exposure to biological select agents and toxins, *Vet. Pathol.*, 2010, **47**(**5**), 779–789.
60. C. J. Roy and L. M. L. Pitt, in Infectious disease aerobiology: aerosol challenge methods in biodefense, *Research Methodology and Animal Models*, ed. J. R. Swearengen, CRC Press, Boca Raton, FL, USA, 2005, ch. 5, pp. 61–76.
61. N. Nicas, A. E. Hubbard, R. M. Jones and A. L. Reingold, The infectious dose of variola (smallpox) virus, *Appl. Biosafety*, 2004, **9**(**3**), 118–127.
62. D. C. Bloom, K. M. Edwards, C. Hager and R. W. Moyer, Identification and characterization of two nonessential regions of the rabbitpox virus genome involved in virulence, *J. Virol.*, 1991, **65**(**3**), 1530–1542.
63. F. De Silva and B. Moss, Vaccinia virus Uracil DNA glycosylase has an essential role in DNA synthesis that is independent of its glycosylase activity: catalytic site mutations reduce virulence but not virus replication in cultured cells, *J. Virol.*, 2003, **77**(**1**), 159–166.

64. F. Fenner, The clinical features of mouse-pox (infectious ectromelia of mice) and the pathogenesis of the disease, *J. Pathol. Bacteriol.*, 1948, **60**(**4**), 529–552.
65. F. Fenner, The pathogenesis of the acute exanthems, *Lancet*, 1948, **2**, 915–920.
66. J. C. N. Westwood, *Virus pathogenicity. Mechanisms of Virus Infection*, ed. W. Smith, Academic Press, London, 1963, p. 255.
67. C. T. Cho and H. A. Wenner, Monkeypox virus, *Bacteriol. Rev.*, 1973, **37**(**1**), 1–18.
68. R. M. Buller and G. J. Palumbo, Poxvirus pathogenesis, *Microbiol. Rev.*, 1991, **55**(**1**), 80–122.
69. H. A. Wenner, F. D. Macasaet, P. S. Kamitsuka, P. Kidd and I. Monkeypox, Clinical, virologic and immunologic studies, *Am. J. Epidemiol.*, 1968, **87**(**3**), 551–566.
70. G. M. Zaucha, P. B. Jahrling, T. W. Geisbert, J. R. Swearengen and L. Hensley, The pathology of experimental aerosolized monkeypox virus infection in cynomolgus monkeys (*Macaca fascicularis*), *Lab. Invest.*, 2001, **81**(**12**), 1581–1600.
71. P. B. Jahrling, L. E. Hensley, M. J. Martinez, J. W. Leduc, K. H. Rubins, D. A. Relman and J. W. Huggins, Exploring the potential of variola virus infection of cynomolgus macaques as a model for human smallpox, *Proc. Natl. Acad. Sci. U. S. A.*, 2004, **101**(**42**), 15196–15200.
72. B. Moss, Poxvirus cell entry: How many proteins does it take?, *Viruses*, 2012, **4**(**5**), 688–707.
73. G. L. Smith, A. Vanderplasschen and M. Law, The formation and function of extracellular enveloped vaccinia virus, *J. Gen. Virol.*, 2002, **83**, 2915–2931.
74. W. G. MacCallum and L. M. Moody, Alastrim in Jamaica, *Am. J. Hyg.*, 1921, **1**(**4**), 388–409.
75. H. E. Michelson and K. Ikeda, Microscopic changes in variola, *Arch. Dermatol. Syphilol.*, 1927, **15**(**2**), 138–164.
76. G. Bras, Observations on the formation of smallpox scars, *Arch. Pathol.*, 1952, **54**(**2**), 149–156.
77. E. A. Boulter, H. B. Maber and E. T. W. Bowen, Studies on the physiological disturbances occurring in experimental rabbit pox: an approach to rational therapy, *Br. J. Exp. Pathol.*, 1961, **42**, 433–444.
78. S. S. Marennikova, Field and experimental studies of poxvirus infections in rodents, *Bull. World Health Organ.*, 1979, **57**(**3**), 461–464.
79. J. Noble Jr and J. A. Rich, Transmission of smallpox by contact and by aerosol routes in Macaca irus, *Bull. World Health Organ.*, 1969, **40**(**2**), 279–286.
80. J. W. LeDuc and P. B. Jahrling, Strengthening national preparedness for smallpox: an update, *Emerging Infect. Dis.*, 2001, **7**(**1**), 155–157.
81. R. Gispen, J. D. Verlinde and P. Zwart, Histopathological and virological studies on monkeypox, *Arch. Gesamte Virusforsch.*, 1967, **21**(**2**), 205–216.

82. S. O. Foster, E. W. Brink, D. L. Hutchins, J. M. Pifer, B. Lourie, C. R. Moser, E. C. Cummings, O. E. Kuteyi, R. E. Eke, J. B. Titus, E. A. Smith, J. W. Hicks and W. H. Foege, Human monkeypox, *Bull. World Health Organ.*, 1972, **46**(**5**), 569–576.
83. S. S. Marennikova, E. M. Shchelukhina, N. N. Maltseva, K. L. Cimiskjan and G. R. Macevic, Isolation and properties of the causal agent of a new variola-like disease (monkeypox) in man, *Bull. World Health Organ.*, 1972, **46**(**5**), 599–611.
84. S. S. Marennikova and E. M. Shchelukhina, Susceptibility of some rodent species to monkeypox virus, and course of the infection, *Bull. World Health Organ.*, 1976, **53**(**1**), 13–20.
85. C. L. Sehgal and S. N. Ray, Laboratory studies on monkeypox virus, *J. Commun. Dis.*, 1982, **14**(**1**), 26–35.
86. L. Khodakevich, M. Szczeniowski, M. D. Manbu, Z. Jezek, S. Marennikova, J. Nakano and D. Messinger, The role of squirrels in sustaining monkeypox virus transmission, *Trop. Geogr. Med.*, 1987, **39**(**2**), 115–122.
87. F. Charatan, US doctors investigate more than 50 possible cases of monkeypox, *Br. Med. J.*, 2003, **326**, 1350.
88. P. B. Jahrling and V. Wahl-Jensen. Orthopoxviruses, in *Biodefense Research Methodology and Animal Models*, ed. J. R. Swearengen, CRC Press, San Diego, CA, USA, 2012, ch. 12, pp. 207–225.
89. G. McFadden, Poxvirus tropism, *Nat. Rev. Microbiol.*, 2005, **3**(**3**), 201–213.
90. S. S. Marennikova, E. M. Shchelukhina and O. A. Zhukova, Experimental infection of squirrels *Sciurus vulgaris* by monkeypox virus, *Acta Virol.*, 1989, **33**(**4**), 399.
91. R. B. Tesh, D. M. Watts, E. Sbrana, M. Siirin, V. L. Popov and S. Y. Xiao, Experimental infection of ground squirrels (*Spermophilus tridecemlineatus*) with monkeypox virus, *Emerging Infect. Dis.*, 2004, **10**(**9**), 1563–1567.
92. J. Knight, Prairie-dog model offers hope of tackling monkeypox virus, *Nature*, 2003, **423**, 674.
93. C. L. Hutson, V. A. Olson, D. S. Carroll, J. A. Abel, C. M. Hughes, Z. H. Braden, S. Weiss, J. Self, J. E. Osorio, P. N. Hudson, M. Dillon, K. L. Karem, I. K. Damon and R. L. Regnery, A prairie dog animal model of systemic orthopoxvirus disease using West African and Congo Basin strains of monkeypox virus, *J. Gen. Virol.*, 2009, **90**, 323–333.
94. S. Y. Xiao, E. Sbrana, D. M. Watts, M. Siirin, A. P. da Rosa and R. B. Tesh, Experimental infection of prairie dogs with monkeypox virus, *Emerging Infect. Dis.*, 2005, **11**(**4**), 539–545.
95. P. B. Jahrling, in Animal models and pathogenesis, *Scientific Review of Variola Virus Research, 1999–2010*, ed. A. S. Khan and G. L. Smith, WHO, Geneva, Switzerland, 2010, ch. 5, pp. 85–101.
96. J. Stabenow, R. M. Buller, J. Schriewer, C. West, J. E. Sagartz and S. Parker, A mouse model of lethal infection for evaluating prophylactics and therapeutics against monkeypox virus, *J. Virol.*, 2010, **84**(**8**), 3909–3920.

97. H. A. Wenner, C. T. Cho, C. R. Bolano and P. S. Kamitsuka, Studies on the pathogenesis of monkeypox. II. Dose-response and virus dispersion, *Arch. Gesamte Virusforsch.*, 1969, **27**(**2**), 166–178.
98. P. B. Jahrling, G. M. Zaucha and J. W. Huggins, in Countermeasures to the reemergence of smallpox virus as an agent of bioterrorism, *Emerging Infections*, ed. W. M. Scheld, W. A. Craig and J. M. Hughes, ASM Press, Washington, DC, 2000, ch. 4, pp. 213–222.
99. K. J. Stittelaar, J. Neyts, L. Naesens, G. van Amerongen, R. F. van Lavieren, A. Holý, E. De Clercq, H. G. Niesters, E. Fries, C. Maas, P. G. Mulder, B. A. van der Zeijst and A. D. Osterhaus, Antiviral treatment is more effective than smallpox vaccination upon lethal monkeypox virus infection, *Nature*, 2006, **439**, 745–748.
100. P. L. Earl, J. L. Americo, L. S. Wyatt, L. A. Eller, J. C. Whitbeck, G. H. Cohen, R. J. Eisenberg, C. J. Hartmann, D. L. Jackson, D. A. Kulesh, M. J. Martinez, D. M. Miller, E. M. Mucker, J. D. Shamblin, S. H. Zwiers, J. W. Huggins, P. B. Jahrling and B. Moss, Immunogenicity of highly attenuated MVA smallpox vaccine and protection against monkeypox, *Nature*, 2004, **428**, 182–185.
101. H. Wei, D. Huang, J. Fortman, R. Wang, L. Shao and Z. W. Chen, Coadministration of cidofavir and smallpox vaccine reduced vaccination side effects but interfered with vaccine-elicited immune response and immunity to monkeypox, *J. Virol.*, 2009, **83**(**2**), 1115–1125.
102. J. Huggins, A. Goff, L. Hensley, E. Mucker, J. Shamblin, C. Wlazlowski, W. Johnson, J. Chapman, T. Larsen, N. Twenhafel, K. Karem, I. K. Damon, C. M. Byrd, T. C. Bolken, R. Jordan and D. Hruby, Nonhuman primates are protected from smallpox virus or monkeypox virus challenges by the antiviral drug ST-246, *Antimicrob. Agents Chemother.*, 2009, **53**(**6**), 2620–2625.
103. A. J. Goff, J. Chapman, C. Foster, C. Wlazlowski, J. Shamblin, K. Lin, N. Kreiselmeier, E. Mucker, J. Paragas, J. Lawler and L. Hensley, A novel respiratory model of infection with monkeypox virus in cynomolgus macaques, *J. Virol.*, 2011, **85**(**10**), 4898–4909.
104. H. S. Bedson and M. J. Duckworth, Rabbitpox: an experimental study of the pathways of infection in rabbits, *J. Pathol. Bacteriol.*, 1963, **85**, 1–20.
105. M. C. Lancaster, E. A. Boulter, J. C. Westwood and J. Randles, Experimental respiratory infection with poxviruses. II. Pathological studies, *Br. J. Exp. Pathol.*, 1966, **47**(**5**), 466–471.
106. M. M. Adams, A. D. Rice and R. W. Moyer, Rabbitpox virus and vaccinia virus infection of rabbits as a model for human smallpox, *J. Virol.*, 2007, **81**(**20**), 11084–11095.
107. H. S. N. Greene, A pandemic of rabbitpox, *Proc. Soc. Exp. Biol. Med.*, 1933, **30**, 892–894.
108. J. Jansen, Todliche infektionen von kanichen durch ein filtrierbares virus, *Zbl. Bakt. Parasit. Infekt.*, 1941, **148**, 65–68.
109. C. J. Roy and T. G. Voss, Use of the aerosol rabbitpox virus model for evaluation of anti-poxvirus agents, *Viruses*, 2010, **2**(**9**), 2096–2107.

110. A. Nalca and D. K. Nichols, Rabbitpox: a model of airborne transmission of smallpox, *J. Gen. Virol.*, 2011, **92**(**1**), 31–35.
111. A. Nalca, J. M. Hatkin, N. L. Garza, D. K. Nichols, S. W. Norris, D. E. Hruby and R. Jordan, Evaluation of orally delivered ST-246 as postexposure prophylactic and antiviral therapeutic in an aerosolized rabbitpox rabbit model, *Antiviral Res.*, 2008, **79**(**2**), 121–127.
112. N. L. Garza, J. M. Hatkin, V. Livingston, D. K. Nichols, P. J. Chaplin, A. Volkmann, D. Fisher and A. Nalca, Evaluation of the efficacy of modified vaccinia Ankara (MVA)/IMVAMUNE against aerosolized rabbitpox virus in a rabbit model, *Vaccine*, 2009, **27**(**40**), 5496–5504.
113. G. Thomas, Sampling rabbitpox aerosols of natural origin, *J. Hyg.*, 1970, **68**(**4**), 511–517.
114. D. G. Brownstein, P. N. Bhatt, L. Gras and T. Budris, Serial backcross analysis of genetic resistance to mousepox, using marker loci for Rmp-2 and Rmp-3, *J. Virol.*, 1992, **66**(**12**), 7073–7079.
115. D. G. Brownstein and L. Gras, Differential pathogenesis of lethal mousepox in congenic DBA/2 mice implicates natural killer cell receptor NKR-P1 in necrotizing hepatitis and the fifth component of complement in recruitment of circulating leukocytes to spleen, *Am. J. Pathol.*, 1997, **150**(**4**), 1407–1420.
116. D. J. Esteban and R. M. Buller, Ectromelia virus: the causative agent of mousepox, *J. Gen. Virol.*, 2005, **86**, 2645–2659.
117. M. M. Stanford, G. McFadden, G. Karupiah and G. Chaudhri, Immunopathogenesis of poxvirus infections: forecasting the impending storm, *Immunol. Cell Biol.*, 2007, **85**(**2**), 93–102.
118. R. J. Jackson, A. J. Ramsay, C. D. Christensen, S. Beaton, D. F. Hall and I. A. Ramshaw, Expression of mouse interleukin-4 by a recombinant ectromelia virus suppresses cytolytic lymphocyte responses and overcomes genetic resistance to mousepox, *J. Virol.*, 2001, **75**(**3**), 1205–1210.
119. M. J. Martinez, M. P. Bray and J. W. Huggins, A Mouse Model of Aerosol-Transmitted Orthopoxviral Disease, *Arch. Pathol. Lab. Med.*, 2000, **124**(**3**), 362–377.
120. A. Phelps, A. J. Gates, M. Hillier, L. Eastaugh and D. O. Ulaeto, Comparative efficacy of replicating smallpox vaccine strains in a murine challenge model, *Vaccine*, 2005, **23**(**27**), 3500–3507.
121. D. C. Quenelle, D. J. Collins, W. B. Wan, J. R. Beadle, K. Y. Hostetler and E. R. Kern, Oral treatment of cowpox and vaccinia virus infections in mice with ether lipid esters of cidofovir, *Antimicrob. Agents Chemother.*, 2004, **48**(**2**), 404–412.
122. M. S. Lee, J. M. Roos, L. C. McGuigan, K. A. Smith, N. Cormier, L. K. Cohen, B. E. Roberts and L. G. Payne, Molecular attenuation of vaccinia virus: mutant generation and animal characterization, *J. Virol.*, 1992, **66**(**5**), 2617–2630.
123. M. Bray and M. E. Wright, Progressive vaccinia, *Clin. Infect. Dis.*, 2003, **36**(**6**), 766–774.

124. A. Mayr, Smallpox vaccination and bioterrorism with poxviruses, *Comp. Immunol. Microbiol. Infect. Dis.*, 2003, **26**(**5–6**), 423–430.
125. A. Alcami and B. Moss, in Smallpox vaccines, *Scientific Review of Variola Virus Research, 1999 – 2010*, ed. A. Khan and G. Smith, WHO, Geneva, Switzerland, 2010, ch. 1, pp. 1–15.
126. G. J. Hatch, V. A. Graham, K. R. Bewley, J. A. Tree, M. Dennis, I. Taylor, S. G. Funnell, S. R. Bate, K. Steeds, T. Tipton, T. Bean, L. Hudson, D. J. Atkinson, G. McLuckie, M. Charlwood, A. D. Roberts and J. Vipond, Assessment of the protective effect of Imvamune and Acam2000 vaccines against aerosolized monkeypox virus in cynomolgus macaques, *J. Virol.*, 2013, **87**(**14**), 7805–7815.
127. C. Samuelsson, J. Hausmann, H. Lauterbach, M. Schmidt, S. Akira, H. Wagner, P. Chaplin, M. Suter, M. O'Keeffe and H. Hochrein, Survival of lethal poxvirus infection in mice depends on TLR9, and therapeutic vaccination provides protection, *J. Clin. Invest.*, 2008, **118**(**5**), 1776–1784.
128. H. Stickl, V. Hochstein-Mintzel, A. Mayr, H. C. Huber, H. Schafer and A. Holzner, MVA Stufenimpfung gegen Pocken, *Dtsch. Med. Wochenschr.*, 1974, **99**(**47**), 2386–2392.
129. J. S. Kennedy and R. N. Greenberg, Imvamune: modified vaccinia Ankara strain as an attenuated smallpox vaccine, *Expert Rev. Vaccines*, 2009, **8**(**1**), 13–24.
130. C. A. Meseda, A. E. Mayer, A. Kumar, A. D. Garcia, J. Campbell, P. Listrani, J. Manischewitz, L. R. King, H. Golding, M. Merchlinsky and J. P. Weir, Comparative evaluation of the immune responses and protection engendered by LC16m8 and Dryvax smallpox vaccines in a mouse model, *Clin. Vaccine Immunol.*, 2009, **16**(**9**), 1261–1271.
131. M. Saijo, Y. Ami, Y. Suzaki, N. Nagata, N. Iwata, H. Hasegawa, M. Ogata, S. Fukushi, T. Mizutani, T. Sata, T. Kurata, I. Kurane and S. Morikawa, LC16m8, a highly attenuated vaccinia virus vaccine lacking expression of the membrane protein B5R, protects monkeys from monkeypox, *J. Virol.*, 2006, **80**(**11**), 5179–5188.
132. M. Kidokoro and H. Shida, Vaccinia Virus LC16m8Δ as a Vaccine Vector for Clinical Applications, *Vaccines*, 2014, **2**, 755–771.
133. M. Kidokoro, M. Tashiro and H. Shida, Genetically stable and fully effective smallpox vaccine strain constructed from highly attenuated vaccinia LC16m8, *Proc. Natl. Acad. Sci. U. S. A.*, 2005, **102**, 4152.
134. G. Andrei and R. Snoeck, Cidofovir activity against poxvirus infections, *Viruses*, 2010, **2**(**12**), 2803–2830.
135. D. F. Smee and R. W. Sidwell, A review of compounds exhibiting anti-orthopoxvirus activity in animal models, *Antiviral Res.*, 2003, **57**(**1–2**), 41–52.
136. D. F. Smee, Progress in the discovery of compounds inhibiting orthopoxviruses in animal models, *Antiviral Chem. Chemother.*, 2008, **19**(**3**), 115–124.

137. J. W. Huggins and N. Tikunova, in Antiviral drug development for the treatment of smallpox – status of small-molecule therapeutics, *Scientific Review of Variola Virus Research, 1999–2010*, ed. A. S. Khan and G. L. Smith, WHO, Geneva, Switzerland, 2010, ch. 6, pp. 103–128.
138. D. F. Smee, K. W. Bailey and R. W. Sidwell, Treatment of lethal vaccinia virus respiratory infections in mice with cidofovir, *Antiviral Chem. Chemother.*, 2001, **21**(**1**), 71–76.
139. M. Bray, M. Martinez, D. F. Smee, D. Kefauver, E. Thompson and J. W. Huggins, Cidofovir protects mice against lethal aerosol or intranasal cowpox virus challenge, *J. Infect. Dis.*, 2000, **181**(**1**), 10–19.
140. M. Bray, M. Martinez, D. Kefauver, M. West and C. Roy, Treatment of aerosolized cowpox virus infection in mice with aerosolized cidofovir, *Antiviral Res.*, 2002, **54**(**3**), 129–142.
141. C. J. Roy, R. Baker, K. Washburn and M. Bray, Aerosolized cidofovir is retained in the respiratory tract and protects mice against intranasal cowpox virus challenge, *Antimicrob. Agents Chemother.*, 2003, **47**(**9**), 2933–2937.
142. D. F. Smee, K. W. Bailey, M. Wong and R. W. Sidwell, Intranasal treatment of cowpox virus respiratory infections in mice with cidofovir, *Antiviral Res.*, 2000, **47**(**3**), 171–177.
143. D. Verreault, S. K. Sivasubramani, J. D. Talton, L. A. Doyle, J. D. Reddy, S. Z. Killeen, P. J. Didier, P. A. Marx and C. J. Roy, Evaluation of inhaled cidofovir as postexposure prophylactic in an aerosol rabbitpox model, *Antiviral Res.*, 2012, **93**(**1**), 204–208.
144. R. M. Buller, G. Owens, J. Schriewer, L. Melman, J. R. Beadle and K. Y. Hostetler, Efficacy of oral active ether lipid analogs of cidofovir in a lethal mousepox model, *Virology*, 2004, **318**(**2**), 474–481.
145. K. Y. Hostetler, Alkoxyalkyl prodrugs of acyclic nucleoside phosphonates enhance oral antiviral activity and reduce toxicity: current state of the art, *Antiviral Res.*, 2009, **82**(**2**), A84–A98.
146. S. Parker, E. Touchette, C. Oberle, M. Almond, A. Robertson, L. C. Trost, B. Lampert, G. Painter and R. M. Buller, Efficacy of therapeutic intervention with an oral ether-lipid analogue of cidofovir (CMX001) in a lethal mousepox model, *Antiviral Res.*, 2008, **77**(**1**), 39–49.
147. A. D. Rice, M. M. Adams, B. Lampert, S. Foster, A. Robertson, G. Painter and R. W. Moyer, Efficacy of MX001 as a prophylactic and presymptomatic antiviral agent in New Zealand white rabbits infected with rabbitpox virus, a model for orthopoxvirus infections of humans, *Viruses*, 2011, **3**(**2**), 63–82.
148. A. D. Rice, M. M. Adams, G. Wallace, A. M. Burrage, S. F. Lindsey, A. J. Smith, D. Swetnam, B. R. Manning, S. A. Gray, B. Lampert, S. Foster, R. Lanier, A. Robertson, G. Painter and R. W. Moyer, Efficacy of CMX001 as a post exposure antiviral in New Zealand White rabbits infected with rabbitpox virus, a model for orthopoxvirus infections of humans, *Viruses*, 2011, **3**(**1**), 47–62.

149. D. W. Grosenbach, R. Jordan and D. E. Hruby, Development of the small-molecule antiviral ST-246 as a smallpox therapeutic, *Future Virol.*, 2011, **6**(**5**), 653–671.
150. G. Yang, D. C. Pevear, M. H. Davies, M. S. Collett, T. Bailey, S. Rippen, L. Barone, C. Burns, G. Rhodes, S. Tohan, J. W. Huggins, R. O. Baker, R. L. Buller, E. Touchette, K. Waller, J. Schriewer, J. Neyts, E. DeClercq, K. Jones, D. Hruby and R. Jordan, An orally bioavailable antipoxvirus compound (ST-246) inhibits extracellular virus formation and protects mice from lethal orthopoxvirus challenge, *J. Virol.*, 2005, **79**(**20**), 13139–13149.
151. A. Berhanu, D. S. King, S. Mosier, R. Jordan, K. F. Jones, D. E. Hruby and D. W. Grosenbach, ST-246 inhibits in vivo poxvirus dissemination, virus shedding, and systemic disease manifestation, *Antimicrob. Agents Chemother.*, 2009, **53**(**12**), 4999–5009.
152. D. W. Grosenbach, A. Berhanu, D. S. King, S. Mosier, K. F. Jones, R. A. Jordan, T. C. Bolken and D. E. Hruby, Efficacy of ST-246 versus lethal poxvirus challenge in immunodeficient mice, *Proc. Natl. Acad. Sci. U. S. A.*, 2010, **107**(**2**), 838–843.
153. D. C. Quenelle, R. M. Buller, S. Parker, K. A. Keith, D. E. Hruby, R. Jordan and E. R. Kern, Efficacy of delayed treatment with ST-246 given orally against systemic orthopoxvirus infections in mice, *Antimicrob. Agents Chemother.*, 2007, **51**(**2**), 689–695.
154. E. Sbrana, R. Jordan and H. De, *et al.*, Efficacy of the antipoxvirus compound ST-246 for treatment of severe orthopoxvirus infection, *Am. J. Trop. Med. Hyg.*, 2007, **76**(**4**), 768–773.
155. R. Jordan, A. Goff, A. Frimm, M. L. Corrado, L. E. Hensley, C. M. Byrd, E. Mucker, J. Shamblin, T. C. Bolken, C. Wlazlowski, W. Johnson, J. Chapman, N. Twenhafel, S. Tyavanagimatt, A. Amantana, J. Chinsangaram, D. E. Hruby and J. Huggins, ST-246 antiviral efficacy in a nonhuman primate monkeypox model: determination of the minimal effective dose and human dose justification, *Antimicrob. Agents Chemother.*, 2009, **53**(**5**), 1817–1822.
156. S. J. Flint, V. R. Racaniello, L. W. Enquist and A. M. Skalka, *Principles of Virology, Third Edition, Volume II: Pathogenesis and Control*, ASM Press, Washington, DC, 2009.

CHAPTER 3

Inhalational Anthrax – Issues in Dose–Response and Hazard Evaluation

AVSHALOM FALK[a] AND ARIK EISENKRAFT*[a,b,c]

[a]NBC Protection Division, Israel Ministry of Defense, HaKirya, Tel-Aviv 61909, Israel; [b]Institute for Research in Military Medicine, Faculty of Medicine, The Hebrew University, Jerusalem, Israel; [c]The IDF Medical Corps, Israel Ministry of Defense, HaKirya, Tel-Aviv 61909, Israel
*E-mail: nbc_pd@mod.gov.il

3.1 Introduction

Bacillus anthracis, the spore-forming bacterium which causes anthrax, is a pathogen of herbivorous mammals, with a natural life history adapted to these hosts and their way of life.[1,2] Humans may contract anthrax by contact with infected animals, their products or a contaminated environment. Until recently, the main recognized forms of human anthrax, depending on the route of infection, were cutaneous, gastrointestinal and inhalational.[1,3,4] A more recent form is injectional anthrax, discovered in illicit drug users who had injected contaminated heroin.[5] Inhalational anthrax, the most devastating and almost always lethal form of anthrax, is caused by inhalation of airborne spores. In general, it is an uncommon occurrence, mainly in the occupational setting,[6] but an imminent threat to the general population is

Issues in Toxicology No. 25
Aerobiology: The Toxicology of Airborne Pathogens and Toxins
Edited by Harry Salem and Sidney Katz

Published by the Royal Society of Chemistry, www.rsc.org

emerging from its use as a biological warfare or bioterror agent.[3,4] It has been weaponized in the past by several states,[7,8] used in an attempted bioattack by the Japanese cult Aum Shinrikyo in 1993[9] and, more recently, used in the high-impact anthrax letter attacks in the USA in 2001.[10] These events, caused by *B. anthracis* spore-laced letters mailed to several individuals and organizations, resulted in five deaths out of 11 inhalational anthrax casualties, prescription of antibiotic prophylaxis for over 30 000 persons and widespread contamination of post offices and targeted facilities, whose remediation required costly (nearly US$1 billion in total) and extensive decontamination operations.[10–12] The responders to these attacks faced knowledge gaps in medical diagnosis and treatment, detection and hazard evaluation, which hampered decision-making and response actions. Most importantly, there were knowledge gaps in dose–response relationships and lack of reliable and established exposure limits that already existed or were in progress in other fields of risk management.[12–14] These knowledge gaps led to the extensive prophylaxis and decontamination responses that, according to some authors, were over-reactive and could have been better planned if more realistic evaluation of exposure hazards and precise methods for estimation of residual contamination levels had been available.[12,14,15]

During and in the aftermath of the 2001 anthrax letter events, extensive research efforts were undertaken to address these and other issues regarding risk evaluation of inhaled *B. anthracis* spores. Although many issues have not been settled and, to our knowledge, human exposure standards have not yet been established, the research efforts yielded many insights that may guide response planning and further research. These research efforts encompassed a variety of approaches: re-analysis of past inhalational anthrax events,[16,17] environmental and scenario simulation studies[18] and modeling of animal exposure data.[17,19,20]

The purpose of this chapter is to review major results of these studies and insights concerning the analysis of the risk to humans from inhalation of *B. anthracis* spores. The major topic discussed is quantitative dose–response relationships, studied by animal exposure, modeling and extrapolation to humans. Some environmental and scenario simulation studies with relevance to hazard evaluation in bioterror scenarios are also discussed.

We review studies published in the open literature or institutional reports available in the public domain. The reviewed literature covers the following topics:

1. Studies and lessons learned from the analysis of major past outbreaks of inhalational anthrax (reviewed in Holty *et al.*[6]), that have been thoroughly analyzed: (i) the outbreak of inhalational anthrax in the goat hair processing mill in Manchester, NH, USA in 1957; (ii) the Sverdlovsk incident of 1979; and (iii) the 2001 anthrax letter attacks in the USA. Results of studies performed in the aftermath of these events continue to be analyzed, incorporating novel knowledge and analytical methodologies.

2. Animal exposure studies yielding dose–response information, regression models, extrapolation to humans for derivation of exposure estimates and other host–pathogen issues relevant to risk management in bioterror events.
3. Field studies employing dispersion of *B. anthracis* spore surrogates such as *B. atrophaeus* and *B. thuringiensis*, and also studies on the natural ecology of *B. anthracis*, which have implications for hazard evaluation.
4. Application of dose–response information in integrative models incorporating biological and environmental factors in risk analysis.

Finally, the implications of these studies and their insights for protective and response measures in bioterror events are discussed.

3.2 Problem Characterization

Human inhalational anthrax, caused by inhalation of airborne *B. anthracis* spores, has been known as an occupational disease contracted by workers exposed to dust or aerosols produced while handling or processing contaminated animal materials, traditionally wool and hair, giving it the common name “woolsorters’ disease”.[21,22] After the introduction of a human vaccine, together with livestock vaccination in the developed world, the frequency of industrial inhalational anthrax has been reduced. However, the potential scope of inhalational exposure risks to anthrax has been extended by the potential for biowarfare/bioterror attacks and new occupational and lifestyle activities. Recently, drum makers contracted inhalational anthrax from contaminated hides used for the manufacture of African-style drums.[23–25] In other cases, the affected individuals were bystanders in the vicinity of industrial locations where spores had been aerosolized.[22,26,27] In addition to these, sporadic naturally acquired cases whose environmental source of exposure cannot be traced have also been reported.[28,29] The largest bioterrorism scenario was experienced by the USA in 2001, with its toll of five deaths among 11 inhalational anthrax cases, high public impact and extensive and costly mitigation efforts.[4,10] The properties that rate anthrax as an effective biological weapon are the ease of acquisition and production, the stability and inhalational infectivity of the spores and the high lethality of the disease. The general cumulative human mortality from inhalational anthrax in the 20th century was higher than 91%, including the 2001 anthrax letter attacks.[6] For this group in particular, the survival was 55% due to improved management together with cooperative and increasingly directed diagnostic and epidemiological survey efforts by the public health system.[6] Even with the current treatment options, the lethality of inhalational anthrax is very high and the survival of patients depends on the timing of treatment.[6,29] This emphasizes the need for early exposure detection, early diagnosis of the illness and rapid mobilization of medical resources in case of a mass attack.

3.2.1 Exposure Scenarios

As a public health problem, inhalational anthrax is a rare event, which implies a low risk level that, as described above, is limited to definite population sectors, the occupational or industrial areas in particular. A more generalized risk is a deliberate release in bioterror or biowarfare scenarios. According to many authors, the risk of these events is even lower, but if they do occur, their consequences may be more devastating from public health, social and economic aspects. From the viewpoint of the release mechanism, the scenarios for inhalational exposure to *B. anthracis* spores are the following: (i) indoor point source release, (ii) outdoor large-scale release and (iii) continuous or repeated exposure to low levels of spores emitted from a persistent source. A summary of these scenarios, relating to the dispersal mechanism, sample events or studies and the main risk features, is presented in Table 3.1. The sample events and studies from which important lessons on exposure assessment and risk identification can be obtained are reviewed in this chapter. The events related to point sources have been analyzed and motivated studies in actual sites and simulations.

The scenarios of deliberate dispersion from stationary or airborne spray devices are frequently analyzed as worst-case scenarios in theoretical studies, which modeled the consequences and effects of mitigation strategies in response to dispersal of 1.0–1.5 kg of anthrax spore preparations containing 10^{14}–10^{15} spores kg^{-1} in a large metropolitan area such as Washington, DC, or New York City.[30–32] One attempt to disperse anthrax spores from a stationary spray device was carried out in Tokyo by the Aum Shinrikyo cult in 1993. Not much can be learnt from this event, as the spores released were of a vaccine strain, the dispersal was not efficient and not much detail has been published about the dispersal device, which frequently malfunctioned.[9] The fact that such an attack was undertaken stresses the reality of this threat scenario, which may have grave consequences under different circumstances. The aerial dispersal of spores of *B. thuringiensis* over large areas in attempts to control the insect pest gypsy moth (*Lymantria dispar*) in Vancouver, Canada, in 1999 and in Washington, DC, USA in 2008 were exploited in order to evaluate the environmental consequences of aerial dispersal of *B. anthracis* in bioterror scenarios.[33,34] The topics studied were dispersal, settling, penetration of houses and cross-contamination. The third scenario is of concern in areas contaminated with low levels of spores either from settling of airborne spores, cross-contamination from heavily contaminated areas or objects and residual contamination due to incomplete decontamination. Minute amounts of spores can be re-aerosolized due to human activity and there are unsettled issues such as the actual risk from low-level exposure, especially to highly susceptible individuals, or if the repetitive exposure to minute doses, which do not pose an appreciable risk individually, have a cumulative risk. This issue is relevant to the definition of decontamination goals and decisions about the repopulation of contaminated sites and the possibility of alternative strategies of risk management if thorough decontamination is not feasible.[31]

Table 3.1 Characterization of exposure scenarios discussed in this chapter.

Scenario type	Mechanism of release and exposure	Sample event/study	Inhalational hazard characterization
I. Point source indoor release	Aerosolization due to mechanical agitation of contaminated or laced object – envelope, animal hair or hidden in an enclosed space	(a) *Industrial* – Goat Hair processing mill, Manchester, NH, 1957 (b) *Occupational* – Manufacturing of drum heads, New York, 2006 (c) *Bioterror* – Anthrax letters, 2001	May be high for individuals in the immediate vicinity of source during aerosolization Collateral risk – low or unknown
II. Large-scale outdoor release	Aerosolization of variable amount of spores from aircraft or stationary devices (deliberate scenarios). Accidental release of large amount of spores	(a) *Environmental surrogate* – Aerial spray of *B. thuringiensis* spores as pest control (b) *Industrial* – Accidental release of *B. anthracis* spores in Sverdlovsk, 1979 (c) *Bioterror* – Release of *B. anthracis* Sterne spores by Aum Shinrikyo, Kameido, Japan, 1993 (d) *Bioterror* – Simulated release of *B. anthracis* spores from aircraft	Risk variable, depending on initial amount of released spores, meteorological conditions and distance from source. May cover a large downwind area
III. Continuous or repeated low-level exposure from a persistent source	Same as I, re-aerosolization from areas with residual contamination due to contact with contaminated objects, settling of airborne spores or incomplete decontamination	(a) *Industrial* – Goat hair processing mills during routine operation (b) *Bioterror* – Residual contamination in postal facilities harboring laced letters and other facilities (*e.g.* Hart building) where laced letters were opened in 2001	Probably low, depends on human activity. High risk to susceptible individuals, but low or none to subclinically infected immunocompetent ones

3.2.2 Issues in Host–Pathogen Relationships – Dose–Response and Incubation Period

The issues relevant to risk assessment and emergency response planning for inhalational anthrax outbreaks are dose–response relationships and the duration of the incubation period. Dose–response relationships are measured quantitatively as the probability of response – infection, illness or death – as a function of the exposure or absorbed dose of the agent. The response probability is the metric of risk, which can be predicted if the dose is known or evaluated using atmospheric dispersion and deposition models. The predicted spatiotemporal evolution of exposure doses and a dose–response function allow one to delimit the dimensions of the attack in order to focus the response efforts and carry out resource allocation. Furthermore, assessment of low-level exposure risk is cardinal in the determination of safety limits and decontamination goals. The dose–response distribution function for inhalational anthrax in humans is not known. Owing to the rarity of inhalational anthrax in humans and the absence of information on the exposure dose in the reported events, most of the information should be based on animal studies. According to the extensive compilation of published effective doses for anthrax in animals,[35] the inhalational median infective dose or ID_{50} value of *B. anthracis* spores for nonhuman primates is in the range 2500–55 000 spores, reflecting the heterogeneity in experimental methodologies, animal species, bacterial strains and other parameters. This range has been presented as a measure to the human infective dose in the *JAMA* reviews by Inglesby *et al.*[3,4] Other works cite an estimate of 8000–10 000 spores for the human inhalational ID_{50} [30,31,36,37] and discrete values of 8000 or 8600 spores used in the analyses of the Sverdlovsk outbreak[16,38] and other modeling studies.[30,31] A similar uncertainty concerns the probit slope, which determines the rate of risk change with the exposure dose. Human estimates used by different authors vary from 0.67 to 2.4 [16,30,31,36,38,39]. Wein and Craft justified the lower value, indicating a broad susceptibility distribution, as suitable for a human population that may include highly susceptible individuals.[37] Higher values were selected for more defined population sectors such as the military[36] or the Sverdlovsk victims, with an age distribution different from that of the general population in Russia at that time.[16] The low-level exposure risk assessment, which is cardinal in determination of exposure limits and decontamination goals, is aggravated by the scarcity of appropriate experimental data, which makes the estimation of low exposure risk dependent on extrapolation from high-level exposure data.[16,17,19] Owing to the small probit slope estimates, extrapolation to low-risk probabilities leads to very low doses, which imply extremely low safe exposure limits for the general population.[40] In addition to the core issue of the dose–response probability function choice and the accuracy of risk prediction, there are additional questions and knowledge gaps. One of them relates to the differential susceptibility with age. Some observations, including the age distribution of the Sverdlovsk victims,[38] suggested that the susceptibility to inhalational infection with *B. anthracis* spores increases with age. Several authors incorporated age-variable estimates of the dose–response parameters in modeling studies of human risk.[30,41] The increasing susceptibility to inhalational anthrax with age merits further research, although it was not found to be the case in an animal study,[42] and incorporation of age-related susceptibility distribution in modeling of the Sverdlovsk outbreak did not improve the accuracy of the incubation period modeling.[16] Other questions relevant to risk assessment are the effect of continuous or repeated low-level exposure, mentioned above, and the possibility of a threshold level, which will be discussed later.

Knowledge and predictions concerning the incubation period are important in public health and attempts at modeling the incubation period have been known since the Spanish Flu pandemic of 1918.[43] In the case of inhalational anthrax, in the same way as other infectious diseases, the distribution of the incubation period duration affects the response policy in many respects (discussed by Wikening[16,18]). First, the left margin of the distribution predicts the time of outbreak detection by the index cases, assuming no prior information or sensors for early warning. In the case of intelligence warning or early detection, it has a confirmatory diagnostic significance. It also determines the time available for efficient prophylactic measures: as antibiotic treatment is effective when started before or shortly after the appearance of disease symptoms, the prediction concerning the time course of the outbreak development will aid in planning the response activities. The right margin of the distribution has implications for the duration of antibiotic prophylaxis.[44] On the basis of the incubation periods of the Sverdlovsk victims[38] and animal studies demonstrating relapse to active disease after cessation of post-exposure antibiotic treatment due to the persistence of viable spores in the lungs of the animals,[45] a 60 day course of antibiotic prophylaxis was prescribed for people exposed in the 2001 anthrax letter attacks.[44] Several studies, reviewed in this chapter, modeled the duration of the incubation period as a dose–response phenomenon,[16,18,44,46] thus allowing prediction of the duration of the incubation period on the basis of exposure dose distribution, which may permit planning of antibiotic drug distribution for post-exposure prophylaxis.

3.2.3 Microbial Risk Assessment and Regulation, with Reference to Inhalational Anthrax

The aim of exposure regulation is to protect the population from harmful exposure to chemical and physical agents and pathogens. The practical goal is the definition of exposure levels that will afford exposure for definitive time periods, or for a whole lifetime, without appreciable compromise of health and quality of life. Exposure standards for toxic chemicals have been defined for intake *via* food, water and air, covering the general population, subgroups such as industrial workers, and for a spectrum of exposure scenarios to include routine life or work, accidental exposure in industrial releases and chemical terror events. The major issues in defining exposure standards include (i) the proper balance between health protection and over-burdening with protective measures, (ii) giving a protective answer to a heterogeneous population, including highly susceptible individuals, in a broad spectrum of circumstances and (iii) bridging the uncertainty gaps between the scientifically known data and the complexities of real life. The data available as points-of-departure in exposure limit calculations are either anecdotal or derived under laboratory conditions, which may not represent fully or accurately the realistic scenarios. The prevalent uncertainty issues are extrapolation from animals to humans and across susceptibility heterogeneity in the human population and transition from acute or subchronic experimental exposure to chronic exposure. The traditional methodology addressing these uncertainties for noncarcinogenic toxicity is based on chemical-specific NOAEL (no observable adverse-effect level) or LOAEL (lowest observable adverse-effect level) divided by a product of uncertainty factors, each addressing another issue.[47] This is acceptable for noncancer effects of toxic chemicals, whose dose–response modeling is based on a deterministic theory, stating individual effective doses in the population so that a population threshold dose can be estimated. For carcinogenic toxicity and pathogens, dose–response models are based on stochastic process modeling, by which no threshold can be defined. The development of exposure limits for carcinogenic chemicals and pathogens employs the "acceptable risk" concept. Accordingly, a disease-specific risk level, whose incremental contribution above the background risk level does not compromise public health significantly, is chosen as a reference. An acceptable risk level for infection by common pathogens is an additional probability of 10^{-4}–10^{-5} per year. A good example is the definition of allowable pathogen/surrogate levels in potable water and remediated wastewater used for irrigation.[48] A similar approach for determining the remediation target level after biological attack has been suggested by Raber *et al.*[49] The acceptable risk level is defined in a consensus-generating process, considering epidemiological parameters and a broad array of other factors. For routine purposes, risk assessment is done for chronic exposure, occurring during a whole lifetime or a significant fraction of the lifetime, depending on lifestyle or occupational parameters such as consumption of food and drinks, residence time in one place or length of a workplace shift. Biological attacks are considered as unique events and residual contamination may decay to negligible risk levels within some time period, making a whole-lifetime risk assessment unnecessary.[15] An issue in the hazard assessment of inhalational anthrax is the threshold dose, below which exposure will not lead to infection or disease.[12,14] Earlier animal studies concentrated on exposure to high doses, leaving the low-dose exposure open or accessible by extrapolation.[14,50] The inability to define low-level risk led to a strict remediation strategy for the spore-contaminated areas, defining the remediation target as no viable spores detected in samples taken after decontamination.[12,14] More recent studies of acute and repeated ("chronic") low-level exposure of animals suggest that an infection threshold dosage may be measurable,[51,52] thus allowing the derivation of evidence-based decontamination goals.

The high environmental persistence of *B. anthracis* spores, compared with other biological agents, pose a special concern.[49,53] The most famous case is Gruinard Island, Scotland, UK, where *B. anthracis* spores, dispersed in biological weapon trials carried out in World War II, remained viable in the soil for 40 years until decontamination. This instance is frequently cited as a paradigm case of the persistent risk from deposited spores, but the low levels of spores recovered in soil sampling undertaken in 1979 presumably represent a minor residue of the large amount of spores, originally deposited in the immediate proximity to the point of release, from the $\sim 10^{14}$ spores released in total.[54,55] This and other studies suggest that after release, the environmental fate of *B. anthracis* spores is variable and the contamination levels may undergo some decay in the absence of deliberate intervention. The decay processes of *B. anthracis* spores in the environment are poorly understood and difficult to predict, as is the resuspension of deposited spores.[49] In view of these uncertainties, some authors suggest establishing a site-specific and event hazard assessment, determination of decontamination goals and conditions for repopulation rather than predetermined fixed standards.[49] It is not our goal here to make recommendations on any strategy. However, a thorough understanding of the factors affecting risk from deposited spores, with a central place for dose–response relationships, is mandatory irrespective of the strategy endorsed.

3.3 Inhalational Anthrax – Pathogenesis and Treatment

3.3.1 The Pathogen

Bacillus anthracis is a Gram-positive, spore-forming aerobic bacterium that is a member of the *Bacillus cereus* group. *B. anthracis* is the only obligate animal pathogen of the group, the other members being the opportunistic pathogen *B. cereus* and the insect pathogen *B. thuringiensis*.[1] Molecular phylogeny studies imply that *B. anthracis* has evolved as a clone of *B. cereus*.[56] The bacterium is distributed worldwide, with endemic foci in all the continents inhabited by wild and domestic herbivorous mammals, which are the natural hosts for the pathogen. The disease has been known since antiquity and molecular studies suggested that it was present in wild and domestic animals since prehistory and probably disseminated by cross-continent migration of humans and their herds.[56] In the external environment, the pathogen survives as dormant spores that upon ingestion by a susceptible host germinate to vegetative bacteria that multiply and cause a lethal disease. After the death of the host and breakdown or opening of the corpse by scavengers, the multitude of bacteria sporulate and are released to the environment to infect a new host.[1,2] Humans may contract anthrax by contact with infected animals, their products or the contaminated environment. The spores, which are the infectious units, can penetrate through skin abrasions, the digestive tract and the respiratory tract.[1,3,4,6]

3.3.2 Mechanism of Infection Establishment and Disease Progression

Upon penetration and encountering suitable conditions, notably specific nutrients, the spores germinate to vegetative bacilli. In the case of cutaneous anthrax in humans and other resistant species, spore germination takes place locally and the bacteria remain in the site to cause a local infection, with seldom systemic dissemination, whereas in more susceptible species such as mice it results in a lethal disease.[57,58] The mechanism of entry and dissemination of the pathogen by the respiratory route is more complex: spores that penetrate deep into the small airways and alveoli are engulfed by resident phagocytes – macrophages and dendritic cells.[59–65] Within phagocytes, the spores are carried through the lymphatics to draining mediastinal lymph nodes, where they germinate, multiply and egress into the bloodstream to disseminate to other organs and start a systemic infection.[66] This pathway, dubbed the "Trojan Horse model," may not be the sole mechanism of spore dynamics: studies using bioluminescence imaging real-time observations in mice showed the possibility of additional trajectories and suggested an alternative model – the "Jailbreak model" – by which spores germinate on the surface of airway epithelia, transgress the epithelia and reach the local lymphatics where they multiply, protected by their capsule and toxins.[67] As the evidence favors the dominance of the "Trojan Horse" trajectory in humans and other animals, the role of other mechanisms such as the "Jailbreak model" has not been ruled out. Understanding the detailed mechanisms of spore and bacterial dynamics is crucial for advanced modeling of the disease course, as will be discussed later.

3.3.3 Virulence Factors

The main virulence factors of *B. anthracis* are the tripartite toxin complex of protective antigen (PA), lethal factor (LF) and edema factor (EF) and the poly-γ-glutamic acid capsule.[1] LF (90 kDa) is a Zn^{2+} metalloproteinase whose cellular substrates are the mitogen-activated protein kinases (MAPKKs). EF (89 kDa) is a Ca^{2+}–calmodulin-dependent adenylyl cyclase. PA (83 kDa) is required for cell-surface receptor binding and cellular entry of the enzymatic proteins and is not toxic by itself.[1,68] Once inside the cell, the toxins perturb signaling and regulatory mechanisms.[69–71] Both lethal and edema toxins interfere with the function of immune cells, activities which aid the bacteria in the establishment of infection in the early phase,[69] while the cardiovascular and hepatic toxicities and meningitis are the leading causes to death of the host in the fulminant phase of the disease.[70–73] Other bacterial products that contribute to the hemodynamic dysfunction at this stage of the disease are the cell-wall peptidoglycan and nontoxin metalloproteinases.[74,75] The genes encoding the toxin proteins and capsule biosynthesis enzymes are located on two virulence plasmids. pXO1 (181.7 kb) and pXO2 (93.2 kb), respectively.[1] Curing the bacteria of one or both plasmids attenuates virulence, as is the case with the Sterne vaccine strain, which is toxigenic but nonencapsulated due to loss of pXO2.[1] However, these vaccine strains retain a certain degree of virulence in susceptible animals. Recent studies, employing deletion of toxin and/or capsule genes by homologous recombination, are concordant with the current views on the differential role of capsule and toxins at the different stages of inhalational anthrax in the guinea pig and rabbit models: expression of either of the toxin genes is required for overcoming the host innate defenses in the establishment of infection, while at the fulminant phase of the disease, the death of the host can be accomplished by nontoxin factors associated with the high bacterial burden.[73,76–78]

3.3.4 Clinical Course, Pathology and Treatment

Inhalational anthrax in humans develop in two phases: a prodromal phase, in which initial, nonspecific symptoms appear, and a fulminant phase, which follows within one or a few days and progresses quickly to sepsis, meningitis and fatal shock.[3,4,6,73] The pre-2001 literature, based on 18 inhalational anthrax cases reported in the USA and the Sverdlovsk incident investigation data, viewed the clinical course of inhalational anthrax as biphasic, with a first stage of mild symptoms lasting from a few hours to a few days and a severe and deteriorating phase.[4,22] The first 10 inhalational anthrax patients in the 2001 attacks became symptomatic in a median of 3.5 days before arriving at hospital. Their symptoms at presentation were fever or chills, sweats, fatigue or malaise, minimal or nonproductive cough, dyspnea and nausea or vomiting.[11] Abnormal chest X-ray findings were seen in all 10 patients and included mediastinal widening (seven patients), infiltrates (seven patients) and pleural effusions (eight patients). Other findings were mild neutrophilia (seven patients), elevated serum transaminases (nine patients) and hypoxemia (six patients).[11] On the basis of the clinical experience gained in the 2001 attacks, Lucey[79] suggested dividing the clinical course of human inhalational anthrax into five stages: (i) asymptomatic or incubation; (ii) early-prodromal, characterized by nonspecific initial symptoms; (iii) intermediate-progressive, characterized by pathological chest X-ray and clinical laboratory findings described above, positive blood cultures and worsening of symptoms; and (iv) late-fulminant, characterized by respiratory failure, meningitis and shock. Symptomatic patients in stages (ii)–(iii) can be cured by prompt antimicrobial and supportive care, unlike fulminant patients, as was the case with the 2001 attack victims.[6] Pathological studies of 42 of the Sverdlovsk victims and the 2001 anthrax letter attacks – five fatalities and two survivors – revealed generally similar results.[80–82] The most conspicuous findings were hemorrhagic lymphadenitis in the mediastinal lymph nodes and pleural effusions, lesions in the lungs, spleen and digestive tract and hemorrhagic meningitis (21 patients in the Sverdlovsk study and the first victim in 2001).[81,82] The pattern of pathological changes in the internal organs was consistent with hematogenous spread of the bacteria from the primary germination sites at the mediastinal lymph nodes. Inflammatory pathological changes in most organs were mild to moderate and even milder in the 2001 patients, who received early and aggressive antibiotic treatment,[82] and consistent with direct cardiovascular toxicity of the toxins and other microbial products as major causes of death. The pathological studies were pivotal in establishing the inhalational route of infection in both cohorts of anthrax cases.

3.3.5 Medical Treatment

Inhalational anthrax is treatable with antibiotics and there is a human licensed vaccine based on alum-adsorbed PA[1,3,4]. The antimicrobial treatment is effective only if given prophylactically during the incubation period or soon after the appearance of symptoms.[3,4] In the 2001 anthrax letter attacks, survival of the inhalational anthrax victims was 55% compared with a few percent known previously, due to the improved management, which included combined antimicrobial therapy and drainage of pleural effusions as part of the supportive measures.[4,6] In these events, as before, the survival of patients depended on the timing of treatment – the survival prospects were higher if treatment was started at the early (prodromal or progressive intermediate) rather than at the fulminant phase of the disease.[6,29] Accordingly, there are still gaps in the ability to increase the survival of inhalational anthrax patients and new therapies are being sought, concomitantly with increasing the understanding of pathogenesis mechanisms and the development of suitable animal model systems.[74,83] Acknowledgment of the role of the toxins in the pathogenesis of late-stage systemic anthrax led to the development of antitoxin therapy as an adjunct to antimicrobials.[74,83] Among the current approaches, passive immunotherapy is available as human anthrax immune globulin (AIG), which has been used as an adjunct to antimicrobial therapy in three inhalational anthrax cases, of which two survived.[23,24,29] A human anti-PA monoclonal antibody, raxibacumab, improved the outcome in spore-challenged animal studies and was found to be safe in healthy volunteers.[84]

3.4 Physical and Biological Factors Affecting Inhalational Risk

This section provides an overview of the physical and biological factors associated with inhalational anthrax risk in humans. In other words, these are the factors that will determine the presence and fate of *B. anthracis* spores in the environment and its bioavailability for inhalational uptake and causing disease. As the main theme of this chapter is dose–response relationships, the topics will be covered in brief, as other chapters present them in greater depth. The physical factors are those affecting primary aerosolization and airborne dispersal, deposition of spores on surfaces and resuspension of settled spores. Spore-related biological factors are environmental stability, germination and replication within the host and surface properties. Host-related biological factors are pulmonary penetration, deposition and clearance and immunological factors. Pulmonary penetration and deposition of anthrax spores, like aerosol particles in general, are affected by the anatomy of the respiratory system, which is one of the sources of variation in pulmonary deposition of particles between animal species and humans.[85]

3.4.1 Pulmonary Deposition of Inhaled *B. anthracis* Spores

The pulmonary deposition of inhaled particles is a physical process, affected by the aerodynamic properties of the particles, flow dynamics in the airways and airway structure.[86] Another factor of importance is the breathing pattern – small animals are obligate nose breathers, whereas nonhuman primates and humans switch irregularly to mouth breathing, which reduces the filtering efficiency of the upper respiratory tract.[85,87] Therefore, species differences in pulmonary deposition of aerosol particles, arising from pulmonary anatomical and functional differences, affect the dose actually delivered to the respiratory bronchioles and alveoli and is a major factor in the determination of the susceptibility of different species to inhalational anthrax.[85,88] The physical parameters that affect pulmonary deposition and retention of bacterial spores are not different from those affecting nonbiological particles, and models developed for risk analysis of radioactive airborne particles, for example, are applicable to biological aerosols, including airborne *B. anthracis* spores.[89] Gutting and co-workers[90–92] measured the pulmonary deposition rate of inhaled *B. anthracis* Ames spores, delivered as a monodisperse aerosol with a mean particle diameter of 0.3 ± 1.0 μm, as 4.63 ± 1.1% of the inhaled dose. This result was similar to that of Raabe *et al.*,[93] who reported that the deposition of ^{169}Yb-labeled aluminosilicate particles of 0.97–1.09 μm aerodynamic diameter in the pulmonary region of rabbits was 4.0–6.6% of the inhaled particles.[92] An important factor affecting the respiratory infectivity and course of disease by inhaled pathogens is the size of the aerosol particles carrying them, which determines the primary site of deposition in the respiratory tract.[94] The first study addressing the significance of particle size in determining the respiratory infectivity of *B. anthracis* spores was the classical study of Druett *et al.*,[95] who employed techniques developed at their institution[96,97] for exposure of animals to pathogens delivered by monodisperse aerosols of definite particle diameters. The study on the influence of particle size on the infectivity of inhaled spores in rhesus macaques and guinea pigs will be discussed in detail later. The published data were used extensively by others for dose–response modeling due to the detailed reports.[17,19,20,98] We shall state briefly that the LD_{50}

for spores delivered in 12 μm diameter particles were 15- and 17-fold higher than spores in 1 μm diameter particles in guinea pigs and rhesus macaques, respectively. In guinea pigs, where particles of intermediate sizes were also tested, they found that the cutoff of particle size effect on infectivity was a 5 μm diameter. The infectivity for particles below this size did not change sharply with particle diameter.[20,95] In their study, Druett *et al.* did not address directly the relation of infectivity and the site of deposition in the respiratory tract and the qualitative effects of particle size on disease pathogenesis. The relationships between infectivity, particle size and site of deposition were inferred by considering the results of a parallel study on the deposition and retention of inhaled ^{32}P-labeled spores of the surrogate bacterium *Bacillus subtilis* in mice, guinea pigs and rhesus macaques.[99] In this study, the distribution of labeled spores delivered in particles of varying sizes from 1 μm and up to 12 μm in diameter in the head, trachea and lung regions of the respiratory tract and their retention with time after inhalation were followed. In both guinea pigs and monkeys, particle retention measured shortly after inhalation in the head region increased with particle size, whereas retention in the pulmonary region was highest for 1 μm particles and decreased sharply at particle sizes of 4–6 μm and higher. In mice, only 1 μm particles were studied and their deposition pattern was similar to that of guinea pigs and monkeys. Retention of 1 μm particles in the lungs of guinea pigs did not change with time, whereas considerable clearance was observed for larger particles deposited in all regions. On the basis of the studies of Young *et al.*,[100] demonstrating an extremely low infectivity of *B. anthracis* spores by the oral route, Druett *at al.* inferred that infection by ingested spores that were cleared from the respiratory tract was unlikely. The issues of sites of deposition and pathogenesis were addressed directly by Thomas *et al.*, in a study of exposure to spore aerosols of 1 and 12 μm diameters in a murine model.[101] The model employed spores of the toxigenic and nonencapsulated *B. anthracis* vaccine strain STI and the mouse inbred strain A/J, which is highly susceptible to infection by this organism due to deficiencies in innate immunity. In this study, 1 μm diameter particles were generated by a Collision nebulizer and the 12 μm diameter particles by a novel system, named flow-focused aerosol generator. The mice were exposed in a nose-only apparatus. As expected, the infectivity of spores delivered in 1 μm particles was higher than that delivered in 12 μm particles: the respective LD_{50}s were 2430 and 7650 cfu (colony-forming units) and the lengths of time to death were 101.6 ± 10.4 and 161.0 ± 16.1 h, respectively. Enumeration of spores deposited at various anatomical regions shortly after exposure showed that 1 μm particles were deposited primarily in the lungs (19.7% pulmonary deposition *versus* 0.2% in the nasal passages, $p = 0.0002$), whereas 12 μm particles were deposited mainly in the nasal passages (40.9% in the nasal passages *versus* 1.7% in the lungs, $p = 0.0001$). Smaller but similar numbers of spores were deposited in the trachea and esophagus and larger numbers were deposited in the stomach, for particles of the two sizes. Bacteriological and histopathological characterization of the infections resulting from the two-size particles showed that the involvement of the upper respiratory tract was similar for both forms of infection. However, in the 1 μm particle infection the involvement of the lung and mediastinal lymph nodes was seen and in the 12 μm particle infection a greater involvement of the stomach, Peyer's patches and mesenteric lymph nodes was observed. These findings suggest that the initial port of entry for both forms of infection was the upper respiratory tract, but there were differences in additional ports of entry. Direct entry of spores through the lungs and mediastinal lymph nodes may take place in the 1 μm particle infection and through the gastrointestinal tract in the 12 μm particle infection. The latter may be caused by ingestion of spores cleared from the respiratory tract by mucociliary action, pass the harsh environment of the stomach, germinate in the intestine and penetrate the Peyer's patches, which were shown to be the ports of entry for *B. anthracis* in a murine model of gastrointestinal anthrax.[102] Dissemination of bacteria to the spleen, kidneys and thymus was the same for both forms of infection, indicating that the terminal disease was similar for both forms. However, the level of terminal bacteremia was higher in the 1 μm particle infection.

3.4.2 *In vivo* Fate of Inhaled *B. anthracis* Spores

Inhaled spores are deposited in the airway walls all along the respiratory tract. Spores that traverse the tracheobronchial region and are deposited in the tracheobronchial tree are generally trapped in the mucus lining the airway epithelia and cleared by mucociliary action, whereas spores deposited in the small airways and alveoli may persist there for long periods of time as dormant spores and are slowly cleared, mainly by phagocytes that transport them to the draining lymph nodes. Early studies on the fate of inhaled spores in the lungs of rhesus macaques, exposed to a high dose (~15 LD_{50}) of the fully virulent Vollum strain and protected from infection by vaccination and antibiotics, revealed retention of spores, of which about 65% remained heat-resistant or dormant, up to 42 days after aerosol exposure.[46] Follow-up for up to 100 days post-exposure revealed that the retained fractions of spores were 15–20, 2, 0.5–1% and traces at 42, 50, 75 and 100 days after exposure, respectively.[46] Recent mouse studies on the mechanism of persistence of intranasally-inoculated *B. anthracis* spores of the attenuated, nonencapsulated 7702 strain revealed that the retained spores, mostly dormant, were in a strong association with terminal airway and alveolar epithelial cells, with some spores found inside epithelial cells.[103] This persistence was not associated with dormancy alone, but to some other factors connected with pathogenicity, as inhaled spores of the *B. subtilis* did not show persistence.[103] Earlier *in vitro* and *in vivo* studies showed that *B. anthracis* spores can adhere to and also become internalized by airway and alveolar epithelial cells, a process which may define an alternative route of systemic dissemination besides trafficking into the lymphatics by phagocytes.[104–106] The interactions of *B. anthracis* and macrophages are important in view of their dual role as agents of spore clearance on the one hand and dissemination of infection on the other.[60,61,67,107] This dichotomy is reflected in the results of *in vitro* studies of macrophages from different origins, either primary alveolar macrophages, primary peritoneal macrophages, bone marrow-derived macrophages or macrophage cell lines. Some studies have shown that phagocytosed spores survive, germinate or multiply intracellularly,[59–62,108] whereas other studies demonstrated the opposite – killing of intracellular germinating *B. anthracis* by macrophages.[109,110] The reason for these conflicting results may be differences in the germination state of the spores at the time of uptake by macrophages – germinating spores are killed whereas dormant spores have increased survival.[110,111] These considerations of different spore fates upon interaction with macrophages and epithelial cells have importance in modeling of *B. anthracis* fate *in vivo*.

3.5 Overview of Dose–Response Regression Models

The dose–response model is a mathematical function relating a biological response parameter and the dose of exposure to a pathogen or toxin, *i.e.* the pathogenic or toxic load. The dose–response function for any host–pathogen system is defined by regression of experimental exposure data and statistical fitting methods.[112–114] The dose–response function is a probability distribution of the response as a function of the dose, determined by a median value and a standard deviation. The median effective dose, ED_{50}, will be referred to, depending on the endpoint of the study or model, as the median lethal dose, LD_{50}, or median infective dose, ID_{50}. The mechanism of pathogen and host interactions, concerning the quantitative outcome as a function of the dose, is described by two opposing views: one assumes that the dose effect is determined by the cooperative action of the infectious units comprising the dose, and the other assumes that each infectious dose is acting independently and one organism is sufficient to cause infection and disease.[113] Furthermore, the first concept assumes that there is a threshold exposure dose that has to be acceded for infection to occur and response heterogeneity in the population is reflected in heterogeneity in the individual dose thresholds. The second concept makes no prior assumptions about the population of the host but assigns a probability for successful colonization of the pathogen for any host–pathogen system.[113] The models of the first type, known as deterministic models, were initially developed for chemical toxins but were adapted for pathogens, whereas the second type of models, known as stochastic or single hit models, are specifically oriented to the mechanisms of host–pathogen interactions.[113] The dose metric for pathogens is measured in counts of discrete biological activity units (colony-forming units or cfu for bacteria, and plaque-forming units or pfu for viruses). In studies with *B. anthracis* spores, the dose metric is heat-resistant cfu (65 °C for 30 min, generally).[90] The response metric is the probability of infection, disease or death, depending on the organism, disease pathogenesis or desired endpoint. In the case of inhalational anthrax, most animal studies measured death, as the untreated disease is nearly always lethal and no other endpoints were amenable for study. Recent studies employed markers such as bacteremia, blood PA or LF levels or increased body temperature as markers of disease.[115–118] In this section, we discuss the "classical" dose–response models used in published studies of *B. anthracis*: the deterministic log-normal or log-probit models and the stochastic exponential, Beta–Poisson and some of their derivatives. Other models, used or tested in the context of microbial risk analysis can be found in the extensive literature[114,119–121] and novel modeling strategies for inhalational anthrax will be discussed in following sections.

3.5.1 The Log-Normal and Log-Probit Models

The log-normal model is based on the assumption that microorganisms act cooperatively and the individual threshold dose to a pathogen (or toxin) in the population is distributed log-normally. Thus the probability $P(D)$ that an individual will be infected by a dose D of microorganisms is defined by

$$P(D)=\frac{1}{\sqrt{2\pi}}\int_{-\infty}^{Z}\exp\left(-\frac{z^2}{2}\right)\mathrm{d}z \tag{3.1}$$

The argument Z in eqn (3.1) is defined as

$$Z=\frac{X-\mu}{\sigma} \tag{3.2}$$

where $X = \ln D_i$, D_i is the dose i in number of microorganisms ($\log_{10} D$ can also be used), $\mu = \ln D_{50}$, which is the natural logarithm of the median infective dose, D_{50} (or ID_{50}), and σ is the logarithmic standard deviation. The integral in eqn (3.1) can be calculated numerically, so it can be rewritten in eqn (3.3), where erf is the error function:

$$P(D)=\frac{1}{2}+\frac{1}{2}\operatorname{erf}\left(\frac{Z}{\sqrt{2}}\right) \tag{3.3}$$

The value of the $P(D)$ can be derived from Z using tables of the normal distribution cumulative probability function. The commonly used log-probit method was introduced in the 1930s,[112] by transforming eqn (3.2) to eqn (3.4):

$$\text{Probit}=5+\text{slope}(\ln d-\ln D_{50}) \tag{3.4}$$

By definition, Probit = $Z + 5$. A graph of Probit ("probability unit") as a function of $\ln D$ (or $\log_{10} D$) will be a straight line with a slope = $1/\sigma$. The reason for introduction of the probit transformation was the need to avoid negative values of Z when the dose is lower than the median, in order to facilitate calculation by electromechanical devices. Subsequently, it became standard in dose–response calculations.

3.5.2 Independent Action/"Single Hit" Stochastic Models: The Exponential and Beta–Poisson Models

The exponential and Beta–Poisson dose–response models are based on two assumptions: (1) that at least one organism capable of causing disease must be absorbed by the host and (2) that not all organisms in the absorbed dose actually cause disease due to decay and defensive actions of the host. Therefore, each organism has a certain probability k of causing infection.[113] The exponential model defines the probability of infection or disease in an individual receiving a dose of D organisms by the following equation:

$$P(D)=1-\exp(-kD) \tag{3.5}$$

In this model, the value of k, which is a measure of the infectivity of the organism, is regarded as constant. When tested experimentally, this model had a good fit with many data sets. Fitting problems with the exponential model can be corrected by treating k not as a constant value but as a distribution of infection probabilities. This assumption underlies the Beta–Poisson model, defining $P(D)$ as follows:

$$P(D)=1-\left[1+\left(\frac{D}{\beta}\right)\right]^{-\propto} \tag{3.6}$$

The parameters α and β are derived by fitting of experimental data; α is a slope parameter, *i.e.* as its value increases the slope of the curve approaches exponential, and β is a scale parameter, *i.e.* modifying β shifts the position of the curve along the x-axis but the shape remains unchanged.

The median infective dose can be calculated by substitution of $P(D) = 0.5$ in eqn (3.7), as follows:

$$D_{50}=\beta(2^{1/\propto}-1) \tag{3.7}$$

and eqn (3.8) can be rewritten in the following form, often found in the literature:[20,50,122]

$$P(D)=1-\left[1+\left(\frac{D}{N_{50}}\right)\left(2^{1/\propto}-1\right)\right]^{-\propto} \tag{3.8}$$

where N_{50} denotes the median infective dose. At very low doses, both the exponential and Beta–Poisson probability functions approach linearity,[123] as characteristic of nonthreshold models. The linear approximation of the exponential function, which holds if $kD \ll 1$, is given by

$$P(D)\sim k*D \tag{3.9}$$

The linear approximation of the Beta–Poisson function, which holds if $\alpha D \ll \beta$, is given by

$$P(D)\sim\frac{\propto}{\beta}*D \tag{3.10}$$

Several fitting studies have shown the superiority of the two-parameter Beta–Poisson over the one-parameter exponential model concerning goodness of fit.[50,113,122] Increasing the number of parameters, such as in the three-parameter Weibull–Gamma model, increases the flexibility of the model and can be used if the two-parameter models fit poorly to the data.[114,119] Also, the lower-parameter models are special cases of high-parameter models, as in the case of the one-parameter exponential and the two-parameter Beta–Poisson models.[114,124] Model comparison studies, employing the abundant published data on human exposure to food- and water-borne pathogens, showed that two-parameter models are generally of adequate fit, but the choice of model remains problematic as there is no ideal model and model uncertainty is seen especially in low-dose risk prediction.[50,114,124] The most straightforward way of dealing with this uncertainty is to select the more conservative estimates, regarding these as the most health protective.[50] Several strategies have been proposed to introduce model uncertainty, in addition to experimental uncertainty, into risk assessment, in order to refine judgment and risk management decisions.[114] As will be discussed in detail later, the exponential and log-probit models are frequently used in modeling studies of inhalational anthrax owing to their applicability and simplicity.

3.5.3 Biological Evidence in Support of Independent Action/"Single Hit" Hypothesis

The validity of the independent action/"single hit" model was demonstrated in the classical studies of Meynell,[125,126] who stated several predictions arising from the model and tested them in lethal infection of mice with *Salmonella paratphii* B or *S. typhimurium* at different doses and infection routes. The two hypotheses tested were the "single hit" hypothesis outlined above and a minimal lethal dose (MLD) hypothesis, stating that in causing a lethal infection, the bacteria act cooperatively, thus there is a minimal lethal dose below which no lethality will take place. This MLD differs among individual animals. The predictions tested concerned the composition of bacterial populations sampled from the blood after the death of mice infected by different inoculation doses, containing equal proportions of readily distinguishable genetic variants. If the MLD hypothesis holds, the progeny of all infections will have a variant composition identical with that of the original inocula. If the independent action/"single hit" holds, then in the bacteria recovered from each dead mouse, one type will predominate. The results were compatible with the independent action/"single hit" model in the following manner: although at very high doses the progeny bacteria had the same composition as the infecting inocula, progeny samples became more heterogeneous between individual mice and skewed towards one type as the dose decreased, and single type populations were observed only in mice infected at sub-LD_{50} doses. This pattern was seen if mice were inoculated by the intraperitoneal or oral route and the distribution of types in the output of high-dose infection was accounted for by a breach in the resistance of the mice, allowing infection by more than one bacterium. Results supporting the independent action/"single hit" hypothesis were obtained in some studies of inhalational exposure to *B. anthracis* spores. Zelle *et al.*[127] infected rats and guinea pigs *via* aerosol with a mixture of three variants of *B. anthracis* that differed in colony morphology. All variants were of equal virulence. Analysis of the bacteria recovered from the blood of dying animals revealed a high proportion (68%) of animals infected with only one type of spores, which was consistent with infection by no more than three bacteria. Two recent studies in mice exposed by aerosol or intranasal instillation to spores of genetically tagged *B. anthracis* mutants revealed the occurrence of bottlenecks in the dissemination of the bacteria from the sites of spore germination to the internal organs, which is consistent with an independent rather than cooperative action in the dissemination of the bacteria and progression of systemic disease.[128,129]

3.6 Dose–Response Studies and Modeling in Animals

All the dose–response data available for modeling come from animal studies. The animal models frequently used in inhalational anthrax studies are mice, guinea pigs, rabbits and nonhuman primates.[130] As the ultimate goal of the dose–response studies are extrapolations to human exposure, the most relevant studies are those carried out on nonhuman primates, whose respiratory anatomy is most similar to that of humans,[85] as is the pathology of inhalational anthrax.[130] Next in line are guinea pigs and rabbits, the latter gaining importance in medical countermeasure research,[131] as well as dose–response studies and advanced modeling that have been carried out recently.[51,52,90] The studies detailed in the following sections involve nonhuman primates and guinea pigs and recent rabbit dose–response studies. The main emphasis is on studies that have been used as raw material for dose–response modeling, but some other studies are also presented for comparison and reference.

3.6.1 Guinea Pig Studies

The earliest detailed studies on aerosol exposure of animals to *B. anthracis* spores were reported by Young *et al.*,[100] who published a series of articles describing studies on the respiratory pathogenesis of *B. anthracis* in various animals, using an aerosol exposure chamber. They exposed various animal species to compare the susceptibility and disease pathology, but their main study model was the guinea pig. In these experiments, animals were exposed to different doses of spores for 5 min each time and observed for mortality for 7 days. The LC_{50} (or 50% lethal respiratory exposure, LRE_{50}, in their terminology) for guinea pigs was 1.9×10^5 spores per liter of air. The LC_{50} was calculated by the probit method, using a probit slope of 2.56 in all calculations. It should be mentioned that normalization of virulence in the comparative animal studies was based on body weight and the respiratory minute volume was not measured or considered in the dosage calculations. In addition, they were the first to observe that the lethality of anthrax spores varied not only among the animal species but also among bacterial strains, medium composition and other factors, indicating the need to control these and other variables in inhalational anthrax exposure studies. Their quantitative and histopathology studies in guinea pigs and other animals established the pathological nature of inhalational anthrax as a systemic infection, with pulmonary involvement as a port of entry rather than a focus of bacterial multiplication and the role of invasiveness as a basis for variation of virulence in different bacterial strains and host species, in addition to the mechanism of "single hit,"[127] as discussed above.

In order to clarify an issue raised by contemporary studies of the aerosol particle size dependence of the outcome of respiratory infections, Druett *et al.*[95] started a series of controlled laboratory studies, starting with *B. anthracis* spores, in guinea pigs and rhesus macaques (*Macaca mulatta*). The spore-containing particles were of four size categories: single spore particles of 1 μm final diameter and particles carrying spore aggregates, with diameters of 3.5, 4.5, 8 and 12 μm. The single-spore particles were generated by the Henderson apparatus, in which liquid droplets were generated by a Collision atomizer.[96] For the larger particles, a different apparatus employing a spinning disc sprayer and wind tunnel was used.[97] Dose metrics were expressed as spore concentration, *N*, in cfu per liter of air, calculated from impinger samples; dosage was defined as the concentration multiplied by exposure time (*Nt*) and the dose was defined as the total number of cfu in the volume inhaled during exposure time. The respiratory minute volume used in these calculations, 0.15 min L^{-1}, was obtained from a literature source and was updated by the authors, who recalculated the doses for modeling. The bacterial strain used in the guinea pig and macaque experiments was a highly pathogenic strain (M3) isolated from a monkey exposed by the respiratory route with spores of an original Vollum strain. The duration of each exposure was 1 min and the animals were observed for mortality for 10 days. A dose–response curve was determined for each particle size category and the median lethal dose for each was calculated by the probit method. A summary of the results is presented in Table 3.2, showing for each particle size the LCt_{50} (or LNt_{50}), the LD_{50} and probit slope. More detailed dose and mortality data for the data sets of single-spore experiments used by Bartrand *et al.*[20] for modeling are shown later in Table 3.4. The data show a sharp fall in lethality above the 4.5 μm size category and that the LD_{50} for 12 μm particles was 17-fold greater than for single spore particles. The dependence of infectivity on particle size was consistent with the preferential pulmonary deposition of spores in particles of diameter smaller than 4 μm, observed by Harper and Morton.[99]

Altboum *et al.*[132] studied the dose–response relationships of two strains of *B. anthracis* – Vollum and ATCC 6605 spores – in intranasal inoculation of guinea pigs in the development of a model system for evaluation of post-exposure antimicrobial treatment protocols. Anesthetized female Hartley guinea pigs were

Table 3.2 Exposure of guinea pigs and rhesus macaques to *B. anthracis* spore aerosols of different particle sizes. Adapted from Druett *et al.*[95]

		Guinea pigs			Rhesus macaques		
Particle size/μm	Spores/ particles[a]	LCt_{50}[b]	LD_{50}[c]	Probit slope	LCt_{50}[b]	LD_{50}[c]	Probit slope
Single spores	1	3.4×10^5	5.0×10^4	2.13	4.5×10^4	5.3×10^4	3.19
3.5	18	3.6×10^5	5.5×10^4	2.44	ND[d]	ND	ND
4.5	36	4.9×10^5	7.4×10^4	1.26	ND	ND	ND
8	235	3.8×10^6	5.7×10^5	2.54	ND	ND	ND
12	680	8.6×10^6	8.6×10^5	1.25	6.4×10^5	7.6×10^5	1.17

[a]The influence of the number of spores per particle was tested by exposing guinea pigs to aerosol particles containing spores in numbers different from those above but of similar diameters, *i.e.* 8 μm particles containing 18 spores per particle and 12 μm particles containing 18 spores per particle, in addition to the particles of the respective sizes containing 235 and 680 spore per particle. Provided that the total inhaled dose was the same, mortality was affected by particle size and not the number of spores per particle.

[b]LCt_{50} (presented as cfu min^{-1} L^{-1} air): the median of the product of spore concentration and exposure time from the log-probit-transformed dose–response curve. The concentration refers to the impinger count of viable cells, divided by the volume (flow rate in L air min^{-1} × sampling time). LCt_{50} values were presented in order to make the recalculations, made by later modeling of these data,[20,163] explicable.

[c]LD_{50} (presented as cfu per animal): calculated by multiplication of the LCt_{50} by the respiratory minute volume (0.15 L min^{-1} for guinea pigs and 1.2 L min^{-1} for rhesus macaques).

[d]ND: not done.

infected with 2×10^2–2×10^7 cfu of Vollum and 3×10–3×10^6 cfu of ATCC 6605. Mortality was seen at doses above 3×10^4 cfu of both strains (see Table 3.5). The LD_{50}s, estimated by the Reed and Muench method, were 4×10^4 and 8×10^4 for Vollum and ATCC 6605, respectively, and mean time to death ranged between 3.5–3 and 9.0–3.5 days for the two strains, with an inverse dose–response trend. The dissemination of bacteria was followed in animals exposed to 30 × LD_{50} of Vollum spores. At 24 h post-infection, no dissemination was observed. Bacteria were observed in the blood and spleen of one animal out of three at 24 h, and at 48 h all animals had a disseminated infection. A guinea pig model of aerosol exposure with spores of the Ames strain was reported recently.[133] In this study, 160 animals were exposed at three iterations to aerosolized spores at doses ranging from 6.4×10^3 to 7.7×10^5 spores per animal in a nose-only apparatus. The LD_{50}, calculated by a fitted probit model, was 5.01×10^4 cfu [95% confidence interval (CI), 3.44×10^4–7.54×10^4 cfu] and the LD_{90} was 4.51×10^5 cfu (95% CI, 2.35×10^5–1.47×10^6 cfu). These results are comparable to those of Druett *et al.*[95] and Altboum *et al.*,[132] in spite of differences in the *B. anthracis* strains and exposure techniques.

3.6.2 Nonhuman Primate Studies

In their early study, Young *et al.*[100] exposed monkeys (species not reported) to aerosols of *B. anthracis* spores, reporting an inhalational LC_{50} of 2×10^5 spores per liter of air.[100] In the study by Druett *et al.*[95] on the particle size dependence of the lethality of inhaled *B. anthracis* spores, groups of rhesus macaques were exposed to different doses of single spore and 12 μm aerosol particles and the dose–response relations were assessed as in the guinea pig experiments. The LD_{50} values calculated by the authors based on a respiratory minute volume of 1.2 L min^{-1} were 5.3×10^4 cfu for the single particle and 7.6×10^5 cfu for the 12 μm aerosol, a 14-fold greater value for the larger particles. The probit slope of the dose–response of the 12 μm particle aerosol was about threefold lower than the single-spore dose–response. Also, animals exposed to the 12 μm aerosol developed massive edema of the head and neck about 1–2 days before death, indicating infection *via* deposition at the upper respiratory tract. A summary of the results is given in Table 3.2 and detailed dose and mortality data are presented later in Table 3.4.

In the wake of the outbreak of inhalational anthrax at the New Hampshire goat hair processing mill in 1957, a group of Centers for Disease Control and Prevention (CDC) researchers performed a field study in a goat hair processing mill in North Carolina, in order to substantiate the risk of continuous exposure to the presumably low concentrations of anthrax spores aerosolized during the processing of contaminated goat hair.[134] The mill processed goat hair imported from India and Pakistan and was often contaminated by anthrax spores. The study design, following the classic study of Riley[135] on airborne transmission of tuberculosis, included continuous exposure of animals to the aerosol effluents diverted at the first processing stage (the "picking" machine) into a trailer where the animal cages were held and following mortality, pathology and airborne *B. anthracis* spore levels, to calculate exposure doses and obtain dose–response information. The study took place in five runs, exposing in total 150 cynomolgus macaques (*Macaca fascicularis*). Of the 91 animals autopsied and included in the analysis, 23 animals died of inhalational anthrax, giving a total mortality rate of 25.3%. The overall course of the study is summarized in Table 3.3. The course of airborne spore doses in <5 μm particles and mortality during runs 3–5, which were analyzed, are shown in Figure 3.1.

In run 3, for example, an exposure peak of 2500 spores on the first day was followed by a cluster of six deaths between the 15th and 19th days, suggesting a mechanism of disease causation by exposure to discrete peaks rather than a cumulative effect of consecutive doses. Quantitatively, the authors concluded that exposure to about 1000 spore-bearing particles in a period of 3–5 days resulted in an approximate mortality of 10% and doses from 3500 to 5000 particles over a 5 day period resulted in mortalities between 20 and 24%. In support of this work, a commentator[39] presented results of exposure studies of cynomolgus macaques from the US Army laboratories at Fort Detrick, MD. In these studies, 1236 animals were acutely exposed (1–10 min) to *B. anthracis* Vollum spore aerosols and the mortalities observed during 10 days post-exposure were analyzed by the probit method. The statistical analysis yielded an LD_{50} of 4130 spores (95% CI, 1980–8630 spores) and a probit slope of 0.669 probits/$\log_{10}$(dose) (95% CI, 0.520–0.818 spores). The shallow nature of this dose–response curve is reflected in an interquartile interval (25–75% mortality rates) of two orders of magnitude in dose.

Table 3.3 Summary of runs, exposure doses and outcome of cynomolgus macaque exposure to industrial aerosols. Adapted from Brachman *et al.*[134]

			Calculated inhaled dose of spores (<5 μm particles)			Outcome	
Run no. and dates[a]	Exposure mode	Cumulative exposure length/days	Total cumulative dose[c]	Daily average	No. of animals	Total anthrax deaths	Anthrax mortality rate/%
3: 14 Oct–29 Nov 1963	Constant	47	19 962	530[d]	32	14	43.8
4: 3 Feb–13 Mar 1964		41	4959	198[e]	31	7	22.6
5a: 1–3 June 1964	Only when "picking" hair[b]	2.29[f]	947	413	28	2	7.1
5b: 29–30 June 1964		1.29[g]	1347	1041	22	0	0

[a]Runs 1 and 2 were not included in the analysis and have therefore been omitted.
[b]Run 5 was carried out in two short exposure and sampling periods, performed when the picking machine was operating at enhanced output, producing discrete exposure peaks. The picking was the first stage of processing, in which the clumps of raw hair were raked apart, producing large amounts of dust. Air from the surroundings of this machine was circulated to a trailer, housing the monkey cages.
[c]Exposure doses were calculated on from air sampler counts, obtained with a two-stage impinger selecting aerosol particles of <5 μm diameter and a respiratory minute volume of 1 L min^{-1}.
[d]Averaging was based on 32 days when exposure actually took place.
[e]Averaging was based on 25 days when exposure actually took place.
[f]Exposure during three separate periods over three successive days.
[g]Exposure during two separate periods over 31 h.

Table 3.4 Selected data used for modeling[a,b] by Haas[19] and Bartrand *et al.*[20]

Study	*B. anthracis* strain	Test animal	Exposure method	Dose (inhaled spores)	No. of animals	No. dead	Mortality/%[d]
Altboum *et al.*[132]	Vollum	Guinea pig	Intranasal	20 000 000	7	7	100
				2 000 000	8	7	87.5
				200 000	12	10	83.3
				20 000	12	6	50
				2000	8	0	0
				200	4	0	0
Altboum *et al.*[132]	ATCC 6605	Guinea pig	Intranasal	3 000 000	10	10	100
				300 000	10	8	80
				30 000	10	3	30
				3000	10	0	0
				300	6	1	16.6
				30	6	0	0
Druett *et al.*[95]	Vollum	Rhesus macaque	Aerosol, 1 μm particles	398 400[c]	8	8	100
				300 000	8	7	87.5
				240 000	8	8	100
				160 800	8	3	37.5
				155 520	8	5	62.5
				137 520	8	6	75
				108 720	8	5	62.5
				70 320	8	1	12.5
Druett *et al.*[95]	Vollum	Guinea pig	Aerosol, 1 μm particles	118 000	32	28	87.5
				76 230	32	21	65.5
				40.838	32	18	56.3
				19 820	32	8	25

[a]Adapted from Haas,[19] Bartrand *et al.*,[20] Altboum *et al.*[132] and Druett *et al.*[95]
[b]Only intranasal spore challenge and aerosol exposure to particles of 1 μm diameter were included in this modeling phase.
[c]Aerosol exposure doses from the Druett *et al.* study were recalculated by using respiratory minute volumes of 2.4 and 0.118 L min^{-1} for rhesus macaques and guinea pigs, respectively, rather than 1.2 and 0.15 L min^{-1} originally.
[d]Mean times to death reported by Altboum *et al.* were 5.5, 4.1, 3.1 and 3 days for guinea pigs exposed to 2×10^4, 2×10^5, 2×10^6 and 2×10^7 Vollum strain spores, respectively. The times to death for guinea pigs exposed to 3×10^4, 3×10^5 and 3×10^6 ATCC 6605 spores were 9.0, 5.3 and 3.5 days, respectively.

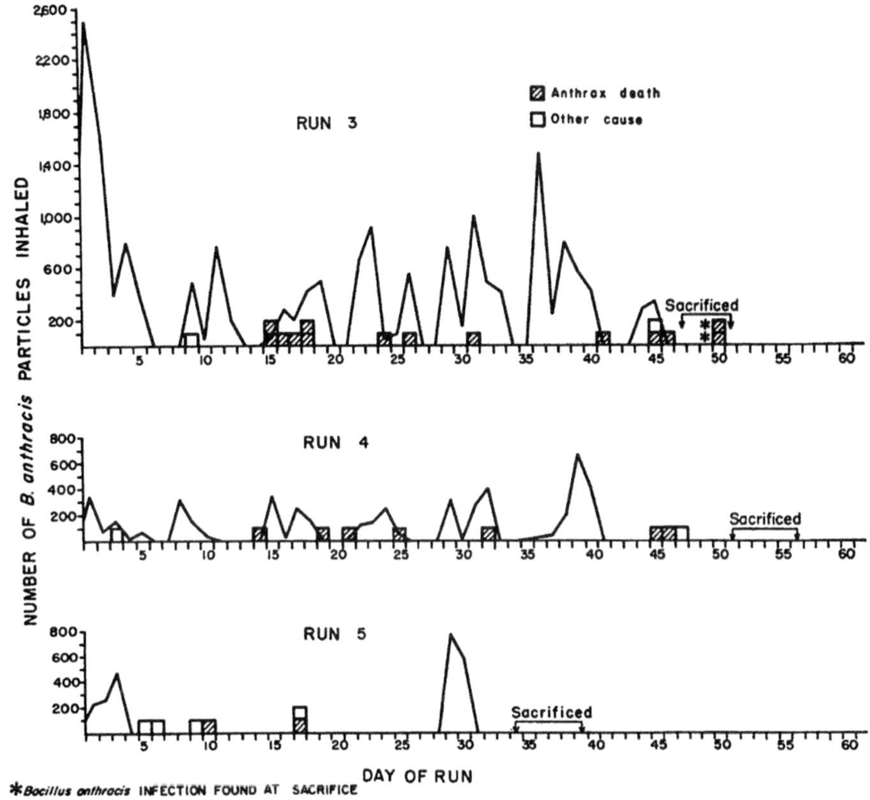

Figure 3.1 Time course of inhaled doses of airborne *B. anthracis* and animal mortality during runs 3, 4 and 5 of the goat hair processing plant study.[134] Exposure doses were calculated from air sampler counts, obtained with a two-stage impinger selecting aerosol particles of <5 μm diameter, and a respiratory minute volume of 1 L min^{-1}. Animals dying from inhalational anthrax or nonanthrax causes are indicated.

The inhalational LD_{50} values for the Vollum and Ames *B. anthracis* spores, determined in nonhuman primates of different species, were higher (reviewed by Phipps *et al.*[131]). A study employing 14 cynomolgus macaques, exposed by aerosol to doses of 4.5×10^4 to 2.94×10^6 cfu of *B. anthracis* Ames spores, determined the inhalational LD_{50} for spores of the Ames strain as 61 800 cfu (95% CI, 34 000–110 000 cfu) and a probit slope of 4.21 [standard error (SE), 2.42][136]. All the animals receiving doses higher than 2×10^5 cfu died within 4 days. The respective LD_{50}s for rhesus macaques were 5.5×10^4 and 5.0×10^4 cfu for Ames and Vollum, respectively.[131,137,138] A study in which rhesus macaques were exposed by aerosol to doses of $>10^4$ *B. anthracis* Vollum spores reported that animals exposed to $<5 \times 10^5$ spores survived 3–20 days (mean 7.5 days) post-exposure and animals exposed to $>5 \times 10^5$ spores survived 2–5 days (mean 3 days).[139] The only study of inhalational anthrax in chimpanzees (*Pan troglodytes*) was reported by Albrink and Goodlow.[140] In this study, four young male chimpanzees (9–11.5 kg body weight) were challenged with aerosolized *B. anthracis* Vollum 1B spores at measured doses of 32 800, 34 350, 39 700 and 66 500 cfu. The two individuals receiving the lower doses survived after mild and transient bacteremia and toxemia, whereas the other two developed fulminant disease and died. In the animals that died, bacteremia and toxemia were seen initially on day 2–3 post exposure and progressed until death on day 6. An LD_{50} estimate of 3.7×10^4 cfu from these data was presented by Phipps *et al.*[131] A recently introduced nonhuman primate inhalational anthrax model, the African green monkey (*Chlorocebus aethiops*), has a similar susceptibility to the macaque models ($LD_{50} = 1.1 \times 10^4$ cfu; 95% CI, 2.98×10^3–8.1×10^4 cfu).[116] The discrepancies between the macaque LD_{50} reported in the 1990s and those reported in the 1960s were accounted for by time elapsed, differences in spore preparation methods, lack of standardized aerosol characterization methods and unreported or poor health of the animals. In fact, the animals of Brachman *et al.*'s study[134] and others[138,139] were often infested with pulmonary parasites, which might have increased their susceptibility due to bacterial entry through pre-existing lesions. A considerable number of animals in the Brachman study died of nonanthrax causes during the experiment.

3.6.3 Dose–Response Modeling

One of the gaps recognized after the 2001 anthrax letter attacks was a methodology for assessing quantitative microbial risk from inhalational exposure to anthrax spores. In order to fill this need, the Haas group embarked on the application of the dose–response model fitting methodology, developed and employed by Haas and others for food- and water-borne pathogens,[48,50,113,122,123] to inhalational anthrax risk assessment. The methodology employed the selection of an appropriate experimental data set, fitting a dose–response model, usually the exponential or Beta–Poisson model, by maximum likelihood estimation and verification of goodness of fit.[50,122] As human data for dose–response analysis of inhalational anthrax are not obtainable, the only available data sources for modeling are animal studies, preferably nonhuman primates. In an initial assessment of the risk of mortality to primates, Haas[19] evaluated two models based on published data sets: the log-probit model of Glassman[39] and an exponential model fitted to recalculated single-spore exposure data for rhesus macaques from Druett *et al.*[95] An exponential model [eqn (3.5)] with the parameter $k = 7.16 \times 10^{-6}$ and an LD_{50} of 96 800 was fitted to recalculated inhaled doses (Tables 3.4 and 3.5). Comparison of this model with Glassman's demonstrated a large difference in the risk predictions: the predicted risk levels converged only at the high doses and when extrapolated to extremely low doses, *i.e.* inhalation of 10 spores, the gap between the predictions spanned three orders of magnitude. The two models were also tested for their ability to predict the results of Brachman *et al.*'s study[134] described above. To facilitate the comparison, an exponential model with the parameter $k = 2.6 \times 10^{-5}$ was fitted to the

Table 3.5 Best-fit models by Bartrand *et al.*[20] for guinea pig (GP) and rhesus macaque (RM) respiratory exposure to *B. anthracis* spores. Adapted from Bartrand *et al.*[20]

Data set	Best-fit model	Parameters	P^a	LD_{50} (95% CI)	LD_{10} (95% CI)	Panel in Figure 3.2
GP, Vollum, 1 μm aerosol	Exponential	$k = 1.65 \times 10^{-5}$	0.699	4930 (32 661–53 346)	6390 (5005–8212)	d
GP, ATCC-6605, intranasal	Beta Poisson	$\alpha = 0.549$; $N_{50} = 28\,472^b$	0.631	25 472 (~10 000–~100 000)	7196 (826–21 700)	b
	Exponential	$k = 7.15 \times 10^{-6}$	0.284	94 320 (74 100–126 060)	14 800 (11 100–19 600)	a
RM, Vollum, 1 μm aerosol	Exponential	$k = 7.16 \times 10^{-6}$	0.188	96 800 (70 700–136 000)	14 123 (10 787–18 200)	c
GP, Vollum, intranasal	Exponential	$k = 7.51 \times 10^{-6}$	0.128	147 411 (17 400–357 700)	23 095 (2730–53 800)	Not shown

[a]Maximum likelihood and goodness of fit: the parameters of a distribution are determined from the data by the method of maximum likelihood estimate (MLE). The MLEs of the parameters are those that minimize deviance Y, defined as $Y = -2 \ln LR$ (LR = likelihood ratio of the estimated and observed outcome probabilities). P or the goodness of fit probability, determined by comparison of the minimized deviance (Y^*) of the model to a critical χ^2 value ($\chi^2_{0.95,\, n-k}$; the 0.95 probability value of a χ^2 distribution with degrees of freedom determined by the difference between the number of doses, n, and the number of parameters of the model, k). If the P value of Y^* is smaller than 0.05, the fit is rejected. The higher the value of P, the better is the fit.[20,50]
[b]N_{50}: median lethal dose in the notation of the article's authors, equivalent to d_{50} of eqn (11) in Bartrand *et al.*[20]

cumulative doses of four exposure runs (Table 3.3), assuming that exposure on each day was an independent event. This exponential model predicted the observed mortalities fairly well, but the Druett-based model underestimated the risk and the Glassman model was dramatically over-predictive. This contrasted Glassman's claim[39] of consistency of his data with the observations of Brachman *et al.* The implications of this work were the following: (i) a "single hit" model, *i.e.* exponential, may be applicable for the risk prediction of inhalational anthrax in nonhuman primates; (ii) the low-dose exposure predicted the risk as nonzero, as found for other microorganisms; and (iii) a large uncertainty in the extrapolation of the available models at the time to doses implying low levels: if an "acceptable" risk of 10^{-3} is sought, the Druett-based model predicted dose is 140 spores and the Glassman-based model predicts a dose of less than one spore for 10 individuals, if applied to humans.

In order to refine the modeling, Bartrand *et al.*[20] searched in the open literature for additional data suitable for pooling in a single model. Of 17 studies reporting dose–response data for *B. anthrax* infections in animal models, only two were found suitable: the aerosolized spore exposure studies of rhesus macaques and guinea pigs of Druett *et al.*[95] and the intranasal exposure of guinea pigs by Altboum *et al.*[132] The data of Druett *et al.* were found to be the only data specifying dose–response for inhalation of <5 μm diameter particles, a fully reported range of doses and lethality and two animal models. The data of Altboum *et al.*, even though a different exposure method was employed, was as detailed and had dose–response data for two *B. anthracis* strains – a fully virulent (Vollum) and an attenuated (ATCC 6605) strain. This scope of data was required as the goal was not only to find the best fitting model but also to see if data from different experiments can be pooled. The dose and mortality data for the selected sets are given in Table 3.4, the parameters and goodness of fit statistics of the best fitted models are given in Table 3.5 and the fitted model curves are shown in Figure 3.2. For three of the four sets, the best fitting models were exponential and the best of these was the guinea pig 1 μm aerosol exponential model ($p = 0.699$).

The only Beta–Poisson fit shown is of intranasal exposure of guinea pigs to the ATCC 6605 strain. An exponential model was also fitted to this experiment, which was used for the pooling studies. A Beta–Poisson model was also successfully fitted for the guinea pig 1 μm aerosol experiment (not shown). The rhesus macaque 1 μm aerosol model of the previous study[50] fitted less well ($p = 0.188$) than other models. The ability to pool data was assessed by the goodness of fit statistics of several interspecies and intraspecies model pairs. Of the two interspecies pairs, only one pair – guinea pig intranasal ATCC 6605 and rhesus macaque 1 μm aerosol exposure – could be pooled with a good fit ($p = 0.4$) (Figure 3.2e), so no definitive conclusions could be drawn on interspecies extrapolation from this study. Another aspect modeled in this study was the dose–response relationship for spore-bearing particles of different diameters on the basis of the experiments of Druette *et al.* For particles with diameters <5 μm, exponential models were generally fitted, while Beta–Poisson models were fitted for the larger particle dose–response relationships. For guinea pigs, the lethal doses for aerosol particles of <5 μm diameter did not show a striking variability with particle diameter and could also be pooled. The same was true for the particles above 5 μm diameter that could be used as a cutoff (Figure 3.3).

3.6.4 Exposure Dose and Delivered or Deposited Dose

As described earlier, particle deposition in the lungs is a major factor in determining species differences in susceptibility to inhalational anthrax. Furthermore, the dose actually delivered to the bronchiolar and alveolar regions determines the probability of infection takeoff and the dynamics of subsequent buildup of pathogen burden. To evaluate the relation between the inhaled and delivered doses for different species, Weir and Haas[88] developed a Markov chain model for calculation of the fluid phase spore concentration in three sub-regions of the respiratory tract: (i) from the nares to the

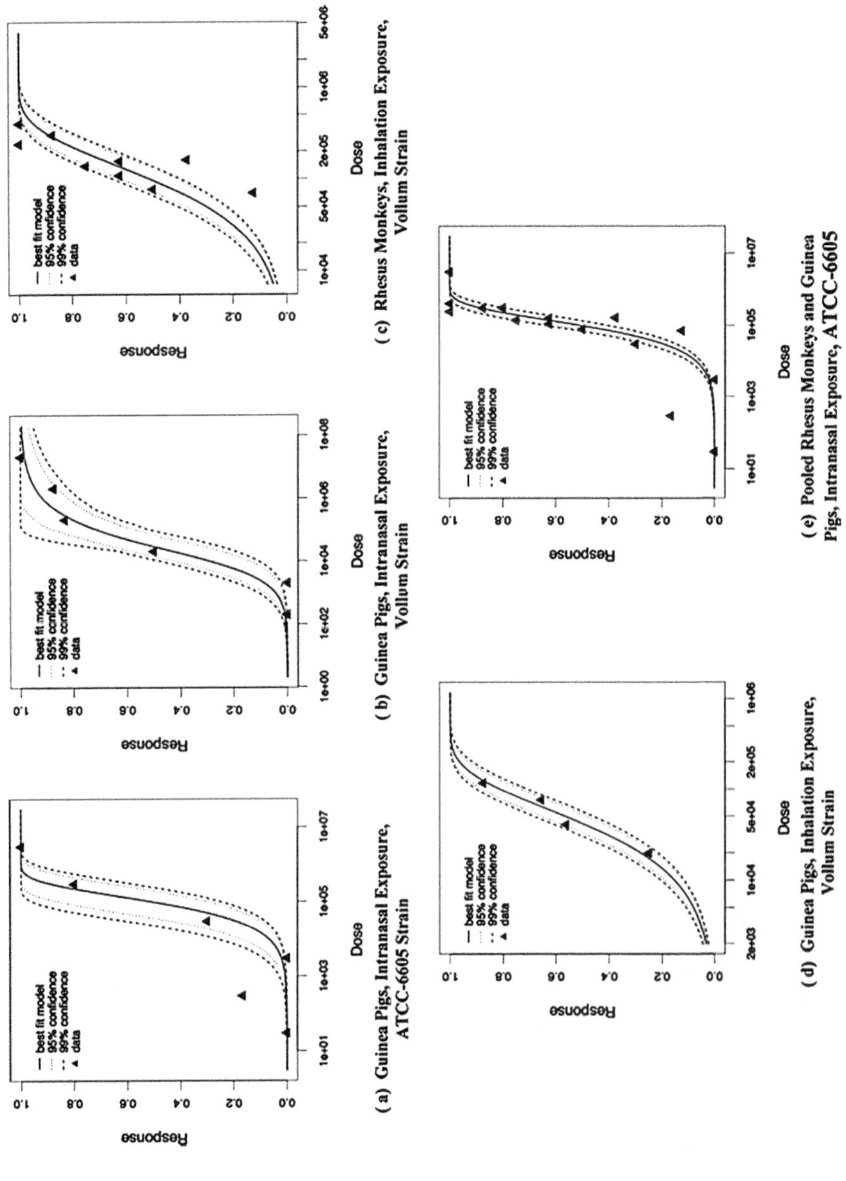

Figure 3.2 Best-fit models, by Bartrand *et al.*,[20] of respiratory exposure of guinea pigs and rhesus macaques to *B. anthracis* spores. Model types and parameters of panels (a)–(d) are shown in Table 3.5 as indicated. Panel (e) shows the pooled exponential model of the data from (a) and (c). Reproduced from Bartrand *et al.*[20]

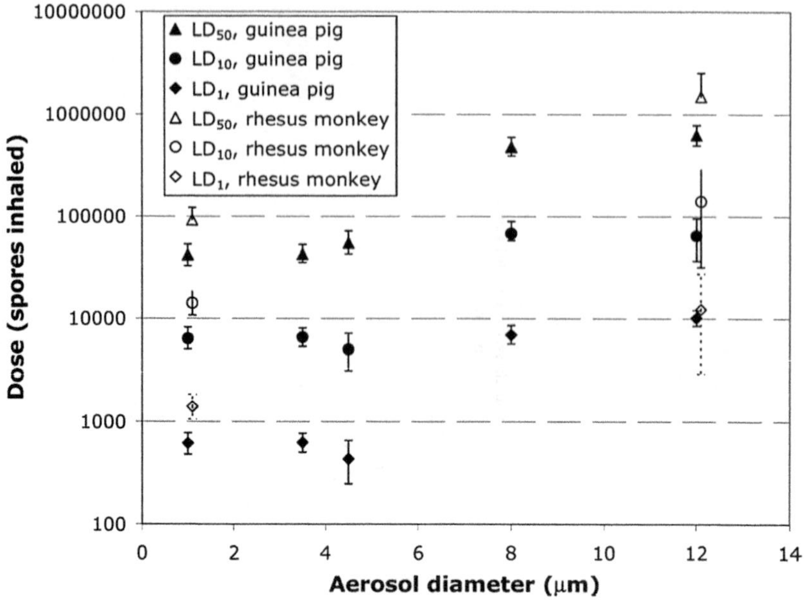

Figure 3.3 Particle size dependence of dose–response relationship for aerosolized *B. anthracis* spore lethality for guinea pigs and rhesus macaques, as modeled by Bartrand *et al.*[20]

larynx; (ii) from the larynx to the main bronchi; and (iii) from the main bronchi to the alveoli. The alveoli themselves were not included in this model of the first stage of the infection process but in a separate model of the second stage, modeling the buildup of bacterial burden. The Markov chain is a stochastic model of the transition probabilities of spores between the fluid phase and surface of the three sub-regions, in addition to losses due to spore inactivation and exhalation. On the basis of estimated fluid phase spore concentrations in the third sub-region after 60 min of breathing, species-specific delivered dose correction factors from the exposure doses were defined as 3.77×10^{-6}, 2.6×10^{-5} and 8.34×10^{-4} for guinea pigs, rhesus macaques and humans, respectively. On the basis of the delivered dose correction factors, a more realistic comparison of interspecies susceptibilities can be made: if the inhaled LD_{50}s for guinea pigs and rhesus macaques are 147 471 (95% CI, 17 000–357 000) and 92 000 (95% CI, 29 400–154 560), respectively,[20] then the calculated delivered doses are 0.556 (95% CI, 0.0644–1.323) for guinea pigs and 2.392 (95% CI, 0.764–4.019) for rhesus macaques.[88] This implies that if the guinea pigs are more susceptible than rhesus macaques on the basis of inhaled dose, then the opposite is judged by the pulmonary delivered doses, being one spore for guinea pigs and four spores for rhesus macaques.[88] For humans, on the basis of the broad estimate of 2500–55 000 spores, the calculated delivered dose is in the range 2.085–45.87 spores.[88] In order to facilitate modeling of subsequent steps in the infection process and interspecies extrapolation, the delivered dose correction factors can be implemented into the exponential or Beta–Poisson models, as shown in eqn (3.11) for the exponential model and eqn (3.12) for the Beta–Poisson model:

$$P(D) = 1 - \exp\left(-k^{*} D\eta_{\mathrm{dd}}\right) \tag{3.11}$$

$$P(D) = 1 - \left[1 + \left(\frac{D\eta_{\mathrm{dd}}}{N^{*}_{50}}\right)\left(2^{1/\alpha} - 1\right)\right]^{-\alpha} \tag{3.12}$$

The delivered dose is the product of the inhaled dose D and the delivered dose correction factor η_{dd}. The delivered dose-specific parameters are k^* and N^*50 for the exponential and Beta–Poisson models.[88]

Gutting *et al.*[141] estimated and compared the inhaled and delivered spore doses in nonhuman primates, rabbits and guinea pigs using a different approach. First, they modeled the dose–response relationship of inhaled doses on the basis of selected data sets. Second, they evaluated the species-specific pulmonary deposition factors. Finally, they modeled the dose–response relationship for the calculated delivered doses. Two of the four nonhuman primate data sets were new – one for rhesus macaques (from USAMRIID) and one for cynomolgus macaques (from Battelle), both exposed to spores of the Ames strain. The unpublished data are reported in their paper[141] and are not reproduced here. The other two data sets are the historical data of Druett *et al.*[95] and Henderson *et al.*[45] for rhesus macaques exposed to the M36 (Vollum) strain delivered in 1 μm diameter aerosol particles. For the nonhuman primate data, exponential models with k parameters ranging from 1.27×10^{-5} to 1.43×10^{-5} (for the Druett *et al.* data) were fitted successfully (MLE p values of 0.1878–0.9821). A single model ($k = 1.37 \times 10^{-5}$) was fitted successfully ($p = 0.901$) for the pooled data of the four sets. For rabbits, three new data sets were used: an unpublished data set from USAMRIID [raw data from the study published by Zaucha *et al.*[142] and the others from Gutting *et al.*[90] and the US Environmental Protection Agency (EPA)[143]]. In all sets the animals were exposed to spores of the Ames strain. Exponential models with k parameters of 6.16×10^{-6}, 2.34×10^{-6} and 7.99×10^{-6} were fitted successfully (p values of 0.7304, 0.8101 and 0.7741) for the USAMRIID, Gutting *et al.* and EPA data sets, respectively. The two guinea pig data sets were the historic set of Druett *et al.*[95] in which the animals were exposed to 1 μm diameter aerosol particles of M36 (Vollum) spores and the new study described above of Savransky *et al.*[133] in which Hartley guinea pigs were exposed to spores of the Ames strain. The exponential model fitted to the Druett data ($k = 9.52 \times 10^{-5}$, $p = 0.1954$) was different from the exponential model fitted previously for that set ($k = 1.65 \times 10^{-5}$; Bartrand *et al.*[20]), because of the different values of the respiratory minute volume used in the inhaled dose calculations (0.150 L min^{-1} by Druett *et al.* and 0.118 L min^{-1} by Bartrand *et al.*). The exponential model fitted for the Savransky *et al.*[133] data set had a parameter $k = 1.09 \times 10^{-5}$ ($p = 0.1158$). In contrast to the primate data, an exponential model for the pooled guinea pig data ($k = 9.79 \times 10^{-5}$) fitted poorly ($p = 0.0893$). The pulmonary deposition factors were estimated for the whole lung, including the tracheobronchial and pulmonary regions. For the nonhuman primates, a pulmonary deposition factor of 22% of the inhaled dose was obtained by averaging the data of Harper and Morton,[99] who measured lung deposition of 17.5% for *B. subtilis* spores in 1 μm particles, and the Henderson *et al.*[45] estimated deposition of 25.64% of the inhaled dose for aerosolized *B. anthracis* spores. As no deposition data were for cynomolgus macaques, the same deposition factor was used for this species, assuming similar physiological parameters. The deposition factors for guinea pigs and rabbits were 30 and 9%, respectively, on the basis of the radiolabeled particle pulmonary deposition studies of Raabe *et al.*,[93] estimations by a lung deposition model and measurement of pulmonary deposition of *B. anthracis* spores in rabbits[90] and *B. subtilis* spores in guinea pigs.[99] Using these species-specific deposition factors, the inhaled doses were converted to deposited doses and exponential models were fitted. Pooled exponential models were fitted to the nonhuman primate ($k = 6.22 \times 10^{-5}$), rabbit ($k = 7.34 \times 10^{-5}$) and guinea pig ($k = 3.29 \times 10^{-5}$) data. The conversion of doses did not affect the goodness of fit determined for the inhaled doses. On the basis of the LD_{50} estimated by the inhaled dose models, the most susceptible animals are the nonhuman primates (LD_{50} = 50 600; bootstrapping 95% CI, 40 400–62 500). Less susceptible are guinea pigs (LD_{50} = 70 800; 95% CI, 62 000–80 300) and the least susceptible are rabbits (LD_{50} = 102 600; 95% CI, 56 000–168 000). Considering the deposited dose LD_{50}, the nonhuman primates and rabbits have a similar susceptibility (nonhuman primate LD_{50} = 11 100; 95% CI, 9250–13 300; rabbit LD_{50} = 9440; 95% CI, 5380–15 400), while guinea pigs are significantly less susceptible from both (LD_{50} = 21 000; 95% CI, 18 600–23 900). These results agree qualitatively with those of Weir and Haas,[88] showing the discrepancy in animals' susceptibility as judged by the inhaled or the delivered doses and reflecting the high susceptibility of guinea pigs and rabbits seen in the course of the disease in these animal models. Other insights from this study are that data from different dose–response studies can be pooled for modeling in the same or closely related species such as the rhesus and cynomolgus macaques, but not for different and more distant mammalian species, as observed by Bartrand *et al.*[20] Similarly, the Beta–Poisson model was inappropriate for modeling the new data sets and therefore the exponential model was chosen.

3.6.5 Competing Risks Model

The competing risks model is a probabilistic model developed for incubation period modeling by Brookmeyer and co-workers.[44,46] The application of this approach in modeling of the incubation period distribution in the Sverdlovsk incident and animal exposure data will be presented later; this section discusses the dose–response aspect. The model is based on the assumption that an inhaled spore may be either cleared by the pulmonary defense mechanisms – mucocilliary action or killing by macrophages – or germinate, produce toxins and multiply to cause disease. The fate of an inhaled spore is determined by competition of the two risks. If the risk, or rate, of spore clearance per time unit is θ and the risk, or rate, of spore germination per time unit is λ, then the probability that a spore germinates before being cleared is defined by

$$\int_0^\infty \lambda e^{-\lambda t} e^{-\theta t}\, dt = \frac{\lambda}{\lambda+\theta} \qquad (3.13)$$

The number of spores that germinate from an inhaled dose D in the lungs has a Poisson distribution with a mean of $D\lambda/(\lambda + \theta)$. Therefore, the probability that at least one spore will germinate and cause an infection, or the infection probability, is defined by the exponential function (3.14):

$$P(D) = 1 - \exp\left(-D\frac{\lambda}{\lambda+\theta}\right) \qquad (3.14)$$

The dose of spores that will cause disease in P% of the population, so that $p = P/100$, is defined as the toxic dose $TD(P)$ or ID_p. $TD(P)$ can be derived from eqn (3.14) by setting $P(D)$ to p and solving for D:

$$TD(P) = \frac{-(\lambda+\theta)\ln(1-p)}{\lambda} \qquad (3.15)$$

Brookmeyer and co-workers[44,46] derived the rate parameters θ and λ from nonhuman primate studies and applied them for modeling of the human incubation period of inhalational anthrax, as will be presented below. By fitting an exponential decay model to the data of Henderson *et al.*[45] on the post-exposure pulmonary retention of spores in rhesus macaques, Brookmeyer *et al.*, estimated θ as 0.07 per day. The spore germination probability λ was estimated on the basis of published primate LD_{50} for inhalational anthrax, between 4000 to over 100 000 spores. By substituting these values into eqn (3.14), setting $P(D) = 0.5$ and $\theta = 0.07$, λ could be estimated to be between 5×10^{-7} and 1×10^{-5}. As λ is very small compared with θ, the equations can be simplified by setting $\lambda + \theta \sim \theta$.

3.6.6 Dynamic Modeling of Cumulative Repeated Exposures

In many situations in real life, exposure to pathogens does not occur as a single high-dose event but as a series of repeated low-level exposure events, with intermissions between them. In the case of inhalational anthrax, settled spores in a contaminated environment may be re-aerosolized when a mechanical disturbance by wind or human activity takes place, and this may result in exposure and infection hazard, as observed in the goat hair processing mills and on some occasions in facilities contaminated by anthrax letters in 2001.[144,145] This may not always be the case.

In sharp contrast to the study of Brachman *et al.*,[134] demonstrating a significant risk to primates continuously exposed to anthrax spore-contaminated air from a goat hair mill, observational studies in such mills showed that employees exposed to airborne spores developed immunity due to subclinical exposure.[146–148] The overall risk from continuous or serial exposures to low levels of pathogens is determined by the interplay of retention and accumulation of pathogens on the one hand and clearance by the innate immune and defense mechanisms on the other, together with recruitment and depletion of immune system elements. Conventional dose–response modeling is insensitive to the timing and dosing pattern of the pathogen, *i.e.* an assessment based on three exposure events to dose x or one exposure event to dose $3x$ will result in the same risk, given the independent action theory. Pujol *et al.*[149] suggested an alternative modeling approach, considering the dynamics of pathogen and innate immune elements and their interrelationships in the course of exposure to pathogens, called by them the cumulative dose model. The model parameters define a series of processes: recruitment of innate immune elements by constitutive and pathogen-induced pathways, constitutive decay and depletion of immune elements through consumption by interactions with the pathogen and "natural" pathways, in addition to recruitment, multiplication, clearance and decay of the pathogen. The feasibility of the cumulative dose model was shown by successful fitting to experimental datasets and demonstration of the effects of dosing dynamics on the evaluated risk, but it is computationally intensive and not practical for use in disease transmission models. Therefore, a simpler dynamic model was developed by Mayer *et al.*[150] and validated using the primate exposure study of Brachman *et al.*[134] The model calculates pathogen level as a negative function of time. In the case of a single exposure, the pathogen level declines until it is cleared completely or establishes an infection. In the case of serial exposures, pathogen level is modeled as a time series of levels that are sums of the current exposure and the residue from the previous one. The pathogen level is introduced into a risk function analogous to the classical exponential model, which is a limiting case of the dynamic model. In addition, a statistical framework for maximum likelihood testing of the model is presented. An outline of the model logic and evaluation is given below.

As mentioned, the pathogen level declines with time until its complete clearance or establishment of infection. In order to capture the pathogen's fate between these two outcomes, the model has to include the depletion of immune elements through interaction with the pathogen, until immune defense is overwhelmed and infection is established. The change in pathogen level (N) with time is represented by the following differential equation:

$$\frac{\mathrm{d}N}{\mathrm{d}t} = -\gamma N^{\alpha} \tag{3.16}$$

where γ is an approximate representation of pathogen clearance rate and α is introduced in order to represent the degree to which the immune system is overwhelmed by the pathogen. For preservation of biological feasibility, the value of γ is constrained between 0 and ∞ and the value of α is constrained between 0 and 1. When $\alpha = 1$, the pathogen elimination is a simple exponential function with a *per capita* rate constant γ, *i.e.* the ability of the immune system to eliminate one pathogen is not affected by the number of pathogens within the host. When $0 < \alpha < 1$, the *per capita* rate γ is attenuated by a factor of $1/N^{1-\alpha}$, which means that the efficiency of the immune system is affected by the pathogen numbers. As α decreases and approaches 0, the shape of the curve becomes more linear and slower for a fixed γ.

To explain the analytics of the calculation of the risk from multiple exposures on the basis of eqn (3.16), we shall first present the calculation of the risk for a single-dose inoculation. The authors analyzed the solutions for a single inoculation with a dose $D = N(0)$ at time 0, as follows:

(i) When $\alpha = 1$, then

$$N(D) = De^{-\gamma t} \tag{3.17}$$

(ii) When $0 < \alpha < 1$, then

$$N(t,D) = \left(t\gamma(\alpha-1) + D^{1-\alpha}\right)^{\frac{1}{1-\alpha}} \tag{3.18}$$

The extinction time, t_e, is the time in which all the pathogen is cleared, so that when $t \geq t_e$, $N(t,D) = 0$. In the case when $\alpha = 1$, then $t_e = \infty$ [eqn (3.19)]. In the case when $0 < \alpha < 1$, then

$$t_e = \frac{D^{1-\alpha}}{\gamma(1-\alpha)} \tag{3.19}$$

The effective dose for a single inoculation is defined in eqn (3.20), by integration of $N(t,D)\mathrm{d}t$ from inoculation to extinction:

$$\int_0^{t_e} N(t,D)\mathrm{d}t = \frac{D^{2-\alpha}}{\gamma(2-\alpha)} \tag{3.20}$$

If the instantaneous probability for a single pathogen, per unit time, to initiate an infection is s, then the probability for an individual harboring N pathogens at a time point to be infected is $sN(t,D)\mathrm{d}t$. By integrating and exponentiating the hazard function over time, the survival probability becomes

$$S(t) = e^{-s\int_0^t N(t,D)\mathrm{d}t} \tag{3.21}$$

By substituting eqn (3.20) for the time period required for complete pathogen clearance, $t = t_e$, and defining the infection probability $P = 1 - s$, the probability of infection as a function of the dose $P(D)$ is

$$P(D) = 1 - \exp\left[-s\frac{D^{2-\alpha}}{\gamma(2-\alpha)}\right] \tag{3.22}$$

When $\alpha = 1$, the model is reduced to the classical exponential model with a constant $k = s/\gamma$. When $0 < \alpha < 1$, the simple exponential relation, based on independent action of pathogens, does not hold as the *per capita* clearance rate becomes dependent on inoculum size.

In multiple exposure scenarios, the input doses are presented as a sequence of inoculations, $\{D_i\}_{i=1}^{k}$, with each dose D_i delivered at a time point t_i. The sequences of inoculations and times correspond, but the final observation time T for each subject generally falls between sequential inoculation events, hence it is considered separately. As the leading dynamic process in this model is the declining pathogen level with time, the effective dose of each inoculation event is the sum of the residual number of pathogens from previous inoculations and the dose of the current inoculation, which also declines with time. The cumulative effective dose at time T is the sum of partial cumulative pathogen numbers, as defined mathematically in the equation

$$\int_0^T N\left(t,\{D_i\}_{i=1}^{k}\right)\mathrm{d}t = \sum_{i=1}^{k-1}\left[\int_0^{t_{i+1}-t_i} N\left(t,n_i+D_i\right)\mathrm{d}t\right] + \int_0^{T-t_k} N\left(t,n_k+D_k\right)\mathrm{d}t \qquad (3.23)$$

where n_i refers to the residual pathogen level at t_i. The last term is the final pathogen increment if $T > t_k$ and becomes equal to 0 if $T = t_k$. The survival function at time T, $S(T)$, is shown in eqn (3.24), and the corresponding risk is $1 - S(T)$.

$$S(T) = \exp\left[-s\int_0^T N\left(t,\{D_i\}_{i=1}^{k}\right)\mathrm{d}t\right] = 1 - P(T) \qquad (3.24)$$

The behavior of the pathogen numbers in two hypothetic multiple exposure scenarios is exemplified schematically in Figure 3.4, showing two different timing patterns resulting in equivalent cumulative doses and exposure duration (40 organisms, 200 min).

Note that under the specified model parameters (α = 0.5, γ = 0.05), the pattern in Figure 3.4b results in longer pathogen persistence. The model was evaluated using run numbers 3 and 4 from Brachman *et al.*'s primate continuous exposure study.[134] The values of the parameters α, s and γ with the optimal maximum likelihood fit were 0.9, 1.81×10^{-7} h^{-1} and 0.0097 h^{-1}, respectively. The predicted risk for runs 3 and 4 was 50 and 16%, respectively, compared with the observed 44 and 23%. The effect of exposure timing on risk was evaluated by comparison of two scenarios – a single bolus and a time-series exposure with a total dose of 15 000 spores. At α = 0.5, the risk levels were 0.64 for the single bolus and 0.59 for the time-series exposure. The difference decreased with α and vanished at α = 1.0. At α = 0.9 it was only 0.01, indicating a negligible effect of exposure timing for Brachman *et al.*'s results. Another test of the feasibility of the model was comparison with other anthrax dose–response models. When setting α = 1.0, the model based on Brachman *et al.*'s data could be reduced to an exponential model with a $k = s/\gamma = 3.57 \times 10^{-5}$. The predicted risk from this model and a model with α = 0.9 were comparable to other exponential models reported in the literature, such as those of Haas[19] and Brookmeyer *et al.*[46] One limitation of the model is the use of abstract parameters whose biological significance is

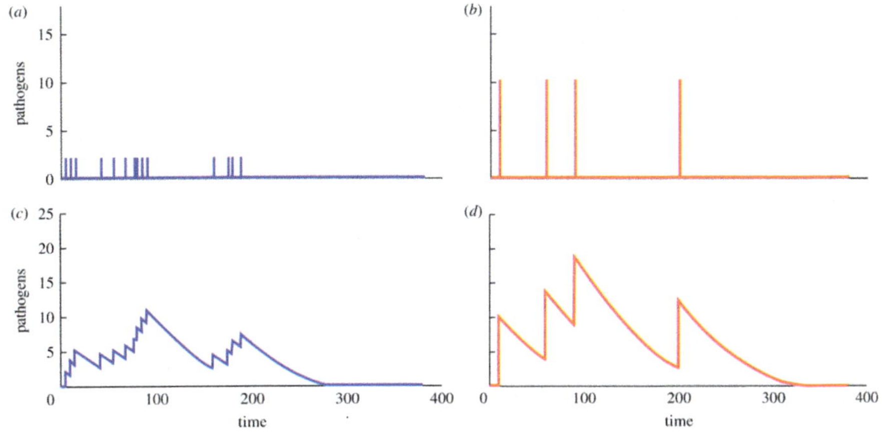

Figure 3.4 Schematic examples of two different inoculation patterns evolution (a and b) of corresponding evolution of within-host pathogen numbers with time (in minutes) (c and d).[150] The timing of pathogen exposure is different in the two patterns, although the total cumulative dosages are the same (40 pathogens over 200 min). Under the specified model parameters (α = 0.5, γ = 0.05), the pathogen persists longer (d) in the four exposures pattern (b). Reproduced from Mayer *et al.*, 2012.[120]

not readily interpretable from the results. The parameter γ is interpretable only for the special case of exponential pathogen clearance, while other pathogen clearance kinetics are feasible. Another limitation is that the model does not take into account the pathogen reproduction within the host, which is also considered in the more complex model of Pujol *et al.*[149]

3.6.7 Benchmark Dose Analysis of Nonhuman Primate Studies

The US Environmental Protection Agency (EPA) is in the process of transition from the classical NOAEL/LOAEL-based method to the benchmark dose (BMD) system, which is based on low-level dose–response relationships in deriving point-of-departure (POD) levels for risk assessment and exposure limit determination.[121] The benchmark dose represents an exposure dose resulting in a certain level of increase in a selected benchmark response (BMR) level above the background level. Thus a BMR_{10} is an extra risk of 0.1 (or 10%) over the background and the respective dose is BMD_{10}. The value taken as POD for low-level risk assessment is the 95% CI lower limit, designated benchmark dose lower bound or BMDL. The current BMD software of the EPA permits the fitting and test of fit and calculating BMD and BMDLs with an array of models for dichotomous (quantal) responses of the type discussed here and continuous responses. The system has been developed for toxicological risk assessment but it is also applicable for microbiological risk analysis.[124]

The applicability of the BMD system has been assessed for dose–response analysis of nonhuman primate exposure to aerosolized *B. anthracis* spores.[98] Of the nonhuman primate dose–response data, the only data found suitable from consideration of data reporting and have been published are Druett *et al.*'s data[95] for exposure to single spore (~1 μm diameter) particles and two US Government-sponsored studies from which requisite experimental details were reported or could be assumed with some confidence. Here we present only the analysis of Druett *et al.*'s data.[95] As one of the advantages of the system is the possibility of unbiased testing of an array of models, seven dichotomous models could be fitted and compared, such as the $\log_e$-probit, Weibull, which is a two-parameter model from which the classical exponential model derived by setting its shape parameter to 1, and a programmed Beta–Poisson, which does not have built-in software. Goodness of fit was tested by three criteria. All seven models tested demonstrated acceptable fit according to similar p values (range 0.21–0.32) and an additional model selection tool, the Akaike Information Criterion (AIC). A sample of the models analyzed, parameters and derived BMD and BMDL doses are given in Table 3.6. The best fitting, with the highest p value and lowest AIC value, was the Weibull-derived exponential model (p = 0.32), with a parameter $k = 1.44 \times 10^{-5}$, a BMD_{50}, or LD_{50}, of 48 000 spores and a $BMDL_{50}$ of 37 000 spores.

The lower dose values, which should be considered as POD for low-level risk assessment, are given in Table 3.6. It is also interesting that the two-parameter Weibull model was not better fitting than its derivative exponential model, and so was the Beta–Poisson, as found by Bartrand *et al.*[20]

3.6.8 Benchmark Dose Analysis of Rabbit Exposure Studies

Rabbits are gaining importance as an animal model for inhalational anthrax medical countermeasure studies because the pathology of the disease is similar to that of the human and nonhuman primate infection[130,131,142] and their handling is less demanding than for nonhuman primates. The two dose–response studies conducted by the EPA and Battelle – the acute low-dose study[143] and the repeated daily dosing study[151,152] – were carried out after careful calibration of the exposure, sampling and assays[90,91] with real-time monitoring of physiological parameters by telemetry, measurement of respiratory rates during exposure by plethysmography, instrumental determination of aerosol particle size distributions in real time and follow-up of levels of bacteremia and toxemia and other biochemical and immunological biomarkers. In the acute dosage study, 30 animals in six dose groups were exposed to ×10 increasing doses from 2×10^0 to 2.75×10^5 cfu and a high-dose control group exposed to 8.75×10^6 cfu. Deaths were observed in dose groups of 2×10^3 cfu and higher and the LD_{50} was determined as 5.2×10^4 cfu ($\log_e$-logistic model), about two times lower than the value of 1.05×10^5 cfu determined by Zaucha *et al.*[142] and used as nominal for this animal model. Moreover, LD_{50} values similar to this were obtained by fitting

Table 3.6 Comparison of sample models fitted by the US EPA benchmark dose software to the Druett *et al.*[95] rhesus macaque respiratory exposure to *B. anthracis* spores. Adapted from Taft and Hines.[98]

Model	Parameters	P	AIC[e]	BMD_{50}	BMD_{10}	BMD_{01}	$BMDL_{50}$	$BMDL_{10}$	$BMDL_{01}$
Log_e probit[a]	$\beta = 1.39; \alpha = -15.1$	0.23	78.1857	56000	14000	2700	40000	6100	580
Weibull[b]	$\beta = 1.64 \times 10^{-7}; \alpha = 1.40$	0.25	77.4941	56000	14000	2700	40000	6100	580
Weibull exponential[c,d]	$k = 1.44 \times 10^{-5}$	0.32	77.1788	48000	7300	700	37000	5600	540
Beta–Poisson[d]	$\alpha = 1.1 \times 10^{10}; N_{50} = 54700$	0.24	85.8595	55000	8300	790	38000	5200	470

[a]Fitted by the generalized form, $P = \Phi(\alpha + \beta \ln X)$ where Φ is the cumulative distribution function, X is the dose, α is the intercept and β is the slope parameter. These parameters can be converted to the parameters calculated by Druett *et al.*,[95] who used the Finney[112] $\log_{10}$-based probit method.

[b]The model equation is $P = 1 - \exp\left[-\left(\beta X^{\alpha}\right)\right]$ where X is the dose, α is the shape and β is the slope parameter.

[c]Weibull model reduced to classical exponential by setting the shape parameter α to 1.

[d]The parameter notations for the exponential and the Beta–Poisson models were changed to those used in the main text. The parameter k in the exponential model is equivalent to the parameter β of the Weibull or Weibull-exponential model equations from Taft and Hines.[98] Likewise, the parameter N_{50} is equivalent to the parameter β of the Beta–Poisson equation there.

[e]The Akaike Information Criterion (AIC) is a model selection criterion that measures the information loss when using a model. The AIC for a given model is defined as $AIC = -2 \log L + 2p$ where $\log L$ is the log-likelihood of the model given the data and p is the number of estimated parameters included in the model. Among a set of specified models, the model with the lowest AIC is the "best."

exponential models, either classical or the competing risks methodology, to this data set pooled with other rabbit exposure data.[91,141] The survivors in the low-dose groups (2.86×10^2 and 2.05×10^3 cfu) were infected, as evident from the finding of mild or moderate inflammation of the lungs, giant cells and minimal bacteria in the lungs of the higher-dose group in microscopic post-euthanasia examination. In the daily low-dose exposure study, three dose groups of seven rabbits each were exposed to *B. anthracis* Ames spore doses of 2.91×10^2, 1.22×10^3 and 1.17×10^4 cfu once daily for 60 consecutive days. All animals in the lowest dose group survived, whereas in the 1.17×10^4 group one animal died after 18 days from the first exposure and in the 1.17×10^4 cfu group four animals died with a mean time to death of 14.80 ± 4.28 days. As in the acute study, all challenged animals that survived were infected, as evident from post-euthanasia microscopic findings in the lungs. The LD_{50} was determined as 8094 cfu per day using logistic regression. The $BMDL_{50}$ and $BMDL_{10}$ for the average daily dose were determined as 2.6×10^3 and 2.9×10^2 cfu, respectively, by the best fitting model ($\log_e$-logistic). The $BMDL_{50}$ and $BMDL_{10}$ for the total cumulative dose were 4.4×10^4 and 4.9×103 cfu, using the same regression model.

3.7 The Sverdlovsk Accidental Release and Incubation Period Modeling

3.7.1 The Sverdlovsk Accidental Release and Initial Modeling

Starting from early April 1979, an unusual outbreak of human anthrax took place in the industrial town of Sverdlovsk in the former USSR (now Ekaterinburg, Russia) with parallel outbreaks in farm animals in the region.[16,38] The human outbreak was explained by the Soviet authorities as resulting from consumption of illegally-marketed meat, but epidemiological and pathological evidence, obtained by US and Russian investigators, confirmed that the outbreak was caused by an accidental release of anthrax spores from a military microbiological facility in the southern part of the town.[38,80,81] The pathological findings in 42 of the victims for whom autopsy reports and paraffin-embedded tissue samples were available to researchers were consistent with known features of inhalational anthrax in humans and nonhuman primates.[80,81] Epidemiological data, collected in Russia by a research team led by Professor Matthew Meselson, formed a data set covering 77 cases, of whom 66 died and 11 survived. In 70 of these patients, the dates of onset of disease and death were recorded, thus permitting modeling of the incubation periods and validation of dose–response models, as discussed below.[16,32,46] Later studies based their modeling on a total of 81 cases, with 70 fatalities and 11 survivors.[16,32,46] The approximate locations of 66 of the victims at the time of the release (~14:00) allowed the location of the hazard zone to be identified, which was matched with atmospheric dispersal modeling using meteorological records and thus adding more information for evaluating the release.[16,38] Molecular epidemiology studies confirmed the presence of

B. anthracis molecular markers in DNA extracted from tissue samples, and also the probable origin of the infecting strain.[153,154]

Epidemiological observations of importance were the following.[38] (i) The recorded onset period of anthrax in patients spanned over 6 weeks, from April 4 to May 15 1979. Based on the inferred day of release, 2 April, the longest incubation period observed was 43 days, with a modal incubation period of 9–10 days. The earlier report of Abramova *et al.*[80] listed 42 victims whose death from inhalational anthrax was confirmed by autopsy. The dates of symptom onset and death were recorded for 30 of these, whose incubation periods, based on 2 April as the exposure date, ranged from 5 to 40 days with a median of 30 days and mean of 16 days.[159] (ii) Of the 77 patients listed in Meselson *et al.*,[38] 55 were men whose mean age was 42 years and the mean age for women was 55 years. No man was younger than 24 years old and only two women, aged 24 and 32 years, were under 40. (iii) Approximately 60% of 33 men with the relevant information were reported as moderate or heavy smokers and about half as moderate or heavy drinkers, but no women were reported as smokers or drinkers. (iv) Few patients were reported to have serious medical problems. (v) Of 35 men whose occupation was known, seven were welders. Items (iii)–(v) may indicate possible risk factors. (vi) Public health preventive measures were distribution of prophylactic antibiotics and later a vaccination campaign, started from mid-March. Other details on the course of the outbreak, patient management and their location and possible circumstances of exposure are detailed in the paper by Meselson *et al.*[38] and a book by Guillemin.[155]

Meselson *et al.* estimated the source strength of the spores on the basis of meteorological data, estimated attack rate and existing models of atmospheric dispersion and dose–response relationship for inhalational exposure to *B. anthracis* spores.[38,156] Meteorological data, recorded at a local airport and stored in a US repository, showed that a northern wind within a sector of 320–350° blew on 2 April 1979 during 04:00–19:00 local time, matching exactly the hazard zone determined by the patients' location data and downwind from the military compound where the dispersal allegedly took place. Other weather data were consistent with a neutral atmospheric stability. Thus, using an atmospheric dispersion model,[157] the downwind relative aerosolized spore dosages and the dosage contours of the human and animal hazard zones could be calculated and plotted. The human attack rate was evaluated as 2%, on the basis of data from several locations downwind of the alleged source: in a ceramic pipe factory 2.8 km downwind from the source on the wind centerline, 10 of the 450 employees working there were infected. Similar infection rates, ranging from 1 to 3%, were estimated from surrounding locations. According to Meselson, the greatest uncertainty in evaluation of the source strength lay in the dose–response relationships.[156] For his calculations, Meselson considered two dose–response models: (i) a log-normal model with a human LD_{50} of 8000 spores and a probit slope of 0.7, based on US Army estimations and Glassman,[39] and (ii) an independent action model using a nonhuman primate

LD_{50} of 45 000 spores determined for rhesus macaques by Druett *et al.*[95] The model equation is the following:[156,158]

$$\text{Attack Rate} = 1 - 0.5^{D/\text{LD}_{50}} = 1 - \exp\left[-0.69\left(D/\text{LD}_{50}\right)\right] \tag{3.25}$$

For the log-normal distribution with LD_{50} = 8000 spores and probit slope = 0.7, an attack rate of 2% at a downwind distance of 2.8 km and a breathing rate of 30 L min^{-1} (moderate work), the calculated exposure dose was nine spores and the source strength was 4×10^9 spores. If the independent model and LD_{50} of 8000 spores are considered, then the exposure dose was 230 spores; a third estimate based on independent action and an LD_{50} of 45 000 spores, resulted in a dose of 1300 spores for the ceramic pipe factory.[38,156] The estimated source strengths for the three doses are 4, 110 and 600 mg of spores, assuming density of 10^9 spores per milligram.[156]

3.7.2 Re-Evaluation of Human Inhalational Anthrax on the Basis of the Sverdlovsk Release

Meselson's early work demonstrated, on the one hand, the potential wealth of the outbreak data for the exploration of low-level exposure to inhalational anthrax and, on the other, the uncertainties in employing the partial and reconstructed raw data. The investigators of the Sverdlovsk release were faced with the uncertainties that occur in *a posteriori* investigations of outbreaks. Epidemiological, clinical and pathological data could be obtained but information concerning release mechanism, exposure doses and physical features such as the quality of the material and the release efficiency of spores in respirable particles (<5 μm diameter) were unavailable and had to be assumed for modeling. Several authors have re-analyzed the Sverdlovsk event using additional human and animal data that became available, novel modeling techniques and different assumptions in order to evaluate dose–response and incubation period distribution models. Thus, the relevance of animal data to human anthrax risk evaluation could be assessed by applying them to models incorporating the Sverdlovsk data. In this section, we discuss the application of the Sverdlovsk data in the evaluation of dose–response models,[16] models of the incubation period[16,32,46] and a recently developed model[17] bridging animal and human data in modeling of dose–response and the incubation period in human inhalational anthrax. These studies, although incomplete in many ways, were a step forward in settling the uncertainties in the analysis of the Sverdlovsk release and development of novel analytical methodologies.

3.7.3 Wilkening's Evaluation of Dose–Response Relationships

For the analysis of the Sverdlovsk outbreak dose–response relationships, Wilkening[16] considered four dose–response models: (i) a log-normal model with an LD_{50} of 8600 spores and a probit slope of 0.67, based on the Glassman/

Jemski model;[39] (ii) a similar log-normal model but with a probit slope of 1.43;[36] (iii) a log-normal model with a probit slope of ~2.4 and an LD_{50} dependent on age distribution, as modeled by Wein *et al.*,[30] which yielded LD_{50} values between 8000 and 10 000 spores depending on the demography assumed; this work analyzed two versions – one using the demography of the Soviet Union as a whole for that period and the other based on the demography of the Sverdlovsk victims (see Figure 3.5); and (iv) an exponential competing risk model [eqn (3.14)], with the parameters $\theta = 0.109$ per day and $\lambda = 8.8 \times 10^{-9}$ per day.

The model graphs of probability of infection *versus* dose are shown in Figure 3.5, labeled in this order as models A–D. Simulation of the spatial distribution of the spore plum using an atmospheric dispersion model and the above dose–response models showed that the dose–risk isopleths based on models A and D, but not models B and C, were consistent with Meselson's modeling and epidemiological analysis.[38] Models A and D were also more consistent than models B and C with the cumulative distribution of the 81 patients with the downwind distance from the source.[16] Model A predicted that 50% of the patients received fewer than two spores (geometric mean 2.4 spores, geometric SD 3.2 spores) and model D predicted that 50% of the patients received fewer than 360 spores (geometric mean 360 spores, geometric SD 4.7 spores). These estimates, meaning that most patients received either 1–10 spores (model A) or 100–2000 spores (model D), agree with Meselson's estimates and show that most patients received doses far below the LD_{50}.

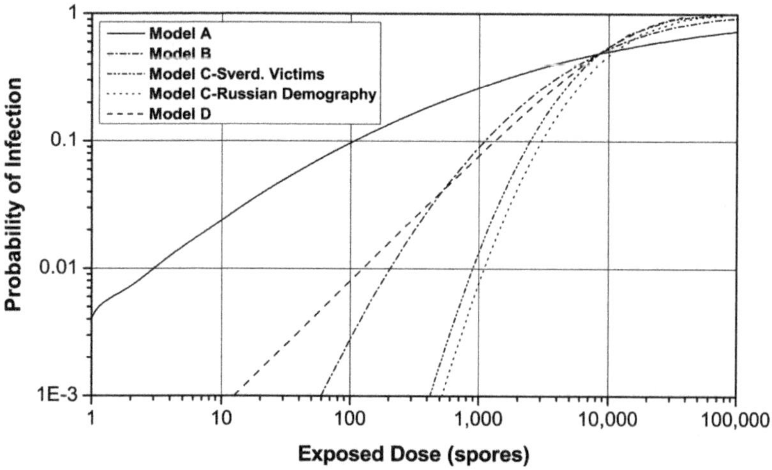

Figure 3.5 Probability of infection *versus* dose plots of the dose–response models defined and analyzed by Wilkening for the Sverdlovsk incident.[16] Model A is a log-normal model with an LD_{50} of 8600 spores and a probit slope of 0.67; model B is a similar log-normal model but with a probit slope of 1.43; model C is a log-normal model with a probit slope of ~2.4 and an LD_{50} dependent on age distribution. Here are shown two versions, one based on the demography of the Sverdlovsk victims and the other using the demography of the Soviet Union as a whole for that period; model D is an exponential competing risk model with the parameters $\theta = 0.109$ per day and $\lambda = 8.8 \times 10^{-9}$ per day. Reproduced from Wilkening.[16]

3.7.4 Modeling of the Incubation Period

As discussed above, information on the incubation period of inhalational anthrax under realistic exposure conditions is scarce. In most of the industrial and environmental exposure cases reported, the exposure time is not known exactly or not at all.[159] Animal studies have two major limitations in this respect. In general, high doses are employed in studies intended to evaluate vaccines and medical countermeasures and the recorded endpoint is death, which occurs after a short time: the mean survival time in rabbits is 2.4 days and in rhesus macaques 4.8 days post-exposure (A. Friedlander, cited in IOM[159]). In recent studies where disease biomarkers are recorded, the time from onset of exceptional markers to death is short,[115–118] whereas in human anthrax there is a relatively long prodromal and symptomatic phase and death occurs 4.7 days post-onset of symptoms (A. Friedlander, cited in IOM[159]). The human incubation period for inhalational anthrax could be inferred in two outbreaks: the anthrax letter attacks of 2001 and the Sverdlovsk incident. In the 2001 anthrax letter attacks, the median incubation period for six patients was 4 days, range 4–6 days,[11] whereas for three others it was estimated as between 5 and 10 days.[159] In the case of the Sverdlovsk incident, a list of the times of symptom onset and death was available, but the date and duration of exposure were uncertain. On the basis of all lines of evidence collected by the Meselson investigation team,[38] the inferred date of release was 2 April. Although this timing is still under debate,[159] all modeling and analyses discussed below refer to this date of release. As for the duration of exposure, it was assumed to be limited to the day of exposure and as the aerosol cloud tends to settle within a short time, the risk from resuspension of settled spores was assumed to be negligible and human-to-human transmission of anthrax is unlikely. Another complicating factor was the public health measures starting in mid-April, which might have truncated the data and affected the incubation period and risk evaluation. This was considered in some[32,160] and ignored in other studies cited below.[46] In the following section, we discuss modeling the incubation period of inhalational anthrax as a function of the dose and model testing using the Sverdlovsk data.

3.7.5 Log-Normal and Log-Linear Models

These models were analyzed by Wilkening[16,32,161] following Sartwell,[162] who analyzed many disease outbreaks whose time courses were recorded and concluded that the probability distribution of the duration of the incubation period can be described as log-normal. Accordingly, the cumulative distribution function $F(t)$ of the length of incubation periods for a given outbreak can be written as

$$F(t)=\left[\frac{1}{\sigma\sqrt{2\pi}}\right]\int_0^t\left(\frac{1}{x}\exp\left\{-\left[\frac{(\ln x-\ln M)^2}{(2\sigma)^2}\right]\right\}\right)\mathrm{dx} \tag{3.26}$$

where M is the median duration and σ is the standard deviation of the incubation periods. The distributions of the median incubation periods and standard deviation with the exposure dose can be modeled by the following log-linear equations:

$$M = \alpha + \beta \log(D) \tag{3.27}$$

$$\sigma = \gamma + \delta(D) \tag{3.28}$$

where α, β, γ and δ are parameters. Wilkening fitted a model with the parameters $\alpha = 10.3, \beta = -1.35, \gamma = 0.804$ and $\delta = -0.079$ on the basis of the Sverdlovsk dose–response model A (M = 2.4 spores) and nonhuman primate high-dose exposure data.[16,161] The prediction of the Sverdlovsk incubation period duration distribution by this model is discussed and shown below.

3.7.6 Brookmeyer's Competing Risks Model

The competing risks dose–response model of Brookmeyer and co-workers[44,46] has been described above. An application of this model to the analysis and prediction of the Sverdlovsk incubation period distribution is presented here. The basic concept of the incubation period model is as follows: the cumulative attack probability of a disease, $F(t)$, is defined as the cumulative probability that a person will become diseased in a time shorter than t post-exposure. If only a proportion p of exposed people will become diseased, then $F(t)$ is a nondecreasing function whose limit is p as t increases. The incubation period distribution $F^*(t)$ is defined as the probability that the incubation period is less than t days among the proportion p of the people who will develop disease. The relation between $F(t)$ and $F^*(t)$ is defined by $F(t) = pF^*(t)$. $F^*(t)$ is the normalized function of $F(t)$ and as time increases it approaches 1.

To incorporate time dependence in the competing risks model, the probability that a spore may germinate within t days was derived as follows:

$$\int_0^t \lambda e^{-\lambda u} e^{-\theta u}\, du = \frac{\lambda}{(\lambda+\theta)\left(1 - e^{-(\lambda+\theta)t}\right)} \tag{3.29}$$

In analogy with eqn (3.14), the probability that at least one spore of a dose D will germinate within t days, or the infection probability, is defined by

$$P(D,t) = 1 - \exp\left[-\frac{D\lambda}{\lambda+\theta}\left(1 - e^{-(\lambda+\theta)t}\right)\right] \tag{3.30}$$

Eqn (3.30) describes the probability that at least one spore of a dose D will germinate at time t, which is actually $F(t)$. Setting the dose in eqn (3.30) as $TD(P)$ of eqn (3.15) gives

$$F(t,p) = 1 - (1-p)^{(1-\exp(-(\lambda+\theta)t))} \tag{3.31}$$

It should be recalled that the parameters λ and θ are the spore germination and spore clearance rates, respectively. By using the approximation $\lambda + \theta \sim \theta$, eqn (3.31) and its normalized form, $F^*(t,p)$ are simplified as follows:

$$F(t,p) = 1-(1-p)^{(1-\exp(-\theta t))} \tag{3.32}$$

$$F^*(t,p) = \frac{1-(1-p)^{(1-\exp(-\theta t))}}{p} \tag{3.33}$$

Eqn (3.33) shows the relationship between the incubation period, clearance rate and proportion of diseased people. For $\theta = 0.07$, based on the pulmonary spore clearance kinetics in the primate study of Henderson *et al.*[45] and an attack rate of 1%, the calculated median incubation period [*i.e.* the median of $F^*(t,p)$] was 9.9 days, which is in good agreement with the observed median incubation period of 10 days from the Sverdlovsk data.[38,46] However, fitting of the model to the Sverdlovsk incubation period data, discussed below, re-evaluated θ as 0.109 per day.[46]

The above derivation did not consider the delay between spore germination and the onset of symptoms. In order to include this time delay, the incubation period distribution can be modeled as a convolution of the probability that a spore will germinate before being cleared at time $t - s$ and the density function of the probability that the delay between spore germination and symptom onset has a duration s, as in eqn (3.34):

$$F^*(t) = \int_0^t F(t-s)g(s)\mathrm{d}s \tag{3.34}$$

A simple model for the probability density of s can be an exponential distribution, $g(s) = \gamma e^{-\gamma s}$, with a hazard rate γ. The combined equation is

$$F^*(t,p) = \frac{1}{p}\int_0^t \left\{1-(1-p)^{(1-\exp[-\theta(t-s)])}\right\}\gamma e^{-\gamma s}\,\mathrm{d}s \tag{3.35}$$

Fitting of the incubation period distribution to several values of θ and fixed values of the delay between spore germination and symptoms yielded several values, of which the value for a delay of 2 days ($\gamma = 0.346$ per day), $\theta = 0.109$ per day, had the best fit with the observed Sverdlovsk data, as shown by the Sverdlovsk incubation period histogram and the predicted density function in Figure 3.6.

3.7.7 Wilkening's Modification of Brookmeyer *et al.*'s Model

Wilkening[16,32,161] modified the incubation period model of Brookmeyer *et al.* by modeling the delay between spore germination and symptom onset as a log-normal distribution rather than a simple exponential distribution. The exponential model of the previous study is not accurate because it does not represent accurately the dynamics of bacterial growth from germination to symptom onset. Wilkening took account of the four phases of bacterial growth

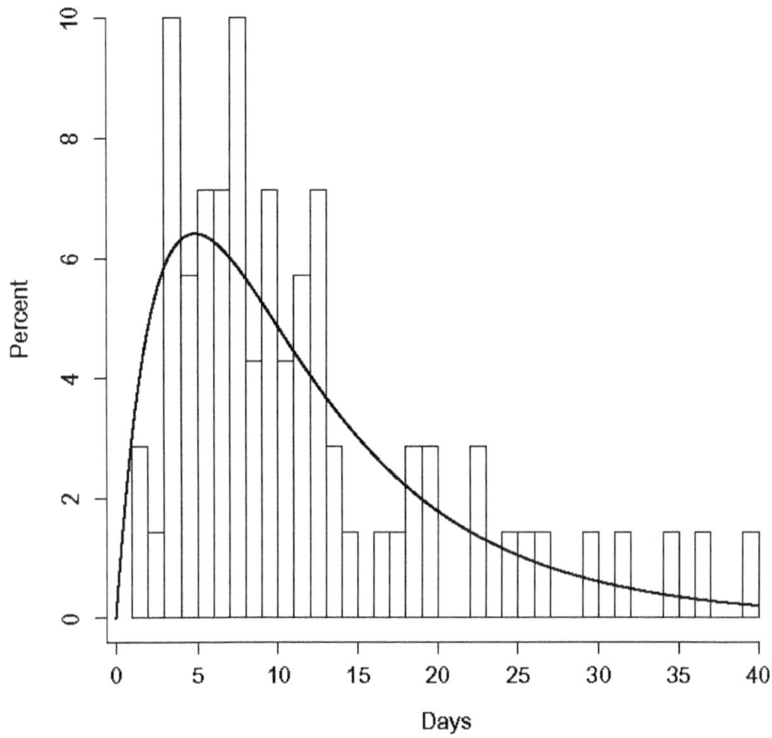

Figure 3.6 Histogram of the incubation periods from the Sverdlovsk outbreak and the predicted incubation period density function, calculated by Brookmeyer *et al.*[46] on the basis of the competing risk model. The derivation and fitting of the density function from eqn (3.31) in the main text are discussed in Brookmeyer *et al.*[46] Reproduced from Brookmeyer *et al.*[46]

dynamics: (i) a lag phase of adaptation to a new environment in which no growth occurs; (ii) a logarithmic phase, in which an exponential or logarithmic increase in bacterial cell number takes place; (iii) a stationary phase, in which the net bacterial growth is zero due to depletion of nutrients; and (iv) a decline phase in cell numbers. Wilkening identified the bacterial stationary phase with the clinical prodromal stage and the bacterial decline phase with the clinical fulminant stage, in which toxins are abundant and may play a major role in the disease pathogenesis. A premise of the model is that the onset of symptoms is associated with a certain threshold level, $N_{\text{threshold}}$ of the bacterial number or load within the host. The time T from spore germination to the onset of symptoms can be calculated by the following equation:

$$T = \left(\frac{t_2}{24\ln 2}\right) * \ln\left(\frac{24\ln 2N_{\text{threshold}}}{\lambda t_2 D}\right) \tag{3.36}$$

where t_2 is the bacterial generation time in hours and D is the exposure dose. The lag phase is of short duration relative to the growth phase and therefore can be ignored. As bacterial growth is affected by many factors, the time delay from spore germination to symptoms onset can be modeled as a log-normal distribution whose probability density function $g(s)$ is

$$g(s)=\left(\frac{1}{\sigma_g\sqrt{2\pi}}\right)\left(\frac{1}{s}\right)\exp\left(-\frac{(\ln(s)-\ln(T))^2}{2\sigma_g}\right) \tag{3.37}$$

where T is the median bacterial growth period from eqn (3.36) and σ_g is the standard deviation of the bacterial growth period distribution.

Wilkening[16] constructed five incubation period models and compared their fits with the Sverdlovsk data. Three models were log-linear models and two were modified Brookmeyer's competing risks models, differing in their underlying dose–response models. Model A1 (Figure 3.7) was a log-linear model with the parameters [eqn (3.27) and (3.28)] $\alpha = 10.3$, $\beta = -1.35$, $\gamma = 0.804$ and $\delta = -0.079$ on the basis of the Sverdlovsk dose–response model A (median dose, $M = 2.4$ spores) and nonhuman primate high-dose exposure data. Two other models of this type were model B ($\alpha = 6.33$, $\beta = -0.725$, $\gamma = 0.17$ and $\delta = 0$, *i.e.* a constant standard deviation) and model C, in which the incubation period is independent of dose ($\alpha = 11.0$, $\beta = 0$, $\gamma = 0.713$ and $\delta = 0$). Model A2 is a competing risks modified model (parameters given in the caption of Figure 3.7), whose dose dependence is modeled by dose–response model A, and model D is a competing risks modified model with the dose dependence of the incubation period derived *ab initio*. The dose dependence of the incubation periods by the five models is shown in Figure 3.7, together with the Sverdlovsk estimated median doses by the respective dose–response models and experimental nonhuman primate high-dose exposure data.

Model B is not consistent with the Sverdlovsk median dose estimates and model C is not consistent with the nonhuman primate data, while the competing risks models D and A2 fit the Sverdlovsk data better than the log-linear model, predicting a median incubation period of 9.5 days (Figure 3.7) and thus giving some credit to the former in modeling the human incubation period as a function of the dose. The predicted incubation period distribution densities of the Sverdlovsk victims by the five models, compared with the observed distribution, are shown in Figure 3.8.

The comparison shows that models A1, A2 and D predict the Sverdlovsk distribution better than models B and C, with A1 (log-linear) and A2 (modified Brookmeyer's competing risks) giving the best fit.[161] The lack of ability to distinguish between the distributions modeled by models A1 and A2 may result from the lack of data covering the middle dose range, which is one of the limitations of the models. Another limitation is the inclusion, in the same model, of the human data, which record the time to onset of symptoms, and the nonhuman primate data, which record the time to death.

The last limitation was addressed in an extended version of the model, published in 2008.[32] In this study, the time to death of nonhuman primates and the human time to symptoms data were fitted separately and the incubation period was modeled by the modified Brookmeyer model D, which

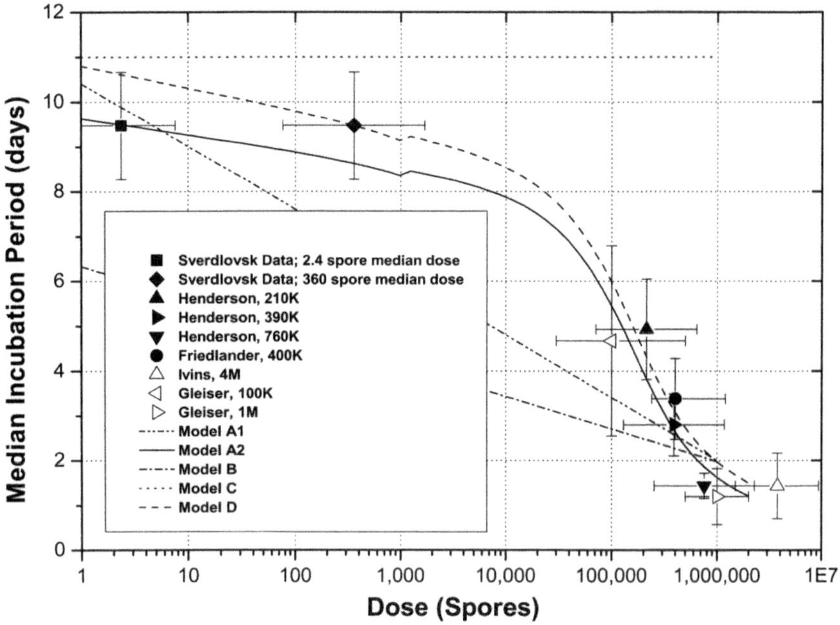

Figure 3.7 Dependence of the median incubation period on the exposure dose, according to five models tested by Wilkening.[16] The curves of models A2 and D are composed of two parts: the doses below 1000 spores are modeled on the basis of the Sverdlovsk estimated LD_{50} of 8600 spores, whereas the higher doses are modeled on the basis of the nonhuman primate LD_{50} of 55 000 spores. The data points to the left of models A2 and D are the median doses for Sverdlovsk according to dose–response models A and D of Figure 3.5, respectively. The data points to the right of the curves are nonhuman primate experimental high-dose exposure data. The open symbols show data points that were not included in the model fit but are consistent with the model. Model A1 is a log-linear model with the parameters $\alpha = 10.3$, $\beta = -1.35$, $\gamma = 0.804$ and $\delta = -0.079$. Model B has the parameters $\alpha = 6.33$, $\beta = -0.725$, $\gamma = 0.17$ and $\delta = 0$, *i.e.* a constant standard deviation. In model C, the incubation period is independent of dose ($\alpha = 11.0$, $\beta = 0$, $\gamma = 0.713$ and $\delta = 0$). The parameters of model A2 are $\theta = 0.110$, $\lambda = 8.84 \times 10^{-6}$, $t_{lag} = 1$ h, $t_2 = 2.06$ h, $N_{threshold} = 1 \times 10^9$, $\sigma_g = 0.542$, $\gamma = 0.804$, $\delta = -0.079$. The parameters of model D are $\theta = 0.109$, $\lambda = 8.79 \times 10^{-6}$, $t_{lag} = 1$ h, $t_2 = 2.07$ h, $N_{threshold} = 1 \times 10^9$, $\sigma_g = 0.542$. Reproduced from Wilkening.[16]

models the dose dependence of the incubation period *ab initio*. The time to death probability distribution for the nonhuman primate data was modeled analogously to the time to symptom onset probability for humans. Like the onset of symptoms in eqn (3.34), the probability $M(t)$ that a host will die after a time t can be modeled as a convolution of the cumulative probability of symptom onset and the probability density of the time to death by

$$M(t) = \int_0^t I(t-x)h(x)\mathrm{d}x \tag{3.38}$$

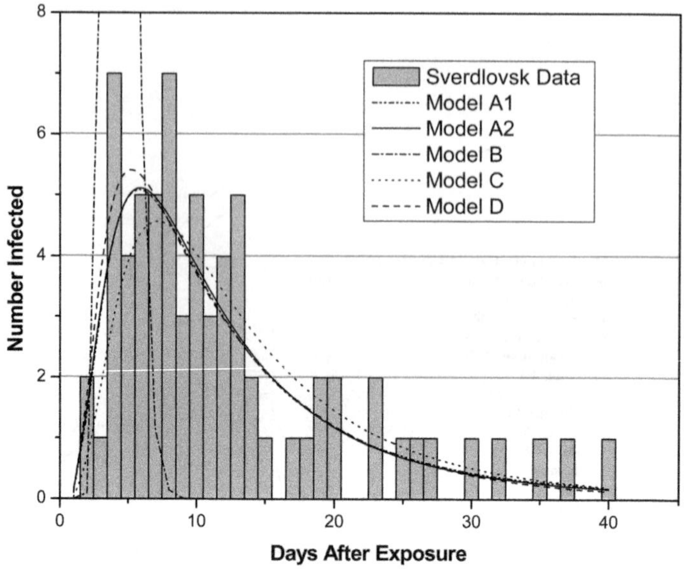

Figure 3.8 Predicted Sverdlovsk incubation periods by models A1–D compared with the observed Sverdlovsk incubation period distribution. Source: supporting Figure 15 of Wilkening;[16] http://www.pnas.org/content/suppl/2006/04/271/0509551103.DC1/09551Fig15.pdf (last accessed 1 March 2015).

where $I(t-x)$ is the cumulative probability of symptom onset at time $t-x$, given by eqn (3.38), and $h(x)$ is the probability density function of the time from symptom onset to death. The time from symptom onset to death can be fitted by a log-normal distribution similar to eqn (3.37) with a median time M_d and a standard deviation σ_d. Separate maximum likelihood fittings were made for the Sverdlovsk human incubation period data and the experimental nonhuman primate exposure studies. As the strain infecting the Sverdlovsk victims was unknown, its LD_{50}/ID_{50} was assumed to be 8600 spores and the median exposure dose 360 spores. The primates, exposed to doses ranging from 214 000 to 760 000 spores in four different experiments, were used as controls in vaccine efficiency studies. The exposure strains were Vollum or Ames strain, with an LD_{50} of 55 000 spores.[45,137] In the maximum likelihood fitting of the Sverdlovsk data, 16 victims who became symptomatic after 15 April, the presumed starting date of prophylactic measures, were right-censored because the effect of treatment on their incubation period and disease course was unknown. So were three animals in one of the Henderson *et al.* studies, who became symptomatic 19 days post-exposure. The parameters fitted for both humans and monkeys were θ (= 0.131 per day), t_2 (= 1.81 h), M_g (= 3.35 days), σ_g (= 0.412) and σ_d (= 0.916) and the parameters M_d and $N_{threshold}$ were held constant at 7 h and 10^{10} bacteria, respectively. The fitted curves for the human incubation period and the nonhuman primate time to death are shown in Figure 3.9. Assuming a median exposure dose of 360 spores, the predicted median incubation period of the Sverdlovsk victims was 9.05 days (95% CI, 8.0–10.3 days).

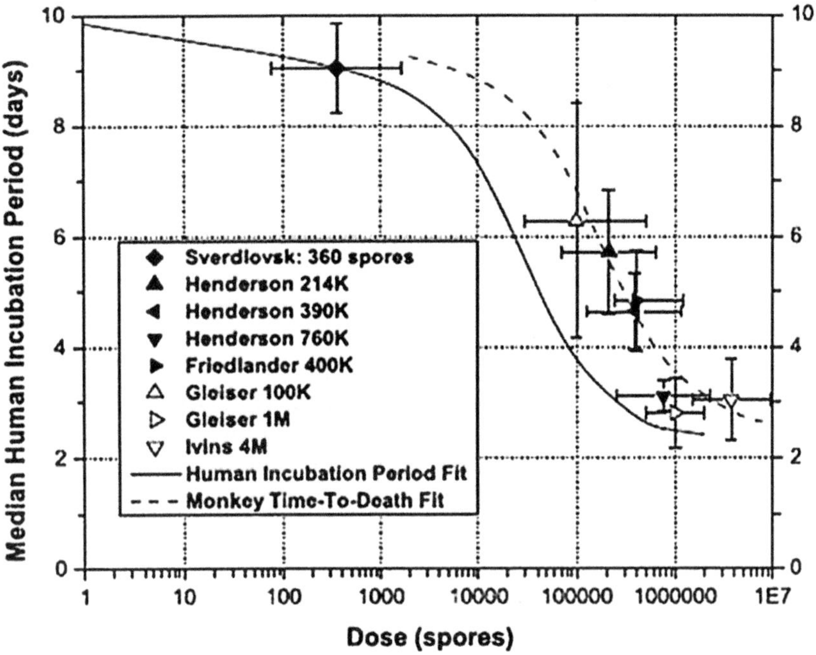

Figure 3.9 Median human incubation period and median nonhuman primate time to death as a function of the exposed dose, modeled in Wilkening.[32] The Sverdlovsk victims are assumed to be exposed to a median dose of 360 spores with an LD_{50} of 8600 spores. The nonhuman primates were exposed to spores of *B. anthracis* Ames strain with an LD_{50} of 55 000 cfu. The open triangles show data points not included in the model fit but which are consistent with the model. Reproduced from Wilkening.[16]

The limitation of the absence of experimental data covering the middle dose range was not resolved and the combined use of human time to symptom onset and monkey time to death data, although addressed by separate maximum likelihood fitting, need some refinement by using human-only and monkey-only data and parameters reflecting species differences in pulmonary spore clearance rates and the critical bacterial loads. The model is computationally heavy, because the integrals in eqn (3.34) and (3.38) can be calculated only numerically. To enhance the usability of this model, Wilkening fitted log-normal distributions of incubation periods for different spore doses, based on the model's predictions (presented as Table 4 in Wilkening's paper[32]).

3.7.8 The Exposure–Infection–Symptomatic Illness–Death Model

Toth *et al.*[17] presented four criteria to evaluate the eligibility of a dose–response model for human inhalational anthrax risk assessment: (i) the parameter values are derived from dose–response data; (ii) the shape of the dose–response curve is consistent with the Sverdlovsk incubation period data; (iii) the model is based on biological mechanistic assumptions; and (iv) estimates the incubation period *ab initio*. Examination of published models showed that none of them (some are listed in Table 3.7, which is based on Table 1 in Toth *et al.*'s paper[17]) met all the criteria – of the 12 models evaluated, 10 did not satisfy criterion (iv) and the one that did (Brookmeyer *et al.*[46]) was not based on dose–response data. They provided a novel model, a hybrid of nonhuman primate dose–response and human historic mortality data, incorporating all of the above elements.

Three models were fitted to the data of Druett *et al.*[95] and were evaluated (Table 3.7): the original log-normal model (model D1, Table 3.7), the exponential model of Haas,[19] model D2 in Table 3.7 and their own fitting (model D3). The models with the best statistical fit were D1 and D3 ($ID_{50} = 48\,000$, $k = 1.43 \times 10^{-5}$). The fit results were very close (deviances of 10.3 and 11.3 for D1 and D3, respectively). Model D3, although somewhat inferior in deviance, was suggested as preferable as it is based on a mechanistic theory and is more parsimonious (one parameter in the exponential rather than two in the log-normal). It is interesting that a biological risk assessment-based planning manual prepared by the Institute for Defense Analysis, Alexandria, VA, for NATO[163] chose as a dose–response model an original fitting of the Druett *et al.*[95] data, yielding a very close model ($k = 1.35 \times 10^{-5}$, $ID_{50} = 51\,000$ spores). As the Druett data-based model did not satisfy all the criteria, they continued with the more complex analysis of Brachman *et al.*'s[134] study, which provided also for evaluating the time course of infection. The modeling strategy included the following steps (Figure 3.10): (i) the dose- and time-dependent risk of spore germination before being cleared at time *t*, using a modified form of eqn (3.30), from Brookmeyer *et al.*[46]:

$$I(D,t) = 1 - \exp\left[-kD(1-e^{-\theta t})\right] \tag{3.39}$$

where $I(D,t)$ is the probability of infection at time t after inhaling a dose of D spores. The parameter k, the probability that one spore will germinate independent of time, had to be optimized and θ, the rate constant of spore clearance, was fixed at 0.07 per day according to Brookmeyer *et al.*[46] (ii) The delay from spore germination to appearance of symptoms, T, modeled as a log-normally distributed variable by Wilkening,[16,32] was optimized in this model as a fixed value. The fraction $S(D,t)$ of infected individuals who become symptomatic after a time t is 0 for $t < T$ and equals $I(D,t-T)$ when $t \geq T$. (iii) The delay from onset of symptoms to death, which was fixed in the modeling of Wilkening (because only time from exposure to death could be inferred from the primate study data and there were uncertainties concerning the effect of medical intervention on the time from symptom onset to death in the Sverdlovsk victims), was considered as a variable with a probability density function $f(\tau)$, τ being the time from symptom onset to death. The distribution function was fitted from the data of Holty *et al.*,[6] covering 82 historical human inhalational anthrax infections. Maximum likelihood

Table 3.7 Comparison of primate and human dose–response models and parameters. Adapted from Toth *et al.*[17]

Model notation[a]	Reference[a]	Model form[b]	Parameters	Estimated infective dose (ID_x)				Model criteria[d]			
				ID_{50}	ID_{10}	ID_1	$ID_{0.1}$	Criterion 1	Criterion 2	Criterion 3	Criterion 4
J	Glassman[39]	Log-normal	ID_{50} = 4130; slope = 0.69	4130	50	1	0.1	✓	✓		
D1	Druett *et al.*[95]	Log-normal	ID_{50} = 53 000; slope = 3.19	53 000	21 000	9900	5700	✓			
D2	Haas[19] Bartrand *et al.*[20]	Exponential	$k = 7.16 \times 10^{-6}$	96 800	14 800	1400	140	✓	✓	✓	
D3	Toth *et al.*[17]	Exponential	$k = 1.43 \times 10^{-5}$	48 000	7400	700	70	✓	✓	✓	
B1	Haas[19]	Exponential	$k = 2.6 \times 10^{-5}$	27 000	4100	390	38	✓	✓	✓	
B2	Mayer *et al.*[150]	Exponential (time independent)	$k = 4 \times 10^{-5}$ ($\alpha = 1$)	18 000	2700	250	25	✓	✓	✓	
B3	Mayer *et al.*[150]	Exponential (extended)	$k = 1.87 \times 10^{-5}$ ($\alpha = 0.9$)	16 000	2300	330	41	✓	✓	✓	
B4	Toth *et al.*[17]	Exponential (time dependent)	$k = 6.4 \times 10^{-5}$	11 000	1700	160	16	✓	✓	✓	✓
E1	Rickmeier *et al.*[36]	Log-normal	ID_{50} = 8940; slope = 1.43	8940	1135	211	62				
E5[c]	Brookmeyer *et al.*[46], Wilkening[16]	Exponential (time dependent)	$k = 8.1 \times 10^{-5}$	8600	1300	120	12		✓	✓	✓

[a]Notation of Toth *et al.*,[17] according to the source of data set: J = Glassman;[39] D = Druett *et al.*;[95] B = Brachman *et al.*;[134] E = expert opinion. The Reference column indicates the modeling study.

[b]Model forms are denoted in the terminology of Toth *et al.*[17]

[c]The value of the parameter *k* was not specified in Brookmeyer *et al.*,[46] but was used by Wilkening.[16] The parameter was calculated using the equation $k = \lambda/(\theta + \lambda)$ given spore clearance rate $\theta = 0.109$ per day and spore germination rate $\lambda = 8.8 \times 10^{-6}$ per day.

[d]Criteria used to evaluate the models are (1) the parameter values are derived from dose response data; (2) the shape of the dose–response curve is consistent with Sverdlovsk data; (3) the model is derived from mechanistic assumptions; and (4) the model estimates the incubation period.[17] A positive answer to a criterion is checked.

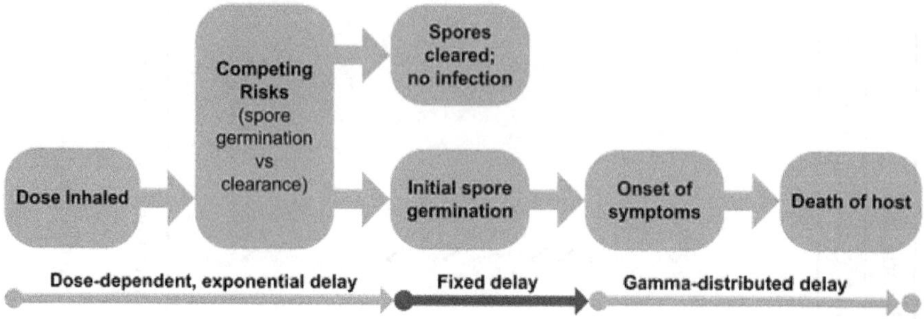

Figure 3.10 Block diagram of the Exposure–Infection–Symptomatic Disease–Death (EISD) model of Toth *et al.*:[17] The scheme summarizes the major assumptions and modeling strategies of the timeline of inhalational anthrax infection and disease progression. Source: Toth *et al.*,[17]; http://www.plospathogens.org/article/info%3Adoi%2F10.1371%2Fjournal.ppat.1003555 (last accessed 1 March 2015).

fitting with the required censoring resulted in the following gamma probability density function:

$$f(\tau)=\frac{1}{\Gamma(a)b^{a}}\tau^{a-1}\,\mathrm{e}^{-\tau/b} \tag{3.40}$$

With the parameters optimized and fixed as a = 5.43 and b = 0.864, the average time was 4.7 ± 2.0 days. The mortality function $M(D,t)$ was defined, in analogy with eqn (3.38), as

$$M(D,t)=\int_{0}^{t}S(D,t-T)f(\tau)\mathrm{d}\tau \tag{3.41}$$

The equations for dose- and time-dependent probabilities of infection $I(D,t)$, symptoms onset $S(D,t)$ and mortality $M(D,t)$ are applicable for analysis of the time course of disease after exposure to a single dose at $t = 0$, which was not the case with the Brachman *et al.* data. One approach for integrating the repeated dosage regime was the model of Mayer *et al.*[150] described above. The approach of Toth *et al.*[17] was different. They assumed that a death on day t resulted from events of exposure, germination and growth taking place until day $t - 1$ and calculated the probabilities $I(D,t)$, $S(D,t)$ and $M(D,t)$ by summation of the daily incremental effective doses until day $t - 1$ and introducing these into exponential model equations as detailed in the *Materials and Methods* section of their paper. Also, the optimization used the data for runs 3, 4 and 5 whereas Mayer *et al.* used only runs 3 and 4. In total, the model contained five biological parameters, of which θ, a and b were fixed and the other two, k and T, were optimized from the Brachman *et al.* data with the values of $k = 6.4 \times 10^{-5}$ and $T = 2.3$ days. The ID_{50} was 11 000 (7200–17 000) spores, ID_{10} 1700 (1100–2000) spores and ID_1 160 (100–250) spores. The median incubation periods were 9.9 (7.7–13.1) days for ID_{50}, 11.8 (9.5–15.0) days for ID_{10} and 21.1 (9.9–15.3) for ID_1. The median incubation period for the low-dose

ID_1 was consistent with the median incubation period, 13 days, of the 30 autopsy-confirmed Sverdlovsk inhalational anthrax victims.[80,159] Hence the model fulfills the four adequacy criteria for human low-dose risk assessment of inhalational anthrax.

3.8 The 1957 Goat Hair Mill Outbreak and Its Implications – "From Mills to Mail"

Goat hair mills, which processed goat hair for the production of coat linings, were often contaminated with anthrax spores, coming from the usual raw material – goat hair imported from Middle Eastern and Far Eastern countries such as Iran, India and Pakistan, which were often contaminated with spores.[21,22,134] This was the cause of occasional cases of anthrax, mostly cutaneous, the occurrence of which was drastically reduced after the introduction of vaccination in the early 1960s.[22] An exceptional anthrax outbreak took place in a goat hair processing mill in Manchester, NH, USA, in 1957. In this outbreak, nine workers contracted anthrax; of these, six cases were inhalational with five fatalities and three had cutaneous anthrax and recovered.[164] The investigation of this outbreak continued with several studies, the importance of which in the research of inhalational anthrax has not waned, including Brachman *et al.*'s study.[134] Although this outbreak preceded the Sverdlovsk outbreak discussed above, it is analyzed here because of implications for matters that arose after the 2001 anthrax letter attacks and conceptual issues that were discussed in earlier sections. The two issues related to this outbreak are the question of the infection threshold in human inhalational anthrax and the occupational risk for mailroom workers.

3.8.1 Main Epidemiological Findings

The epidemiological investigation of the outbreak attributed the source to one highly contaminated bale of goat hair imported from India.[164] Workers might have been exposed to aerosols produced during processing maneuvers in the early stages of the manufacturing line, which involved raw hair and extremely dusty conditions. The main contributing risk factors were the following. (i) *Vaccination and health status*: the mill workers participated in a field study of an alum-precipitated PA vaccine, enrolling employees on a voluntary basis. Of the outbreak patients, five were not vaccinated and four were inoculated with placebo. No other medical predisposing factors were indicated. (ii) *Prior exposure to anthrax*: all 19 workers (of 950 in total) who had a history of cutaneous anthrax before August 1957 did not develop clinical anthrax during the outbreak. Also, serological testing, described in more detail below, revealed a significant proportion of anti-PA positives among unvaccinated workers, tested especially in the carding and combing departments. This suggests that prior exposure had a protective effect. (iii) *Occupational and environmental factors*: two inhalational and two cutaneous anthrax patients worked in the carding department and two inhalational patients were employed in the combing department, the most dusty but also poorly ventilated sites. It is worth noting that a cluster of worker's complaints on dust-related upper respiratory symptoms was recorded by the mill's nurse in the period 14–21 August, 6 days before the outbreak onset. Air sampling studies of *B. anthracis* are discussed below. (iv) *Virulence of the outbreak causal organism*: three strains, isolated from two inhalational cases and one cutaneous case, were tested for virulence to animals in the US Army laboratories and found to be as virulent as the most virulent strains they had. The strain isolated from the first patient of inhalational anthrax had a respiratory LD_{50} of 6000 and 50 000 spores for monkeys and guinea pigs, respectively. The respiratory virulence for guinea pigs was enhanced by the detergent used in the plant for hair scouring,[164] as observed in previous studies of the effect of detergent on virulence of *B. anthracis* spores.[165]

3.8.2 Exposure and Risk Assessment Studies in American Goat Hair Mills

In the wake of the outbreak, several studies were undertaken by research teams from the CDC and the US Army laboratories in order to assess exposure to spores and quantify risk of inhalational anthrax in the New Hampshire mill outbreak and other similar mills. The studies included the collection of evidence of physical contamination of workers, serological studies in order to assess exposure and infection and aerosol sampling for detection and quantification of airborne anthrax spores in the mills. A study, carried out before the New Hampshire mill outbreak, in two goat hair mills reported on the isolation of *B. anthracis* from 33–36% pharyngeal washes and nasal swabs taken from apparently healthy workers employed in the picking, carding and combing departments, whereas in samples from workers in the cleaner drawing and spinning departments the isolation rates were only 0–5%.[146] Serological specimens were collected from apparently uninfected volunteers among the New Hampshire plant workers in three rounds.[148] The serum donors included previously infected, vaccinated and unvaccinated workers. Among the 72 unvaccinated workers, 11 were anti-PA seropositive, and seven of these worked in the carding and combing departments. Similarly, among 56 sera collected in three other goat hair plants, eight were positive, and of these, four were from the carding–combing areas.

Aerosol sampling was carried out in the New Hampshire outbreak mill during February 1958 and in a similar mill in Pennsylvania during May 1958.[147] In the Pennsylvania mill, no inhalational anthrax was reported but the occurrence of cutaneous anthrax cases indicated contamination with spores. The questions that these studies were designed to answer concerned the numbers of anthrax spores inhaled by workers during an 8 h work shift and if these were in particles of a size smaller than 5 μm, which are retained in the lower respiratory tract. In each plant, sampling was carried out in the combing and weaving zones, in order to compare a high- and a low-risk zone.

In each mill, samples were collected along an 8 h shift on two consecutive days, either continuously or intermittently with a 30 min interval between each collection, depending on the sampler type. Total *B. anthracis* organisms and organisms in <5 μm particles were calculated by averaging of liquid samplers and particle-collecting impactor counts. Eight-hour inhaled doses were based on an assumed daily respiration rate of 7 m^3 (~15 L min^{-1}). In both mills, the airborne total and <5 μm particle-carried *B. anthracis* organisms were significantly higher in the combing than in the weaving zones, but not significantly different between the mills. In the New Hampshire mill, a worker in the carding department inhaled 620 and 2200 total *B. anthracis* organisms during an 8 h shift on the first and second sampling days and 140 and 690 organisms in <5 μm particles on the respective days. The total doses for the weaving department were 21 and 77 organisms for the first and second days, respectively, and the respective <5 μm particle doses were 12 and 24 organisms. The inhaled doses calculated for the Pennsylvania mill were the following: a worker in the carding department inhaled 1300 and 620 total *B. anthracis* organisms on the first and second days. The respective <5 μm particle doses were 510 and 410 organisms. The doses for the weaving department were 19 and 44 total *B. anthracis* organisms and 19 and 25 in <5 μm particles on the first and second days, respectively. The fact that such levels of exposure did not result in inhalational anthrax show that they could have been too low to cause the disease in generally healthy workers. Taking together these and the serological studies, it is possible that such exposures could cause subclinical infections that resulted in the development of protective immunity in some of the workers. The causes of the unusual outbreak in the New Hampshire mill became less clear in view of these results, making the explanations offered no more than hypothetical.

3.8.3 Anthrax Risk in a Present-Day Belgian Hair and Wool Processing Plant

Studies on the prevalence of *B. anthracis* and the human response were carried out in recent years in a plant for cleaning sheep wool and goat hair in Belgium.[166,167] The plant has been operating since 1890 and complies with modern standards of industrial hygiene and safety, including proper ventilation and use of personal protective equipment. As human vaccination against anthrax is not mandatory in Belgium and other EU countries, it was of interest to study the occurrence of pathogens, including *B. anthracis*, in raw materials, in several stations in the cleaning process, in dust, waste effluents and airborne aerosol particles, using modern pathogen tracking and characterization techniques that did not exist in the late 1950s. *B. anthracis*, comprising ~1% of the microbiological flora of the plant, was isolated from hair, dust and effluent samples from the sites where goat hair, but not sheep wool, was processed. Goat hair was imported from Middle Eastern and Central Asian countries in which anthrax is endemic, whereas sheep wool was imported from European countries where anthrax is not prevalent among sheep. Molecular typing of the isolates revealed that they were from different sources and distinct from indigenous Belgian *B. anthracis* isolates from cattle. *B. anthracis* was also associated with <5 μm airborne particles sampled in zones where raw hair was handled and the evaluated levels were comparable to those reported previously for the American goat hair mills.[166] A survey of immune responses to *B. anthracis* among the employees of the plant revealed a few unvaccinated individuals with antibody- and cell-mediated immune responses to PA.[167] In a later study,[168] the prevalence of seropositivity among the workers was found to be 12.1% in general and was higher in workers processing and sorting raw goat hair (20–30%) than in "cleaner" occupations (3%). It is interesting that seropositivity was negatively associated with mask use, which was protective by epidemiological criteria (adjusted prevalence ratio of 0.3, $p = 0.015$), indicating a role for inhalational exposure to anthrax in the plant and encouraging the use of respiratory protection in high-risk occupations.

3.8.4 Risk Threshold for Human Inhalational Anthrax

The question of risk threshold was addressed by Cohen and Whalen[169] in view of the extreme rarity of inhalational anthrax in goat hair processing mills in spite of substantial amounts of airborne respirable spores and evidence for subclinical exposure of workers. The aerosol sampling studies in the New Hampshire and Pennsylvania mills were indicative of a possible no-risk exposure level of hundreds of spores and above, but they were carried out on only two days at each plant, whereas longer studies were needed in order to establish the existence of a safe exposure level. The study of Brachman *et al.*[134] was carried out for periods culminating in 57 days of continuous sampling, in which the spore source, in this case the picking machine, was active for 31 days. To evaluate the "no-risk" dose, the authors preferred to use the airborne spore levels rather than the monkey dose–mortality data, because (i) no human anthrax infection occurred as workers were exposed to those doses, (ii) the low reliability of human extrapolation and (iii) the poor health status of the monkeys, which did not suit the target population of healthy workers. The reported dosages of airborne spores were converted to human equivalents on the basis of an estimated respiratory rate of 7 m^3 per 8 h. The cumulative distribution of daily exposure doses showed that on 17 of the 31 days of exposure the daily dose was less than 600 spores, with only three days with exceptionally high doses of >2000 spores. Accordingly, the authors recommended a dose of 600 spores as a lower boundary for the maximum human no-risk inhalational dose. For comparison, the ID_{02} for the Sverdlovsk incident was evaluated by Wilkening[16] as 2.4 or 360 spores and by Meselson[156] as 9260 or 1300 spores, depending on the dose–response model used. According to model B4 of Toth *et al.*,[17] exposure to 600 spores corresponds to an infection risk of 2–6%, which is an unacceptable risk level for the general population.[17] It is worth noting that these probabilities are based on the data of the Brachman *et al.* study, hence it is of interest to compare the no-risk dose of 600 spores with models based on the Druett *et al.* data. The original log-probit model (model D1 in Table 3.7 and Figure 3.12) predicts an extremely low risk for a dose of 600 spores. The exponential models based on recalculated data by Toth *et al.*[17] (model D3 in Table 3.7 and Figure 3.12) and the BMD analysis by Taft and Hines[98] (Table 3.6) predict the ID_{01} or BMD_{01} as 700 spores and the $BMDL_{01}$ as 540 spores, which are close to the proposed threshold of Cohen and Whalen.[168,169] The proposed threshold seems to agree with the Druett-based models, which may be more suitable than the Brachman data for predictions concerning healthy individuals. However, this apparent agreement requires careful scrutiny in view of the relevant response endpoints (infection, disease and death), as the animals in the experiments were infected and succumbed but the American mill workers, although they did not become ill, might have developed subclinical infections at a possibly substantial rate.

3.8.5 Modeling Indoor Risk from Airborne Anthrax Spores

The 2001 anthrax letter attacks did not affect the intended victims, *i.e.* government institutions and media, as badly as it affected postal workers, exposing them to a novel occupational risk: seven of 11 inhalational anthrax patients and victims were US Postal Service mail sorters,[10,11] of whom two died. An "exercise" in modeling the risk of anthrax infection in the Brentwood, Washington, DC, and Trenton, NJ, mail distribution centers, where four and two employees, respectively, contracted inhalational anthrax by exposure to spores suspended from mail sorting machines that handled contaminated mail[11,144] and risk prediction for typical American home and office rooms, was carried out by Fennelly *et al.*,[170] using the Wells–Riley model. The model, which was developed for the prediction of measles transmission in schools,[171] links the probability of infection or attack rate (P_{infect}) in a closed space to the number of sources, infectious potency of the agent, exposure time and physiological and environmental parameters, according to the basic equation

$$P_{\text{infect}} = \frac{C}{S} = 1 - \exp\left(-\frac{Iqtp}{Q}\right) \tag{3.42}$$

where C is the number of cases, S is the number of susceptible individuals, I is the number of sources, *e.g.* coughing people in the case of contagious respiratory diseases, q is the number of infectious "quanta" (a quantum is defined as the number of pathogens per source and per unit time), t is the exposure time, p is the respiratory minute volume and Q is the number of air volume changes per hour (ACH) in the room. The model is still in use to predict infection risk in closed spaces such as hospital wards,[120,172] it is simple and flexible and it can incorporate additional parameters such as the use of respirators, killing by UV light or airborne pathogen loss due to deposition.[120] For the case of an anthrax envelope, $I = 1$. To evaluate the exposures at the mail distribution centers, the physical and engineering parameters (space dimensions, air exchange rates) could be obtained from the US Postal Services, attack rates (*C*/*S*) were evaluated as the numbers of cases divided by the total number of employees, being 0.25% (2/750) for Trenton and 0.66% (4/610) for Brentwood. Exposure time was assumed as an 8 h work shift and respiratory rate as 14.6 L min^{-1}, for moderate work. The infectious quantum is defined statistically as the dose required for an infection probability of 0.63 or 1.25 × ID_{50}. Fennelly *et al.*[170] estimated the infectious dose of the basis of the measurements of *B. anthracis* exposure doses in goat hair mills by Dahlgren *et al.*[147] and the Glassman/Jemski log-probit model.[39] Estimating the attack rate in the 1957 outbreak at the New Hampshire mill as 25% (four of 16 workers in the carding–combing exposure zones) and an estimated daily dose of 140–690 spores in <5 μm particles and 620–2200 total spores,[147] this was the ID_{25}. According to the 1966 Glassman model, where the ID_{50} is 10 × ID_{25}, the corresponding ID_{50}s for the New Hampshire mill are 1400–6900 spores in <5 μm particles and 6.200–22 000 total spores and the quanta are 1750–8625 spores in <5 μm particles and 7250–27 500 total spores. By applying the model for the Trenton postal facility, the exposure rate was calculated as 0.238–2.08 infectious quanta per minute or 114–998 infectious quanta generated per 8 h day. In spore numbers, the cumulative daily amount generated was estimated as 2×10^5–8.61×10^6 spores in <5 μm particles and 8.8×10^5–2.75×10^7 total spores. The respective estimates for Brentwood were 0.96–6.0 quanta per minute or 461–2550 quanta per 8 h day and in spore numbers they were 8.06×10^5–2.48×10^7 in <5 μm particles and 3.84×10^6–7.92×10^7 total spores. These estimates were criticized as erroneous[169] as the measured doses at the New Hampshire mill were taken when no inhalational anthrax occurred at the mill, so they could not be connected to the outbreak attack rate. However, considering the "common wisdom" human ID_{50} of 8000–10 000 spores and the recent estimates cited above on the basis of primate dose–response data, the upper estimates of Fennelly *et al.*[170] are fairly reasonable. The risk levels in numbers of cases per 10^4 were calculated for exposure levels from 0.001 to 1000 quanta, hypothetical rooms of different sizes and different air exchange rates (0.5–2.0 ACH) and respiratory minute volumes (4–10 L min^{-1}). For an exposure level of one quantum, the risk varies between 6 and 936 cases per 10^4 depending on the scenario and for a 0.1 quantum it decreases to 0.6–98. Of the factors analyzed, the room size and removal rate by ventilation affected risk more strongly than respiratory rates.

3.8.6 Experimental Testing of Spore Emission from Spore-Laden Letters and Mail-Sorting Machines

Experimental studies to assess the risk of anthrax letters were performed by the Defense Research and Development Canada (DRDC) in a model of simulated letters containing dried spores of *Bacillus atrophaeus*, known previously as *B. globigii* or BG. The motivation for the first experiments that started in 2001 before the anthrax letter attacks in the USA came from a series of hoax anthrax letters sent to various institutions and individuals, first in the USA and later in Canada. The first experiments characterized the dispersion of spores from letters opened in a room simulating a mailroom or an office.[18,173] Later experiments included an animal exposure study in the model room to assess pulmonary deposition of spores emitted from letters,[174] a study of dispersion from a spore-laden letter in a running mail-sorting machine[18] and an experiment in a real office environment.[175] These model studies demonstrated the emission and pulmonary intake of high numbers of spores that may far exceed the inhalational LD_{50} human estimates. The first experiments were conducted in a 45 m^3 aerosol chamber configured to simulate a mailroom or office.[18,173] The experimenter sat behind a desk and was assigned to open 10 "letters," one of them laced with either 1 or 0.1 g of dried *B. atrophaeus* spore powder containing 10^{11} spores per gram. Dispersed spores were sampled for 10 min, starting from the opening of a spore-laden letter, by two arrays of slit samplers at different locations to follow the time course and spread of the cloud, a personal filter attached to the inlet of the experimenter's protective mask canister and an ambient filter sampler. A typical cloud, as characterized by the slit sampler, reached within ~2 min a sharp concentration peak that declined slowly to background levels within 10 min of sampling. The concentration metric was ACPLA (agent-containing particles per liter of air) because the aerosol particles, 0.75–9 μm in size, were mainly spore aggregates and each colony in the slit sampler counts represented a particle and not a single spore. Typical peak heights were 300 or 80 ACPLA for the letters that contained 1 or 0.1 g of spore powder, respectively. The average yields of spores from the personal filters were 6.7×10^6 cfu for the 1 g and 8.0×10^5 cfu for the 0.1 g letters. Assuming a human inhalational LD_{50} of 10 000 spores, the inhaled dose equivalents were 670 and $80 \times LD_{50}$. Based on the data for the ambient filter samplers, a human opening an anthrax letter could inhale 770 or $120 \times LD_{50}$ from a 1 or 0.1 g powder-containing letter, respectively, when breathing at a minute volume of 10 L min^{-1} and threefold higher doses if stressed and breathing at 30 L min^{-1}. To assess further the risk from opening a spore-laden letter, pulmonary deposition was studied in swine exposed to a spore cloud released by opening a 1 g letter in the model room. Quantitative analyses of pulmonary spore deposition in lungs from euthanized animals showed a mean content of 21 000 spores (interquartile range 7100–46 100 cfu) in both lungs after exposure to a cloud peaking at >200 ACPLA and lasting for 30 min. The total exposure doses of the animals could not be determined precisely and were estimated as >200 000.[174] Another study of letter opening in an office building wing was conducted in order to follow cloud spread and the effects of various activity patterns of the experimenter on the cloud spread and risk.[175] In the control scenario, the experimenter remained seated after opening the letters, whereas in five other scenarios he responded by leaving the room, trying to stop cloud spread by shutting the door of the room where he opened the letter or turning the

HVAC off, performing mitigating activities such as wetting the powder spill to prevent suspension, simulating panic behavior or performing the immediate response actions recommended by the CDC. Analysis of slit samplers deployed in several rooms showed that the spore cloud spread in the premises within minutes of letter opening irrespective of the scenario, and the spore levels in all scenarios were higher than in the control scenario. The calculated exposure doses, in which 0.1 g powder-containing letters were opened, ranged from 3.9×10^5 cfu in the control to 1.6×10^6 in the worst scenario and the human lethal dose equivalents were 39–160 × LD_{50}. Taken together, these experiments demonstrated that contrary to previous belief, even the apparently passive dispersal from opening a spore-laden letter may release spore clouds that are more than enough for a lethal exposure and that actions taken in attempts to reduce risk are ineffective, if not deleterious. For the mail-sorting machine studies, the emitted spore particles were measured with a fluorescence aerosol particle sizer because slit sampler arrays could not be used. Data from these studies could be compared with those from the letter opening experiments because (i) the fluorescence-based counts were well correlated with the slit sampler colony counts and (ii) in order to facilitate quantitative comparisons, the slit sampler colony counts were computed as log-normal frequency density functions and using the median $\log_{10}$ACPLA as the concentration metric of each experiment. The median aerosol particle concentration from a sorting machine handling a 1 g powder-containing letter was 2 $\log_{10}$ACPLA (SD = 0.3 $\log_{10}$ACPLA) compared with 1.83 $\log_{10}$ACPLA and 1.15 $\log_{10}$ACPLA for 1 and 0.1 g powder-containing letters, respectively, opened manually.[18]

3.8.7 Modeling of the Spore Release at the Brentwood Postal Facility in 2001

On the basis of the dispersal from a letter containing 1 g of spores in the sorting machine, Ho and Duncan[18] modeled the aerosolized spore release that caused the fatalities at the Brentwood Postal Service facility in the 2001 anthrax letter attacks. The two extensions that had to be made were setting the spore powder content in the letter to 3 g and considering the time of spore cloud decay after the initial ~30 s peak concentration. Using a scale factor of 3 and a 10% compensation for sampler efficiency, the initial median cloud concentration was set as 2.7 $\log_{10}$ACPLA. A hypothetical decay curve for the Brentwood cloud was constructed on the basis of the decay phase of the cloud generated by opening a 1 g powder-containing letter. According to this model, the mean concentration decayed to 1.0 $\log_{10}$ACPLA in 23 min. This final level is convertible to an aerosol inhaled dose of 4.5×10^3 spores, which may pose a substantial risk to a susceptible individual. It was important to consider the time dimension of the release as the two victims were not mail sorters and as they were located "off target" from the contaminated machine. The spore cloud might have traveled for some time until they were exposed. It is noteworthy that the operator of that machine did not contract anthrax.[18] To compare this analysis with that of Fennelly *et al.*,[170] both types of analysis predict an overall low risk, and the Fennelly *et al.* analysis does not distinguish between an overall low risk for the postal facility workers as a whole and a local and momentary risk for a few susceptible individuals from a short-term highly concentrated cloud, a risk on which the model of Ho and Duncan is focused. This scenario is also relevant to the 1957 outbreak in the New Hampshire goat hair mill, where a few susceptible individuals might have encountered a momentary spore cloud of high concentration.

3.9 Evaluation, Conclusions and Future Directions

Dose–response relationship analysis is a central theme in any assignment of risk assessment. It is not confined to chemical or biological risk assessment, as general approaches from these fields can be applied to any natural or human-generated calamity, provided that a quantitative metric for agent dose or intensity and a model connecting it to the effect are at hand.[176] The case of microbial risk assessment, the quantification, the risks of infection, illness or lethality caused by contact with pathogens, is unique because the models are underlain by assumptions concerning biological properties of the host, pathogen or both. This applies to the family of log-normal models, which are based on assumptions concerning the distribution of individual susceptibilities in the host population or cooperative action of pathogenic organisms, and to the family of exponential models, which assume independent action of the pathogen in the infecting dose.[113]

Inhalational anthrax in humans is an extremely highly lethal but a rare disease, occurring mainly in the occupational setting and among workers in the hair and wool industries in activities that produce large amounts of dust and aerosolization of *B. anthracis* spores, where an unfortunate encounter between a susceptible individual and highly contaminated raw material takes place. It is also rare or, to our knowledge, nonexistent in animal anthrax. In animal anthrax, the main route of infection is intestinal and even in places such as the Etosha Park in Namibia, where aerosolization and windborne spread were measured, the inhalational risk was estimated to be extremely low.[177] The other major past concern regarding inhalational anthrax was in biological warfare, in view of the high bioweapon potential of the agent and the more recently emergent interest in bioterror threat and response.

3.9.1 Applicability of Data Sets for Dose–Response Modeling

In contrast to the dose–response modeling of food- and water-borne infections, where many human experimental and epidemiological data provide raw material for modeling, dose–response modeling of inhalational anthrax can rely only on animal studies and, among these, the data sets available and technically suitable for modeling are few in number and, as the studies were carried out a number of years ago, they may have uncertainties in data quality and reporting. Several reviews have evaluated the suitability of published or institutional data sets for dose–response modeling.[20,52,98,163] The criteria of Bartrand *et al.*[20] for selection of data sets were the following: (i) exposure by inhalation of bolus dose; (ii) specified doses of inhaled spores; (iii) specified response; (iv) information on aerosol particle size and *B. anthracis* strain used. The selection methodology of Taft and Hines,[52,98] based on guidelines of the National Research Council and the EPA, was more rigorous. The study was rated by the following criteria: (i) dose–response data available; (ii) particle size distributions, or identification of particles as single spore or spore aggregates, were based on measurements; (iii) reporting of dose groups with low lethality (<50%) or an overall lethality rate of <50% in doses for individuals; (iv) use of real-time methods during exposure to derive respiratory rates; (v) group number (≥5) or total number tested or individual dose measurements (≥12) sufficient for dose–response modeling. High usability is scored if all five criteria are fulfilled, medium usability is scored upon fulfillment of three or four criteria and low usability is scored if fewer than three criteria are satisfied or there is a negative answer to criterion (i) or (v). As stated above, the rhesus macaque aerosol exposure study of Druett *et al.*[95] and the guinea pig intranasal challenge study of Altboum *et al.*[132] were the only studies selectable for modeling from those evaluated by Bartrand *et al.*[20]

The Druett *et al.* study was also selected by Curling *et al.*[163] as a basic dose–response model for the development of military medical response planning guidelines for NATO. The most conspicuous limitation of this study is uncertainty in the respiratory minute volume used for inhaled dose calculations, which had to be redone by more recent modelers. The merits of the study are the detailed dose and response reporting, sufficient numbers of dose groups and subjects per group, fair coverage of response range (12.4–100%) and thorough physical characterization of aerosol particles, as the effect of particle size on infectivity was a major goal of the study. The evaluation of Taft and Hines graded the Druett *et al.*[95] and Altboum *et al.*[132] studies as of medium utility, the former because of the inaccuracy of respiratory rate data and the latter because of lack of particle information, as it was an intranasal challenge study. The studies complying with all their criteria and graded as high utility involved New Zealand White rabbit aerosol exposure to *B. anthracis* Ames spores. They were performed in collaboration of the US Department of Defense, EPA and the Battelle Memorial Institute. Parts of the studies were published[90,91,151] and reported fully in available EPA documents.[143,152] The studies, which are of high experimental quality and a breakthrough in dose–response assessment, are described in brief above. The meticulous and resource-intensive nonhuman primate exposure to goat hair mill aerosols of Brachamn *et al.*[134] was evaluated as inadequate by Curling *et al.*[163] and criticized by other studies (*e.g.* Cohen and Whalen[169]), but in spite of many uncertainties and difficulties in interpretation of the data, this study was used to test modeling strategies as unique of its kind with nonhuman primates. The major obstacle in interpretation of the data was the unclear dose–response relationships or attribution of airborne spore peaks to animal deaths. This and other hurdles could be tackled with more sophisticated modeling approaches.[17,150] The recent EPA[151,152] study of repeated multiple dosing of rabbits was designed to answer the question of repeated low doses in a direct and analyzable manner.

3.9.2 Comparison of the Nonhuman Primate and Human Dose–Response Models

Toth *et al.*[17] made a thorough comparative analysis of the current dose–response models based on published primate exposure data and human estimated parameters. The comparison lists the parameters, estimated infective doses and satisfaction of the four model evaluation criteria (Table 3.7). The models designated J, D1–D3 and B1–B3 are based purely on nonhuman primate data sets. Model B4 is a human model based on nonhuman primate dose–response data and human time to death distribution and models E1 and E5 are human estimate models. Comparison of the nonhuman primate models fitted to Brachman *et al.*'s data is shown in columns B1–B3 in Table 3.7 and Figure 3.11. The two models of Mayer *et al.*,[150] the time-independent exponential B2 and extended, time-dependent exponential B3 do not deviate much from each other and from the B1 exponential model of Haas,[19] as observed also by Mayer *et al.*[150]

All three B models fulfill criteria 1–3, *i.e.* based on dose–response data, their shape is consistent with the Sverdlovsk incubation period data, are based on mechanistic assumptions but do not estimate the incubation period (criterion 4). A comparison of the low-dose parts of the human and nonhuman primate models discussed is presented in Figure 3.12.

Druett *et al.*'s data[95] were also modeled by Toth *et al.*[17] and their exponential model (D3) differed from that of Haas[19] (D2) in a twofold higher k but twofold lower infectious dose values, resulting from the use of a higher respiratory minute volume by Haas (2.4 L min^{-1}) instead of the original 1.2 L min^{-1}, which was retained in the calculations by Toth *et al.*[17] It is noteworthy that model D3 is very close to the exponential model fitted by Taft and Hines[98] for this data set and the recent model fitted by Gutting *et al.*[141] to data pooled from four exposure studies employing two primate species (rhesus and cynomolgus macaques) and two *B. anthracis* strains (Vollum/M36 and Ames), with a parameter $k = 1.37 \times 10^{-5}$ and LD_{50} = 50 600 spores. Except for model D1, which satisfies only criterion 1, models D2 and D3 satisfy criteria 1–3 but not criterion 4. Comparisons of the low-range dose–response curves are shown in Figure 3.12. The low-dose parts of the two log-normal models (J) of Glassman[39] and Druett *et al.*[95] (D1) are extremely inconsistent with each other and with the other models cited, as was shown earlier in Haas.[19] Table 3.7 and Figure 3.12 also show the human dose–response models E1 of Rickmeier *et al.*[36] and B4 of Toth *et al.*[17] and the exponential time-dependent competing risks of Brookmeyer *et al.*[46] and Wilkening.[16] The model E1, advised by an expert panel convened by Rickmeier *et al.*.[36] for human risk analysis to

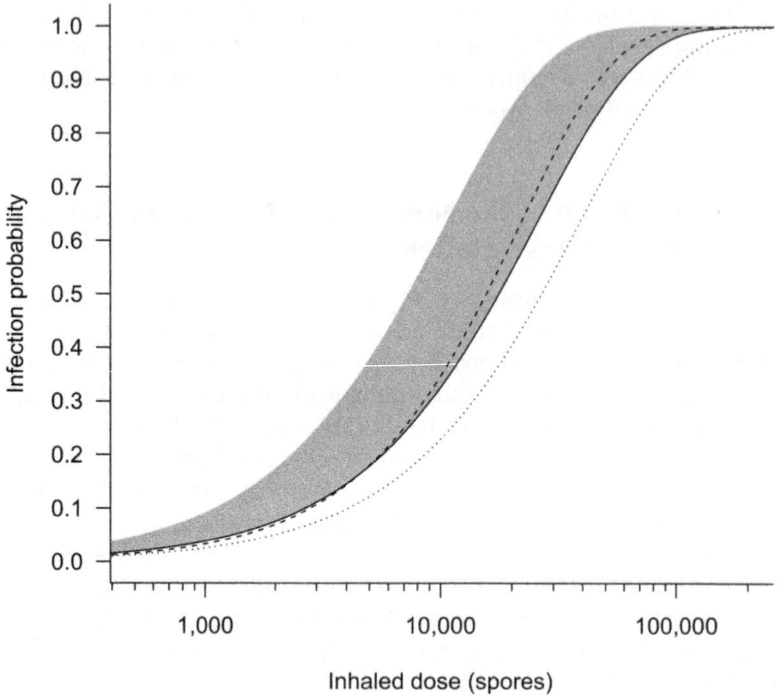

Figure 3.11 Comparison of dose–response models fits to the Brachman *et al.*[134] nonhuman primate exposure study, from Toth *et al.*[17] The shaded area is the 95% CI of model B4 (Table 3.7), which is the authors' best fit of the Brachman *et al.*[134] data. The dashed line is Mayer *et al.*[150] extended exponential model **B3** ($\alpha = 0.90$, $r = 1.87 \times 10^{-5}$); the solid line is Mayer *et al.*[150] exponential model **B2** ($r = 3.95 \times 10^{-5}$); the dotted line is Haas[19] exponential model **B1** ($r = 2.6 \times 10^{-5}$). Source: Toth *et al.*,[17] ; http://www.plospathogens.org/article/info%3Adoi%2F10.1371%2Fjournal.ppat.1003555 (last accessed 1 March 2015).

biological threat agents, based its probit slope of 1.42 on the human ID_{50} estimates ranging between 8000 and 10 000 spores and ID_{10}, ranging between 1000 and 2000 spores. Although they are human estimates whose empirical bases were not presented, they are mentioned here for comparison with the other models and they are fairly consistent with model B4. Model E5, used by Brookmeyer *et al.*[46] and Wilkening,[16] assumed an LD_{50} of 8600 spores and initially based k on nonhuman primate spore clearance kinetics, but was finally fitted to the Sverdlovsk data.[46] Although the model was not based on dose–response data, it satisfies criteria 2–4. The compatibility of the incubation period distribution predicted on the basis of models B4 and E5 with the Sverdlovsk incubation period data gives them additional credibility for human inhalational anthrax risk assessment. Another property of importance is the consistency in prediction of risk upon extrapolation to low doses, a problem raised by Haas[19] in his first attempt to model published nonhuman primate

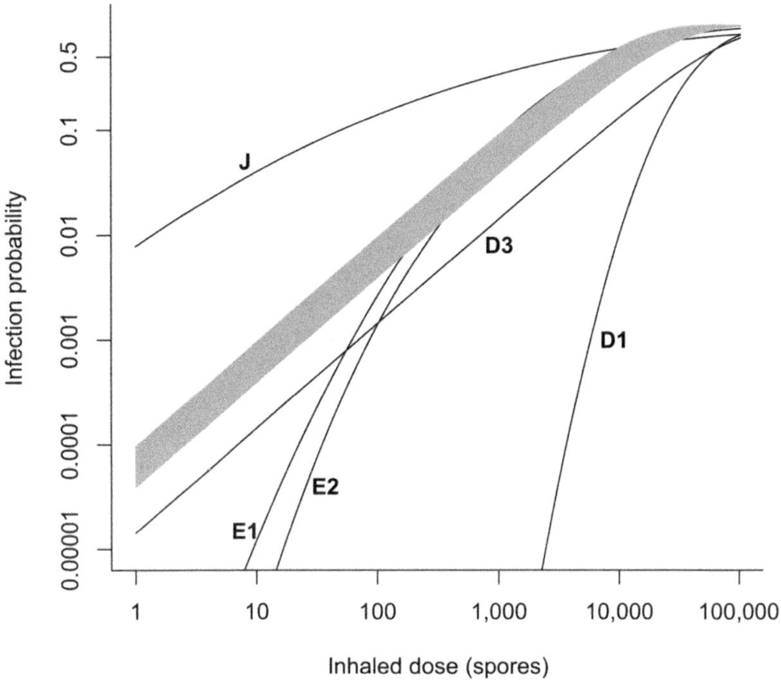

Figure 3.12 Comparison of nonhuman primate and human dose–response models for inhalational anthrax, from Toth *et al.*[17] The model notations are listed in Table 3.7. The shaded area is the 95% CI of model B4, which is the authors' best fit of the Brachman *et al.*[134] data. Model D2 was omitted from the figure by the authors. Model E2 is not shown in Table 3.7. Model E5 is included completely within the shaded area. Source: Toth *et al.*,[17] ; http://www.plospathogens.org/article/info%3Adoi%2F10.1371%2Fjournal.ppat.1003555 (last accessed 1 March 2015).

exposure data. Contrary to the log-probit models of Glassman (J) and Druett *et al.*[95] (D1), all the other models are fairly consistent in the prediction of low-risk infective doses (ID_{10}, ID_1 and $ID_{0.1}$) and extrapolate to doses that may not underestimate the risk like model D1 or overestimate the risk like model J.

3.9.3 Single High-Dose (Bolus) and Multiple Low-Dose Exposure Patterns

As discussed above, a more realistic exposure pattern in some settings, *e.g.* the industrial environment of goat hair mills or areas with residual surface contamination from an airborne release or incomplete decontamination, may undergo multiple exposures to low doses of resuspended spores. The question raised was whether the risk measured in single high-dose (bolus) experiments is the same as that in the more realistic multiple exposures to low doses. The nonhuman primate studies and models summarized in

Table 3.7 were of two sorts, as exemplified by Druett *et al.*,[95] which was a bolus exposure study, and by Brachman *et al.*,[134] which was the only multiple low-dose primate study published. The two models fitted for the Brachman *et al.* data by Mayer *et al.*[150] were consistent, showing small differences in the estimated risk if using a time-dependent or time-independent modeling strategy, leading to the conclusion that in this case the dose-timing effects were small.[150] In the exponential model by Toth *et al.*[17] for the same study, which was different in the selection of data and risk estimation methodology, the time-dependent exponential model parameter was threefold higher but the lower extrapolated infectious doses were similar. The low-dose acute and multiple daily exposure studies in rabbits did not show dramatic differences in benchmark doses – the multiple low-dose BMDs were only about twofold higher than in the acute exposure study[152] and this difference can be accounted for, at least in part, by model choice. In view of the low rates of spore clearance and germination, the results showing a small or no difference in the dose–response models for single-dose or multiple low-dose exposures are not surprising. Contrary to the earlier interpretation of the Brachman study and as modeled by Mayer *et al.*[150] and Toth *et al.*,[17] the risk may be related to the cumulative spore numbers in the alveoli. It should be noted that in the case of multiple exposures with different individual doses and timing, the risk may be different from that for a single exposure to the total dose. In the broad sense, the risk in single *versus* multiple exposures varies for different pathogens with different within-host dynamics.[149]

3.9.4 Modeling of the Incubation Period

The studies described earlier modeled the distribution of the incubation period duration as a random variable, as in the classical studies of Sartwell.[162] The first incubation period model of the Sverdlovsk incident estimated the distribution parameters using a likelihood method developed for the analysis of truncated data,[160] taking into account the fact that about 10 days after the outbreak started, prophylactic antibiotic administration might have changed the course of the outbreak in an unknown way. The parameters evaluated by estimation of instantaneous intervention were a median incubation period of 11 days and Sartwell's dispersion factor, exp σ, of 2.04.[160] Later, the mechanistic models described earlier linked the incubation period duration to exposure dose.[16,17,32,44,46] Both Brookmeyer *et al.*'s original competing risks model and Wilkening's extended models showed that the duration of the incubation period at low doses was relatively insensitive to the exposure dose whereas at doses from nearly the LD_{50} and higher, the duration of the incubation period shortened sharply with the dose. This was similar to earlier observations with other pathogens and classical birth-death infection models.[178,179] The classical birth-death models were based on quantitative studies with viruses and nonspore-forming bacteria. The underlying mechanistic assumptions of the anthrax models were similar to those of the classical models but were adapted to the biology of *B. anthracis*, in which the infective unit is the dormant spore. The most important feature was the stochasticity of spore germination and outgrowth.[44,46]

The incubation period modeling has implications for evaluating the size and timing of inhalational anthrax attacks and the optimization of post-exposure antibiotic prophylaxis. The application of incubation models in situational awareness of inhalational anthrax attacks is discussed later. In this section, we present studies on the optimal duration of prophylaxis and the timing of the prophylaxis campaign. Earlier studies with animal models have shown that premature cessation of post-exposure therapy results in disease eruption due to persistent, dormant and viable spores in the lungs.[45,137] The pre-2001 recommendation for a 60 day duration of post-exposure antibiotic prophylaxis (PEP) for inhalational anthrax[3] was strengthened by the modeling of Brookmeyer *et al.*,[160] which implied that 1% of the incubation periods in the Sverdlovsk distribution model were longer than 58 days. The optimal duration of antibiotic prophylaxis was modeled by Brookmeyer *et al.*[44] as a function of the exposure dose, using their competing risks model. If the time of cessation of antibiotic prophylaxis is τ, integrating eqn (3.29) from τ to infinity and derivation of the probability that a single spore will germinate after τ as a function of dose and time as in eqn (3.30) allows the calculation of τ, the optimal duration of post-exposure prophylaxis, as a function of the risk-related dose, $TD(P)$. Thus, for reducing the risk to less than 10^{-3}, a person exposed to $TD(1)$, $TD(10)$ or $TD(50)$ should remain on antibiotics for at least 33, 67 and 93 days, respectively; to reduce the risk to $<10^{-4}$, and the respective durations of antibiotic prophylaxis should be at least 66, 99 and 126 days. The same relationships of risk reduction, duration of prophylaxis and exposure dose were calculated by Toth *et al.*[17] Using their dose–response model B4, they presented the prophylaxis durations required for dose reduction from 0.5 down to 10^{-4} for doses of 10 000, 100 and 100 spores. They showed that a 60 day course of antibiotics is adequate in reducing the risk to <0.1% for those exposed to 1000 spores (ID_6) or less and <0.01% for those exposed to 100 spores ($ID_{0.7}$) or less. A sample study of the effect of timing of a prophylaxis campaign on population protection after a large-scale anthrax attack is that of Hupert *et al.*[180] This study evaluated the effect of timing of a prophylaxis campaign on hospital surge, using as a basic scenario the CDC's Cities Readiness Initiative (CRI) prophylaxis campaign goal of starting the campaign at a delay of 2 days from exposure and completion within 48 h. The effects of complying with or deviating from this goal on population protection, measured by reduction of hospital surge, was modeled using either Brookmeyer *et al.*'s[46] or Wilkening's[32] incubation period models. The results showed that under baseline conditions of complying with the CRI goal and assuming 90% antibiotic effectiveness and 95% progression to diseases in the absence of treatment, between 86% (Wilkening model, W) and 87% (Brookmeyer model, B) will be protected. At the extreme end of the study timeframe, which is a delay of 1 week and a campaign duration of 7 days, protection will decrease to 39.5% (Wilkening) or 49.3% (Brookmeyer).

Each additional campaign day beyond the 48 h CRI goal, added, on average, 2.4% (model B) or 2.9% (model W) to the exposed population requiring hospitalization (range 0.5–3.6%, depending on delay in campaign initiation). Every additional day of delay in starting the campaign beyond the CRI goal added, on average, 5.2% (model B) or 6.5% (model W) to the hospitalization demand (range 2.0–7.5%, depending on the campaign's duration).

3.9.5 Animal-to-Human Extrapolation of Infectious Doses

As discussed above, human data on which to base dose–response modeling are lacking owing to the rarity of inhalational anthrax in humans, the absence of quantitative exposure data from historical and recent human cases and the ethical imperative against human experimental exposure to dangerous pathogens. Therefore, human risk assessment of inhalational anthrax is dependent on animal exposure data. The only human data that could be exploited in the quantitative risk assessment studies were the epidemiological investigation data from the 1979 Sverdlovsk incident, environmental and epidemiological studies carried out in goat hair mills in the wake of the 1957 industrial inhalational anthrax outbreak in New Hampshire and environmental studies undertaken in sites contaminated by anthrax letters in 2001.[144,145] These data can be used as indirect or supporting evidence only. In the Sverdlovsk case, not all patient data were available or accurate, there was insufficient information on the effects of treatment given to hospitalized patients and public health measures for exposed population and, finally, attempts at exposure modeling were based on assumptions concerning source strength and meteorological conditions at the time of the release.[16,32,38,155] When the environmental studies in the goat hair mills were undertaken, no anthrax cases were reported in those mills, so no direct positive connection between spore levels and human morbidity could be established.[134,147,169] The environmental studies at the Hart Senate building[145] and the Brentwood postal facility[144] were performed late after facility closure. The use of animal data in human risk assessment is complicated by interspecies susceptibility differences, even in the animal models in which the overall pathogenesis of inhalational anthrax is similar to the human disease – the rhesus, cynomolgus and African green monkey nonhuman primates, New Zealand White rabbits and Hartley guinea pigs.[130] Non-human primates are more readily taken as surrogates when human risk is assessed, because of the evolutionary relatedness, anatomical and physiological similarities and similar pathogenesis of inhalational anthrax. Estimates of dose–response relationships from primate studies were used directly in human risk assessment without dose corrections[38,88,155,163] given the broad range of 2500–55 000 spores as the inhalational LD_{50} of anthrax spores in primates, with inclusion of humans within this range.[3,4,88] Combined modeling of primate experimental dose–response data with human epidemiological data showed some success in achieving consistent results, especially in extrapolation to low-level exposure;[17,32] results are summarized in Table 3.7. The Bayesian hierarchical meta-analysis of a large number of experimental results of studies with a large number of spore preparations in one animal species, the guinea pig, gave some support to the narrower estimate of 8000–10 000 spores as a conservative estimate of the human ID_{50} within a broad range of possible values.[181] This estimate is also supported by the modeling studies discussed above and summarized in Table 3.7. However, a more rigorous methodology is needed, especially if studies using animal models other than nonhuman primates is considered and in the assessment of low-dose exposure for the determination of safe exposure limits and decontamination goals. In this case, relevant physiological species differences should be considered for the determination of dose-adjustment factors from animals to humans. An example of a plausible methodology for extrapolation from animal-determined BMDs and BMDLs to humans, with its strengths and limitations, is given below.

3.9.6 Application of the EPA Methodology of Animal-to-Human Extrapolation in Inhalational Anthrax

The methodology for the derivation of human reference concentrations (RfCs) for inhaled gaseous or particulate toxicants, when based on animal studies, includes, as a first step, the calculation of a human equivalent concentration (HEC), which is used as a point of departure for further dose adaptation to the requisite exposure scenario (reviewed by Jarabek *et al.*[182]). The conversion of the animal concentration to HEC is performed in the following steps: (i) calculation of the animal effective internal dose using an animal dosimetry model, (ii) conversion of the animal internal dose to a human effective internal dose and (iii) calculation of the HEC from the human internal dose, using a human dosimetry model.[182] The dosimetry models for transition from exposure concentration to internal dose and *vice versa* bring into consideration the differences in exposure duration between the animal study and human scenario, respiratory minute volumes of animals and humans and in the case of particulate materials – the species-specific pulmonary deposition fraction for the relevant particle sizes.[182] The target doses in these examples will be the human equivalent $BMDL_{10}$-derived concentrations. The first example is based on the BMD analysis of nonhuman primate data.[51] The animal BMD selected was from a recent nonhuman primate study, for which the best-fitting model was a $\log_e$-logistic model, yielding a BMD_{50} of 10 000, a BMD_{10} of 1100 and a $BMDL_{10}$ of 550 cfu. Applying a nonhuman primate pulmonary deposition rate of 12% for 2 μm diameter particles,[183] the animal $BMDL_{10}$ was converted to a deposited or effective internal dose, of 66 spores deposited in the alveolar region. Assuming the same internal dose for humans and a pulmonary deposition rate of 20% for 1–2 μm diameter particles, the internal dose can be converted to an inhaled dose of 330 spores. For a respiratory minute volume of 0.017 $m^3\ min^{-1}$ for an adult male performing light activity, the derived 1 minute HEC is 1900 spores m^{-3}. A second example is based on the BMD analysis of the multiple low-dose exposure of study in New Zealand White rabbits.[152] The rabbit critical value for this example is a $BMDL_{10}$ of 4900 spores, derived from the best-fitting $\log_e$-logistic model ($p = 0.62$, AIC = 20.3447) for the total cumulative dose (ref. 152, Appendix V). If the alveolar deposition rate for ~1 μm diameter particles in rabbits is 0.56%, then the deposited dose is 270 spores, which is taken as the same for humans. Using the human deposition rate of 20%, the human cumulative inhaled dose is 1400 spores. Determination of HEC is more difficult in this case as it has to be adjusted for specified human exposure timing. The transformed values may be useful as the POD for the derivation of health protective exposure limits and decontamination goals, provided that properly selected uncertainty factors are applied. To decrease uncertainty, cross-validation with studies with other animal species and application of more elaborate dosimetry models should be undertaken.

As the pulmonary deposition rate is a major determinant in interspecies differences in susceptibility to infection by inhaled pathogens, interspecies extrapolation of infective inhalational doses depends on its accurate estimation. Gutting *et al.*[141] had at their disposal several sources of empirical and model-based information for the parameterization of pulmonary deposition in rabbits and guinea pigs, but not in nonhuman primates. The most accurate lung deposition models in use for risk assessment – the LUDEP/ICRP66 models – is intended for humans,[89,182] and multiple-path deposition (MPPD) models are available for humans and rats.[182,184] A radio-imaging study of pulmonary deposition in cynomolgus macaques determined, for particles 2.3 μm in size, an extrathoracic deposition rate of 39% and a pulmonary deposition of 12%. For particles of 5.1 μm, the respective deposition rates were 58 and 8%.[183] The particle deposition rates for humans, derived by the LUDP/ICRP66 model, were 41.1 and 18.2% for extrathoracic and pulmonary deposition of 2 μm particles and 64.8 and 16.3% for 5.1 μm particles, respectively.[183] Recently, a lung morphometric model applicable to pulmonary deposition modeling was developed for the rhesus macaque.[184] The pulmonary deposition rate for 2 μm diameter particles, modeled for nose-breathing young adult rhesus macaques, was comparable to the value of 12% deposition measured by Cheng *et al.*[183] in cynomolgus. Using this deposition estimate for calculation of the nonhuman primate deposited dose in the Gutting *et al.*[141] model roughly halves the primate deposited dose, making it closer to the rabbit model estimate. The deposition rate of 5.1 μm diameter particles measured by Cheng *et al.*[183] in cynomolgus macaques was 8%, being threefold greater than the model result for young adult rhesus macaques.[184] This emphasized the concerns raised by Asgardian *et al.* for the validity of their model for cynomolgus macaques, which challenges the assumption of similarity in pulmonary deposition, made by Gutting *et al.*,[141] in the two species. The ICRP66 model, developed initially for inhalation hazard assessment from radioactive particles, has been adapted recently for inhaled pathogen risk assessment.[89] The deposition equations model the deposition of particles on the basis of their size, shape and other physical properties. For *B. anthracis* spores, the literature properties of single spores were incorporated in the model as a default, stating that more accurate results are dependent on realistic parameters of the bioaerosol particles. The model also addresses the variability in deposition rates among humans, using height as a covariate.[89] This discussion demonstrates that even though an extrapolation methodology is biologically feasible, more work is needed in refining of model inputs in order to obtain more reliable human estimates, with possible use of existing models from other fields of human risk assessment.

3.9.7 Retrospective Risk Assessment from Epidemiological and Environmental Studies in Biological Attack Scenarios

The response to biological attack scenarios is challenging owing to uncertainties in essential data for situational awareness, planning and resource allocation.[30,37] Assuming absence or partial coverage of biological agent detection and warning systems, a suspected biological attack will be detected by irregular morbidity in the affected population, probably with unique spatiotemporal patterns.[3,4,185] Detection of unusual biological events is augmented by syndromic surveillance systems, which monitor emergency departments and community physician visits with emphasis on relevant syndromes, *e.g.* respiratory symptoms for the case of inhalational anthrax and other biological agents.[186,187] Once an unusual epidemiological event has been recognized, a major issue is the discrimination of a deliberate biological attack from an unusual natural cause.[188,189] After an attack confirmation, information will be required on the size, location and timing of the attack and a forecast of the progression of the disease outbreak. As the only data available for situational awareness at this stage may be the serial accumulation of cases, the size and timing of an attack can be estimated with the help of an incubation period distribution model, even from partial observations of case accumulation at different times from the start of the outbreak.[160] A Bayesian algorithm for back-calculation of the size of an anthrax bioterror attack from case numbers and times of onset of symptoms and a known incubation-period distribution model was published by Walden and Kaplan.[190] The algorithm, which calculates posterior probability density functions for the size and timing of an attack, was tested with a simulated inhalational anthrax attack of 100 cases, using the log-normal incubation period distribution of the Sverdlovsk incident from Brookmeyer *et al.*[160] as a reference. On the basis of the first five cases occurring on the first day of the attack, the algorithm estimated the size of the attack as 850 cases, but as more cases accumulated, the accuracy improved and on day 4 (18 cases) the attack size was estimated correctly (100 cases), determined to take place 1.8 days before the first case (comparable to the Sverdlovsk estimated release time of 2 days before the first case). Other examples of Bayesian algorithms for the detection of aerosolized anthrax attacks and back-calculation of attack parameters from biosurveillance data are presented below.

A more elaborate Bayesian algorithm for the detection of an aerosol biological attack by biosurveillance data – the Bayesian Aerosol Release Detector

(BARD) – was developed by Hogan *et al.*[185] The development of this algorithm was inspired by the work of Meselson's group, who inferred the cause of the Sverdlovsk incident as an aerosol release on the basis of combined data of case timing, geographical location at the alleged time of release, meteorological data and dispersion modeling.[38,155] The algorithm, which integrates medical surveillance data and an atmospheric dispersion model, calculates posterior probabilities of an airborne anthrax release and crucial attack parameters using readily available inputs of emergency department (ED) visits and meteorological data. The attack detection is based on probabilistic discrimination between two hypotheses: the null hypothesis H_0, which relates the ED visits for respiratory complaints in the recent 24 h period to background causes only, and the alternative hypothesis H_1, which relates the ED visits to both background causes and an aerosolized *B. anthracis* spore attack. The posterior probability of H_1 is calculated, by Bayes' equation as follows:

$$P(H_1|\mathbf{b},\mathbf{G},\mathbf{M}) = \frac{P(\mathbf{b}|H_1,\mathbf{G},\mathbf{M})P(H_1)}{P(\mathbf{b}|H_1,\mathbf{G},\mathbf{M}) + P(\mathbf{b}|H_0,\mathbf{G})P(H_0)} \tag{3.43}$$

In other words, $P(H_1|\mathbf{b},\mathbf{G},\mathbf{M})$ is the conditional probability of H_1 given the vector **b** of ED visits in the last 24 h for each Zip-code sub-region of the surveillance region, the matrix **G** of geographical data (*e.g.* Zip codes) from which the surveillance data were taken and the matrix **M** of meteorological data. The above equation contains the two prior probabilities $P(H_0)$ and $P(H_1)$ and two conditional probabilities – $P(\mathbf{b}|H_0,\mathbf{G})$ and $P(\mathbf{b}|H_1,\mathbf{G},\mathbf{M})$. $P(H_1)$, the probability of an aerosolized anthrax attack in one of the 30 largest US cities in 10 years, was estimated at $1/(10 \times 365 \times 30) \sim 10^{-5}$ and $P(H_0)$ is $1 - P(H_1)$. The detailed method of calculation of the conditional probabilities can be found in the paper by Hogan *et al.*[185] In brief, the conditional probability $P(\mathbf{b}|H_0,\mathbf{G})$ is calculated by integrating the surveillance and geographical data in a binominal probability model. The conditional probability $P(\mathbf{b}|H_1,\mathbf{G},\mathbf{M})$ is calculated by integrating the surveillance data with meteorological data and infection probabilities, which are calculated by a Gaussian dispersion model, dose–response and incubation period distribution models. In addition to the posterior probability of H_1, the BARD algorithm produces also the likelihood ratio of the posterior probabilities $P(\mathbf{b}|H_1,\mathbf{G},\mathbf{M})/P(\mathbf{b}|H_0,\mathbf{G})$ or Bayes' factor and the posterior distributions and the expectations of the release location, source strength, time of release and a list of Zip codes of sub-regions with decreasing probabilities that some individuals may be infected. The dose–response model used in the article was Glassman's log-probit model (ID_{50} = 8625 and probit slope = 0.668)[39] and Wilkening's incubation period distribution model A1 (Figure 3.7).[16] However, the algorithm is flexible and can accommodate other models. The algorithm performance was tested with 225 simulations of aerosolized anthrax spore attacks with varying parameters of release location, quantity of released spores, size of attacks and also ID_{50}, probit slope and incubation period characteristics in order to test the sensitivity of the algorithm to dose–response parameter uncertainties. The released quantities were 0.01, 0.1 and 1 kg with a spore content of 10^{14} spores kg^{-1} and the attacks were from 8 to 169 190 cases in size with a mean size of 19 386 cases. At a sensitivity of 86.2% and a false alarm rate of one every 3 years, the mean time to detection was 3.37 days from release (95% CI: 3.23–3.51 days) or 1.40 days from the time of first case presentation at an ED (95% CI: 1.31–1.52 days). At a sensitivity of 90.2% and a false alarm rate of one per month, the respective times to detection were 3.16 days (95% CI: 3.04–3.29 days) and 1.11 days (95% CI: 0.99–1.22 days). The threshold posterior probability for alarm was 0.0014. In sensitivity testing, the parameter changes that resulted in a larger outbreak size and a shorter incubation period [*e.g.* ID_{50} = 2000 spores, probit slope = 0.3 and incubation period of 6.33 days at $\log_{10}$(dose) = −0.725] resulted in shorter detection times, whereas parameter changes that resulted in a smaller outbreak size and a longer incubation period [*e.g.* release quantity 10^{12} spores, ID_{50} = 55 000 spores, probit slope = 1.43, incubation period of 12 days at $\log_{10}$(dose) = −1.628] increased the detection times. The most detrimental parameter changes were increased probit slope and reduced release quantity. The critical detection threshold for this analysis was an attack size of 900 cases: at a false alarm rate of one every 3 years, for the time to detect an attack of 900 cases or more, 100% of the outbreaks were detected by 3.32 days on average, whereas for outbreaks <900 cases in size, only 12% of the outbreaks were detected in 12 days on average. It should be stated that this testing was performed under several strong assumptions, such as an atmospheric dispersion model applicable to a flat terrain or that the affected population remained in their homes when the attack took place. Easing of these assumptions and inclusion of more realistic scenarios such as mobility of the affected population or alternative dispersion models were included in later detection algorithm studies.[191]

Another Bayesian algorithm for inference of size, timing and dose (N, τ and $\log_{10}D$) from short time series (3–7 days long) of cases was developed at Sandia National Laboratories.[192] The algorithm was tested under several "idealized" and more challenging simulations that reflect realistic situations. In the first test, the scenarios were constructed using the same incubation period model as used for inference, Wilkening's model A2 (Figure 3.7).[16] In this way, the differences between the input and inferred parameters could be attributable only to stochastic noise. In this analysis, simulated attacks of 100 or 10 000 cases and doses of 1, 100 and 10 000 spores could be back-calculated with fair accuracy with respect to size and time from data accumulated over 5 days, whereas dose was more difficult to evaluate owing to the broad distribution density curves. In some cases, bimodal density distributions of size or time were produced, but this was not considered a fallacy as the correct decision could be reached using additional information. More realistic scenarios were produced by simulation of arrays of different exposure doses per attack, generated by a Gaussian plume dispersion model and using different models in construction of the input data (Wilkening's model D) and inference (Wilkening's model A2). The latter simulation was designed to test the effect of structural disparity between the real-life data and the analytical model on the algorithm's performance. These challenging scenarios somehow compromised the accuracy of the estimates, but the overall utility of the algorithm in optimizing medical resource allocation in response to an attack was preserved. Retrospective analysis of the Sverdlovsk incident estimated N as 55 cases (90% CI: 41.15–66.94 cases) and $\tau = -2$ days (90% CI: −3.22 to −1.38 days) on the basis of 9 days of data, ending at the presumed start of the prophylaxis campaign and 2 days before the start of vaccination. A back-calculation algorithm developed in the UK was based on a model combining aerosol dispersion, pathogen within-host dynamics based on the Brookmeyer *et al.* and Wilkening models and population mobility.[191] The algorithm could detect aerosol anthrax attacks of small sizes (about 25–35 cases) and predict attack parameters (release quantity, location and time of attack) accurately after 10–15 cases, but even the less accurate predictions after about five cases were adequate in assisting response planning. The Bayesian algorithms, especially those using the advanced pathogen dynamic, dose–response and incubation period models, were efficacious in the detection of small-sized attacks and prediction of attack parameters at operationally relevant times. In a study evaluating the detection of an anthrax aerosol attack by a "conventional" syndromic surveillance system, detection of simulated anthrax attacks of 15 000–49 000 cases in size by syndromic surveillance was obtained after 3.1–3.6 days, compared with 3.7–4.1 days on average by laboratory-confirmed clinical diagnosis, only if the specificity was low (one false alarm every 10 days). Increasing the specificity to one false alarm every 40 days impaired the performance badly, increasing the time to detection and reducing the proportion of attacks detected prior to clinical diagnosis.[186] If the value of current syndromic surveillance in the case of anthrax is more of a support to other means of attack detection, then the ability to predict attack parameters and assist with operational planning of the response should be a considerable benefit.

The above algorithms were developed and tested against scenarios of outdoor releases of aerosolized spores. In the case of indoor releases, several authors have developed frameworks for the evaluation of risk from environmental samples, integrating dose–response, particle dynamics and surface sampling performance data.[193–196] Hong and co-workers[194–196] developed an integrated model of indoor pathogen-carrying particle dynamics for evaluation of retrospective risk, *i.e.* the risk to people who are present on-site during the release, and prospective risk, *i.e.* the risk to people re-entering the site prior to decontamination or agent decay, if relevant, from the quantitative analysis of environmental samples collected any time after the release. The first paper[194] focused on the inhalational risk from anthrax spores, and a later paper[195] extended the analysis to other CDC category A pathogens (*Francisella tularensis*, *Yersinia pestis* and the viral agents of smallpox and Lassa fever, in addition to *B. anthracis*) and additional modes of exposure – dermal contact in the case of *B. anthracis* and ingestion in the case of other pathogens. The model's object in these studies was a hypothetical simple office room with an HVAC system. Pathogen-carrying particle dynamics were modeled by a matrix of first-order differential equations of particle movement between different compartments. In the 2010 paper,[194] these included room air, tracked floor surfaces, untracked floor surfaces, ceiling, walls, the HVAC system and human nasal passages of people present in the room. The main particle transfer processes were deposition to all surfaces and resuspension from tracked surfaces due to human activity. In the 2012 paper,[195] transfer to human hands and from hand to mouth were added to address contact or ingestion risks. A methodology was developed for linking spore (or any other pathogen) concentration on environmental media – tracked and untracked floors, ceilings, walls, HVAC filters and nasal passages – to risk. Retrospective risk was evaluated by back-calculation of the concentration in the room air compartment at the time of release and onwards, as long as people were present in the room, from the surface concentrations measured at different times of sampling after the initial deposition by converting them to time-integrated exposure doses that were used as inputs into dose–response models. Prospective risk was similarly evaluated by using the relevant transport models, *e.g.* resuspension for inhalational risk, for calculation of time-integrated exposure doses. The target of these calculations was to understand the sampling effort needed, in terms of minimal sampling area for surface sampling, recovery from HVAC filters or analytical detection limits in order to assess a required level of risk. As particle size is an important factor in determining both inhalational risk and particle movement dynamics, analyses were performed for particle size categories of 1, 3, 5 and 10 μm aerodynamic diameters. Particles of 1 μm diameter pose the highest inhalational risk, but on the other hand their deposition rate to surfaces is low and, once deposited, they are not readily resuspended. In addition, their content in the released aerosol may be low, as judged from analyses of the 2001 anthrax letters captured by the FBI.[196] The results of the analysis showed that estimation of risk at the generally accepted levels of 10^{-4}–10^{-6} from surface or HVAC filter sampling may be challenging.[194,195] For a retrospective risk of 10^{-6} inferred from deposited 1 μm particles in the Hong *et al.* model,[194] the concentration on floors was calculated to be as low as 0.0035 spores m^{-2} and requires the sampling of an area of 485 m^2 if the hypothetical sampling recovery is 100%. For a more acceptable risk level of 10^{-4}, a sampling area of 4.9 m^2 is required if 100% sampling recovery is achieved and 36 m^2 for a more realistic recovery.[194] Sampling such areas may be feasible using suitable equipment (large wipes and vacuum samplers) and collection of multiple samples. At the other extreme, 10 μm particles behave differently: although they are more relevant to dermal contact or ingestion rather than inhalational risk, they are more readily deposited on surfaces and adhere less firmly once deposited, and may also be more abundant in the aerosol material.[194–196] The sampling effort for assessment of risk from analysis of surface-deposited

10 μm particles is more feasible. For a retrospective inhalational risk of 10^{-4}, a sampling area of 0.1 m^2 will be required if the recovery rate is 100% and 0.485 m^2 for a more realistic recovery rate.[194] For a retrospective risk level of 10^{-4} from 1 μm particles deposited in HVAC filters, the spore concentrations entrapped in the filters was calculated as 9.2×10^4 spores m^{-2}. The sampling areas required are 1.8×10^{-4} m^2 if the recovery efficiency is 100% and 1.4×10^{-3} m^2 for a more realistic recovery rate.[194] Converting these sampling surfaces to those required for risk level of 10^{-6} (multiplying by a factor of 1000 on assuming linearity of risk with dose) yields 0.18 and 1.4 m^2, respectively. The respective sampling areas for the same risk level related to 10 μm particles are 0.031 and 0.15 m^2 if sampling recoveries are 100% or less, as defined above.

These studies are important in developing a methodology for risk assessment on the basis of environmental sampling, which may be useful in the future for the evaluation of repopulation of affected sites and risk-based determination of decontamination goals. It is also a way for evaluating novel sampling strategies such as extraction or vacuum sampling of HVAC filters.[197] However, in its present state the methodology is subject to restrictive assumptions and model uncertainties.[194,195] One such assumption is that the particles in the room are homogeneously mixed from the start of dispersion. For sites in the vicinity of the source, such as a room where an anthrax letter has been opened, this is not true, and until final equilibrium a significant high local risk may be present in the first minutes. Therefore, the models are more applicable to sites remote from the source, where the arriving aerosol is more evenly mixed. The uncertainties affecting retrospective and prospective risk evaluation from environmental samples have been mapped and prioritized for further research.[195] The statistical evaluation showed that inhalational risk evaluation is mostly affected by uncertainties in air exchange rates, breathing rates, particle densities (for retrospective risk) and resuspension rates (for prospective risk). The uncertainty in inputs such as air exchange rates are mainly subject to variability that can be mapped by data collection. Other uncertainties stem also from epistemic gaps, of which the most relevant to inhalational risk are particle resuspension rates and dose–response relations.

3.10 Summary

Inhalational anthrax, the gravest and nearly 100% lethal form of human anthrax without treatment, has been, and still is, a rare occurrence in the industrial or normal public health context, but a real threat in the context of biological terrorism and warfare. The 2001 anthrax letter attacks in the USA revealed the enormous knowledge gaps in the risk evaluation, which is needed for efficient planning of the response to such an attack and mitigation of the long-term consequences. These invoked large investments by governments and other institutions in research directed towards the development of models and methods for risk evaluation and their application in the analysis of response strategies to bioterror anthrax attacks. Dose–response experimental and modeling studies have been reviewed in this chapter with emphasis on their relevance to risk analysis, response planning and mitigation of the effects of these attacks. They were focused on the following themes:

1. *Dose–response modeling of experimental animal data.* The determination of risk as a function of exposure dose was the main theme of the studies reviewed here. The most important issues are the risk from exposure to low doses and extrapolation of human risk from animal experimental data. Risk of low-dose exposure is important in risk analysis-based determination of safe or negligible risk exposure limits and decontamination goals. The most successful modeling strategies were the "single hit" exponential dose–response model and the similar "competing risks" model.[44,46,91,198] The more empirical log-probit model was also found useful for general modeling of human risk, but is regarded by some authors as overestimating low-level risk.[16,17,19] Most of the modeling studies used nonhuman primate data as reasonable default models for human risk assessment, although more studies are required to understand inter-species differences in susceptibility.
2. *Dynamic modeling of multiple exposures.* The exposure pattern to a series of low-level exposure events, rather than a single high-dose exposure, is more realistic in an environment contaminated by low levels of settled spores that may be re-aerosolized when a mechanical disturbance by wind or human activity takes place. Therefore, evaluation of risk for this mode of exposure is an important additive to the low-level risk determination outlined above. Two experimental studies, one with nonhuman primates[164] and the other with rabbits,[152] and two modeling strategies[17,150] have shown that a single-dose exposure may be a limiting case for a more general pattern of repeated doses. Furthermore, the predicted risks of equivalent doses in the two patterns of exposure were not extremely different in the two studies, but in general, the difference may be strongly dependent on the within-host pathogen dynamics and timing and intensity of individual exposures.
3. *Modeling of the incubation period distribution.* The incubation period of inhalational anthrax in the Sverdlovsk incident was found to be of variable length (2–43 days) and successfully modeled as dose dependent. The ability to predict the incubation period and dose is important in predicting the time of appearance of the first few patients, subsequent patient accumulation and hospital surge after an attack and the required duration of prophylactic treatment. Some algorithms for back-calculation of the size, timing, agent quantity and location of an attack were developed on the basis of these models.

4. *Risk analysis in special environments.* Environmental studies carried out in goat hair processing mills, where workers have been constantly exposed to low levels of aerosolized spores, and studies with surrogate bacterial spore dispersion in a simulated office environment, assessing the dynamics of airborne spores and risk following opening of an anthrax letter, were discussed in detail, in order to evaluate risk in those environments. Also, integrative modeling of spore dynamics in an office environment to assess retrospective (at the time of event) and prospective (exposure upon re-entry to a contaminated office) risk from analysis of delayed environmental samples[194–196] were discussed.

The studies reviewed, although demonstrating impressive progress since Haas's exposition of the limited ability to evaluate inhalational exposure risk to anthrax spores from published primate data,[19] have limitations. The first is that they are based on a small number of published primate data sets, are mostly historical and some are of limited utility owing to lack of proper experimental control.[152] However, recent publications of modeling studies incorporated previously unpublished nonhuman primate, rabbit[91,92,141] and guinea-pig[92,133,141] studies of better experimental quality. Exposure studies with well-established small animal models such as rabbits and guinea pigs are easy to perform using state-of-the-art experimental methodologies, but their utility for extrapolation to humans depends on a thorough understanding of the factors that affect interspecies susceptibility. A crucial parameter in determining interspecies susceptibility is alveolar deposition rates of spores, which can be evaluated by using existing particle deposition models, even though the literature points to further research needs.[184] Other issues in differential susceptibility can be addressed using biokinetic and deterministic models.[92,198]

Issues not addressed in this chapter are inter-individual variations in susceptibility to inhalational anthrax and specific risk factors. There is an anecdotal observation on the possible role of background morbidity in susceptibility to inhalational anthrax,[26] but no thorough research on this point is available. Other proposed covariates with susceptibility to inhalational anthrax are age[41] and lifestyle habits such as smoking and drinking.[38] Although it is possible that pulmonary conditions may affect spore clearance and innate immunity and thus increase susceptibility to inhalational anthrax, most of the discussion on this point has been speculative. This is a topic for further study and rather important owing to implications concerning population sectors with special needs.

This chapter was centered on *Bacillus anthracis* and inhalational anthrax, as a major bioterror threat with unique biological properties, relating to the spore as the infectious agent and its outstanding environmental persistence. As stated above, the theory and practice principles in the field of quantitative microbial risk assessment (QMRA), as coined by Haas, have originated from work on food- and water-borne pathogens. They have been applied to *B. anthracis* and extended later to other bioterror high-priority agents.[36,163,195,199–201]

References

1. M. Mock and A. Fouet, Anthrax, *Annu. Rev. Microbiol.*, 2001, **55**, 647–671.
2. M. Hugh-Jones and J. Blackburn, The ecology of Bacillus anthracis, *Mol. Aspects Med.*, 2009, **30**, 356–367.
3. T. V. Inglesby, D. A. Henderson, J. G. Bartlett, M. S. Ascher, E. Eitzen, A. M. Friedlanderr, J. Hauer, J. McDade, M. T. Osterholm, T. O'Toole, G. Parker, T. M. Perl, P. K. Russell and K. Tonat, Anthrax as a biological weapon. Medical and public health management, *J. Am. Med. Assoc.*, 1999, **281**, 1735–1745.
4. T. V. Inglesby, T. O'Toole, D. A. Henderson, J. G. Bartlett, M. S. Ascher, E. Eitzen, A. M. Friedlanderr, J. Geberding, J. Hauer, J. Hughes, J. McDade, M. T. Osterholm, G. Parker, T. M. Perl, P. K. Russell and K. Tonat, Anthrax as a biological weapon, 2002. Updated recommendations for management, *J. Am. Med. Assoc.*, 2002, **287**, 2236–2252.
5. T. Berger, M. Kassirer and A. Aran, Injectional anthrax – new presentation of an old disease, *Euro Surveill.*, 2014, **19**, 20877, http://www.eurosurveillance.org/ViewArticle.aspx?Articleid=20877, last accessed on February 16, 2015.
6. J.-E. C. Holty, D. M. Bravata, H. Liu, R. A. Olshen, K. M. McDonald and D. K. Owens, Systematic review: a century of inhalational anthrax cases from 1900 to 2005, *Ann. Intern. Med.*, 2006, **144**, 270–280.
7. R. A. Zilinskas, Iraq's biological weapons, *J. Am. Med. Assoc.*, 1997, **278**, 418–424.
8. K. Alibek and S. Handelman, *Biohazard: The Chilling True Story of the largest Covert Biological Weapons Program in the World*, Random House, New York, NY, 1999.
9. H. Takahashi, P. Keim, A. F. Kaufmann, C. Keys, K. L. Smith, K. Taniguchi, S. Inouye and T. Kurata, Bacillus anthracis incident, Kameido, Tokyo, 1993, *Emerging Infect. Dis.*, 2004, **10**, 117–120.
10. L. A. Cole, *The Anthrax Letters: A Bioterrorism Expert Interrogates the Attacks that Shocked America*, Skyhorse Publishing, New York, NY, 2009.
11. J. A. Jernigan, D. S. Stephens, D. A. Ashford, C. Omenaca, M. S. Topiel, M. Galbraith, M. Tapper, T. L. Fisk, S. Zaki, T. Popovic, R. F. Meyer, C. P. Quinn, S. A. Harper, S. K. Fridkin, J. J. Sejvar, C. W. Shepard, M. McConnell, J. Guarner, W.-J. Shieh, J. M. Malecki, J. L. Gerberding, J. M. Hughes, B. A. Perkins and members of the Anthrax Bioterrorism Investigation Team, Bioterrorism-related inhalational anthrax: the first 10 cases reported in the United States, *Emerging Infect. Dis.*, 2001, **7**, 933–944.
12. D. A. Canter, Addressing residual risk issues at anthrax cleanups: How clean is safe?, *J. Toxicol. Environ. Health, Part A*, 2005, **68**, 1017–1032.
13. A. S. Sabelnikov, V. Zhukov and C. R. Kempf, Airborne exposure limits for chemical and biological warfare agents: Is everything set and clear?, *Int. J. Environ. Health Res.*, 2006, **16**, 241–253.
14. M. E. Coleman, B. Thran, S. Morse, M. Hugh–Jones and S. Massulik, Inhalation anthrax: dose response and risk analysis, *Biosecur. Bioterror.*, 2008, **6**, 147–159.

15. E. Raber, The challenge of determining the need for remediation following a wide- area biological release, *Biosecur. Bioterror.*, 2011, **9**, 257–261.
16. D. A. Wilkening, Sverdlovsk revisited: Modeling human inhalation anthrax, *Proc. Natl. Acad. Sci. U. S. A.*, 2006, **103**, 7589–7594.
17. D. J. A. Toth, A. V. Gundlapalli, W. A. Schell, K. Bulmahn, T. E. Walton, C. W. Woods, C. Coghill, F. Gallegos, M. H. Samore and F. R. Adler, Quantitative models of the dose–response and time course of inhalational anthrax in humans, *PLoS Pathog.*, 2013, **9**, e1003555.
18. J. Ho and S. Duncan, Estimating aerosol hazards from anthrax letters, *J. Aerosol Sci.*, 2005, **36**, 701–719.
19. C. N. Haas, On the risk of mortality to primates exposed to anthrax spores, *Risk Anal.*, 2002, **22**, 189–193.
20. T. A. Bartrand, M. H. Weir and C. N. Haas, Dose–response models for inhalation of *Bacillus anthracis* spores: interspecies comparison, *Risk Anal.*, 2008, **28**, 1115–1124.
21. P. S. Brachman and F. R. Fekety, Industrial anthrax, *Ann. N. Y. Acad. Sci.*, 1958, **70**, 574–584.
22. P. S. Brachman, Inhalation anthrax, *Ann. N. Y. Acad. Sci.*, 1980, **353**, 83–93.
23. J. J. Walsh, N. Pesik, C. P. Quinn, V. Urdaneta, C. A. Dykewicz, A. E. Boyer, J. Guarner, P. Wilkins, K. J. Norville, J. R. Barr, S. R. Zaki, J. B. Patel, S. R. Reagan, J. L. Pirkle, T. A. Treadwell, N. R. Messonier, L. D. Rotz, R. F. Meyer and D. S. Stephens, A case of naturally acquired inhalation anthrax: Clinical care and analysis of anti-protective antigen immunoglobulin G and lethal factor, *Clin. Infect. Dis.*, 2007, **44**, 968–971.
24. S. Anaraki, S. Addiman, G. Nixon, D. Krahé, R. Ghosh, T. Brooks, G. Lloyd, R. Spencer, A. Walsh, B. McCloskey and N. Lightfoot, Investigation and control measures following a case of inhalation anthrax in East London in a drum maker and drummer, October 2008, *Euro Surveill.*, 2008, **13**, 19076, http://www.eurosurveillance.org/ViewArticle.aspx?Articleid=19076, last accessed on February 16, 2015.
25. T. Q. Nguyen, N. Clark and The 2006 NYC Anthrax Working Group, Public health and environmental response to the first case of naturally acquired inhalational anthrax in the United States in 30 years: Infection of a New York City resident who worked with dried animal hides, *J. Public Health Manag. Pract.*, 2010, **16**, 189–200.
26. P. S. Brachman, J. S. Pagano and W. S. Albrink, Two cases of fatal inhalation anthrax, one associated with sarcoidosis, *N. Engl. J. Med.*, 1961, **265**, 203–208.
27. F. M. LaForce, F. H. Bumford and J. C. Feeley, Epidemiologic study of a fatal case of inhalation anthrax, *Arch. Environ. Health*, 1969, **18**, 798–805.
28. J. Griffith, D. Blaney, S. Shadomy, M. Lehman, N. Pesik, S. Tostenson, L. Delaney, R. Tiller, A. DeVries, T. Gomez, M. Sullivan, C. Blackmore, D. Stanek, R. Lynfield and The Anthrax Investigation Team, Investigation of inhalation anthrax case, United States, *Emerging Infect. Dis.*, 2014, **20**, 280–283.

29. M. D. Sprenkle, J. Griffith, W. Marinelli, A. E. Boyer, C. P. Quinn, N. T. Pesik, A. Hoffmaster, J. Keenan, B. A. Juni and D. D. Blaney, Lethal factor and anti-protective antigen IgG levels associated with inhalation anthrax, Minnesota, USA, *Emerging Infect. Dis.*, 2014, **20**, 310–314.
30. L. M. Wein, D. L. Craft and E. H. Kaplan, Emergency response to an anthrax attack, *Proc. Natl. Acad. Sci. U. S. A.*, 2003, **100**, 4346–4351.
31. L. M. Wein, Y. Liu and T. J. Leighton, HEPA/vaccine plan for indoor anthrax remediation, *Emerging Infect. Dis.*, 2004, **11**, 69–76.
32. D. A. Wilkening, Modeling the incubation period of inhalational anthrax, *Med. Decis. Making*, 2008, **28**, 593–605.
33. D. B. Levin and G. Valdares de Amoram, Potential for aerosol dissemination of biological weapons: lessons from biological control of insects, *Biosecur. Bioterror.*, 2003, **1**, 37–42.
34. S. Van Cuyk, A. Deshpande, A. Hollander, N. Duval, L. Ticknor, J. Layshock, L. Gallegos-Graves and K. Omberg, Persistence of *Bacillus thuringiensis* subsp. *kurstaki* in urban environments following spraying, *Appl. Environ. Microbiol.*, 2011, **77**, 7954–7961.
35. A. Watson and D. Keir, Information on which to base assessment of risk from environments contaminated with anthrax spores, *Epidemiol. Infect.*, 1994, **113**, 479–490.
36. G. L. Rickmeier, G. E. McClellan and G. A. Anno, Biological warfare human response modeling, *Mil. Oper. Res.*, 2001, **6**, 35–56.
37. L. M. Wein and D. L. Craft, Evaluation of public health interventions for anthrax: A report to the secretary's council on public health preparedness, *Biosec. Bioter.*, 2005, **3**, 348–356.
38. M. Meselson, J. Guillemin, M. Hugh-Jones, A. Langmuir, I. Popova, A. Shelokov and O. Yampolskaya, The Sverdlovsk anthrax outbreak, *Science*, 1994, **266**, 1202–1208.
39. H. N. Glassman, Discussion on industrial inhalational anthrax, *Bacteriol. Rev.*, 1966, **30**, 659.
40. C. L. Peters and D. M. Hartley, Anthrax inhalation and lethal human infection, *Lancet*, 2002, **359**, 710–711.
41. G. F. Webb and M. J. Blaser, Mailborne transmission of anthrax: Modeling and implications, *Proc. Natl. Acad. Sci. U. S. A.*, 2002, **99**, 7027–7032.
42. C. R. Lyons, J. Lovchik, J. Hutt, M. F. Lipscomb, E. Wang, S. Heninger, L. Berliba and K. Garrison, Murine model of pulmonary anthrax: kinetics of dissemination, histopathology and mouse strain susceptibility, *Infect. Immun.*, 2004, **72**, 4801–4809.
43. I. Nishiura, Early efforts on modeling the incubation period of infectious diseases with an acute course of illness, *Emerg. Themes Epidemiol.*, 2007, **4**, 2, DOI: 10.1186/1742-7622-4-2.
44. R. Brookmeyer, E. Johnson and R. Bollinger, Modeling the optimum duration of antibiotic prophylaxis in an anthrax outbreak, *Proc. Natl. Acad. Sci. U. S. A.*, 2003, **100**, 10129–10132.

45. D. W. Henderson, S. Peacock and F. C. Belton, Observations on the prophylaxis of experimental pulmonary anthrax in the monkey, *J. Hyg.*, 1956, **54**, 28–36.
46. R. Brookmeyer, E. Johnson and S. Barry, Modelling the incubation period of anthrax, *Stat. Med.*, 2005, **24**, 531–542.
47. M. L. Dourson, S. P. Felter and D. Robinson, Evolution of science-based uncertainty factors in noncancer risk assessment, *Regul. Toxicol. Pharmacol.*, 1996, **24**, 108–120.
48. H. Shuval, Y. Lampert and B. Fattal, Development of a risk assessment approach for evaluating wastewater reuse standards for agriculture, *Water Sci. Technol.*, 1997, **35**, 15–20.
49. E. Raber, A. Jin, K. Noonan, R. McGuire and R. D. Kirvel, Decontamination issues for chemical and biological agent: how clean is clean enough?, *Int. J. Environ. Health Res.*, 2001, **11**, 128–148.
50. C. N. Haas, A. Thayyar-Madabusi, J. B. Rose and C. P. Gerba, Development of a dose–response relationship for *Escherichia coli* O157:H7, *Int. J. Food Microbiol.*, 2000, **1748**, 153–159.
51. U.S. Environmental Protection Agency (US EPA), *Benchmark dose analysis for Bacillus anthracis inhalation exposures in the nonhuman primate and application to risk-based decision making*, EPA 600/R-10/138, December 2010.
52. U.S. Environmental Protection Agency (US EPA), *Review of Bacillus anthracis (Anthrax) Studies for Dose-Response Modeling to Estimate Risk*, EPA/600/S-12/681, October 2012.
53. R. Sinclair, S. A. Boone, D. Greenberg, P. Keim and C. P. Gerba, *Appl. Environ. Microbiol.*, 2008, **74**, 555–563.
54. R. J. Manchee, M. G. Broster, J. Melling, R. M. Henstridge and A. J. Stagg, *Bacillus anthracis* on Gruinard Island, *Nature*, 1981, **294**, 254–255.
55. R. J. Manchee, M. G. Broster, A. J. Stagg and S. E. Hibbs, Formaldehyde solution effectively inactivates spores of *Bacillus anthracis* on the Scottish island of Gruinard, *Appl. Environ. Microbiol.*, 1994, **60**, 4167–4171.
56. P. S. Keim and D. M. Wagner, Humans, evolutionary and ecologic forces that shaped the phylogeography of recently emerged diseases, *Nat. Rev. Microbiol.*, 2009, 7, 813–821.
57. W. J. Cromartie, W. L. Bloom and D. W. Watson, Studies on infection with Bacillus anthracis I. A histopathological study of the skin lesions produced by *B. anthracis* in susceptible and resistant animal species, *J. Infect. Dis.*, 1947, **80**, 1–13.
58. J.-P. Corre, A. Piris-Gimenes, M. Moya-Nilges, G. Jouvion, A. Fouet, I. J. Glomski, M. Mock, J. C. Sirard and P. L. Goosens, In vivo germination of *Bacillus anthracis* spores during murine cutaneous infection, *J. Infect. Dis.*, 2013, **207**, 450–457.
59. F. Shafa, B. J. Moberly and P. Gerhardt, Cytological features of anthrax spores phagocytosed in vitro by rabbit alveolar macrophages, *J. Infect. Dis.*, 1966, **116**, 401–413.

60. C. Guidi-Rontani, M. Weber-Levy, Labruyère and M. Mock, Germination of *Bacillus anthracis* spores within alveolar macrophages, *Mol. Microbiol.*, 1999, **31**, 9–17.
61. C. Guidi-Rontani, M. Levy, H. Ohayon and M. Mock, Fate of germinated *Bacillus anthracis in* primary murine macrophages, *Mol. Microbiol.*, 2001, **42**, 931–938.
62. T. C. Dixon, A. A. Fadl, T. M. Koehler, J. L. Swanson and P. C. Hanna, Early *Bacillus anthracis*-macrophage interactions: Intracellular survival and escape, *Cell. Microbiol.*, 2000, **2**, 453–463.
63. K. C. Brittingham, G. Ruthel, R. G. Panchal, C. L. Fuller, W. J. Ribot, T. A. Hoover, H. A. Young, A. O. Anderson and S. Bavari, Dendritic cells endocytose *Bacillus anthracis* spores: Implications for anthrax pathogenesis, *J. Immunol.*, 2005, **174**, 5545–5552.
64. A. Cleret, A. Quesnal-Hellmann, A. Vallon-Eberhard, B. Verrier, S. Jung, D. Vidal, J. Mathieu and J.-N. Tournier, Lung dendritic cells rapidly mediate anthrax spore entry through the pulmonary route, *J. Immunol.*, 2007, **178**, 7994–8001.
65. L. S. Shetron-Rama, A. C. Herring-Palmer, G. B. Huffnagle and P. C. Hanna, Transport of *Bacillus anthracis* from the lungs to the draining lymph nodes is a rapid process facilitated by CD11c+ cells, *Microb. Pathog.*, 2010, **49**, 38–46.
66. R. E. Lincoln, D. R. Hodges, F. Klein, B. G. Mahlandt, W. I. Jones Jr., B. W. Haines, M. A. Rhian and J. S. Walker, Role of the lymphatics in the pathogenesis of anthrax, *J. Infect. Dis.*, 1965, **115**, 481–494.
67. Z. P. Wiener and I. J. Glomski, Updating perspectives on the initiation of Bacillus anthracis growth and dissemination through its host, *Infect. Immun.*, 2012, **80**, 1626–1633.
68. J. A. T. Young and R. J. Collier, Anthrax toxin: Receptor binding, internalization, pore formation and translocation, *Annu. Rev. Biochem.*, 2007, **76**, 243–265.
69. J.-N. Tournier, S. Rossi Paccani, A. Quesnel-Hellman and C. T. Baldari, Anthrax toxins: A weapon to systematically dismantle the host immune defenses, *Mol. Aspects Med.*, 2009, **30**, 456–466.
70. C. W. Hicks, X. Cui, D. Sweeney, Y. Li, A. Barochia and P. Q. Eichaker, The potential contributions of lethal and edema toxins to the pathogenesis of anthrax associated shock, *Toxins*, 2011, **3**, 1185–1202, DOI: 10.3390/toxins3091185.
71. A. Guichard, V. Nizet and E. Bier, New insights into the biological effects of anthrax toxins: linking cellular to organismal responses, *Micob. Infect*, 2012, **14**, 97–118.
72. S. Liu, Y. Zhang, M. Moayeri, D. Crown, R. J. Fattah, A. N. Wein, Z.-X. Yu, T. Finkel and S. H. Leppla, Key tissue targets responsible for anthrax-toxin-induced lethality, *Nature*, 2013, **501**, 63–68.
73. H. Levy, I. Glinert, S. Weiss, E. Bar-David, A. Sittner, J. Schlomovitz, Z. Altboum and D. Kobiler, The central nervous system as target of *Bacillus anthracis* toxin independent virulence in rabbits and guinea pigs, *PLoS One*, 2014, **9**, e112319, DOI: 10.1371/journal.pone.0112319.

74. D. A. Sweeney, C. W. Hicks, X. Cui, Y. Li and P. O. Eichacker, Anthrax infection, *Am. J. Respir. Crit. Care Med.*, 2011, **184**, 1333–1341.
75. K. E. Remy, P. Qiu, Y. Li, X. Cui and P. Q. Eichacker, *B. anthracis* associated cardiovascular dysfunction and shock: the potential contribution of both nontoxin and toxin components, *BMC Med.*, 2013, **11**, 217, DOI: 10.1186/1741-7015-11-217.
76. J. A. Lovchik, M. Drysdale, T. M. Koehler, J. A. Hurt and R. C. Lyons, Expression of either lethal toxin or edema toxin by *Bacillus anthracis* is sufficient for virulence in a rabbit model of inhalational anthrax, *Infect. Immun.*, 2012, **80**, 2414–2425.
77. H. Levy, S. Weiss, Z. Altboum, J. Schlomovitz, I. Glinert, A. Sittner, A. Shafferman and D. Kobiler, Differential contribution of Bacillus anthracis toxins to pathogenicity in two animal models, *Infect. Immun.*, 2012, **80**, 2623–2631.
78. H. Levy, I. Glinert, S. Weiss, A. Sittner, J. Schlomovitz, Z. Altboum and D. Kobiler, Toxin-independent virulence of *Bacillus anthracis* in rabbits, *PLoS One*, 2014, **9**, e84947, DOI: 10.1371/journal.pone.009947.
79. D. Lucey, Anthrax, in *Mandel, Douglas and Bennett's Principles and Practice of Infectious Diseases*, ed. G. L. Mandell, J. E. Bennett and R. Dolin, Elsevier Churchill Livingstone, Philadelphia, 6th edn, 2005, ch. 324, pp. 3618–3624.
80. F. A. Abramova, L. M. Grinberg, O. V. Yampolskaya and D. H. Walker, Pathology of inhalational anthrax in 43 cases from the Sverdlovsk outbreak in 1989, *Proc. Natl. Acad. Sci. U. S. A.*, 1993, **90**, 2291–2294.
81. L. M. Grinberg, F. A. Abramova, O. V. Yampolskaya and D. H. Walker, Quantitative pathology of inhalational anthrax I: quantitative microscopic findings, *Mod. Pathol.*, 2001, **14**, 482–495.
82. J. Guarner, J. A. Jernigan, W.-J. Shies, K. Tatti, L. M. Flannagan, D. S. Stephens, T. Popovic, D. A. Ashford, B. A. Perkins, S. R. Zaki and The Inhalational Anthrax Pathology Working Group, Pathology and pathogenesis of bioterrorism-related inhalational anthrax, *Am. J. Pathol.*, 2003, **163**, 701–709.
83. A. W. Artenstein and S. M. Opal, Novel approaches to the treatment of systemic anthrax, *Clin. Infect. Dis.*, 2012, **54**, 1148–1161.
84. T.-S. Migone, G. M. Subramanian, J. Zhong, L. M. Healey, L. Corey, M. Develeraja, L. Lo, S. Ullrich, J. Zimmerman, A. Chen, M. Lewis, G. Meister, K. Gillum, D. Sanford, J. Mott and S. D. Bolmer, Taxibacumab for the treatment of inhalational anthrax, *N. Engl. J. Med.*, 2009, **361**, 135–144.
85. C. J. Roy, D. S. Reed and J. A. Hutt, Aerobiology and inhalation exposure to biological select agents and toxins, *Vet. Pathol.*, 2010, **47**, 779–789.
86. H. C. Yeh, R. F. Phalen and O. G. Raabe, Factors influencing the deposition of inhaled particles, *Environ. Health Perspect.*, 1976, **15**, 147–156.
87. M. Lippmann, D. B. Yeates and R. E. Albert, Deposition, retention and clearance of inhaled particles, *Br. J. Ind. Med.*, 1980, **37**, 337–362.
88. M. H. Weir and C. N. Haas, A model for in-vivo delivered dose estimation for inhaled *Bacillus anthracis* spores in humans with interspecies extrapolation, *Environ. Sci. Technol.*, 2011, **45**, 5828–5833.

89. S. Guha, P. Hariharan and M. R. Myers, Enhancement of ICRP's lung deposition model for pathogenic bioaerosols, *Aerosol Sci. Technol.*, 2014, **48**, 1226–1235.
90. B. W. Gutting, T. L. Nichols, S. R. Channel, J. M. Gearhart, A. A. Andrews, A. E. Berger, R. S. Mackie, B. J. Watson, S. C. Taft, K. A. Overheim and R. L. Sherwood, Inhalational anthrax (Ames aerosol) in naïve and vaccinated New Zealand White rabbits: characterizing the spread of bacteria from lung deposition to bacteremia, *Front. Cell. Infect. Microbiol.*, 2012, **2**, 87, DOI: 10.3389/fcmb.2012.00087.
91. B. W. Gutting, D. R. Marchette, G. A. Andrews, A. Director-Myska, S. R. Channel, D. Wolfe, A. E. Berger, R. S. Mackie, B. J. Watson and A. Ruhkin, Modeling low-dose mortality and disease incubation period of inhalational anthrax in the rabbit, *J. Theor. Biol.*, 2013, **329**, 20–31.
92. W. B. Gutting, Deterministic models of inhalational anthrax in New Zealand White rabbits, *Biosec. Bioterror.*, 2014, **12**, 29–41.
93. O. G. Raabe, M. A. Al-Bayati, S. V. Teague and A. Rasolt, Regional deposition of inhaled monodisperse coarse and fine aerosol particles in small laboratory animals, *Ann. Occup. Hyg.*, 1988, **32**(**suppl. 1**), 53–63.
94. R. J. Thomas, Particle size and pathogenicity in the respiratory tract, *Virulence*, 2013, **8**, 847–858.
95. H. A. Druett, D. W. Henderson, L. Packman and S. Peacock, Studies on respiratory infection I. The influence of particle size on respiratory infection with anthrax spores, *J. Hyg.*, 1953, **51**, 359–371.
96. D. W. Henderson, An apparatus for the study of airborne infection, *J. Hyg.*, 1952, **50**, 53–68.
97. H. A. Druett and K. R. May, A wind tunnel for the study of airborne infections, *J. Hyg.*, 1952, **50**, 69–81.
98. S. C. Taft and S. A. Hines, Benchmark dose analysis for *Bacillus anthracis* inhalation exposures in nonhuman primates, *Risk Anal.*, 2012, **32**, 1750–1768.
99. G. J. Harper and J. D. Morton, The respiratory retention of bacterial aerosols: Experiments with radioactive spores, *J. Hyg.*, 1953, **51**, 372–378.
100. G. Young, M. R. Zelle and R. E. Lincoln, Respiratory pathogenicity of Bacillus anthracis spores I. Methods of study and observations of pathogenesis, *J. Infect. Dis.*, 1946, **79**, 233–246.
101. R. Thomas, C. Davies, A. Nunez, S. Hibbs, H. Flick-Smith, L. Eastaugh, S. Smither, A. Gates, P. F. Oyston and S. Eley, Influence of particle size on the pathology and efficacy of vaccination in a murine model of inhalational anthrax, *J. Med. Microbiol.*, 2010, **59**, 1415–1427.
102. I. J. Glomski, A. Piris-Gimenez, M. Huerre, M. Mock and P. L. Goossens, Primary involvement oif pharynx and Peyer's patches in inhalational and intestinal anthrax, *PLoS Pathog.*, 2007, **3**, e76, DOI: 10.1371/journal.ppath.003076.
103. S. A. Jenkins and Y. Xu, Characterization of *Bacillus anthracis* persistence in vivo, *PLoS One*, 2013, **8**, e66177, DOI: 10.1371/journal.pone.0066177.

104. B. H. Russell, R. Vasan, D. R. Keene, T. M. Kohler and Y. Xu, Potential dissemination of *Bacillus anthracis* utilizing human lung epithelial cells, *Cell. Microbiol.*, 2008, **10**, 945–957.
105. B. H. Russell, Q. Liu, S. A. Jenkins, M. J. Tuvim, B. F. Dickey and Y. Xu, In vivo demonstration and quantification of intracellular *Bacillus anthracis* in lung epithelial cells, *Infect. Immun.*, 2008, **76**, 3975–3983.
106. Q. Xue, C. Gu, J. Rivera, M. Höök, X. Chen, A. Pozzi and Y. Xu, Entry of Bacillus anthracis spores into epithelial cells is mediated by the spore surface protein BclA, integrin α2β1 and complement component C1q, *Cell. Microbiol.*, 2011, **13**, 620–634.
107. C. K. Cote, N. Van Roojten and S. L. Welkos, Roles of macrophages and neutrophils in the early host response to Bacillus anthracis spores in a mouse model of infection, *Infect. Immun.*, 2006, **74**, 469–480.
108. G. Ruthel, W. J. Ribot, S. Bavari and T. A. Hoover, Time-lapse confocal imaging of development of *Bacillus anthracis* in macrophages, *J. Infect. Dis.*, 2004, **189**, 1313–1316.
109. T. J. Kang, M. J. Fenton, M. A. Wiener, S. Hibbs, S. Basu, L. Baillie and A. S. Cross, Murine macrophages kill the vegetative form of *Bacillus anthracis*, *Infect. Immun.*, 2005, **73**, 7495–7501.
110. H. Hu, Q. Sa, T. M. Koehler, A. I. Aronson and D. Zou, Inactivation of *Bacillus anthracis* spores in murine primary macrophages, *Cell. Microbiol.*, 2006, **8**, 1634–1642.
111. I. M. Gut, B. Tamilselvan, A. M. Prouty, B. Stojkovic, S. Czeschin, W. A. van der Donk and S. R. Blanke, Bacillus anthracis spore interactions with mammalian cells: Relationship between germination state and the outcome of in vitro, *BMC Microbiol.*, 2011, **11**, 46, DOI: 10.1186/1471-2180-11-46.
112. D. J. Finney, Probit Analysis, *A Statistical Treatment of the Sigmoid Response Curve*, Cambridge University Press, 2nd edn, 1962.
113. C. N. Haas, Estimation of risk due to low doses of microorganisms: A comparison of alternative methodologies, *Am. J. Epidemiol.*, 1983, **118**, 573–582.
114. R. L. Kodell, S.-H. Kang and J. J. Chen, Statistical models of health risk due to microbial contamination of foods, *Environ. Ecol. Stat.*, 2002, **9**, 259–271.
115. D. Kobiler, S. Weiss, H. Levy, M. Fisher, A. Machhaly, A. Pass and Z. Altboum, Protective antigen as a correlative marker for anthrax in animal models, *Infect. Immun.*, 2006, **74**, 5871–5876.
116. C. A. Rossi, M. Ulrich, S. Norris, D. S. Reed, L. M. Pitt and E. Leffel, Identification of a surrogate marker for infection in the African green monkey model of inhalation anthrax, *Infect. Immun.*, 2008, **76**, 5790–5801.
117. A. E. Boyer, C. P. Quinn, A. R. Hoffmeister, T. R. Kozel, E. Saile, C. K. Marston, A. Percival, B. D. Plikaytis, A. R. Woolfit, M. Gallegos, P. Sabourin, L. G. McWilliams, J. L. Pirkle and J. R. Barr, Kinetics of lethal factor and poly-D-glutamic acid antigenemia during inhalation anthrax in thee rhesus macaques, *Infect. Immun.*, 2009, **77**, 3432–3441.

118. J. E. Comer, B. D. Ray, L. N. Hennig, G. V. Stark, R. E. Barnewell, J. M. Mott and G. T. Meister, Characterization of a therapeutic model of inhalational anthrax using an increase in body temperature in New Zealand White rabbits as a trigger for treatment, *Clin. Vaccine Immunol.*, 2012, **19**, 1517–1525.
119. R. L. Buchanan, J. L. Smith and W. Long, Microbial risk assessment: Dose–response relations and risk characterization, *Int. J. Food Microbiol.*, 2000, **58**, 159–172.
120. G. N. Sze To and C. Y. H. Chao, Review and comparison between the Wells-Riley and dose–response approaches to risk assessment of infectious respiratory diseases, *Indoor Air*, 2010, **20**, 2–16.
121. U.S. Environmental Protection Agency (US EPA), *Benchmark Dose Technical Guidance*, EPE/100/R-12-001, June 2012.
122. C. N. Haas, J. B. Rose, C. Gerba and S. Regli, Risk assessment of virus in drinking water, *Risk Anal.*, 1993, **13**, 545–552.
123. P. F. M. Teunis, N. J. D. Nagelkerke and C. N. Haas, Dose response models for infectious gastroenteritis, *Risk Anal.*, 1999, **19**, 1251–1260.
124. H. Moon, J. J. Chen, D. W. Gaylor and R. L. Kodell, A comparison of microbial dose–response models fitted to human data, *Reg. Toxicol. Pharmacol.*, 2004, **40**, 177–184.
125. G. G. Meynell and B. A. D. Stocker, Some hypotheses on the aetiology of fatal infections in partially resistant hosts and their application to mice challenged with *Salmonella pratyphi*-B or *Salmonella typhimurium* by intraperitoneal injection, *J. Gen. Microbiol.*, 1957, **16**, 38–58.
126. G. G. Meynell, The applicability of the hypothesis of independent action to fatal infections in mice given *Salmonella typhimurium* by mouth, *J. Gen. Microbiol.*, 1957, **16**, 396–404.
127. M. R. Zelle, R. E. Lincoln and G. Young Jr., Respiratory pathogenicity of *Bacillus anthracis* spores II. Genetic variation in respiratory pathogenicity and invasiveness of colonial variants in *B. anthracis*, *J. Infect. Dis.*, 1946, **79**, 247–253.
128. R. D. Plaut, V. K. Kelly, G. M. Lee, S. Stibitz and T. J. Merkel, Dissemination bottlenecks in a murine model of inhalational anthrax, *Infect. Immun.*, 2012, **80**, 3189–3193.
129. D. E. Lowe, S. M. C. Ernst, C. Zito, J. Ya and I. J. Glomski, Bacillus anthracis has two independent bottlenecks that are dependent on the portal of entry in an intranasal model of inhalation infection, *Infect. Immun.*, 2013, **81**, 4408–4420.
130. N. A. Twenhafel, Pathology of inhalational anthrax in animal models, *Vet. Pathol.*, 2010, **47**, 819–830.
131. A. J. Phipps, C. Pemanandan, R. E. Barnewall and M. Lairmore, Rabbit and nonhuman primate models of toxin-targeting human anthrax vaccines, *Microbiol. Mol. Biol. Rev.*, 2004, **68**, 617–629.
132. Z. Altboum, Y. Gozes, A. Barnea, A. Pass, M. White and D. Kobiler, Postexposure prophylaxis against anthrax: Evaluation of various treatment regimes in intranasally infected guinea pigs, *Infect. Immun.*, 2002, **70**, 6231–6241.

133. V. Savransky, D. C. Sanford, E. Syar, J. L. Austin, K. P. Tordoff, M. S. Anderson, G. V. Stark, R. E. Barnewall, C. M. Briscoe, L. Lemiale-Biérinx, S. Park, B. Ionin and M. H. Skiadopoulos, Pathology and pathophysiology of inhalational anthrax in a guinea pig model, *Infect. Immun.*, 2013, **81**, 1152–1163.
134. P. S. Brachman, A. F. Kaufman and F. G. Dalldorf, Industrial inhalation anthrax, *Bacteriol. Rev.*, 1966, **30**, 646–659.
135. R. L. Riley, Airborne pulmonary tuberculosis, *Bacteriol. Rev.*, 1961, **25**, 243–248.
136. D. Vasconceslos, R. Barnewall, M. Babin, R. Hunt, J. Estep, C. Nielsen, R. Carnes and J. Carney, Pathology of inhalational anthrax in cynomolgus monkeys (*Macaca fascicularis*), *Lab. Invest.*, 2003, **83**, 1201–1209.
137. A. M. Friedlander, S. L. Welkos, M. L. M. Pitt, J. W. Ezzell, P. L. Worsham, K. J. Rose, B. E. Ivins, J. R. Lowe, G. B. Howe, P. Mikesell and W. B. Lawrence, Postexposure prophylaxis against experimental inhalation anthrax, *J. Infect. Dis.*, 1993, **167**, 1239–1242.
138. D. L. Fritz, N. K. Jaax, W. B. Lawrence, K. J. Davis, M. L. M. Pitt, J. W. Ezzell and A. M. Friedlander, Pathology of experimental inhalation anthrax in the rhesus monkey, *Lab. Invest.*, 1995, **73**, 691–702.
139. C. A. Gleiser, C. C. Berdjis, H. A. Hartman and W. S. Gochenour, Pathology of experimental respiratory anthrax in Macaca mulatta, *Br. J. Exp. Pathol.*, 1963, **44**, 416–426.
140. W. S. Albrink and R. J. Goodlow, Experimental inhalation anthrax in the chimpanzee, *Am. J. Pathol.*, 1959, **35**, 1055–1065.
141. W. B. Gutting, A. Ruhkin, R. S. Mackie, D. R. Marchette and B. Thran, Evaluation of inhaled versus deposited dose using the exponential dose–response model for inhalational anthrax in nonhuman primate, rabbit and guinea pig, *Risk Anal.*, 2015, **35**, 811–827.
142. G. M. Zaucha, M. L. M. Pitt, J. Estep, B. E. Ivins and A. M. Friedlander, The pathology of experimental anthrax in rabbits exposed by inhalation and subcutaneous inoculation, *Arch. Pathol. Lab. Med.*, 1998, **112**, 982–992.
143. U.S. Environmental Protection Agency (US EPA), *Acute Low Dose Bacillus anthracis Ames Inhalational Exposure in the Rabbit*, EPA 600/R-11/075, September 2011.
144. P. K. Dull, K. E. Wilson, B. Kournikakis, E. A. Whitney, C. A. Boulet, J. Y. W. Ho, J. Ogston, M. R. Spence, M. M. Mckenzie, M. A. Phelan, T. Popovic and D. Ashford, *Bacillus anthracis* aerosolization associated with a contaminated mail sorting machine, *Emerging Infect. Dis.*, 2002, **8**, 1044–1047.
145. C. P. Weis, A. J. Intrepido, A. K. Miller, P. G. Cowin, M. A. Durno, J. S. Gebhardt and R. Bull, Secondary aerosolization of viable *Bacillus anthracis* spores in a contaminated U.S. senate office, *J. Am. Med. Assoc.*, 2002, **288**, 2853–2858.
146. E. R. Carr Jr. and R. R. Rew, Recovery of *Bacillus anthracis* form the nose and throat of apparently healthy workers, *J. Infect. Dis.*, 1957, **100**, 169–171.

147. C. N. Dahlgren, L. M. Buchanan, H. M. Decker, S. W. Freed, C. R. Phillips and P. S. Brachman, *Bacillus anthracis* aerosols in goat hair processing mills, *Am. J. Hyg.*, 1960, **72**, 24–31.
148. P. S. Norman, J. G. Ray Jr., P. S. Brachman, S. A. Plotkin and J. S. Pagano, Serologic testing for anthrax antibodies in workers in a goat hair processing mill, *Am. J. Hyg.*, 1960, **72**, 32–37.
149. J. M. Pujol, J. E. Eisenberg, C. N. Haas and J. S. Koopman, The effect of ongoing exposure dynamics in dose–response relationships, *PLoS Comput. Biol.*, 2009, **5**, e1000399, DOI: 10.1371/journal.pcbi.1000399.
150. B. T. Mayer, J. S. Koopman, E. L. Ionides, J. M. Pujol and J. N. S. Eisenberg, A dynamic dose–response model to account for exposure patterns in risk assessment: a case study in inhalation anthrax, *J. R. Soc., Interface*, 2011, **8**, 506–517, see also supplement available at http://rsif.royalsocietypublishing.org/ccontent/suppl/2010/11/09/rsif.2010.0491.DC1.html, last accessed on February 16, 2015.
151. R. E. Barnewall, J. Comer, B. D. Miller, B. W. Gutting, D. N. Wolfe, A. E. Director-Myska, T. L. Nichols and S. C. Taft, Achieving consistent multiple daily low-dose *Bacillus anthracis* spore inhalation exposures in the rabbit model, *Front. Cell. Infect. Microbiol.*, 2012, **2**, 71, DOI: 10.3389/ficmb.2012.00071.
152. U.S. Environmental Protection Agency (US EPA), *Multiple Daily Low-Dose Bacillus anthracis Ames Inhalation Exposures in the Rabbit*, May 2012, http://nepis.epa.gov/EPA/html/DLwait.htm?url=/Exe/ZyPDF.cgi/P100H0J4.PDF?Dockey=P100H0J4.PDF, accessed 16 December 2014.
153. P. J. Jackson, M. E. Hugh-Jones, D. M. Adair, G. Green, K. K. Hill, C. R. Kuske, L. M. Grinberg, F. A. Abramova and P. Keim, PCR analysis of tissue samples from the 1979 Sverdlovsk anthrax victims: The presence of multiple *Bacillus anthracis* strains in different victims, *Proc. Natl. Acad. Sci. U. S. A.*, 1998, **95**, 1224–1229.
154. R. T. Okinaka, M. Henrie, K. K. Hill, K. S. Lowery, M. Van Ert, T. Pearson, J. Schupp, L. Kenefic, J. Beaudry, S. A. Hofstadler, P. J. Jackson and P. Keim, Single nucleotide polymorphism typing of *Bacillus anthracis* from Sverdlovsk tissue, *Emerging Infect. Dis.*, 2008, **14**, 653–656.
155. J. Guillemin, *Anthrax: The Investigation of a Deadly Outbreak*. University of California Press, 1999.
156. M. Meselson, Note regarding source strength, *The ASA Newsl.*, 1995, http://www.asnltr.com/newsletter/01-6/articles/016a.html, last accessed on February 16, 2015.
157. S. R. Hanna, G. A. Briggs and R. P. Hosker Jr, *Handbook on Atmospheric Diffusion*. U.S. Department of Energy, Technical Information Center, DOE/TIC-11223, Washington, DC, 1982.
158. H. A. Druett, Bacterial invasion, *Nature*, 1952, **170**, 288.
159. Institute of Medicine of the National Academies (IOM), *Prepositioning Antibiotics for Anthrax*, ed. C. Stroud, K. Viswanathan, T. Powell and R. R. Bass, The National Academies Press, Washington, DC, 2012, http//www.nap.edu/catalog.php?record_id=13218, last accessed on February 16, 2015.

160. R. Brookmeyer, N. Blades, M. Hugh-Jones and D. A. Henderson, A statistical analysis of truncated data: application to the Sverdlovsk anthrax outbreak, *Biostatistics*, 2001, **2**, 233–247.
161. D. A. Wilkening, Supporting Text for Wilkening, 2006, http://www.pnas.org/content/suppl/2006/04/77/0509551103.DC1/09551Supp Text.pdf, last accessed on February 16, 2015.
162. P. S. Sartwell, The distribution of incubation periods of infectious diseases, *Am. J. Hyg.*, 1950, **51**, 310–318.
163. C. A. Curling, J. K. Burr, L. Danakian, D. S. Disraelly, L. A. LaViolet, T. J. Walsh and R. A. Zirkle, Technical Reference Manual: NATO Planning Guide for Estimation of Chemical, Biological, Radiological and Nuclear (CBRN) Casualties, Allied Medical Publication-8(C), 2010. Institute for Defense Analysis, Alexandria, VA. IDA Document D-4082, Log H 10–00898, August 2010. Avalable from DTIC, Accession Number: ADA536889. Last accessed on February 16, 2015.
164. P. S. Brachman, S. A. Plotkin, F. H. Bumford and M. M. Achison, An epidemic of inhalation anthrax: the first in the twentieth century: II. Epidemiology, *Am. J. Hyg.*, 1960, **72**, 6–23.
165. J. M. Barnes, The development of anthrax following the administration of spores by inhalation, *Br. J. Exp. Pathol.*, 1947, **28**, 385–394.
166. P. Wattiau, S. R. Klee, D. Fretin, M. Van Hessche, M. Ménart, T. Franz, C. Chasseur, P. Butaye and H. Imbrechts, Occurrence and genetic diversity of *Bacillus anthracis* strains isolated in an active wool-cleaning factory, *Appl. Environ. Microbiol.*, 2008, **74**, 4005–4022.
167. P. Wattiau, M. Govartes, D. Frangoulidis, D. Fretin, E. Kissling, M. Van Hescche, B. China, M. Poncin, Y. Pirenne and G. Hanquet, Immunologic response of unvaccinated workers, exposed to anthrax, *Belgium Emerging Infect. Dis.*, 2009, **15**, 1637–1640.
168. E. Kissling, P. Wattiau, B. China, M. Poncin, D. Fretin, Y. Pirenne and G. B. Hanquet, Anthracis in a wool-processing factory: seroprevalence and occupational risk, *Epidemiol. Infect.*, 2012, **140**, 879–886.
169. M. L. Cohen and T. Whalen, Implications of low level human exposurecto respirable B. anthracis, *Appl. Biosaf.*, 2007, **12**, 109–115.
170. K. P. Fennelly, A. L. Davidow, S. L. Miller, N. Connell and J. J. Ellner, Airborne infection with Bacillus anthracis: from mills to mail, *Emerging Infect. Dis.*, 2004, **10**, 996–1001.
171. E. C. Riley, G. Murphy and R. L. Riley, Airborne spread of measles in a suburban elementary school, *Am. J. Epidemiol.*, 1978, **107**, 421–432.
172. C. J. Noakes and P. A. Sleigh, Mathematical models for assessing the role of airflow on the risk of airborne infection in hospital wards, *J. R. Soc., Interface*, 2009, **6**, S791–S800.
173. B. Kournikakis, S. J. Armour, C. A. Boulet, M. Spence and B. Parsons, *Risk Assessment of Threat Anthrax Letters. Defence R&D Canada, Technical Report, DRES-TR-2001-048*, September 2001, http://www.dtic.mil, ADA399955, last accessed on February 16, 2015.

174. E. J. S. Duncan, B. Kournikakis, J. Ho and I. Hill, Pulmonary deposition of aerosolized Bacillus atrophaeus in a swine model to exposure from a simulated anthrax letter incident, *Inhalation Toxicol.*, 2009, **21**, 141–152.
175. B. Kournikakis, J. Ho and S. Duncan, Anthrax letters: personal exposure, building conramination and effectiveness of immediate mitigation measures, *J. Occup. Environ. Hyg.*, 2010, **7**, 71–79.
176. W. W. Piegorsch, S. L. Cutter and F. Hardisty, Benchmark analysis for quantifying urban vulnerability to terrorist incidents, *Risk Anal.*, 2007, **27**, 1411–1425.
177. P. C. Turnbull, P. M. Lindeque, J. Le Roux, A. M. Bennett and S. R. Parks, Airborne movement of anthrax spores from carcass sites in the Etosha National Park, Namibia, *J. Appl. Microbiol.*, 1998, **84**, 667–676.
178. G. G. Meynell and E. W. Meynell, The growth of microorganisms *in vivo* with particular reference to the relation between dose and latent period, *J. Hyg.*, 1958, **56**, 323–346.
179. G. Shortley and J. R. Wilkins, Independent-action and birth-death models in experimental microbiology, *Bacteriol. Rev.*, 1965, **29**, 102–141.
180. N. Hupert, D. Wattson, J. Cuomo, J. Hollingsworth, K. Neukermans and W. Xiong, Predicting hospital surge after a large-scale anthrax attack: a model-based analysis of CDC's Cities Readiness Initiative prophylaxis recommendations, *Med. Decis. Making*, 2009, **29**, 424–437.
181. J. Mitchell-Blackwood, P. L. Gurian, R. Lee and B. Thran, Variance in Bacillus anthracis virulence assessed through Bayesian hierarchical dose–response modeling, *J. Appl. Microbiol.*, 2012, **113**, 265–275.
182. A. M. Jarabek, B. Asgharian and F. J. Miller, Dosimetric adjustments for interspecies extrapolation of inhaled poorly soluble particles (PSP), *Inhalation Toxicol.*, 2005, **17**, 317–334.
183. Y. S. Cheng, H. Irshad, P. Kuehl, T. D. Holmes, R. Sherwood and C. L. Hobbs, Lung deposition of droplet aerosols in monkeys, *Inhalation Toxicol.*, 2008, **20**, 1029–1036.
184. B. Asgharian, O. Price, G. McClellan, R. Corley, D. R. Einstein, R. E. Jacob, J. Harkema, S. A. Carey, E. Schelegle, D. Hyde, J. S. Kimbell and F. J. Miller, Development of a rhesus monkey lung geometry model and application to particle deposition in comparison to humans, *Inhalation Toxicol.*, 2012, **24**, 869–892.
185. W. R. Hogan, G. F. Cooper, G. L. Wallstrom, M. W. Wagner and J.-M. Depinay, The Bayesian aerosol release detector: an algorithm for detecting and characterizing outbreaks caused by an atmospheric release of Bacillus anthracis, *Stat. Med.*, 2007, **26**, 5225–5252.
186. D. L. Buckeridge, D. K. Owens, P. Switzer, J. Frank and M. A. Musen, Evaluating detection of an inhalational anthrax outbreak, *Emerging Infect. Dis.*, 2006, **12**, 1942–1949.
187. R. D. Fricker Jr., Some methodological issues in biosurveillance, *Stat. Med.*, 2011, **28**, 403–415.

188. R. Grunow and E.-J. Finke, A procedure for differentiating between the intentional release of biological warfare agents and natural outbreaks of disease: its use in analyzing the tularemia outbreak in Kosovo in 1999 and 2000, *Clin. Microbiol. Infect.*, 2002, **8**, 510–521.
189. Z. F. Dembek, M. G. Kortpeter and J. A. Pavlin, Discernment between deliberate and natural infectious disease outbreaks, *Epidemiol. Infect.*, 2007, **135**, 353–371.
190. J. Walden and E. H. Kaplan, Estimating time and size of bioterror attack, *Emerging Infect. Dis.*, 2004, **10**, 1201–1205.
191. J. Legrand, J. R. Egan, I. M. Hall, S. Cauchemez, S. Leach and N. M. Ferguson, Estimating the location and spatial extent of a covert anthrax release, *PLoS Comput. Biol.*, 2009, **5**, e1000356, DOI: 10.1371/joutnal.pcbi.1000356.
192. J. Ray, Y. M. Marzouk, P. T. Boggs, D. M. Gay and H. N. A. Najm, Bayesian approachforestimatingbioterrorattacksfrompatientdata,*Stat.Med.*,2011, **30**, 103–120.
193. P. N. Price, M. D. Sohn, K. S. H. Lacommare and J. A. McWilliams, A framework for evaluating anthrax risk in buildings, *Environ. Sci. Technol.*, 2009, **43**, 1783–1787.
194. T. Hong, P. L. Gurian and N. F. D. Ward, Setting risk-informed environmental standards for *Bacillus anthracis* spores, *Risk Anal.*, 2010, **30**, 1602–1622.
195. T. Hong, P. L. Gurian, Y. Huang and C. N. Haas, Prioritizing risks and uncertainties from intentional release of selected category A pathogens, *PLoS One*, 2012, **2**, e32732, DOI: 10.1371/journal.pone.003732.
196. T. Hong and P. L. Gurian, Characterizing bioaerosol risk from environmental sampling, *Environ. Sci. Technol.*, 2012, **46**, 6714–6722.
197. M. W. Calfee, L. J. Rose, J. Tufts, S. Morse, M. Clayton, A. Touati, N. Griffin-Gatchalian, C. Slone and N. McSweeney, Evaluation of sampling methods for Bacillus spore-contaminated HVAC filters, *J. Microbiol. Methods*, 2014, **96**, 1–5.
198. B. W. Gutting, S. R. Channel, E. A. Berger, J. M. Gearhart, G. A. Andrews, R. L. Sherwood and T. L. Nichols, Mathematically modeling inhalational anthrax, *Microbe*, 2008, **3**, 78–85.
199. Y. Huang and C. N. Haas, Time-dose–response models for microbial risk assessment, *Risk Anal.*, 2009, **29**, 648–661.
200. Y. Huang and C. N. Haas, Quantification of the relationship between bacterial kinetics and host response for monkeys exposed to aerosolized Francisella tularensis, *Appl. Environ. Microbiol.*, 2011, **77**, 485–490.
201. S. S. Teske, Y. Huang, S. B. Tamrakar, T. A. Bartrand, M. A. Weir and C. N. Haas, Animal and human dose–response models for *Brucella* species, *Risk Anal.*, 2011, **31**, 1576–1596.

CHAPTER 4

Toxicity of Botulinum Neurotoxin by Inhalation: Implications in Bioterrorism[†]

MICHAEL ADLER*[a] AND DAVID R. FRANZ*[b]

[a]Neurobehavioral Toxicology Branch, Analytical Toxicology Division, US Army Medical Research Institute of Chemical Defense, 3100 Ricketts Point Road, APG, MD 21010-5400, USA; [b]Former Commander, US Army Medical Research Institute of Infectious Diseases, 1425 Porter Street, Fort Detrick, MD 21702-5011, USA
*E-mail: michael.adler2.civ@mail.mil, davidrfranz@gmail.com

4.1 Introduction

Clostridium botulinum and related clostridial species constitute a family of spore-forming, anaerobic, Gram-positive bacteria found in soils, aquatic sediments and decaying animal carcasses.[1,2] Toxigenic strains of these bacteria elaborate seven antigenically distinct protein neurotoxins (serotypes A, B, C1, D, E, F and G). Botulinum neurotoxins (BoNTs) inhibit the release of the neurotransmitter acetylcholine (ACh) at nerve terminals that innervate skeletal muscle, autonomic ganglia and parasympathetic target organs.[3–5]

[†]The opinions expressed in this chapter are the private views of the authors and are not to be construed as an official Department of the Army position unless so designated by other authorizing documents. This chapter has been approved for public release.

Issues in Toxicology No. 25
Aerobiology: The Toxicology of Airborne Pathogens and Toxins
Edited by Harry Salem and Sidney Katz

Published by the Royal Society of Chemistry, www.rsc.org

There are five recognized forms of botulism. In adults, intoxication by BoNT generally results from ingestion of preformed toxin in contaminated food or beverage (foodborne botulism),[6,7] colonization of penetrating wounds by *Clostridium botulinum* (wound botulism)[8] or overdose of toxin used in clinical or esthetic procedures (iatrogenic botulism).[9,10] In infants up to the age of 1 year, *Clostridium botulinum* can colonize the intestinal tract and produce a form of the disease termed infant botulism.[11] In rare cases, the intestinal tract of adults can undergo a similar colonization, generally after bowel surgery or prolonged antibiotic usage (adult intestinal toxemia).[12]

The basic syndrome of BoNT intoxication is similar for all naturally occurring forms and does not differ appreciably among serotypes,[2,13] although there can be marked differences in latency, severity and duration that are serotype dependent.[14–17] In humans, the earliest signs of botulism generally include visual disturbances, followed by dysphagia, dysphonia and dysarthria, reflecting an especially high susceptibility of cranial efferent terminals to BoNT action.[18,19] A descending generalized skeletal muscle weakness may then develop, progressing from the upper to lower extremities. Involvement of the diaphragm and intercostal muscles can lead to ventilatory failure and death. Survival of botulism requires intensive care, artificial ventilation and timely infusion of antitoxin.[20–23] Among the distinguishing characteristics of botulism are the symmetrical nature of the muscle weakness, the descending progression of signs and the general absence of central nervous system or sensory involvement.[2,13,16,19,20]

In addition to natural outbreaks, intoxication by BoNT can occur following inhalation of BoNT-containing aerosols.[24–27] Inhalation exposure is considered to be the likely route of intoxication in a bioterrorist attack with BoNT in common with other biological and chemical threat agents.[13,25,28–31]

For intoxications that involve exposure to preformed BoNT (foodborne, iatrogenic, inhalation), clinical signs of botulism are usually observed 24–48 h after exposure, although onset times from 2 h to 16 days have been reported.[20,32,33] For botulism acquired secondary to bacterial colonization (infant, adult intestinal toxemia, wound), the onset times are generally longer than those following exposure to preformed toxin.[11–13] The preponderance of signs of botulinum intoxication, including the potentially lethal ventilatory failure, stems from inhibition of neuromuscular transmission. Gastrointestinal disturbances that often accompany foodborne botulism (constipation, nausea, diarrhea, abdominal cramps) may be of parasympathetic origin.[3,5] The latter could also be a consequence of ingestion of other clostridial products or substances resulting from anaerobic spoilage unrelated to *Clostridium botulinum.*[34–36]

The BoNTs are secreted in an anaerobic environment as relatively inactive 150 kDa single-chain protoxins (range 140–167 kDa), surrounded by a complex of neurotoxin-associated proteins (NAPs).[37] The protoxin is subsequently cleaved by trypsin-like enzymes in the bacterial culture or in the host to form the active disulfide-linked dichain neurotoxin consisting of a ~100 kDa heavy chain (HC) and a ~50 kDa light chain (LC).[4,14,38] The BoNTs have three functional domains. The carboxy-terminal half of the HC interacts with polysialogangliosides and protein receptors (synaptotagmin, SV2) of motor neuron terminals and autonomic varicosities.[39,40] This domain also binds to receptors on the apical surface of airway and intestinal epithelial cells, leading to transcytosis of toxin across these barriers.[41–43] The amino-terminal region of the HC promotes translocation of the LC into the cytosol of cholinergic nerve terminals.[44] The entry of the LC into the cytosol completes a multistep process that begins with the encounter of BoNT with the host and is followed by two distinct transport steps: the first conveys the intact holotoxin across an epithelial barrier (gut or airway) to reach the bloodstream and the second transports the toxin across nerve terminal membranes to enable the LC to reach the transmitter secretory apparatus.[4,16,17]

The LC is a zinc-dependent endoprotease that cleaves serotype- and substrate-specific sites in the SNARE (soluble *N*-ethylmaleimide-sensitive factor attachment protein receptor) complex, which leads to cessation of transmitter release.[45] There are no signs or symptoms of exposure during toxin entry when botulism could be treated most successfully with antitoxin.[16] All manifestations of botulism are believed to stem from SNARE protein cleavage and the first specific indication of intoxication (cranial nerve dysfunction) occurs when a critical fraction of SNARE protein is cleaved.[14,17]

Substantial work in the 1990s led to the identification of the targets for the protease activity of the BoNT LC. Serotypes A and E cleave the last nine and last 26 residues, respectively, from the nerve terminal active zone protein SNAP-25 (synaptosomal-associated protein of 25 kDa).[45,46] Serotypes B, D, F and G cleave different sites on synaptobrevin/VAMP (vesicle-associated membrane protein), a small transmembrane protein located on the surface of small synaptic vesicles.[47,48] Serotype C1 cleaves the active zone proteins syntaxin and SNAP-25 and is the only serotype known to target two SNARE proteins.[49] As a consequence of BoNT-mediated cleavage of SNARE proteins, synaptic vesicles are unable to bind to active zone membranes, leading to inhibition of ACh release. Inhibition can last for weeks to months depending on the dose, serotype and nature of the synapse involved.[15,50]

4.2 BoNT As a Biowarfare and Bioterrorist Threat

BoNT and the bacteria that produce BoNT (*C. botulinum*, *C. butyricum*, *C. baratii* and *C. argentinense*) are designated by the US Health and Human Services as Tier 1 select agents. Toxins and pathogens classified as Tier 1 "present the greatest risk of deliberate misuse with significant potential for mass casualties or devastating effect to the economy, critical infrastructure or public confidence and pose a severe threat to public health and safety" (http://www.selectagents.gov/).

The potential use of BoNT as a biological agent has long been recognized, but it was not until World War II that large-scale attempts were made to weaponize BoNT for offensive use.[25,51,52] The biological weapons program was terminated in the United States in 1969 following an executive order by President Nixon, but it continued in the former Soviet Union in spite of its prohibition by the Biological and Toxin Weapon Convention.[53] In more recent times, BoNT was weaponized and field tested by Iraqi scientists during the period preceding the Persian Gulf War of 1991.[26,51] With the presumed destruction of Iraqi biological weapon stockpiles during 1991–2003, the concern over possible deployment of BoNT in biological warfare by an organized military has diminished. However, owing to the rise of global terrorism and terrorist cults, the more immediate threat for the foreseeable future is a terrorist strike involving BoNT. In fact, in the early to mid-1990s, several attempts were made by the Japanese religious cult Aum Shinrikyo to carry out a biological attack using aerosolized BoNT.[54] In March 1995, cult members targeted a Tokyo subway station using sprayers containing BoNT concealed in briefcases. Although these attempts were unsuccessful, during the same month the cult carried out a coordinated attack on the Tokyo subway system using the nerve agent sarin; this attack resulted in 3800 injuries and 12 deaths.[54] Had the BoNT strike succeeded, the casualties could have been considerably higher.[26]

It is likely that a terrorist strike with BoNT would involve release of the neurotoxin as an aerosol[25,30] or deliberate contamination of food or beverage.[55] Aerosol exposure leads to a form of the disease designated inhalation botulism, which is not found in nature.[21] Ingestion of BoNT following malevolent contamination of food or beverage would result in foodborne botulism, indistinguishable from a natural outbreak with the possible exception of an unusually large cluster of cases, multiple simultaneous outbreaks or the occurrence of serotypes not commonly associated with foodborne botulism.[26]

The majority of toxins and chemical agents exhibit greater potency and a more rapid onset by inhalation than when consumed in food or administered parenterally. Moreover, some toxins and pathogens such as *Yersinia pestis* can elicit a different and more severe clinical manifestation of the disease (pneumonic plague *versus* bubonic plague) when the route of exposure is inhalation, a phenomenon that may confound diagnosis and delay treatment.[24] The following questions are fundamental in planning for the consequences of a terrorist attack with aerosolized BoNT. Are the BoNTs more potent by aerosol than by the traditional routes? Are there signs and symptoms of botulism that are unique to aerosol exposure? Are currently used medical countermeasures effective against inhalation botulism? These questions will be addressed in the remainder of this chapter.

4.3 Potency of BoNT by Inhalation

Relative to the extensive literature on naturally occurring botulism in humans, there are comparatively few publications on inhalation botulism apart from a laboratory accident in Germany over 50 years ago[56] and a reported exposure in France following inhalation of BoNT-contaminated cocaine.[33] Each case of inhalation botulism must therefore be viewed with suspicion and considered as a potentially hostile act.[21]

Most of what we know about inhalation botulism has come from studies in nonhuman primates (NHPs) and in small animal models. In the mid-1960s, investigators at Fort Detrick, MD, examined the lethality of aerosolized BoNT/A in three NHP species: squirrel monkeys, rhesus monkeys and chimpanzees. Intravenous (i.v.) exposures, which allow unrestricted entry of toxin to the circulation, were also carried out to determine the inherent sensitivity of these NHPs to BoNT/A. The median lethal dose (LD_{50})[‡] of BoNT/A by inhalation or i.v. injection, expressed as multiples of the mouse intraperitoneal (i.p.) LD_{50} unit ($MIPLD_{50}$) per kilogram body weight, is shown in Table 4.1. In each species, BoNT/A was significantly more potent by the i.v. route than by inhalation: 11-fold in squirrel monkeys, 4-fold in rhesus monkeys and 32-fold in chimpanzees.

The lower potency of BoNT/A by inhalation suggests that only a fraction of the aerosolized toxin is absorbed by the airways and delivered to the bloodstream.[30] Consistent with this notion, Park and Simpson[41] observed a peak plasma concentration of 13 pM in mice 2 h after intranasal (i.n.) administration of 36.4 μg kg^{-1} [^{125}I]BoNT/A, which corresponds to 0.5% of the instilled dose. Although absorption of BoNT by inhalation is relatively inefficient, lethal BoNT aerosols can be readily generated.[24–28]

Among the earliest studies to examine the lethality of BoNT by inhalation relative to other routes of intoxication was that of Cardella,[57] who compared the sensitivity of guinea pigs to i.p., aerosol and oral challenge with five serotypes of BoNT (A–E). For all serotypes tested, aerosol delivery of BoNT was intermediate in lethality between i.p. and oral administration, and was generally closer to oral than to i.p. values (Table 4.2). Intoxication by the oral and inhalation routes involves penetration of BoNT through an epithelial barrier: intestinal for the former and airway for the latter. For oral intoxication, BoNT must also survive the low pH of the stomach and proteolytic enzymes

Table 4.1 Lethality of BoNT/A by aerosol or intravenous administration in NHPs.[a]

	Inhalation		Intravenous	
NHP species	$MIPLD_{50}$[b] kg^{-1}	95% CI[c]	$MIPLD_{50}$[b] kg^{-1}	95% CI[c]
Squirrel monkey	915	262–3192	81	56–105
Rhesus monkey	297	130–681	74	66–83
Chimpanzee	890	N/A[d]	28	N/A[d]

[a]Unpublished data, J. V. Jemski (Technical Evaluation Division, US Army Medical Unit, Fort Detrick, MD). $n = 8–10$ in each group. BoNT/A was delivered as the complex (neurotoxin with associated NAPs). Determination of the inhalation LD_{50} was derived from respiratory minute volume and percentage of the inhaled aerosol retained in the airways during exposure.
[b]Lethality is expressed in $MIPLD_{50}$ units to control for differences in toxin purity and potency.
[c]CI = confidence interval.
[d]N/A = not available.

‡In this chapter, lethality is expressed as LD_{50}, LCt_{50} or multiples of the mouse intraperitoneal LD_{50} unit ($MIPLD_{50}$) based on the expression of lethality in the source material.

Table 4.2 Lethality of BoNT (A–E) by various routes of administration in the guinea pig.[a,b]

	Route of exposure					
Serotype	i.p.	Aerosol[c]	Aerosol/i.p.	Oral	Oral/i.p.	Aerosol/ oral
A	5.2	141	27.1	717	137.9	0.2
B	4.2	350	83.3	306	72.9	1.1
C1	1.6	87	54.4	177	110.6	0.5
D	4.1	186	45.4	436	106.3	0.4
E[d]	34.3	778	22.7	178000	5189.5	0.0044

[a]Data modified from Table 7 of Cardella[57] and reproduced with permission from the US Army Medical Research Institute of Infectious Diseases.
[b]Lethality is expressed as $MIPLD_{50}$ kg^{-1}.
[c]Guinea pigs were exposed (whole body) to static aerosols of BoNT complexed with NAPs.
[d]Serotype E was less potent in guinea pigs than in mice by all routes of administration, especially by oral gavage. The low oral toxicity of BoNT/E may be a result of the different protective NAPs that are associated with BoNT/E.[58]

of the small intestine.[16,37] These harsh conditions presumably account for the lower potency of BoNT by ingestion.

In NHPs (Table 4.1) and in guinea pigs (Table 4.2), BoNT was considerably more toxic by parenteral administration than by inhalation. This difference was also observed in mice where BoNT was tested by i.v. or i.n. administration for differences in potency. Both BoNT/A and BoNT/E showed a 500–1000-fold greater potency by i.v. injection than by i.n. instillation.[43] The lower potency of BoNT by inhalation may seem surprising since many toxins, such as T-2 mycotoxin, tetrodotoxin and saxitoxin, and also organophosphorus chemical warfare agents, are more toxic by inhalation than by parenteral delivery.[24,59–61] However, it must be borne in mind that for inhalation botulism to occur, BoNTs must first penetrate airway epithelial cells to reach the bloodstream, whereas injected toxin has a more direct access to the bloodstream.[4,16,30,41]

Two distinct mechanisms have been proposed for BoNT entry across epithelial membranes: transcytosis, which involves specific binding to the apical surface of epithelial cells, and paracellular transport, in which the epithelial tight junctions are disrupted to enable toxin entry across a damaged epithelial barrier.[30] The available evidence appears to favor the transcytosis pathway.[16,41–43] In general, toxin delivery into the bloodstream is determined by the concentration of BoNT in the lungs, surface area of the airways, duration of aerosol exposure and net apical to basal flux.[16] The transport of BoNT by airway epithelial cells is discussed in the following section.

4.4 Mechanism of Entry of Aerosolized BoNT

The first step in inhalation botulism involves contact of aerosolized BoNT with airway epithelial cells. From studies with differentiated polarized epithelial cells, Simpson and colleagues[41,42] demonstrated that transport of toxin into the bloodstream requires binding of BoNTs to (as yet unidentified) receptors on the apical surface of airway epithelial cells, followed by receptor-mediated endocytosis and, finally, transcytosis across the epithelial monolayer. Using [^{125}I]BoNTA, Park and Simpson[41] demonstrated that transcytosis is bidirectional with a net transport in the apical to basolateral direction. In Calu-3 cells, the net influx of toxin was determined to be 0.22 fmol h^{-1} cm^{-2} in cells exposed to 10 nM BoNT/A, which is consistent with the relatively slow rates of toxin entry observed *in vivo* following i.n. installation of BoNT/A in mice.[41]

The binding domain of the BoNT/A HC can undergo epithelial transcytosis in the absence of the toxin's translocation and catalytic (LC) domains, although at a twofold slower rate than observed with the complete holotoxin.[41] Further, the NAPs that protect the neurotoxin from the low pH and proteolytic enzymes of the gastrointestinal tract[30,37] do not appear to play a prominent role in inhalation toxicity, presumably because the conditions in the airways are more benign than those in the gastrointestinal tract.[43] Consequently, both the pure neurotoxin and toxin complex are transported across airway epithelia at similar rates and emerge after transcytosis as structurally intact, fully active neurotoxins.[41] The BoNTs are exceptionally stable in the bloodstream and do not undergo appreciable binding to plasma proteins, so that most of the BoNT that enters the bloodstream is available to intoxicate target cells in the host organism.[16]

4.5 Human Case Reports of Inhalation Botulism

4.5.1 Laboratory Exposure

To date, only two outbreaks of inhalation botulism have been reported in humans. The first of these was published in 1962 and involved three laboratory investigators in Germany who exposed rabbits and guinea pigs (whole body) to an aerosol of BoNT/A that was stabilized by addition of colloids.[56] They performed necropsies on the animals the following day wearing protective gloves but no mask or respirator. BoNT/A was apparently inhaled when toxin residues that were deposited on the skin and fur became airborne during post-mortem procedures. Each laboratory investigator became symptomatic after a 3 day latent period, with the most prominent signs and symptoms being dysphagia, dizziness, malaise, dysphonia, headache and blurred vision. All three individuals required hospitalization (9 days) and showed positive serum tests for the presence of BoNT by the mouse bioassay.

The positive serum test in these individuals is surprising since laboratory confirmation is rare in patients whose symptoms are largely restricted to cranial nerve palsies.[20] Two of the patients received infusions of equine antitoxin, but whether this led to a significant improvement in their condition is unclear, since antitoxin was administered outside its normal therapeutic window of ≤48 h after exposure.[13] None of the patients experienced difficulties in breathing or frank muscle paralysis. The nature of the symptoms and the relatively rapid recovery described in this case report are consistent with botulism involving a low dose of toxin.

4.5.2 Inhalation of BoNT-Contaminated Cocaine

No new cases of human inhalation botulism were reported for the next four decades until a case of suspected inhalation botulism was described in France in 2006 by Roblot *et al.*[33] The incident involved two drug users who inhaled cocaine that was suspected of being contaminated by BoNT/B. Signs and symptoms in Patient 1 appeared 2 h after inhalation of the contaminated cocaine and consisted of general confusion and diplopia. Six days after onset of intoxication his confusion resolved, but he developed characteristic signs of botulism (loss of accommodation, blurred vision, dry mucosa, constipation and absence of a gag reflex), leading to hospitalization. Interestingly, dysphagia, which was prominent in the 1962 case of inhalation botulism, was not observed in this outbreak, possibly because of differences in serotype. Blood samples were taken 3 and 6 days after the onset of intoxication and were positive for BoNT/B by the mouse bioassay. Patient 1 recovered after 10 days of hospitalization and received antibiotic treatment (macrolide) but no antitoxin. For Patient 2, the onset of intoxication occurred 6 h after inhalation of BoNT/B-contaminated cocaine that was purchased from the same dealer. Patient 2 had a milder case (loss of visual accommodation, blurred vision, constipation), due presumably to inhaling a smaller amount of cocaine, and his symptoms regressed over a 3 day period. BoNT was not detected in the serum of Patient 2.

As in the 1962 inhalation exposure, neither patient in the latter outbreak experienced breathing difficulties or generalized muscle weakness. The unusually rapid onset times of 2 and 6 h after inhalation of cocaine may have been due to a more rapid onset of serotype B.[62] Alternatively, some of the early symptoms (confusion, diarrhea) may have been due to cocaine itself or to a combination of cocaine and BoNT.[63]

In concluding that this outbreak represented true inhalation exposure, food-borne botulism was ruled out, since no common food items were consumed. This still left the possibility of wound botulism, secondary to cocaine-induced sinusitis, as was suggested in previous cases of botulism following intranasal cocaine abuse.[64] Although wound botulism was considered unlikely from the absence of radiographic evidence, sinus infections in inhalation cocaine users are common and small abscesses may be difficult to detect.[65]

Despite the fact that documented cases of inhalation BoNT are exceedingly rare and have been relatively mild, BoNT can be highly lethal following inhalation exposure, as evidenced by numerous animal studies.[27,29,43,57,66–68] The exceptional potency of BoNT and its ability to enter by the respiratory route (albeit slowly) make this neurotoxin a formidable military and bioterrorist threat.

4.6 Histopathological Alterations Following BoNT Exposure

Since inhalation of BoNT would be expected to generate high initial concentrations of toxin in the airways, local tissue damage may reasonably be assumed to result from inhalation of aerosolized BoNT. In animal experiments using a variety of species, no gross pulmonary lesions have been reported following inhalation of BoNT. In a study on rhesus monkeys, Franz *et al.*[66] noted mild to moderate subacute, submucosal tracheitis in 3 of 6 animals that received a lethal dose of BoNT/A by inhalation. These alterations were considered to be secondary to the intoxication, due possibly to aspiration pneumonia as a consequence of toxin-induced relaxation of esophageal sphincter muscles. The findings of Franz *et al.*[66] were confirmed and extended in a more recent study by Sanford *et al.*,[27] who noted a relative absence of gross pulmonary lesions in rhesus monkeys exposed to aerosolized BoNT/A or BoNT/B.

Following an i.n. administration of a 4 $MIPLD_{50}$ dose of BoNT/A in mice, Taysse *et al.*[68] reported a thickening of the alveolar septa, a generalized spreading interstitial edema and moderate intra-alveolar/intra-bronchiolar hemorrhage. These data suggest that some fraction of the i.n. dose may be absorbed in the lungs, although the olfactory nasal epithelium is also capable of absorbing toxicants.[69] Of particular interest was the finding that animals exhibited a similar histopathology whether protected by prior vaccination with the pentavalent toxoid vaccine (PBT) or unvaccinated. This is consistent with suggestions that alterations of airway tissue in the NHPs were not caused directly by BoNT. Taysse *et al.*[68] attributed the histological changes in mice to the action of cytokines, such as tumor necrosis factor alpha (TNFα), released non-specifically in response to the aerosol challenge. TNFα and other cytokines may increase pulmonary vascular permeability and activate inflammatory mediators. Both cytokines and aspirational pneumonia likely play a role in the observed airway histopathology.

In contrast to the relatively benign consequences of inhaled or nasally instilled BoNT, inhalation of many biological toxins leads to severe and potentially fatal necrotic lesions manifested predominantly in the airways. Thus, Poli *et al.*[70] demonstrated that inhalation of ricin in mice led to diffuse airway epithelial necrosis, characterized by severe interstitial edema and inflammation. Inhalation of anthrax spores led to severe tissue damage consisting of splenomegaly, enlargement of lymph nodes and hemorrhage in various organs, particularly the lungs, mediastinum, meninges and gastrointestinal tract.[71] Similarly, inhalation of aerosolized staphylococcal enterotoxin B resulted in pulmonary lesions that included severe, diffuse interstitial and alveolar pulmonary edema, leukocytic infiltrates, perivascular edema and alveolar fibrin deposition.[72] In the context of histopathology, the major difference between BoNT and the above threat agents is that BoNT is highly selective in targeting cholinergic nerve endings and uses the airways as a conduit to reach its desired target.[16]

4.7 Detection and Diagnosis

Based on inhalation studies in NHPs and exposure of laboratory personnel,[56] the typical incubation period for inhalation botulism is projected to be 2–4 days, which is comparable to that reported for foodborne botulism.[20,22] In addition, the signs and symptoms of inhalation botulism are expected to be similar to those of naturally occurring forms, except for the general lack of early gastrointestinal symptoms that often accompany foodborne botulism[23] and the absence of penetrating wounds that are the hallmark of wound botulism.[33] Accordingly, the diagnosis of inhalation botulism would need to be based on signs and symptoms and medical treatment (intensive care, artificial ventilation, antitoxin infusion) should not be delayed for laboratory confirmation, which could take up to 48 h or longer with the standard mouse bioassay.[73] This is in accordance with CDC guidelines for other forms of botulism.[13] Laboratory confirmation involves testing of clinical samples such as blood and stool for the presence of BoNT. In a retrospective study of >200 foodborne botulism patients, BoNT was confirmed in only a fraction of clinical samples tested: 37% of patients tested positive for BoNT in sera and even fewer (23%) had positive stool samples. Moreover, the ability to detect BoNT was further reduced if the time between intoxication and collection of samples was delayed.[73]

In inhalation botulism, the ability to identify BoNT in clinical samples is largely unknown, owing to the rarity of inhalation botulism in humans and the marked differences in reports of BoNT detection between intoxicated humans and NHPs. Of the five humans exposed to BoNT by inhalation, a remarkably high percentage were reported to have detectable levels of BoNT in sera: all three victims of the laboratory incident described in Section 4.5.1 and one of the two cocaine users described in Section 4.5.2 tested positive for BoNT.[33,56] This represents confirmation of toxin in 80% of the inhalation botulism cases compared with 37% in a CDC study of foodborne botulism.[73] Moreover, the human inhalation cases were relatively mild, which should have minimized the probability of detecting BoNT in serum. Consequently, these data must be viewed as atypical and may not apply to future inhalation botulism outbreaks.

In contrast to the findings in humans, when rhesus monkeys were exposed to aerosolized BoNT/A or BoNT/B, investigators were unable to detect BoNT/A in sera using the mouse bioassay even with the highest challenge dose tested (>5000 $MIPLD_{50}$ kg^{-1}).[27] In the same study, high doses of BoNT/B (>25 000 $MIPLD_{50}$ kg^{-1}) were successfully detected in sera but only in animals that eventually died.[27] According to the authors, their bioassay had a detection limit of 2 $MIPLD_{50}$ mL^{-1} serum. The difficulty in identifying BoNT in the above study is in accord with earlier reports that aerosolized BoNT is generally not detectable in serum or stool.[28] The marked discrepancy between humans and NHPs in the detection of BoNT in serum is clearly in need of further investigation.

In conducting epidemiological investigations of natural BoNT outbreaks, testing includes not only clinical material, but also environmental samples. For foodborne botulism, testing would include all suspicious food items recently consumed by the victim. For suspected inhalation botulism, environmental testing would be limited largely to air samples and surfaces on which aerosols may have settled. Depending on particle size, the toxin in an aerosol cloud may be difficult to detect, because particles >5 μm would precipitate rapidly from the aerosol.[30] Since it is not known how long BoNT would remain active on environmental surfaces under different environmental conditions, a sensitive chemical assay such as mass spectrometry would need to be used in conjunction with a functional assay, such as the mouse bioassay, to confirm an aerosol release of BoNT.[74]

4.8 Medical Countermeasures for Inhalation Botulism

In recent times, deaths from BoNT exposure have fallen to <5%, due mainly to improvements in intensive care, including assisted ventilation and the availability of antitoxin.[13,20,25] For severe BoNT intoxication, especially by the more persistent serotypes such as BoNT/A or BoNT/B, recovery can take several months and consume a significant fraction of critical care resources.[32] The current medical and public health capabilities are generally sufficient to deal with sporadic natural outbreaks of botulism, but in a large terrorist strike with aerosolized BoNT, the number of casualties would likely overwhelm the healthcare system and many more fatalities could result.

A fundamental question for the medical management of botulism following a terrorist attack is whether the new vaccine candidate and recently approved antitoxin (currently used for natural outbreaks) would be effective in treating patients following an inhalation exposure of BoNT. Since inhalation botulism does not occur routinely in nature, it is necessary to determine the efficacy of these countermeasures in appropriate animal models, especially in the context of a large bioterrorist attack, where high exposure may be expected.

4.8.1 Efficacy of Vaccines in Inhalation Botulism

4.8.1.1 Protection by Traditional Pentavalent Vaccine

BoNT serotypes A–E were converted into toxoid for immunization by incubation for several weeks in 0.6% formalin at 35–37 °C under mildly acidic conditions (pH 5.5–6.5).[57] This vaccine, designated as pentavalent (ABCDE) botulinum toxoid or PBT, had been available for immunization of high-risk individuals, such as laboratory workers who handle BoNT and military personnel, for over 50 years. Although the PBT vaccine was discontinued in 2011 because of reductions in immunogenicity coupled with a progressive rise in reactogenicity (http://www.cdc.gov/mmwr/preview/mmwrhtml/mm6042a3.htm), it is useful to consider historical data on the efficacy of PBT in inhalation challenge by BoNT. In experiments in which guinea pigs were immunized with PBT and challenged by BoNT, the vaccine was found to be as effective against an inhalation challenge as against exposure by the other routes of administration. Approximately 90% of guinea pigs that were immunized with PBT were able to survive an aerosol challenge of 50–150 respiratory LD_{50} units of BoNT/A 40 days later.[57]

4.8.1.2 Protection by Recombinant Subunit Vaccine Against Aerosol Exposure of BoNT/A

The botulinum vaccine that is currently in clinical trials uses the 50 kDa binding domains of BoNT/A (subtype A1) and BoNT/B (subtype B1) as immunogens and is expressed recombinantly in the yeast *Pichia pastoris* (rBV A/B).[75] The serotype A component of this vaccine was studied in an inhalation model of botulism in rhesus monkeys.[29] Vaccine containing the 50 kDa binding domain of BoNT/A HC was administered intradermally on days 0, 28 and 56. Monkeys were challenged with BoNT/A aerosol for 10 min in a head-only exposure chamber 1 month after the final vaccination and monitored for survival for 2 months after the BoNT/A challenge. All vaccinated animals survived challenge with a high dose of BoNT/A with no signs of intoxication (mean dose = 65 LD_{50}, range = 24–99 LD_{50}). All non-vaccinated monkeys died within 2 days of challenge. Al-Saleem *et al.*[43] tested a similar vaccine for protection in mice. Animals immunized with the 50 kDa binding domain of BoNT/A HC were able to survive i.n. challenge of 7000–10 000 $MIPLD_{50}$ of BoNT/A with no signs of intoxication.[43]

4.8.1.3 Protection by Recombinant Subunit Vaccine Against Aerosol Exposure of BoNT/B

In a separate study, the BoNT/B component of the bivalent vaccine was tested for protection against challenge by BoNT/B in rhesus monkeys.[76] The 50 kDa binding domain of BoNT/B HC was administered i.m. to animals three times at 4 week intervals. Six weeks after the final dose, monkeys were challenged with aerosolized BoNT/B for 10 min in a head-only chamber using a target dose of 50 LD_{50} of BoNT/B. Non-vaccinated animals died of BoNT/B intoxication. Vaccinated animals, however, were fully protected from lethality and also from signs of botulism: six of six monkeys survived challenge with 51–60 LD_{50} doses of BoNT/B. These data suggest that rBV A/B will be protective against multiple lethal doses of aerosolized BoNT/A and BoNT/B.

Despite the efficacy of these vaccines, it is unlikely that the general population will be vaccinated against BoNT, even when rBV A/B becomes licensed by the US Food and Drug Administration (FDA).[75] Two compelling factors argue against large-scale vaccination. First, the incidence of botulism from natural sources is rather low. During the last decade, the total number of cases of botulism in the United States has averaged ~140 per year, with infant botulism accounting for the majority of outbreaks (~100) (http://www.cdc.gov/nationalsurveillance/PDFs/Botulism_CSTE_2011.pdf). This would hardly justify vaccination of large segments of the population. Second, BoNT serotypes A and B are used extensively in clinical medicine; they are often the treatment of choice for dystonia, movement disorders, urinary incontinence, hyperhidrosis, migraine and pain, and immunized individuals would be precluded from being able to use this treatment, presumably for their lifetime.[19,77–79]

4.8.2 Efficacy of Botulinum Antitoxins in Inhalation Botulism

Since a post-symptomatic therapeutic agent for botulinum intoxication is not currently available,[79] the most effective medical countermeasure for BoNT intoxication is infusion of antitoxin as soon as possible after exposure is established.[23] Human- and equine-derived antitoxins have been tested for protection against inhalation challenge of BoNT and both were found to be highly effective.

4.8.2.1 Efficacy of Human and Equine Antitoxins in Rhesus Monkeys Against Aerosol Challenge by BoNT/A: Prophylactic and Presymptomatic Use

Rhesus monkeys were injected with 8–16 international units (IU) kg^{-1} of human hyperimmune globulin or 14–143 IU kg^{-1} of equine "despeciated" $F(ab')_2$ antitoxin, yielding plasma titers of <0.02–0.6 IU mL^{-1}. The animals were challenged head-only with 3.4–10.7 LD_{50} kg^{-1} of aerosolized BoNT/A 48 h later.[66] All animals survived this challenge with no signs of BoNT intoxication. Animals even survived challenge by BoNT/A 6 weeks after a single administration of 16 IU kg^{-1} of human hyperimmune globulin. Control animals, on the other hand, died 2–4 days after BoNT/A challenge and exhibited clinical signs of intoxication 12–18 h before death. These signs were similar to those observed with other forms of botulism in NHPs and consisted of, in order of onset, muscle weakness, intermittent ptosis, poor head control, dysphagia and finally lateral decumbency.

Additional experiments were carried out in which a high dose of equine $F(ab')_2$ antitoxin (143 IU kg^{-1}) was given 46 h after an aerosol exposure of BoNT/A (8.0–10.3 LD_{50} kg^{-1}). Animals that were highly symptomatic at the time of antitoxin administration died during the next 2 days, whereas those with minimal or no signs at the time of antitoxin infusion recovered over the next 10 days and survived. These data indicate that the window for antitoxin administration in inhalation botulism is similar to that for other forms of botulism and that therapeutic efficacy is highly dependent on the severity of symptoms before therapy is initiated. This is reasonable since antitoxin acts by neutralizing BoNT that is free in circulation or bound to nerve terminals but not yet internalized.[16,20]

4.8.2.2 Efficacy of Equine $F(ab')_2$ Antitoxin in Rhesus Monkeys Against Aerosol Challenge by BoNT/A: Infusion Delayed Until First Sign of Intoxication

Rhesus monkeys were exposed to a high dose of BoNT/A (36 LD_{50} kg^{-1}). One group of animals was treated with equine $F(ab')_2$ botulinum antitoxin 24 h after aerosol challenge, whereas treatment was delayed in the second group until the onset of distinct clinical signs of BoNT intoxication (ptosis, dysphagia). A control group received 1 human dose equivalent of *Escherichia coli*

Table 4.3 Efficacy of equine $F(ab')_2$ antitoxin given 24 h after challenge or at onset of signs.[a]

Antitoxin			Inhaled BoNT/A	
Type	Dose/IU[b] kg^{-1}	Time	(LD_{50} kg^{-1})[e]	Alive/total
E. coli $F(ab')_2$	1[c]	24 h	34, 44	0/2
E. coli $F(ab')_2$	1[c]	FS[d]	26, 68	0/2
Equine $F(ab')_2$	14	24 h	27 ± 8	4/4
Equine $F(ab')_2$	14	FS[d]	33, 47	0/2
Equine $F(ab')_2$	143	24 h	37 ± 9	3/4
Equine $F(ab')_2$	143	FS[d]	31, 32	0/2
Equine $F(ab')_2$	429	FS[d]	40 ± 4	0/4

[a]Results from an unpublished study by D. Franz and M. L. M. Pitt (US Army Medical Research Institute of Infectious Diseases, Fort Detrick, MD).
[b]14 IU kg^{-1} corresponds to a 0.1 human dose of antitoxin.
[c]Dose for control *E. coli* $F(ab')_2$ antitoxins was 1 human equivalent.
[d]FS = first signs of BoNT intoxication.
[e]Values represent mean ± SEM of LD_{50} doses in each group or individual values where replicates are <3.

antitoxin 24 h after challenge or at the first signs of botulism. The results are shown in Table 4.3. When equine $F(ab')_2$ BoNT/A antitoxin was given 24 h after challenge, seven of eight monkeys survived exposure of up to a 63 LD_{50} dose of BoNT/A (Table 4.3). If equine $F(ab')_2$ treatment was initiated only after first signs appeared, all eight animals died, even those receiving the highest dose of antitoxin (429 IU kg^{-1}). In this study, the first signs of BoNT/A intoxication were observed at 38.9 ± 3.5 h (mean ± SE) after BoNT/A exposure (range 29–64 h). Viewed collectively, the results of the NHP studies indicate that the time of initiation of antitoxin treatment and magnitude of the BoNT dose appear to be the major determinants of treatment efficacy.

4.8.2.3 *Efficacy of Human Hyperimmune Globulin Antitoxins in Guinea Pigs Challenged by BoNT/A Aerosol: Fixed Schedule*

A similar study was performed in passively immunized guinea pigs, with comparable results.[67] Animals were injected i.p. with human botulinum immune globulin obtained from volunteers vaccinated with PBT to generate a serum titer of 0.25 IU ml^{-1}, which is considered to be protective by the US Centers for Disease Control and Prevention (CDC). Guinea pigs were challenged 24 h later with aerosolized BoNT serotype A, B, C1, D or E at doses corresponding to 25 × LCt_{50} (LCt_{50} = the product of concentration and exposure time that is lethal to 50% of the exposed population). Approximately 90% of guinea pigs survived aerosol challenge by serotypes A, B and E and all animals survived challenge by serotypes C and D. Moreover, these guinea pigs were asymptomatic. Thus, as with active immunization (Section 4.8.1), passive immunization prevented all BoNT-related signs. Since the antibodies were derived from human volunteers, these data further suggest that the donors would have survived an inhalation challenge by BoNT.

As of March 2013, an equine heptavalent (A–G) antitoxin (BAT) has replaced all previously licensed antitoxins in the United States. Although the licensed BAT has not been specifically evaluated for efficacy against inhalation challenge, it can be assumed to have a similar efficacy to the antitoxins that have previously been tested. Moreover, the despeciated F(ab′)$_2$ antitoxin that was found to be protective by Franz *et al.*[66] against aerosol challenge by BoNT/A in NHPs is analogous to the type A antitoxin in the licensed BAT manufactured by Cangene.[7]

4.9 Summary and Conclusions

BoNT is the most toxic substance known and has the potential for both alleviating a variety of muscle and autonomic disorders[19] and causing massive loss of life if deployed by terrorists or hostile nations.[25,26,28] BoNT also has its traditional role as the causative agent in foodborne, intestinal and wound botulism. Deployment of BoNT by terrorists may involve food contamination but the more probable route is aerosol exposure. Battlefield use of BoNT would likely be carried out by aerosol exposure from munitions or aerial spraying. Although inhalation BoNT has only been sporadically encountered in humans, data from animal studies are sufficient to indicate that the symptoms would be similar to foodborne botulism (with the possible exception of diarrhea and nausea) and that the new vaccine candidate and recently approved antitoxin would be as effective for inhalation botulism as for other forms. Moreover, the latency from exposure to onset of signs and symptoms would also be similar to that observed for other forms of botulism and the potency would be intermediate between parenteral and oral exposure to toxin.

Since the vaccine under development is not currently available in the United States (and even after approval, may only be indicated for an at-risk group), and antitoxin has a limited therapeutic window, it is essential to develop additional therapeutic agents to supplement antitoxin and intensive care for the treatment of botulism.[75] These include inhibitors of LC protease activity and agents that can accelerate clearance of light chain from the nerve terminal.[79] Much progress has already been made in this direction and serotype-selective inhibitors effective in symptomatic patients are expected to be available by the next decade for improved treatment of BoNT intoxication both from inhalation exposure and from natural outbreaks.

Abbreviations

ACh	Acetylcholine
BoNT	Botulinum neurotoxin
HC	Heavy chain
IU	International unit
i.m.	Intramuscular

i.n.	Intranasal
i.p.	Intraperitoneal
i.v.	Intravenous
LC	Light chain
LCt_{50}	Median lethal concentration × time
LD_{50}	Median lethal dose
$MIPLD_{50}$	Mouse intraperitoneal LD_{50}
NAP	Neurotoxin-associated protein
PBT	Pentavalent (ABCDE) botulinum toxoid
SNAP-25	Synaptosomal-associated protein of 25 kDa
SNARE	Soluble *N*-ethylmaleimide-sensitive factor attachment protein receptor
TNFα	Tumor necrosis factor alpha

Acknowledgements

The authors would like to thank Drs Susan Maslanka, Angelo Scorpio, Leonard Smith, Sean O'Malley, James Apland and Frank Lebeda for helpful suggestions and enlightening discussions.

References

1. V. Sathyamoorthy and B. R. DasGupta, *J. Biol. Chem.*, 1985, **260**, 10461.
2. C. L. Hatheway, in *Botulinum Neurotoxin and Tetanus Toxin*, ed. L. L. Simpson, Academic Press, San Diego, 1989, p. 3.
3. N. Ambache, *J. Physiol.*, 1951, **113**, 1.
4. L. L. Simpson, *Annu. Rev. Pharmacol. Toxicol.*, 2004, **44**, 167.
5. R. Topakian, C. Heibl, K. Stieglbauer, B. Dreer, M. Nagl, P. Knoflach and F. T. Aichner, *J. Neurol.*, 2009, **256**, 803.
6. A. N. Sheth, P. Wiersman, D. Atrubin, V. Dubey, D. Zink, G. Skinner, F. Doerr, P. Juliao, G. Gonzalez, C. Burnett, C. Drenzek, C. Shuler, J. Austin, A. Ellis, S. Maslanka and J. Sobel, *Clin. Infect. Dis.*, 2008, **47**, 1245.
7. T. H. Hammer, S. Jespersen, J. Kanstrup, V. C. Ballegaard, A. Kjerulf and A. Gelvan, *Infect. Dis.*, 2015, **47**, 190.
8. H. Yuan, G. Inami, J. Mohle-Boetani and D. J. Vugia, *Clin. Infect. Dis.*, 2011, **52**, 862.
9. T. R. Cote, A. K. Mohan, J. A. Polder, M. K. Walton and M. M. Braun, *J. Am. Acad. Dermatol.*, 2005, **53**, 407.
10. E. Yiannakopoulou, *Pharmacology*, 2015, **95**, 65.
11. H. A. Dabritz, K. K. Hill, J. R. Barash, L. O. Ticknor, C. H. Helma, N. Dover, J. R. Payne and S. S. Arnon, *J. Infect. Dis.*, 2014, **210**, 1711.
12. Y. D. Sheppard, D. Middleton, Y. Whitefield, F. Tyndel, S. Haider, J. Spiegelman, R. H. Swartz, M. P. Nelder, S. L. Baker, L. Landry, R. Maceachern, S. Deamond, L. Ross, G. Peters, M. Baird, D. Rose, G. Sanders and J. W. Austin, *Emerging Infect. Dis.*, 2012, **18**, 1.

13. J. Sobel, *Clin. Infect. Dis.*, 2005, **41**, 1167.
14. G. Schiavo, M. Matteoli and C. Montecucco, *Physiol. Rev.*, 2000, **80**, 717.
15. J. E. Keller, *Neuroscience*, 2006, **139**, 629.
16. L. Simpson, *Toxicon*, 2013, **68**, 40.
17. O. Rossetto, M. Pirazzini and C. Montecucco, *Nat. Rev. Microbiol.*, 2014, **12**, 535.
18. E. Habermann and F. Dreyer, *Curr. Top. Microbiol. Immunol.*, 1986, **129**, 94.
19. M. F. Brin and A. Blitzer, in *Botulinum Toxin*, ed. A. Carruthers and J. Carruthers, Saunders Elsevier, London, 2013, p. 6.
20. R. L. Shapiro, C. Hatheway and D. L. Swerdlow, *Ann. Intern. Med.*, 1998, **129**, 221.
21. R. F. Robinson and M. C. Nahata, *Ann. Pharmacother.*, 2003, **37**, 127.
22. J. Sobel, N. Tucker, A. Sulka, J. McLaughlin and S. Maslanka, *Emerging Infect. Dis.*, 2004, **10**, 1606.
23. S. E. Hill, R. Iqbal, C. L. Cadiz and J. Le, *Ann. Pharmacother.*, 2013, **47**, e12.
24. D. R. Franz, in *Medical Aspects of Chemical and Biological Warfare: Textbooks of Military Medicine*, ed. F. R. Sidell, E. T. Takafuji and D. R. Franz, Borden Institute, Washington, DC, 1997, p. 603.
25. S. S. Arnon, R. Schechter, T. V. Inglesby, D. A. Henderson, J. G. Bartlett, M. S. Ascher, E. Eitzen, A. D. Fine, J. Hauer, M. Layton, S. Lillibridge, M. T. Osterholm, T. O'Toole, G. Parker, T. M. Perl, P. K. Russell, D. L. Swerdlow and K. Tonat, *JAMA*, 2001, **285**, 1059.
26. Z. F. Dembek, L. A. Smith and J. M. Rusnak, *Disaster Med. Public Health Prep.*, 2007, **1**, 122.
27. D. C. Sanford, R. E. Barnewall, M. L. Vassar, N. Niemuth, K. Metcafe, R. V. House, I. Henderson and J. D. Shearer, *Clin. Vaccine Immunol.*, 2010, **17**, 1293.
28. D. R. Franz, P. B. Jahrling, A. M. Friedlander, D. J. McClain, D. L. Hoover, W. R. Bryne, J. A. Pavlin, G. W. Christopher and E. M. Eitzen, Jr., *JAMA*, 1997, **278**, 399.
29. G. L. Morefield, R. F. Tammariello, B. K. Purcell, P. L. Worsham, J. Chapman, L. A. Smith, J. B. Alarcon, J. A. Mikszta and R. G. Ulrich, *J. Immune Based Ther. Vaccines*, 2008, **6**, 5.
30. C. J. Roy, D. S. Reed and J. A. Hutt, *Vet. Pathol.*, 2010, **47**, 779.
31. V. Savransky, D. C. Sanford, E. Syar, J. L. Austin, K. P. Tordoff, M. S. Anderson, G. V. Stark, R. E. Barnewall, C. M. Briscoe, L. Lemiale-Bierinx, S. Park, B. Ionin and M. H. Skiadopoulos, *Infect. Immun.*, 2013, **81**, 1152.
32. D. S. Chertow, E. T. Tan, S. E. Maslanka, J. Schulte, E. A. Bresnitz, R. S. Weisman, J. Bernstein, S. M. Marcus, S. Kumar, J. Malecki, J. Sobel and C. R. Braden, *JAMA*, 2006, **296**, 2476.
33. F. Roblot, M. Popoff, J. P. Carlier, C. Godet, P. Abbadie, S. Matthis, A. Eisendorn, G. Le Moal, B. Becq-Giraudon and P. Roblot, *Clin. Infect. Dis.*, 2006, **43**, e51.
34. B. M. Lund, *Lancet*, 1990, **336**, 982.
35. J. Sobel, M. Malavet and S. John, *Clin. Infect. Dis.*, 2007, **45**, e14.

36. Y. Fujinaga, *J. Biomed. Biotechnol.*, 2010, 974943, DOI: 10.1155/2010/974943, epub 2010 Feb 14.
37. S. Gu, S. Rumpel, J. Zhou, J. Strotmeier, H. Bigalke, K. Perry, C. B. Shoemaker, A. Rummel and R. Jin, *Science*, 2012, **335**, 977.
38. B. R. DasGupta and H. Sugiyama, *Biochem. Biophys. Res. Commun.*, 1972, **48**, 108.
39. M. Dong, H. Liu, W. H. Tepp, E. A. Johnson, R. Janz and E. R. Chapman, *Mol. Biol. Cell*, 2008, **19**, 5226.
40. L. Peng, W. H. Tepp, E. A. Johnson and M. Dong, *PLoS Pathog.*, 2011, **7**(**3**), e1002008.
41. J. B. Park and L. L. Simpson, *Infect. Immun.*, 2003, **71**, 1147.
42. M. Elias, F. Al-Saleem, D. M. Ancharski, A. Singh, Z. Nasser, R. M. Olson and L. L. Simpson, *J. Pharmacol. Exp. Ther.*, 2011, **336**, 605.
43. F. H. Al-Saleem, D. M. Ancharski, S. G. Joshi, M. Elias, A. Singh, Z. Nasser and L. L. Simpson, *Infect. Immun.*, 2012, **80**, 4133.
44. L. K. Koriazova and M. Montal, *Nat. Struct. Biol.*, 2003, **10**, 13.
45. G. Schiavo, O. Rosetto, S. Catsicas, P. Polverino de Laureto, B. R. DasGupta, F. Benfenati and C. Montecucco, *J. Biol. Chem.*, 1993, **268**, 23784.
46. T. Binz, Y. Blasi, S. Yamasaki, A. Baumeister, E. Link, T. C. Sudhof, R. Jahn and H. Niemann, *J. Biol. Chem.*, 1994, **269**, 1617.
47. G. Schiavo, F. Benfenati, B. Poulain, O. Rosetto, P. Polverino de Laureto, B. R. DasGupta and C. Montecucco, *Nature*, 1992, **359**, 832.
48. S. Yamasaki, A. Baumeister, T. Binz, J. Blasi, E. Link, F. Cornille, B. Roques, E. M. Fykse, T. C. Sudhof and R. Jahn, *J. Biol. Chem.*, 1994, **269**, 12764.
49. J. Blasi, E. R. Chapman, E. Link, T. Binz, H. Niemann and R. Jahn, *EMBO J.*, 1993, **12**, 4821.
50. M. Adler, J. E. Keller, R. E. Sheridan and S. S. Deshpande, *Toxicon*, 2001, **39**, 233.
51. J. L. Middlebrook and D. R. Franz, in *Medical Aspects of Chemical and Biological Warfare: Textbook of Military Medicine*, ed. F. R. Sidell, E. T. Takafuji and D. R. Franz, Borden Institute, Washington, DC, 1997, p. 643.
52. J. M. Rusnak, M. G. Kortepeter, R. J. Hawley, A. O. Anderson, E. Boudreau and E. Eitzen, Jr., *Biosecur. Bioterror.*, 2004, **2**, 281.
53. H. J. Jansen, F. J. Breeveld, C. Stijnis and M. P. Grobusch, *Clin. Microbiol. Infect.*, 2014, **20**, 488.
54. K. B. Olson, *Emerging Infect. Dis.*, 1999, **5**, 513.
55. L. M. Wein and Y. Liu, *Proc. Natl. Acad. Sci. U. S. A.*, 2005, **102**, 9984.
56. V. E. Holzer, *Med. Klin.*, 1962, **57**, 1735.
57. M. A. Cardella, in *Botulism: Proceeding of a Symposium*, ed. K. H. Lewis and K. Cassel, Jr., U.S. Dept. of Health, Education and Welfare, Public Health Service Publication 999-FP-1, Cincinnati, OH, 1964, p. 113.
58. L. W. Cheng and T. D. Henderson II, *Toxicon*, 2011, **58**, 62.
59. D. A. Creasia, J. D. Thurman, L. J. Jones III, M. L. Nealley, C. G. York, R. W. Wannemacher Jr. and D. L. Bunner, *Fundam. Appl. Toxicol.*, 1987, **8**, 230.
60. D. A. Creasia, J. D. Thurman, R. W. Wannemacher, Jr. and D. L. Bunner, *Fundam. Appl. Toxicol.*, 1990, **14**, 54.

61. K. M. Busl and T. P. Bleck, *Neurotherapeutics*, 2012, **9**, 139.
62. T. C. Flynn and R. E. Clark II, *Dermatol. Surg.*, 2003, **29**, 519.
63. B. Mégarbane and L. Chevillard, *Chem.-Biol. Interact.*, 2013, **206**, 444.
64. D. B. Kudrow, D. A. Henry, D. A. Haake, G. Marshall and G. E. Mathisen, *Ann. Intern. Med.*, 1988, **109**, 984.
65. K. L. McDonald, G. W. Rutherford, S. M. Friedman, J. R. Dietz, B. R. Kaye, G. F. McKinley, J. H. Tenney and M. L. Cohen, *Ann. Intern. Med.*, 1985, **102**, 616.
66. D. R. Franz, L. M. Pitt, M. A. Clayton, M. A. Hanes and K. J. Rose, in *Botulinum and Tetanus Neurotoxins*, ed. B. R. DasGupta, Plenum Press, New York, 1993, p. 473.
67. T. R. Gelzleichter, M. A. Myers, R. G. Menton, N. A. Niemuth, M. C. Matthews and M. J. Langford, *J. Appl. Toxicol.*, 1999, **19**, 835.
68. L. Taysse, S. Daulon, J. Calvet, S. Delamanche, D. Hilaire, B. Bellier and P. Breton, *Toxicol. Pathol.*, 2005, **33**, 336.
69. J. R. Harkema, *Toxicol. Pathol.*, 1991, **19**, 321.
70. M. A. Poli, V. R. Rivera, M. L. Pitt and P. Vogel, *Toxicon*, 1996, **34**, 1037.
71. D. Vasconcelos, R. Barnewall, M. Babin, R. Hunt, J. Estep, C. Nielsen, R. Carnes and J. Carney, *Lab. Invest.*, 2003, **83**, 1201.
72. R. D. LeClaire, R. E. Hunt, S. Bavari, J. E. Estep, G. O. Nelson and C. L. Wilhelmsen, *Toxicol. Pathol.*, 1996, **24**, 619.
73. B. A. Woodruff, A. M. Griffin, L. M. McCroskey, J. F. Smart, R. B. Wainwright, R. G. Bryant, L. C. Hutwagner and C. L. Hatheway, *J. Infect. Dis.*, 1992, **166**, 1281.
74. D. Wang, J. Krilich, J. Baudys, J. R. Barr and S. R. Kalb, *Anal. Biochem.*, 2014, **468C**, 15.
75. L. A. Smith and J. M. Rusnak, *Crit. Rev. Immunol.*, 2007, **27**, 303.
76. J. Boles, M. West, V. Montgomery, R. Tammariello, L. M. Pitt, P. Gibbs, L. Smith and R. D. LeClaire, *Toxicon*, 2006, **47**, 877.
77. F. J. Erbguth, *Mov. Disord.*, 2004, **19**(**suppl. 8**), S2.
78. I. Matak and Z. Lacković, *Prog. Neurobiol.*, 2014, **119–120**, 39.
79. M. Adler, N. Gul, E. Eitzen, G. Oyler and B. Molles, in *Molecular Aspects of Botulinum Neurotoxin, Current Topics in Neurotoxicity 4*, ed. K. A. Foster, Springer, New York, 2014, p. 291.

CHAPTER 5

The Structural Biology and Biochemistry of the Ricin Toxin and the Military Use and Inhalation Toxicology of Ricin Aerosols†

TERRY J. HENDERSON*[a], RUSSELL M. DORSEY[a], GEORGE EMMETT[b], AND HARRY SALEM[a]

[a]Research and Technology Directorate, US Army Edgewood Chemical Biological Center, Aberdeen Proving Ground, MD 21010, USA; [b]Battelle Memorial Institute, Chemical Security Analysis Center, Aberdeen Proving Ground, MD 21010, USA
*E-mail: terry.j.henderson.civ@mail.mil

5.1 Introduction

In the wake of recent terrorist-related incidents across the globe, biological agents have become known for the threat they pose to US and allied militaries and to the civilian populations that they defend. The agents are attractive

†The opinions expressed in this chapter are the private views of the authors and are not to be construed as an official Department of the Army position unless so designated by other authorizing documents. This chapter has been approved for public release.

Issues in Toxicology No. 25
Aerobiology: The Toxicology of Airborne Pathogens and Toxins
Edited by Harry Salem and Sidney Katz

Published by the Royal Society of Chemistry, www.rsc.org

to both foreign states and terrorists because they are relatively inexpensive to produce and require minimal technical infrastructures. Towards such ends, aerosolization is recognized to be the most effective means of agent dissemination for inflicting mass casualties. Governments with active offensive biological warfare programs recognized early that pathogens and toxins formulated for protection against environmental stress, sized for optimal pulmonary delivery and dispersed in a manner that maximizes homogeneity and high-dose exposures, were relatively inexpensive to produce and highly effective weapons against military and civilian targets.[1] The governments soon realized that the airborne delivery of appropriately formulated pathogens and toxins was exceedingly difficult to defeat or prevent,[2] largely because they directly target the most susceptible route of entry – the respiratory tract.

The biological agents commonly associated with military and terrorist use are aptly classified as biological select agents and toxins (BSATs). Under US law, BSATs are biological agents that have been declared by the US Department of Health and Human Services (HHS) or the US Department of Agriculture to have the potential to pose a severe threat to public health and safety. BSATs include several bacteria such as *Coxiella burnetii*, *Francisella tularensis* and *Yersinia pestis*, numerous viruses such as the corona viruses, encephalitis viruses, influenza viruses and viral hemorrhagic fever viruses, and finally, a very large number of biologically derived molecular toxins. Biological toxins are the metabolic products of living organisms and, in contrast to bacteria and viruses, cannot reproduce. A common feature to all such toxins is that minute quantities will exert a pronounced effect on their intended target individuals. The toxins are produced by a vast number of different life forms, ranging from the simplest to the most complex, and each toxin has a distinctive mode of action in conjunction with a characteristic molecular structure and biochemistry. Algal toxins constitute a very diverse group of compounds ranging from simple ammonia to complex polypeptides and polysaccharides. Dinoflagellates are a source of some potent, nonprotein toxins such as saxitoxin and tetrodotoxin; one such toxin is the cause of red tides. The mycotoxins are a wide variety of nonprotein substances produced by molds and fungi. Many molds produce more than one toxin and, in several cases, combinations of mycotoxins synergize to enhance toxicity.[3] Bacteria produce some of the most potent toxins known, including botulinum neurotoxin from *Clostridium botulinum* and shiga toxin from *Shigella dysenteriae*. The leaves, roots or seeds from some plants can be poisonous. Examples include the protein toxins ricin derived from castor beans and abrin from the jequirity pea. Finally, there are a number of toxins produced by animals, including the nonprotein batrachotoxin found in certain species of frogs (poison dart frog) and a wide variety of peptides and proteins from marine snails (the conotoxins in particular), scorpion venoms and snake venoms.

Undoubtedly, the most widely recognized BSAT is the ricin protein toxin from the seeds of the castor bean plant *Ricinus communis*. Both the ricin toxin and the castor bean plant have a long history. Some of the earliest records contain references to the castor bean plant, which was known and cultivated by the ancient Egyptians. Oil pressed from the beans was used as a lubricant and fuel for oil lamps and the beans were sometimes ingested to treat constipation. Even today, castor oil is still commercially produced in many countries across the globe; ~1 Mt of castor beans are harvested annually for castor oil production. Castor oil is currently used in bath oil products, detergents, lubricants, dyeing agents and, most recently, in the production of biodiesel fuel. The ricin toxin is easily produced from the mash remaining after the castor beans have been pressed to extract castor oil.[4] Initially, the ricin protein was known for its ability to aggregate red blood cells and platelets and later the field of immunology was founded by the pioneering research of Ehrlich with the toxin.[5] By introducing small amounts of the toxin to mice, Ehrlich was able to induce specific immunity to ricin. This seminal work demonstrated conclusively that in response to a challenge, certain serum proteins (antibodies) are produced, which affords the host protection.

By virtue of its toxicity, availability and ease of production, ricin was the first BSAT considered for weaponization by the US [6] and other governments. Today, ricin is one of the few BSATs used with known success, including the 1978 attack of the Bulgarian exile Georgi Markov,[7] the 1980 attack of Boris Korczak in a Tysons Corner, VA, parking lot[8] and the 2008 incident involving Roger Von Bergendorff in Las Vegas, NV.[9] There are also over a dozen other documented events that have occurred in the recent past.[10] Ricin is an HHS select toxin covered under 7 CFR (Title 7 of the US Code of Federal Regulations) Part 331, 9 CFR Part 121 and 42 CFR Part 73. In this chapter, we discuss the ricin toxin in detail, with special attention to the military use and inhalation toxicology of its aerosols. As an introduction to the toxin, we begin by reviewing its structural biology and the biochemistry relevant to its toxicity. The balance of the review, however, is dedicated to ricin aerosols, especially the most recent research focused on the inhalation toxicology of the toxin.

5.2 Structural Biology and Biochemistry of the Ricin Ribosome Inactivating Protein

One of the most extensively studied protein families is that of the ribosome-inactivating proteins (RIPs), a large family of protein toxins that includes ricin. The expression "ribosome-inactivating protein" was introduced to designate plant proteins that could inactivate animal ribosomes. Such protein toxins, however, were found sometime later in certain bacteria and fungi.[11] The RIPs have been classified into two distinct types, referred to as types I and II, based on their quaternary structures. Type I RIPs are the most numerous and are synthesized as a single-chain enzyme of ~30 kDa. Their type II counterparts, in contrast, are synthesized as larger precursors and contain a ~30 kDa A-chain linked by a disulfide bond to a lectin B-chain of similar size.[12] The A-chain is responsible for the enzymatic activity of the toxin, while the B-chain binds to a specific carbohydrate receptor on the host cell membrane to allow the A-chain to cross the cell membrane. B-chains contain two carbohydrate-binding sites[13,14] that allow them to aggregate with red blood cells and platelets. Type II RIPs were discovered more than a century ago, when Stillmark[15] discovered that the beans of the castor plant contained a toxic protein, which he named ricin. The toxicity of ricin was initially attributed to its agglutination activity for red blood cells and not its ribosome inactivating activity. Historically, understanding the ricin toxin was complicated by the presence of a homologous protein also isolated from castor beans, which was later identified as the *R. communis* agglutinin. The agglutinin shares considerable sequence identity with the ricin toxin but is much less toxic and is a homodimer of two A- and two B-chains. The coexistence of two highly similar proteins from castor beans, one a potent cytotoxin and the other an effective hemagglutinin, came to light by improved separation methods and the identification of their respective genes.[16–18]

Plant RIPs, including all type I toxins and the A-chains of type II toxins, are RNA *N*-glycosidases capable of hydrolyzing the nitrogen–carbon glycosidic bond of a specific adenosine located in the sarcin–ricin domain of the largest ribosomal RNA.[19] The B-chain binds the D-galactose-terminated receptors on the membranes of animal cells, which facilitates the internalization of the enzymatic A-chain. To date, more than 50 type I and ~15 type II RIPs have been identified.[12] While there is some variation at the N- and C-termini of their respective polypeptide chains, the active sites of the type I RIPs and the A-chains of type II RIPs are well conserved, as are their three-dimensional structures.[20]

Currently, the role of RIPs in plant physiology is not entirely clear. Based on their variable activity towards heterologous and autologous plant ribosomes, several possible roles have been proposed, including anti-viral activity, anti-fungal activity, herbivore defense, a role in the arrest of cellular metabolism during periods of senescence and, finally, a role as storage proteins.[21] Conclusive evidence has been discovered that RIPs not only deadenylate ribosomal RNA but also remove adenine residues from DNA and several other polynucleotide substrates.

5.2.1 Crystal Structure of the Ricin Toxin

Ricin is a type II RIP of 64–65 kDa isolated exclusively from the seeds of *R. communis*. A cartoon rendering of the ricin protein backbone based on X-ray crystallographic data[22–25] is presented in Figure 5.1 to illustrate the toxin's overall molecular architecture. The A-chain contains considerable secondary structure whereas, in contrast, the B-chain is devoid of any regular secondary structural elements. As indicated in the figure, the key features of the B-chain are the two galactose-binding sites and two bound mannose oligosaccharides. The two chains are linked together by a single disulfide bond (yellow in the figure) between Cys259 of the A-chain and Cys4 of the B-chain.

Significantly more X-ray crystallography data have been reported on the A-chain largely because of investigations into the chain's catalytic activity.

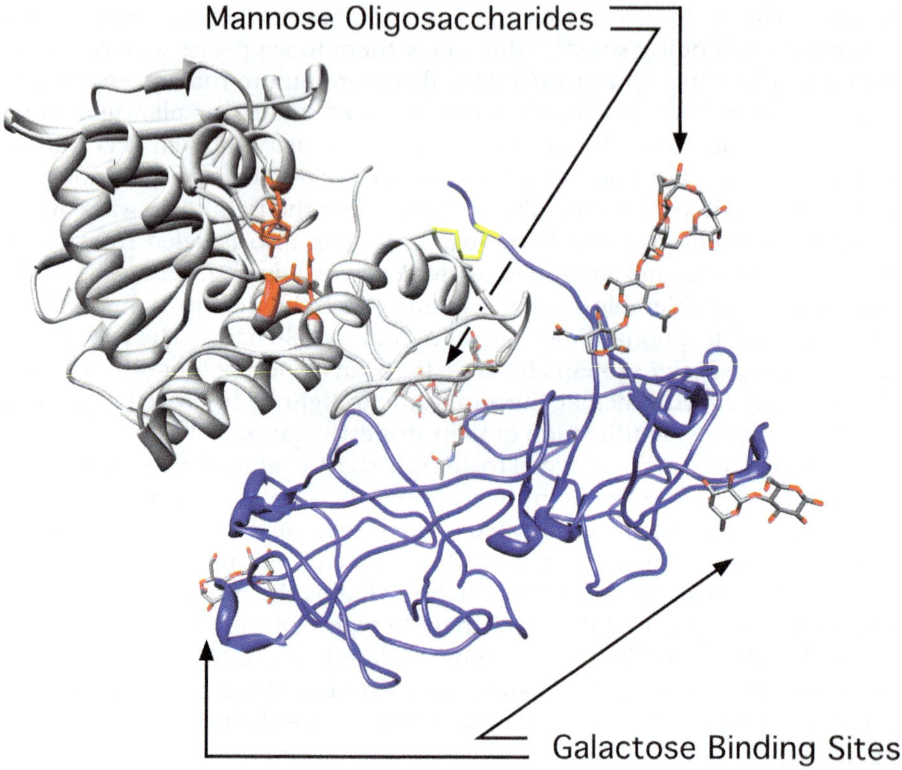

Figure 5.1 Cartoon rendering of the ricin backbone in three-dimensional crystallographic space.[22–25] Key features of the toxin are illustrated, including the A- and B-chains in gray and blue, respectively, the active site residues in red, the cysteine residues forming the disulfide bond between the two chains in yellow and carbohydrates bound to the B-chain colored by element (oxygen is red, nitrogen is blue, carbon is gray and hydrogen is not shown). The bound mannose oligosaccharides and the galactose binding sites of the B-chain are labeled. Figure created with the program USCF Chimera-1.10.1 [125,126] on a computer running Darwin 64 and the Aqua graphical user interface (Apple, Cupertino, CA, USA).

The A-chain is a 267 amino acid, mushroom-shaped α/β globular protein whose structure is largely unaffected by the absence of the B-chain.[26] The chain contains eight α-helices and eight β-strands[26,27] and can be divided into three domains, shown in Figure 5.2 as domains 1, 2 and 3. Domain 1 contains a six-stranded β-sheet with parallel and anti-parallel strands, whereas domain 2 contains five helices, the longest of which forms the central core of the protein chain. It is kinked near its C-terminus to allow two key residues for enzymatic activity, Glu177 and Arg180, to reach the active site. Domain 3 contains the amino acids involved in the A-chain–B-chain interface and contains one α-helix and two β-strands. The location of the active site (red amino acid residues in Figures 5.1 and 5.2) is in a cleft on the A-chain surface and is

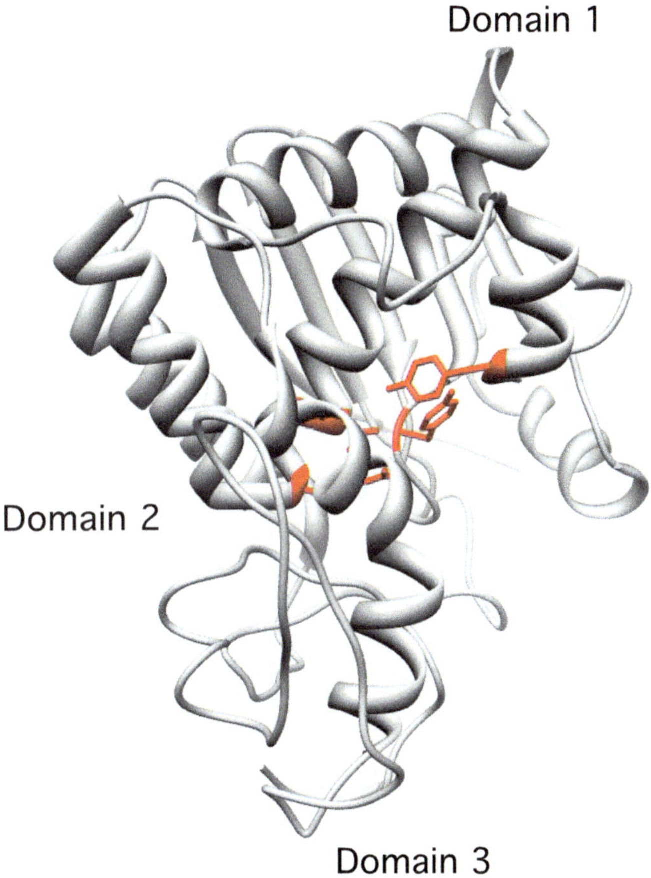

Figure 5.2 Cartoon rendering of the ricin A-chain backbone in three-dimensional crystallographic space[22,23,25] illustrating its three domains (labeled) and active site residues in red. Figure created with the program USCF Chimera-1.10.1 (see the caption of Figure 5.1 for more details).

characterized by several conserved amino acid residues. Glu177 and Arg180 are critical to the rate-limiting step of the toxin's catalytic activity rather than substrate binding, as mutation of these residues gives a reduction in k_{cat} but not k_m. As shown in Figure 5.3, the invariant residues Tyr80 and Tyr123 are principally involved in substrate binding. X-ray crystallography reveals that adenosine is converted to adenine when complexed with the ricin A-chain and intercalates with the aromatic rings of the two tyrosine residues (Figure 5.3a). The active site lies at the back of a deep pocket on the A-chain surface (Figure 5.3b). Recognition of the rRNA substrate contributes substantially to the efficiency of the depurination reaction. Activity is reduced ~10 000-fold when the rRNA proteins are removed[28] but substrate binding is not reduced. Mutation of Ser203 outside the active site to an asparagine reduces A-chain activity,[29] demonstrating clearly that contacts between the rRNA substrate

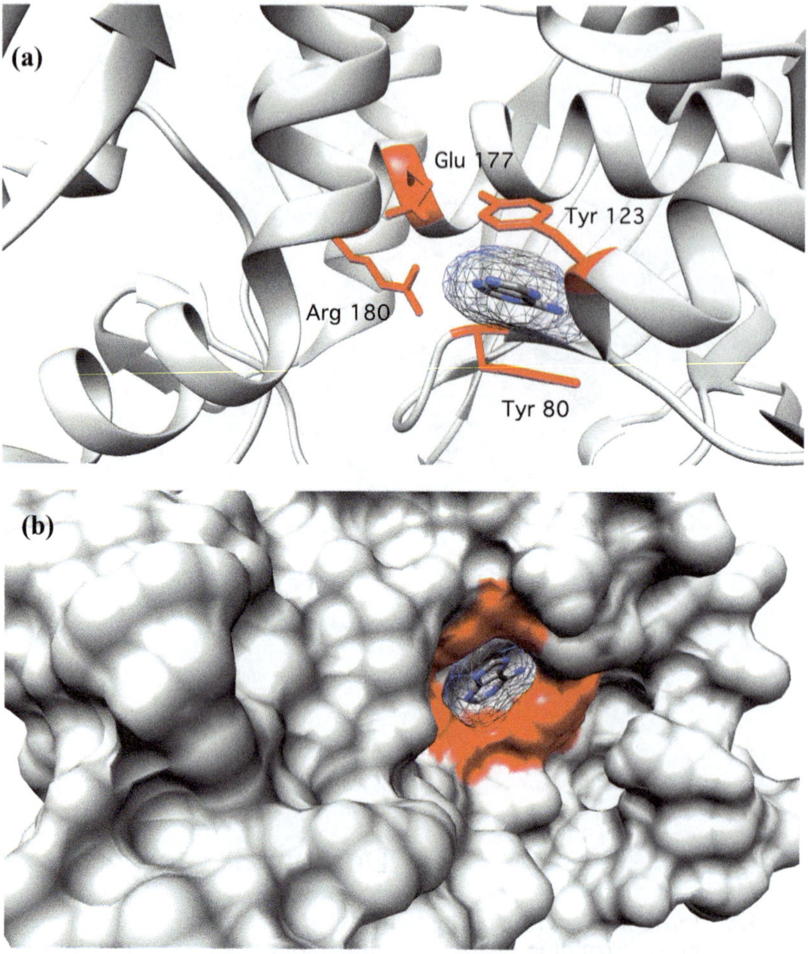

Figure 5.3 The binding of adenine in the ricin active site. Adenosine is converted to adenine when complexed with the ricin A-chain and (a) is held in the active site by intercalation with the aromatic rings of Tyr80 and Tyr123. Glu177 and Arg180 are critical for the catalytic activity of the toxin and (b) line the active site located within a deep pocket of the A-chain. The ricin backbone is represented as a cartoon rendering or solvent-excluded molecular surface in gray with active site amino acid residues shown in red. The structure of adenine is shown without hydrogen atoms and is colored by element (nitrogen is blue and carbon is gray) along with its solvent-excluded molecular surface illustrated with a mesh surface. The figures were created from the crystallographic coordinates of Carra *et al.*[127] using the program USCF Chimera-1.10.1 (see the caption of Figure 5.1 for more details).

and amino acid residues outside the active site are critical to A-chain catalytic activity. There are no known inhibitors of ricin, presumably because of its exceptional affinity for its natural substrate.

At 262 amino acids residues, the ricin B-chain protein, sometimes referred to as the ricin-type beta-trefoil lectin, folds into a bilobal, barbell-shaped architecture.[23] The B-chain has no regular secondary structure but does display several Ω-loops, nonregular secondary structures characterized by the peptide backbone following a loop-shaped course (shown in Figure 5.1 as short, poorly formed blue helices). The folded structure has two homologous domains, each with a galactose-binding site to allow hydrogen bonding to the galactose and *N*-acetylgalactosamine carbohydrates typically found on cell surfaces. Each domain is comprised of three copies of a primitive 40 amino acid folding sub-domain packed around a pseudo-threefold symmetry axis.[30] The sub-domains contribute conserved tryptophan, leucine and isoleucine residues in a compact hydrophobic core and it is this tight, threefold binding that likely drives the folding of the B-chain and stabilizes its structure.[23] The amino acid residues of the sub-domains collectively account for ~92% of all the B-chain amino acids. Galactose binds each domain in a shallow cleft formed by a three amino acid kink on the bottom and an aromatic ring on the top. At the back of the cleft, an aspartate residue forms hydrogen bonds to the C-3 and C-4 hydroxyl groups of galactose and a glutamine bonds to the to the C-4 alcohol. The mannose oligosaccharides of the B-chain are able to bind cells that express mannose receptors,[31] including cells of the reticuloendothelial system.[32]

5.2.2 Cultivars and Isotoxins of Ricin

A number of varieties, or cultivars, of *R. communis* have been identified; some share similar morphology but others have a different appearance. These properties are also found for the seeds produced by the different cultivars, where ricin is synthesized and stored. Ricin isotoxins were discovered nearly 40 years ago when they were first identified as ricin D and ricin E.[33] The ricin D isotoxin was purified and identified 13 years before this discovery by Ishiguro *et al.*[34] The two isotoxins have slightly different amino acid compositions and isoelectric points that differ by 1.5 pH units, but their toxicities were reported to be the same by Mise *et al.*[33] Other, early investigations of the two isotoxins with various cell lines, however, made reference to a ~50-fold higher activity for ricin D in murine cells. The isotoxins were reported to be somewhat specific to the large and small castor bean cultivars, with the ricin D isotoxin being found in the large bean whereas both ricin D and E are found in the small bean cultivars. Other researchers have identified the ricin isotoxins as RCLIII and RCLIV, corresponding to ricin D and ricin E, respectively, with RCLI and RCLII referring to two different *R. communis* agglutinins[35] and not toxins.

5.2.3 Cellular Internalization of Ricin

The enzymatic component of ricin is not active until it is released from the native toxin comprising both subunits. Isolated A-chains have enzymatic activity but lack the cell binding and entry capabilities, whereas isolated B-chains may bind to target cells and even block the binding of the native toxin, but are not toxic. The entry of ricin into a eukaryotic host cell begins with the reversible binding of its B-chain to cell-surface glycolipids and glycoproteins bearing terminal *N*-acetylgalactosamine or β(1 → 4)-linked galactose residues.[36] It has been demonstrated that 10^6–10^8 B-chains can bind a single cell at one time.[37] The toxin molecules are internalized into cells *via* endocytosis in coated pits and vesicles, or smooth pits and vesicles.[36] Endocytosis can be by clathrin-dependent or -independent mechanisms that are somewhat reliant on the cell type and their polarization status.[36,38–42] Once in the cytosol, the vesicles fuse with endosomes that can return some of the toxin molecules to the cell surface by exocytosis or, as shown in Figure 5.4, endosomes can fuse with lysosomes that will destroy the ricin molecules.[43,44] Alternatively, the endosomes can carry ricin molecules to the Golgi complex and endoplasmic reticulum (ER) by retrograde transport[44] (see Figure 5.4). Ricin molecules reaching the trans Golgi network can penetrate its membrane and reach the cytosol or, in some cases, return to the cell surface.[44] For ricin molecules reaching the ER lumen, the disulfide bond connecting their A- and B-chains will be reduced, facilitating partial unfolding of the A-chain.[45] The unraveling toxin protein will be translocated across the ER membrane *via* the Sec61p translocon, following the same pathway as that for incorrectly folded endoplasmic reticulum proteins targeted to ER-associated degradation. Ricin molecules escaping this degradation may find their way into the cytosol; these ricin proteins, and those crossing the trans Golgi network membrane into the cytosol, can refold into a protease-resistant, enzymatically active structure.[46] With a k_{cat} of 1500 min^{-1}, just a single ricin molecule entering the cytosol of a living cell can inactivate enough ribosomes to result ultimately in cell death.[47]

5.2.4 *N*-Glycosidase Activity of Ricin

The catalytic A-chain recognizes and binds a highly conserved region in the large 28S rRNA referred to as the sarcin/ricin loop,[48] a very short stem–loop structure in domain VII of the rRNA molecule ~400 nucleotides from its 3′-end. Within the stem–loop, the ring of a single adenine, A_{4324}, becomes intercalated between the aromatic rings of Tyr80 and Tyr123 in the catalytic cleft of the ricin A-chain (see Figure 5.3a) and is hydrolyzed at the carbon–nitrogen glycosidic bond by *N*-glycosidase action.[47] Although the detailed mechanism of this event is not known, Monzingo and Robertus[49] have proposed a mechanism based on their high-resolution X-ray crystallographic data for substrate analogs in the A-chain active site; the mechanism is summarized in Figure 5.5. The mechanism begins with the sarcin–ricin loop substrate binding the A-chain active site with the A_{4324} adenine ring system intercalating with the aromatic rings of Tyr80 and Tyr123 (Figure 5.3a). Arg180 donates a proton to N-3 of the adenine to break the bond between N-9 of the adenine ring and C-1′ of ribose (Figure 5.5a). This bond cleavage leaves cation character on the ribose ring in the form of an oxycarbonium ion and gives anion character to the adenine ring system (Figure 5.5b); the oxycarbonium ion should have sp^2 character at C-1′ (indicated in Figure 5.5b by the partial double-bond character with O-4′). The bond-forming process

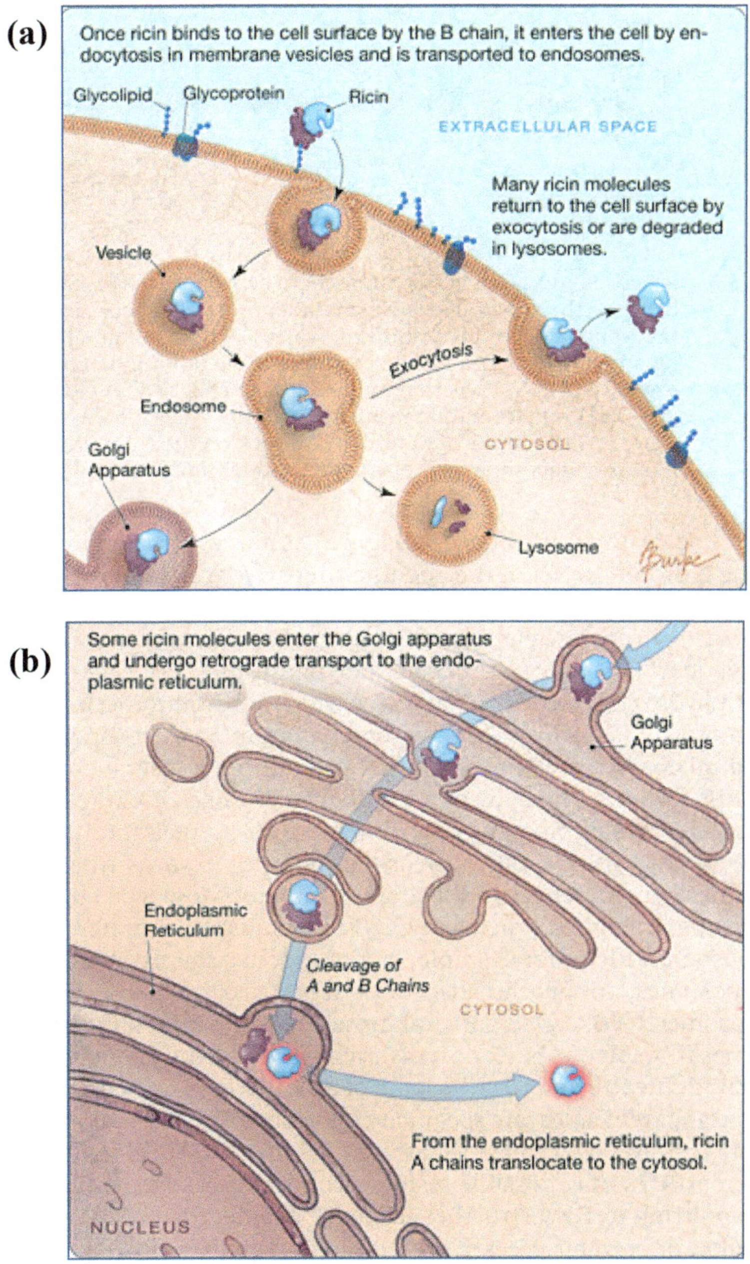

Figure 5.4 Example of the cellular internalization of ricin. The process involves (a) endocytosis by coated pits and vesicles or smooth pits and vesicles, followed by (b) vesicle–endosome fusion. Ricin molecules can then return to the cell surface by exocytosis, or the vesicles may fuse to lysosomes for toxin destruction. Figure from Audi *et al.*[119]

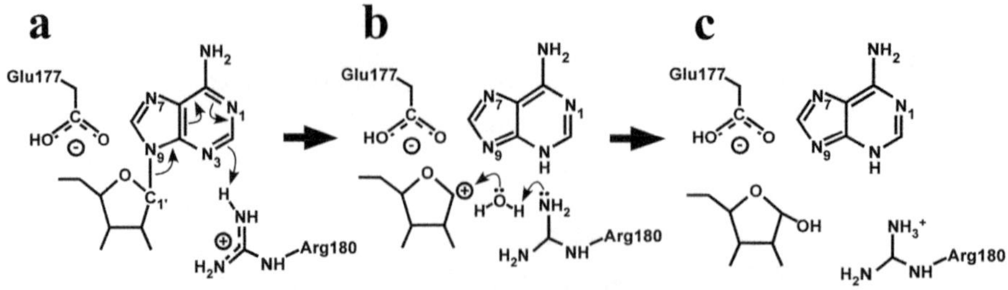

Figure 5.5 Proposed mechanism of ricin A-chain activity. (a) → (b) Bond breaking between the substrate ribose and adenine. Cation character builds on the ribose ring and is stabilized by Glu177 (E177) and anion character on adenine becomes stabilized through the donation of a proton from Arg180 to the N-3 position of the adenine ring system of the substrate. (b) → (c) Bond formation resulting from the attack on the ribose ring by water. Arg180 activates water for its attack on the ribose ring to give the final products shown in (c). Adapted from Monzingo and Robertus.[49]

(represented in the path between Figure 5.5b and c) begins with Arg180 donating a proton to N-3 of the adenine ring system to increase the basicity of the side chain. This, in turn, acts to pull a proton from the water molecule, creating hydroxide character to assist solvent attack on ribose and, ultimately, a neutral ribose (Figure 5.5c). The figure shows the adenine product to have a hydrogen atom at N-3, but this will readily tautomerize in solution to the more stable form with a hydrogen atom at N-9. Site-specific RNA *N*-glycosidase activity is a feature common to all previously identified type I and II RIP toxins.[47] The activity prevents the binding of elongation factors EF-1 and EF-2, resulting in the cessation of mRNA translation.

Although all RIPs exhibit *N*-glycosidase activity towards ribosomes, they display marked differences in substrate specificity. Most type I toxins exhibit very broad specificities, whereas type II toxins display a preference for animal ribosomes. Ricin, for example, is highly active against mammalian and yeast ribosomes, but poorly active, or even inactive, against plant and bacterial ribosomes.[50] Pokeweed antiviral protein, on the other hand, depurinates ribosomes from plants, bacteria, yeasts and various evolutionally diverse animals.[51] Both the RIP and the ribosome of an RIP–ribosome complex contribute to the apparent substrate specificity.[52] Because the rRNA sarcin–ricin loop structure is universally conserved, the broad range of specificities displayed by different RIPs may possibly reside within the associated ribosomal proteins conferring tertiary structure to the ribosome;[52] these differences may either allow or prevent access of the RIPs to the sarcin–ricin loop. Vater *et al.*[53] identified the rat liver ribosomal proteins L9 and L10e as the binding target of the ricin A-chain and yeast ribosomal protein L3 as the binding factor for pokeweed antiviral protein. In addition to the highly specific action on ribosomes, ricin and related RIPs have a less specific action *in vitro* on single- and double-stranded DNA, RNA substrates releasing multiple adenine residues and guanine residues in some instances.[54] Ricin A-chain also catalyzes the hydrolysis of synthetic oligonucleotides as short as six base pairs, provided that a specific tetra loop of guanosine and adenosine residues is present.[55]

5.3 Weaponization of the Ricin Toxin

Although many pathogens and toxins cause disease, toxicity, or other deleterious effects, relatively few of the naturally occurring agents can be adapted for use as a biological weapon. The ability to survive the mechanical rigors of large-scale aerosol dissemination is of paramount importance for the weaponization of a biological agent. In the case of BSATs, an "aerosol" is most suitably described as a system of solid or liquid particles of sufficiently small diameter to maintain stability as a suspension in air.[56] An aerosol includes both the particles and the suspending gas, which almost always is air. Other important characteristics include availability, ease of production and storage and lethality, or its ability to incapacitate through morbidity.[57] Whereas some who develop BSATs pursue agents of death, the philosophy of others is that the most attractive agents are those that incapacitate an adversary. An ideal agent is one that renders an adversary force defenseless while the disseminating force can be easily protected at the same time.

Agent stability after production is another important factor to consider when selecting a BSAT for weaponization. Environmental factors, including temperature, relative humidity, atmospheric pollution and sunlight, can significantly affect agent viability. The most feasible options for storing agents include their production in a liquid or solid form that is easily stored at room temperature or under refrigeration. The most desirable option is dry sample storage at room temperature, as this eliminates the need for refrigeration, allows for particle sizing, or milling, prior to dissemination and minimizes the environmental effects upon dissemination.

When all factors are considered, toxins are almost certainly the most suitable biological agents for use as a biological weapon. However, in contrast to agents capable of replication, the use of toxins as biological agents requires that a lethal amount must be injected, ingested or inhaled. With this in mind, a sizable amount of toxin must be available for dispersion in large-scale attacks, severely limiting the effective toxins to those that are the most lethal and easily produced. Ricin is one of only a handful of toxins that meet these requirements, along with botulinum toxin, staphylococcal enterotoxin B and the trichothecene mycotoxins. Owing principally to its suitability as a biological warfare agent, the ricin toxin has been one of the biological weapons of choice for state-sponsored organizations.[58]

5.3.1 Brief History of Ricin Weaponization

After becoming aware of Germany's involvement with biological weapons during World War I, the US government initiated its own biological weapons program towards the end of the war, which included the weaponization of the ricin toxin.[59] The US Bureau of Mines studied the offensive potential of ricin at the American University Experimental Station in Washington, DC, after limited experimental research on animals demonstrated that it was possible to weaponize ricin. Two methods of ricin dispersal were examined at the time: the adherence of ricin to shrapnel and bullets and the production of a ricin dust cloud. Not only was the shrapnel and bullets concept a direct violation of the Hague Convention of 1899[60], but also the heat generated while firing the coated bullets destroyed a significant amount of the toxin.[61] The dust cloud approach, on the other hand, was considered less promising owing to the limited amounts of purified ricin on hand together with its less efficient delivery *via* the respiratory tract. Although both approaches were laboratory tested to some extent, neither was perfected for battlefield use prior to the end of the war. Sometime during the early 1920s, US military leadership determined that conducting further research into the use of biological weapons was not an efficient use of resources.

Interest in the weaponization of ricin re-emerged during World War II, when the toxin was given the military designation "W" (and "WA" sometime later). During this time, the British and Canadian governments investigated the use of ricin in cluster bombs.[62] The US military's interest in ricin was reinvigorated sometime around 1942 as a project of the National Defense Research Committee[63] and led to chamber and field tests at Dugway Proving Ground, UT, in 1944.[64] Although definite plans for mass producing ricin cluster bombs existed at times and several field trials with different ricin bomblet concepts were conducted, Allied governments concluded that the use of ricin on the battlefield was no more economical than using the chemical agent phosgene (carbonyl dichloride, military designation "CG"), which was already weaponized at the time. This conclusion, however, was based on comparisons of the final weapons rather than the toxicity of ricin. Interest in developing ricin weapons continued for a short time after World War II, but soon diminished when the US Army Chemical Corps initiated their program to weaponize the nerve agent sarin [(*R*,*S*)-propan-2-yl methylphosphonofluoridate, military designation "GB"]. With the exception of one US patent awarded in 1965 for a small bomb that can be used to disperse toxic agents such as ricin in aerosol form,[65] there has not been any reported research or development of ricin as an offensive weapon since World War II. Present-day research focused on ricin by the US government and its allies is exclusively for defense against the toxin used by adversary forces or terrorists.

5.3.2 Aerosolization of Ricin

By a considerable margin, the best option for the use of ricin as an effective military weapon is aerosolization. Research clearly suggests that the toxicity of ricin by inhalation is a function of the aerodynamic equivalent diameter (the diameter of a spherical particle of standard unit density with the identical settling velocity in the same gas) of ricin aerosol particles.[66] The aerodynamic equivalent diameter, d_a, is an indicator of the aerodynamic behavior of a particle, which depends in its overall size, particle shape and density[67] according to the equation

$$d_a = d_e \left(\frac{\rho_p}{\rho_0 \chi} \right)^{\frac{1}{2}}$$

where d_e is the equivalent volume diameter (the diameter of a sphere of the same volume as that of an irregular particle), ρ_p and ρ_0 are the particle density and standard particle density (1 g cm^{-3}), respectively, and χ is the dynamic shape factor (ratio of the resistive force of an irregular particle to that of a spherical particle of the same volume and velocity). The equation is a simple expression for d_a derived by neglecting the slip correction and can be used to illustrate the difficulties in preparing ricin toxin aerosols. For an offensive military weapon, the d_a of ricin aerosol particles must be sufficiently small to reach the human lower respiratory tract to cause intoxication (discussed in more detail in Sections 5.4–5.4.4). Larger particles of >5 μm can be cleared from the upper respiratory tract by a number of different mechanisms (sneezing and swallowing, for example, among other mechanisms; see Section 5.4.4) before ever reaching the lower respiratory tract, and particles of <0.5 μm may not deposit in the respiratory tract at all, as they move by Brownian motion and settle very slowly and can be exhaled before deposition. However, strategies can be devised for engineering the d_a of larger aerosol particles to bring them into the aerodynamic respirable range of 1–5 μm. The equation reveals that particle d_a can be decreased by decreasing the particle size, decreasing ρ_p or increasing χ. Larger particles, therefore, can be made to fall into the aerodynamic respirable range by producing them as low-density, porous particles, producing them as very long, very thin particles (needle-shaped, for example) or a combination of both. The former strategy has been used in the pharmaceutical industry to bring particles with geometric sizes of ~20 μm into the aerodynamic respirable range.[68] For ricin in particular, isolation of the toxin from castor beans can have a considerable effect on the final ρ_p of individual particles because it can be very difficult to isolate pure toxin even with well-equipped laboratory facilities; very pure ricin preparations can be expected to have a ρ_p of ~1.4 gm cm^{-3}.[69] Crystallization techniques might be able to produce very long, very thin ricin particles as in the pharmaceutical industry.

5.3.3 Polydispersity of Ricin Aerosols

When prepared using sophisticated means by individuals with the appropriate knowledge, artificially produced formulations of biological aerosols are altogether different from their naturally produced counterparts. Such formulations can be designed for the generation of aerosol particles with extended residence times and to target directly the most susceptible route of entry. As with most aerosols, ricin aerosols are likely to be polydisperse colloidal systems with a range of particle sizes. The smaller the particle size d_a, the longer it will stay airborne and the more likely it can be re-aerosolized following settling. Particles of <10 μm can stay airborne for extended periods of time,[70–72] whereas a large fraction of aerosol particles of >15 μm simply drop harmlessly to the ground.[73] Because they are not volatile, ricin particles settling to the ground will no longer be a threat, even to unprotected warfighters and civilians. Although there are few data, it is unlikely that secondary aerosol formation caused by vehicular or personnel movement over ground previously exposed to ricin aerosol would even generate a significant threat.[73]

Mathematical modeling based on robust field tests predicted that under ideal meteorological conditions, the mass of ricin necessary to generate a cloud exposing individuals within a 100 km^2 area to a lethal dose is greater than several metric tons.[73] When considering actual battlefield or terrorism situations, however, much variability can be encountered that can impact predictions from the model. Examples of such variability include, for example, increased breathing rates resulting from exercise and dispersion of the aerosol into smaller volumes. No matter what theoretical models predict, ricin must first be extracted from castor beans, formulated and then milled to the appropriate dimensions to achieve a critical d_a size range.[74] The preparation of ricin to an appropriate d_a for inflicting injury or death requires technical skills in addition to access to adequately equipped laboratories.[75] The technical skills necessary for producing large quantities of ricin powder for aerosolization in a sufficiently hazardous form are well beyond the reach of most terrorists lacking state funding and resources.

5.4 Inhalation of Aerosolized Ricin

As mentioned earlier, aerosol is the most likely route of dissemination of BSATs in bioterrorist attacks regardless of the natural route of exposure to a biological agent. With its expansive surface area, the respiratory tract typically serves as a primary site of disease or toxicity, or as an entry portal for the pathogen or toxin stated in a more general sense, with the major manifestation of disease or toxicity realized at more distant sites within the individual. In addition to the characteristics of the aerosol particles that affect the delivered dose of the inhaled BSAT described above, the inherent infectivity, virulence and toxicity of the particular agent are relevant to the outcome after respiratory exposure. BSATs often have extremely different pathogenic effects following inhalation relative to their other routes of exposure and the relative importance of different virulence mechanisms used by BSATs may vary with exposure route.[1]

5.4.1 The Aerodynamics of Inhaled Ricin Particles

BSAT deposition in the respiratory tract is largely dependent upon particle size, but other factors can also become important. Deleterious environmental stressors such as ultraviolet radiation can significantly reduce the viability of agents contained in an aerosol particle. Particle size, however, does determine the initial deposition pattern within the respiratory tract and, ultimately, the ability of the deposited BSAT to initiate disease or intoxication. Particle sizing can be of paramount importance if the BSAT affects only particular cell types found in specific regions of the respiratory tract. Fine particle distributions of 1–3 μm have been found to be the optimal size for pulmonary deposition of most BSATs,[76] whereas larger particle sizes tend to be deposited higher in the respiratory tract (see Figure 5.6). For aerosolized ricin in particular, the deposition pattern is a critical determinant for the response and outcome of the individual. Mice exposed to ricin aerosols containing particles of >5 μm showed minimal effects when administered an inhaled dose of 20 μg kg^{-1}. If this aerosol contained particles of <1 μm, this same dose would be equivalent to roughly four lethal doses[77] because the smaller particles would have found their way into the lower respiratory tract. Similar decreases in lethality with increasing particle size have been found for guinea pigs exposed to aerosols of anthrax spores,[78] *Yersinia pestis*[79] and *Brucella suis*[80] and for guinea pigs[81] and rhesus macaques[82] exposed to aerosols of *Francisella tularensis*. Additionally, region-specific clearance mechanisms that facilitate the spread of an agent

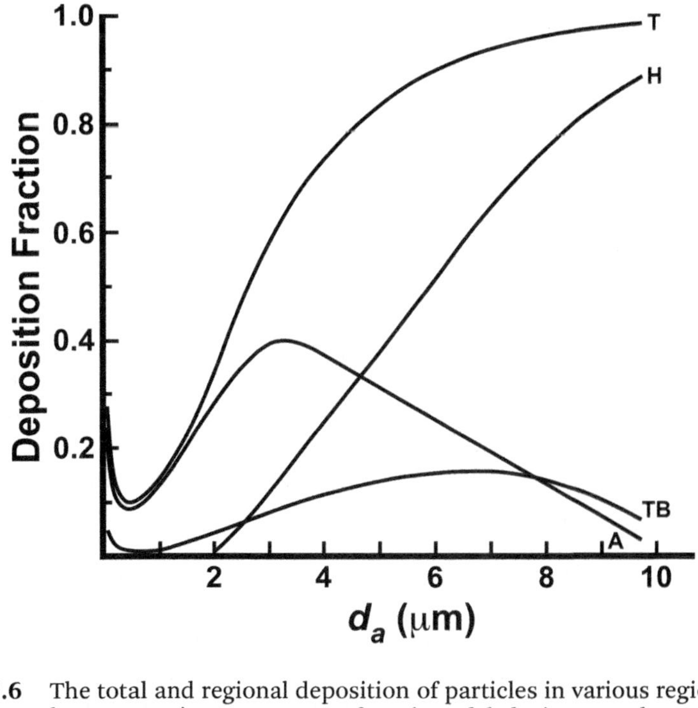

Figure 5.6 The total and regional deposition of particles in various regions of the human respiratory tract as a function of d_a during a moderate workload at 30 respirations per minute and a 750 cm^3 tidal volume. The lines represent calculated values for the head (H), tracheobronchial (TB), alveolar (A) and total (T) deposition. Adapted from Yu and Diu.[128]

to nonrespiratory tissues also contribute to variations in agent pathogenesis and toxicity. For example, Doebler *et al.*[83] found some of the ricin delivered to rodents by aerosol dissemination in the stomach, most likely because the test animals swallowed the toxin after its initial deposition in the nasopharynx.

5.4.2 The Fate of Aerosolized Ricin in the Respiratory Tract

Incidences involving human exposure to BSAT aerosols are rare; however, the fate of many different inhaled particles in humans and laboratory animals has been investigated extensively in the recent past (see Salem and Katz,[84] for example). Results from such investigations and recent computer simulations of aerosol deposition based on computational fluid dynamics in a three-dimensional, anatomically based, human airway model[85] have been used to infer the fate of BSAT aerosols in the human respiratory tract following inhalation. We briefly discuss this fate below, with special attention to the ricin toxin.

As with all particles, inhaled BSAT aerosol particles initially become deposited on the surfaces of the respiratory tract or oral cavity if they are not removed by exhalation. Sites of deposition within the respiratory tract are crucial, as the BSAT particles must reach vulnerable cell or tissue types to induce their disease or toxic effects effectively. The severity of disease or intoxication, therefore, can vary dramatically with the site of initial deposition. Deposited BSAT particles that do not dissolve quickly are physically removed by one of several anatomic clearance mechanisms (discussed in more detail below). The particles that are not cleared become retained in the respiratory tract and collectively constitute the effective BSAT dose that can cause disease or intoxication. These particles, however, can be subjected to defense mechanisms such as the innate (mucociliary clearance mechanisms and sneezing and coughing reflexes) and adaptive immune systems (immunoglobulins and T and B lymphocytes).

Aerosolized BSAT particles may first encounter the hairs within the anterior nares upon inhalation that will impede or intercept the larger particles. The aerosol is then drawn through the posterior nares, becoming warmed and humidified as it passes over the conchae and reaches the nasopharyngeal region (see the illustration of the human respiratory tract in Figure 5.7). This region is characterized by stratified squamous epithelium that extends from the oral pharynx to the larynx. Especially for smaller particles, aerosol flow will pass through the trachea and the branches of the right and left bronchi, which separate into secondary lobar branches and sub-segmental branches, finally dividing into >800 smaller quaternary branches.[86] This branching extends into even smaller airways, where goblet cells producing mucus and secretory glands gradually disappear into the terminal bronchioles. The epithelial lining becomes completely nonciliated at the terminal bronchioles before they separate into ~150 000 respiratory bronchioles that are also nonciliated.[87]

The respiratory bronchioles further divide into ~26 million alveolar ducts[88] that are almost entirely lined with alveoli. Collectively, somewhere between 50 and 100 million alveolar sacs are found in the alveoli of a human adult[89] to provide 30–100 m^2 of gas exchange surface area, depending upon individual body size and fitness level.[90] The alveolar walls are built from a network of reticular and elastic fibers that support fine pulmonary capillaries. Lymphatic channels are found throughout the pleura and septa and constitute means for clearing insoluble deposited particles. Hilar lymph nodes may become secondary reservoirs of deposited material, causing prolonged release of hazardous inhaled materials to the systemic circulation.[91]

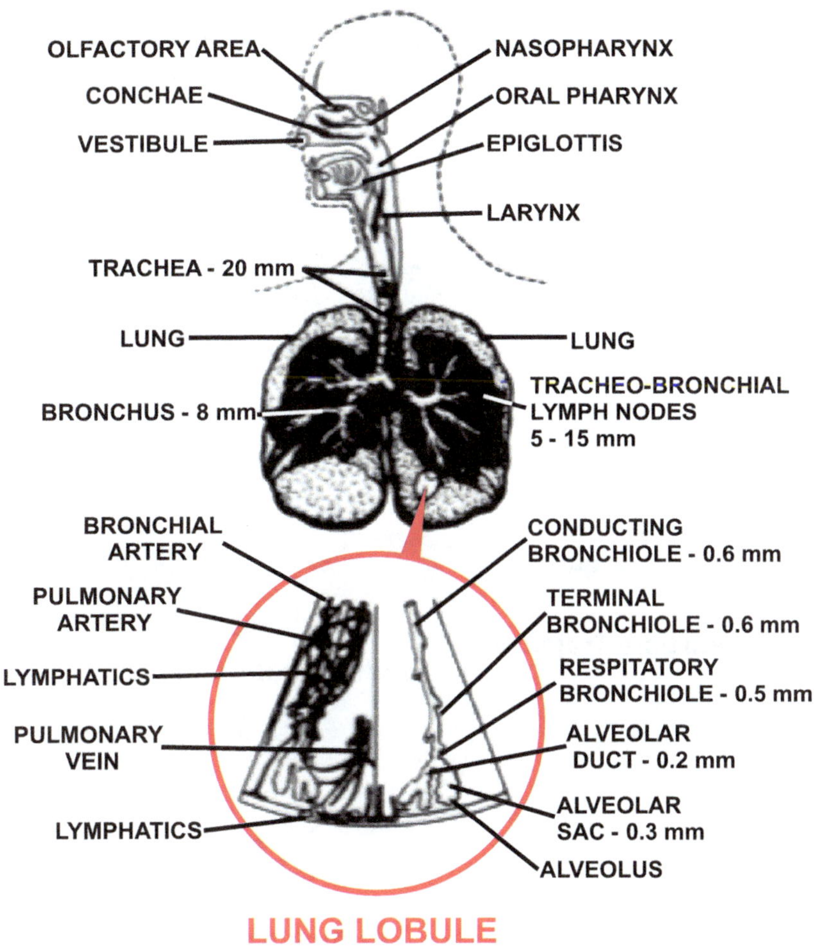

Figure 5.7 Anatomical regions of the respiratory tract. Dimensions in millimeters represent the d_a for particles that can penetrate the respiratory to the corresponding anatomical region. Figure from Stuart.[96]

5.4.3 Deposition of Aerosol Particles in the Respiratory Tract

Airflow rates for aerosols entering the respiratory tract range from 0 to 120 L min^{-1} during normal respiration, depending upon the amount of work performed.[92] During both inspiration and expiration, particle deposition processes that follow physical laws will occur. The deposition of nonfibrous particles in the respiratory tract results from the three physical processes of impaction, sedimentation and diffusion,[84,93–95] which are illustrated in Figure 5.8. The three processes are influenced by (1) the physical and chemical characteristics of the particle, such as its size and shape, density, surface charge, hygroscopicity, solubility and chemical reactivity, and also (2) factors specific to the individual host animal, including the respiratory route as a result of nasal or mouth breathing, respiratory rate and tidal volume and respiratory tract anatomy.[1,96] Impaction is the dominant mechanism for the deposition of larger particles with $d_a > 1–2$ μm and usually occurs in the nasopharyngeal and tracheobronchial regions,

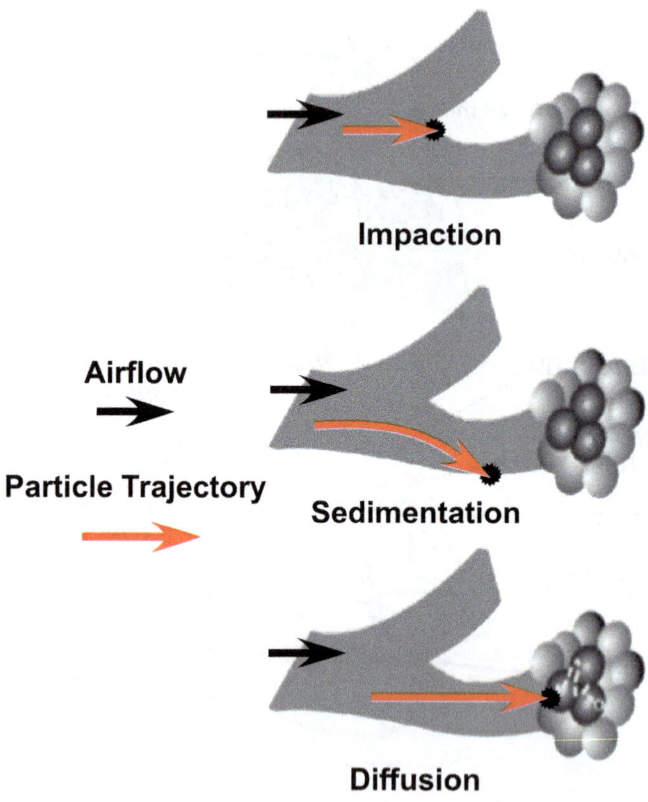

Figure 5.8 Depiction of the deposition of nonfibrous particles in the respiratory tract. Deposition is primarily governed by impaction, sedimentation and diffusion. Adapted from Roy *et al.*[1]

especially at airway branching points where bulk airflow is faster and more turbulent. On the other hand, sedimentation becomes the major mechanism for particles with $0.5 < d_a < 10$ μm, typically occurring in the smaller conducting airways where the air velocity is slower. Also, in the small bronchioles and alveoli where the air velocity is slowest, the diffusion of particles with $0.1 < d_a < 2$ μm becomes the predominant deposition mechanism. Rapid, shallow breathing increases air velocity and also the likelihood of particle deposition by impaction, whereas slow, deep breathing and pauses in respiration reduce air velocity and increase the likelihood of deposition by sedimentation and diffusion.

5.4.4 Clearance of Inhaled Particles from the Respiratory Tract

Clearance from the respiratory tract is the physical removal of insoluble and poorly soluble particles and also the dissolution and absorption of soluble particles.[84,93–95] Particle deposition is the critical first step in determining the subsequent clearance processes. The respiratory tract can be divided into three unique regions based on clearance mechanisms. Inhaled particles deposited in the posterior nares (region 1) can become caught up in mucus and transported by mucociliary action through the nasopharynx and into the gastrointestinal tract.[96] Other mechanisms that can also become involved include sneezing, coughing, swallowing, nose blowing and other actions used to expel inhaled aerosol particles. Particles reaching the larynx and passing through successive generations of conducting airways to the terminal bronchioles (region 2) can also be deposited upon mucus that is propelled upwards by underlying cilia for clearance to the gastrointestinal tract. Finally, inhaled insoluble particles that are deposited below the ciliated epithelium, on the surfaces of the respiratory bronchioles, alveolar ducts, alveolar sacs and alveoli (region 3), will be cleared more slowly to the gastrointestinal tract, or will remain with only gradual dissolution or removal following phagocytosis by pulmonary macrophages. Soluble particles can be absorbed directly into the blood or lymphatic channels after dissolution in respiratory fluids.

5.5 Signs, Symptoms and Toxicity of Ricin Exposure

As with most BSATs, ricin remains biologically active through several different modes of entry into an animal, including ingestion and injection in addition to inhalation. Symptoms and especially toxicity, however, can vary dramatically with these modes of entry,[66,97] with inhalation of the aerosolized toxin considered the most likely route of exposure for causing serious health effects.[98] Response to cutaneous contact with ricin, on the other hand, is not particularly noteworthy. An urticarial, IgE-mediated, allergic reaction may occur after handling the intact castor bean plant or exposure to castor bean dust or pumice. Irritation and the development of pseudo-membranous conjunctivitis after ocular exposure to very low ricin concentrations are also commonly reported. In humans, general signs and symptoms of ricin exposure include fever, fatigue, weakness, muscle pain and dehydration.

5.5.1 Inhalation of Ricin

The pulmonary toxicology of aerosolized ricin has been investigated in a number of different animal models primarily to support the ongoing development of therapeutics and vaccines against the toxin, with the mouse used extensively as an inexpensive but effective model of its aerosol-induced pathogenesis.[83,99–102] Once in the lung, ricin can bind with a variety of different cell types and inhibit ribosomal activity, ultimately leading to its toxic and lethal effects in the animal. As mentioned above in greater detail, this toxicity and lethality are largely a consequence of particle size and its influence on lung deposition. Smaller particles deposit deeper into the respiratory tract (see Figure 5.6) and result in higher mortality, whereas larger particles deposit higher in the respiratory tract and can be cleared by a number of different mechanisms.[1,77] In one study,[77] aerosolized ricin delivered to mice as particles with a 1 μm MMAD (mass median aerodynamic diameter – the diameter at which 50% of the particles by mass are larger and 50% are smaller in a population of airborne particles) deposited extensively in the pulmonary parenchyma and induced mortality from necrotizing alveolitis and pulmonary edema, whereas ricin aerosols with particles of >5 μm MMAD induced minimal effects using comparable or higher presented doses when also delivered to mice. The LD_{50} (lethal dose for 50% of a sample population of a specific test animal) for inhaled ricin toxin delivered to mice has been reported to be between 2.8 and 12.2 μg kg^{-1} for different mouse strains.[66,77,83,103] In rats, the LD50 for the inhalation of two different *R. communis* cultivars was found to be 3.7 and 9.8 μg kg^{-1}, with the differences in toxicity assumed to be due to variations in ricin isotoxins.[104–106] There have also been a number of different investigations on the inhalation toxicology or ricin in nonhuman primates. Wilhelmsen and Pitt[101] found that in rhesus macaques, inhalation of 1.2 μm MMAD ricin particles at 22–42 μg kg^{-1} resulted in respiratory distress 36–48 h post-inhalation. Wannemacher and Anderson[103] found the LD_{50} for inhaled ricin delivered to the same species to be 15 and 5.8 μg kg^{-1} for African green monkeys. One caveat to bear in mind is that the susceptibility of different animal species to ricin toxicity clearly varies, and considerably in some cases.[106] This is almost certainly due to differences in the anatomy and cytology of the respiratory tract to some extent.

The post-mortem findings of exposure to aerosolized ricin toxin typically include diffuse pulmonary edema with multifocal areas of necrosis and inflammation, and injury tends to be significantly worse in the distal airways and alveoli.[101,107–109] Toxicity results directly from the inhibition of protein synthesis, release of cytokine mediators and direct injury to the epithelial membrane, with the primary targets of toxicity being the type I and II pneumocytes.[77,101,107,108] Induced cytokines include TNF-α, IL-6, IL-1β, CXCL1 and CXCL2[110–112] as part of a generalized inflammatory response. Analysis of post-exposure lungs and spleens indicated the upregulation of genes associated with defense processes, response to pathogens, the innate immune response and the inflammatory and stress response.[113] Significant systemic absorption is usually not observed after inhalation and toxicity is primarily limited to the respiratory tract.[101,109] Signs of vascular leak syndrome, characterized by alveolar flooding and extravascular lung water, become evident. Arterial hypoxemia and acidosis develop as fluids accumulate. Wet lung weights in nonhuman primates are generally found to be twice that considered normal.[113] Respiratory failure is likely to be the primary cause of morbidity and mortality in humans and nonhuman primates after ricin inhalation[101] and symptoms include cough, chest tightness, dyspnea, hypoxemia and noncardiogenic pulmonary edema.[73] Animals display outward clinical signs of intoxication including depression, shortness of breath and anorexia as early as 12–16 h post-exposure, with fever developing within 24 h. The clinical status of animals continues to worsen, with an identifiable rapid decline 30 h post-exposure. Animals usually succumb to intoxication 1–3 days following exposure to ricin aerosols. Immunohistochemical analysis of post-mortem tissue showed that ricin binds ciliated bronchiolar lining cells, alveolar macrophages and alveolar lining cells.[107]

There has been one investigation that addresses the pathology of sub-lethal exposure of aerosolized ricin in nonhuman primates.[114] Three rhesus macaques given different nonlethal doses of aerosolized ricin at ~1 μm MMAD initially exhibited dyspnea, tachypnea, labored breathing and anorexia, similarly to the other macaques in the study exposed to lethal doses of the same ricin aerosol. These signs decreased substantially within 48–60 h for the group receiving nonlethal doses, but mild, sporadic signs persisted until the end of the study. Indicative of ongoing reparative processes, these animals experienced fibrosing interstitial pneumonia and type II pneumocytes hyperplasia with minimal edema and variable histocytic and lymphoplasmacytic infiltration. The initial injury gave way to a sub-acute chronic phase with an influx of macrophages, clearing of alveolar spaces and subsequent repair of pulmonary architecture by a robust fibrotic response and type II pneumocyte hyperplasia. The reparative processes correlated directly with the nonlethal dose of ricin aerosol administered. The authors noted that most of the lesions in the lung were remarkably similar to those seen in bleomycin-induced lung injury, the most widely used model for pulmonary fibrosis in laboratory animals.[115] It is almost certain that long-term lung function of animals surviving exposure to inhaled ricin would be greatly diminished from the extensive fibrosis resulting from acute lung injury. This investigation is particularly noteworthy because numerous exposures to sub-lethal doses of ricin aerosols can be anticipated in terrorist attacks involving the protein toxin in aerosolized form.

5.5.2 Ingestion and Injection of Ricin

We briefly address the signs, symptoms and toxicity of ricin exposure by the ingestion and injection routes for comparison with those from inhalation. Most of this information can be found in the report of Worbs *et al.*,[18] which is a compilation of information from 887 cases involving ricin ingestion and injection by humans since 1900. In general terms, the symptoms induced by both routes were observed to be very similar and the severity of symptoms increased with increases in ricin dose. Symptoms arose after 3–20 h after ingestion or injection and included abdominal pain, emesis, diarrhea with or without blood, muscular pain and cramps in the limbs, circulatory collapse, dyspnea and dehydration. Both muscular pain and circulatory collapse were more commonly observed with the injection route, in addition to pain at the injection site. Biochemical analyses often revealed increases in white blood cell count, blood urea nitrogen and both aspartate aminotransferase and alanine aminotransferase levels, which indicate damage to the liver and kidneys, respectively. Autopsy in fatal cases typically found hemorrhagic necrosis in the intestines and heart and also edema in the lungs. The LD_{50} for mice challenged with ricin has been reported as 20 000 μg kg^{-1} for the ingestion route[116] and 2.8–3.3 μg kg^{-1} for the injection route.[117,118]

5.6 Conclusions

R. communis is an economically important plant for the production of castor oil and numerous industrial, medical and cosmetic products. The ricin toxin is easily produced in relatively large quantities from its seeds, which was in part responsible for the development of the toxin as a biological weapon by the United States and its allies during World Wars I and II. More recently, the toxin has gained popularity as a weapon of terror because of its ease of access and extraction from castor beans in addition to its stability. Although the use of ricin can certainly lead to fatal consequences, it does not appear well suited as an agent of bioterrorism. By the ingestion route, ricin erodes the intestinal mucosa, which may lead to massive fluid loss and hypovolemic shock (the severe loss of blood and other fluids that make the heart unable to supply blood to the body's tissues). Both of these symptoms, however, are very manageable under the appropriate medical attention. The contamination of a municipal drinking water facility to generate lethal concentrations of ricin would require impossibly large amounts of ricin powder. By the injection route, ricin may be ideally suited as a weapon of assassination, as clearly evidenced in the cases of Georgi Markov in 1978[7], Boris Korczak in 1980[8] and the 2008 incident involving Roger Von Bergendorff,[9] but it cannot realistically be used for causing mass mortality. Further, there is no evidence of adsorption of the ricin toxin from direct cutaneous contact. In contrast, delivery by the inhalation route does have potential for the use of ricin as an agent for bioterrorism, but using this route for such ends has its own difficulties and problems. To target a large urban population, for example, literally tons of ricin powder must be isolated from castor beans and milled to sufficiently small d_a values to target the regions of the lungs where fatal consequences can result, and this is before its successful aerosolization in a bioterrorism attack. The manufacture of such specially formulated ricin powders in ton quantities is far beyond the reach of almost all terrorists.

An approved therapy or antidote against ricin intoxication is not available. Treatment for ricin exposure involves supportive medicine, the use of intravenous fluids and the suppression of hypertension, and to prevent further toxin absorption, activated charcoal and gastric lavage (commonly referred to as stomach pumping) have been used after toxin ingestion.[119] There has been interest, however, in developing a vaccine against ricin intoxication to protect warfighters, first responders and other emergency personnel. Although there is no such vaccine that is readily available, one is currently in an advanced stage of legislative approval.[120–123] Based on an inactivated ricin A-chain, the vaccine RiVax (Soligenix, Princeton, NJ, USA) has been subjected to toxicological and clinical trial studies. One study demonstrated that nonhuman primates vaccinated with the vaccine were completely protected against a lethal aerosol of ricin with a high survival benefit.[124] The primates were exposed to 3–5 times the amount of ricin known to cause mortality in untreated animals and all survived with no apparent signs of gross lung damage. The vaccine regimen was safe and well tolerated by all animals, with antibodies found in their sera that could neutralize ricin activity. Conversely, all nonhuman primates in the unvaccinated treatment group died within ~36 h of exposure to the aerosolized ricin while developing severe lung damage that included hemorrhaging. A pilot clinical trial of the vaccine in humans also demonstrated that the vaccine induced the production of functionally active antibodies.[122] The most significant benefit of the RiVax vaccine is that it protects against ricin exposure to the vulnerable tissues of the lungs, the most dangerous route of exposure and most likely for military and terrorist scenarios.

Acknowledgements

We gratefully acknowledge the gracious and invaluable efforts of Adana L. Eilo for ensuring that the text was devoid of all classified content and grammatical errors and both Mrs Donna Hoffman and Joy L. Henderson for their expert assistance in preparing this chapter.

References

1. C. J. Roy, D. S. Reed and J. A. Hutt, Aerobiology and inhalation exposure to biological select agent toxins, *Vet. Pathol.*, 2010, **47**, 779–789.
2. Z. F. Dembek, M. G. Kortepeter and J. A. Palvin, Disconnect between deliberate and natural infectious disease outbreaks, *Epidemiol. Infect.*, 2007, **135**, 353–371.

3. F. Weber, R. Freudinger, G. Schwerdt and M. Gekle, A rapid screening method to test apoptotic synergisms of ochratoxin A with other nephrotoxic substances, *Toxicol. In Vitro*, 2005, **1**, 135–143.
4. R. W. Wannemacher, J. F. Hewetson and P. V. Lemley, Comparison of detection of ricin in castor bean extracts by bioassays, immunoassays and chemistry procedures, in *Recent Advances in Toxinology Research*, ed. P. Gopalakrishnakone and C. Tan, Venom and Toxin Research Group, National University of Singapore, Singapore, 1992, pp. 108–119.
5. P. Ehrlich, Experimentelle Unterschungen über Immunität, I: Euber Ricin, *Deutsch Med. Wochenschr.*, 1891, **17**, 976–979.
6. J. Cookson and J. Nottingham, *A Survey of Chemical and Biological Warfare*, Monthly Review Press, New York, 1969, p. 424.
7. R. Crompton and D. Gall, Georgi Markov: Death in a Pellet, *Med. Leg. J.*, 1980, **48**, 51–62.
8. J. D. Douglas, Jr. and N. C. Livingston, *America the Vulnerable: The Threat of Chemical/Biological Warfare. The New Shape of Terrorism and Conflict*, Lexington Books, Lexington, MA, 1987, p. 256.
9. A. Goldman, *Meet the mysterious Roger Von Bergendorff: He files for bankruptcy, had a history of health problems and kept a vial or ricin*, Las Vegas Sun, 2014, http://www.lasvegassun.com/news/2008/mar/05/meet-mysterious-roger-von-bergendorff, retrieved 31 December.
10. James Martin Center for Nonproliferation Studies, *Chronology of incidents involving ricin*, 2014, http://cns.miis.edu/reports/ricin_chron.htm, retrieved 31 December.
11. Y. Endo, K. Tsurugi, T. Yutsudo, Y. Takeda, T. Ogasawara and K. Igarashi, Site of action of a Vero toxin (VT2) from Escherichia coli O157:H7 and of Shiga toxin on eukaryotic ribosomes. RNA *N*-glycosidase activity of the toxins, *Eur. J. Biochem.*, 1988, **171**, 45–50.
12. F. Stirpe and L. Barbieri, Ribosome-inactivating proteins up to date, *FEBS Lett.*, 1986, **195**, 1–8.
13. J. Robertus, The structure and action of ricin, a cytotoxic *N*-glycosidase, *Semin. Cell Biol.*, 1991, **2**, 23–30.
14. H. M. Berman, J. Westbrook, Z. Feng, G. Gilliland, T. N. Bhat, H. Weissig, I. N. Shindyalov and P. E. Bourne, The Protein Data Bank, *Nucleic Acids Res.*, 2000, **28**, 235–242.
15. P. H. Stillmark, Ueber ricen: Arbeiten des Pharmacologischen Institutes zu Dorpat, iii. Cited in Flexner, J., The histological changes produced by ricin and abrin intoxications, *J. Exp. Med.*, 1897, **1889**(2), 197–216.
16. S. Olsnes, E. Saltvedt and A. Phil, Isolation and comparison of galactose-binding lectins from *Abrus precatorius* and *Ricinus communis*, *J. Biol. Chem.*, 1974, **249**, 803–810.
17. L. M. Roberts, F. I. Lamb, D. J. Pappin and J. M. Lord, The primary sequence of *Ricinus communis* agglutinin, comparison with ricin, *J. Biol. Chem.*, 1985, **260**, 15682–15686.
18. S. Worbs, K. Köhler, D. Pauly, M.-A. Avondet, M. Schaer, M. B. Dorner and B. G. Dorner, *Ricinus communis* intxoications in human and veterinary medicine – a summary of real cases, *Toxins*, 2011, **3**, 1332–1372.

19. A. Bolognesi, F. Olivieri, M. G. Battelli, L. Barbieri, A. I. Falasca, A. Parente, B. F. Vecchio and F. Stirpe, Ribosome-inactivating proteins in edible plants and purification and characterization of a new ribosome-inactivating protein from *Cucurbita moschata*, *Biochim. Biophys. Acta*, 2006, **1760**, 783–792.
20. A. Gasperi-Campani, L. Barbieri, M. G. Battelli and F. Stirpe, On the distribution of ribosome-inactivating proteins amongst plants, *J. Nat. Prod.*, 1985, **48**, 446–454.
21. A. Bolognesi, L. Polito, C. Lubelli, L. Barbieri, A. Parente and F. Stirpe, Ribosome-inactivating and adenine polynucleotide glycosylase activities in *Mirabilis jalapa* L. tissues, *J. Biol. Chem.*, 2002, **277**, 13709–13716.
22. W. Montfort, J. E. Villafranca, A. F. Monzingo, S. R. Ernst, B. Katzin, E. Rutenber, N. H. Xuong and J. D. Robertus, The three-dimensional structure of ricin at 2.8 Å, *J. Biol. Chem.*, 1987, **262**, 5398–5403.
23. E. Rutenber and J. D. Robertus, Structure of ricin B-chain at 2.5 Å resolution, *Proteins*, 1991, **10**, 260–269.
24. B. J. Katzin, E. J. Collins and J. D. Robertus, Structure of ricin A-chain at 2.5 Å, *Proteins*, 1991, **10**, 251–259.
25. E. Rutenber, B. J. Katzin, S. Ernst, E. J. Collins, D. Mlsna, M. P. Ready and J. D. Robertus, Crystallographic refinement of ricin to 2.5 Å, *Proteins*, 1991, **10**, 240–250.
26. S. A. Weston, A. D. Tucker, D. R. Thatcher, D. J. Derbyshire and R. A. Pauptit, X-ray Structure of Recombinant Ricin A-Chain at 1.8 Å, Resolution, *J. Mol. Biol.*, 1994, **244**, 410–422.
27. H. T. Wright and J. D. Robertus, The intersubunit disulfide bridge of ricin is essential for cytotoxicity, *Arch. Biochem. Biophys.*, 1987, **256**, 280–284.
28. Y. Endo and K. Tsurugi, The RNA *N*-glycosidase activity of ricin A-chain: The characteristics of the enzymatic activity of ricin A-chain and with RNA, *J. Biol. Chem.*, 1987, **263**, 8735–8739.
29. J. H. Gould, M. R. Hartley, P. C. Welsh, D. K. Hoshizaki, A. Frankel and R. M. Roberts, Alteration of an amino acid outside the active site of the ricin A-chain reduces its toxicity towards yeast ribosomes, *Mol. Gen. Genet.*, 1991, **230**, 81–90.
30. E. Rutenber, M. P. Ready and J. D. Robertus, Structure and evolution of ricin B chain, *Nature*, 1987, **326**, 624–626.
31. S. Magnusson, R. Kjeken and T. Berg, Characterization of two distinct pathways of endocytosis of ricin by rat liver endothelial cells, *Exp. Cell Res.*, 1993, **205**, 118–125.
32. B. M. Simmons, P. D. Stahl and J. H. Russell, Mannose receptor-mediated uptake of ricin toxin and ricin A chain by macrophages. Multiple intracellular pathways for A chain translocation, *J. Biol. Chem.*, 1986, **26**, 7912–7920.
33. T. Mise, G. Funatsu, M. Ishiguro and M. Funatsu, Isolation and characterization of ricin E from castor beans, *Agric. Biol. Chem.*, 1977, **41**, 2041–2046.

34. M. Ishiguro, T. Takahashi, G. Funatsu, K. Hayashi and M. Funatsu, Biochemical studies in ricin I. Purification of ricin, *J. Biochem.*, 1964, **55**, 587–592.
35. T. T.-S. Lin and S. S.-L. Li, Purification and physicochemical properties of ricins and agglutinins from *Ricinus communis*, *Eur. J. Biochem.*, 1980, **105**, 453–459.
36. M. Moya, A. Dautry-Varsat, B. Goud, D. Louvard and P. Boquet, Inhibition of coated pit formation in Hep2 cells blocks the cytotoxicity of diphtheria toxin but not that of ricin toxin, *J. Cell Biol.*, 1985, **101**, 548–559.
37. N. Sphyris, J. M. Lord, R. Wales and L. M. Roberts, Mutational analysis of the *Ricinus* lectin B-chains. Galactose-binding ability of the 2 gamma subdomain of *Ricinus communis* agglutinin B-chain, *J. Biol. Chem.*, 1995, **270**, 20292–20297.
38. M. R. Jackman, W. Surety, J. A. Ellis and J. P. Luzio, Inhibition of apical but not basolateral endocytosis of ricin and folate in Caco-2 cells by cytochalasin D, *J. Cell Sci.*, 1994, **107**, 2547–2556.
39. W. Surety, N. A. Bright and J. P. Luzio, The effects of cytochalasin D and phorbol myristate acetate on the apical endocytosis of ricin in polarized Caco-2 cells, *J. Cell Sci.*, 1996, **109**, 2927–2935.
40. A. Llorente, A. Rapak, S. L. Schmid, B. van Deurs and K. Sandvig, Expression of mutant dynamin inhibits toxicity and transport of endocytosed ricin to the Golgi apparatus, *J. Cell Biol.*, 1998, **140**, 553–563.
41. M. R. Jackman, J. A. Ellis, S. R. Gray, W. Surety and J. P. Luzio, Cell polarization is required for ricin sensitivity in a Caco-2 cell line selected for ricin resistance, *Biochem. J.*, 1999, **341**, 323–327.
42. T.-G. Iversen, G. Skretting, A. Liorente, P. Nicoziani, B. Van Deurs and K. Sandvig, Endosome to Golgi transport of ricin is independent of clathrin and of the Rab9- and Rab11-GTPases, *Mol. Biol. Cell*, 2001, **12**, 2099–2107.
43. K. Sandvig and S. Olsnes, Effect of temperature on the uptake, excretion and degradation of abrin and ricin by HeLa cells, *Exp. Cell Res.*, 1979, **121**, 15–25.
44. K. Sandvig, B. Spilsberg, S. U. Lauvrak, M. L. Torgersen, T. G. Iversen and B. van Deurs, Pathways followed by protein toxins into cells, *Int. J. Med. Microbiol.*, 2004, **293**, 483–490.
45. S. B. Kornfeld, J. E. Leonard, M. D. Mullen and R. Taetle, Assessment of ligand effects in intracellular trafficking of ricin A chain using anti-ricin hybridomas, *Cancer Res.*, 1991, **51**, 1689–1693.
46. M. J. Lord, N. A. Jolliffe, C. J. Marsden, C. S. Pateman, D. C. Smith, R. A. Spooner, P. D. Watson and L. M. Roberts, Ricin. Mechanisms of cytotoxicity, *Toxicol. Rev.*, 2003, **22**, 53–64.
47. Y. Endo, K. Mitsui, M. Motizuki and K. Tsurugi, The mechanism of action of ricin and related toxic lectins on eukaryotic ribosomes. The site and the characteristics of the modification in 28 S ribosomal RNA caused by the toxins, *J. Biol. Chem.*, 1987, **262**, 5908–5912.

48. F. Rajamohan, C. Mao and F. M. Uckun, Binding interactions between the active center cleft of recombinant pokeweed antiviral protein and the alpha-sarcin/ricin stem loop of ribosomal RNA, *J. Biol. Chem.*, 2001, **276**, 24075–24081.
49. A. F. Monzingo and J. D. Robertus, X-ray analysis of substrate analogs in the ricin A-chain active site, *J. Mol. Biol.*, 1992, **227**, 1136–1145.
50. S. Yoshinari, S. Koresawa, S. Yokota, H. Sawamoto, M. Tamura and Y. Endo, Gypsophilin, a new type 1 ribosome-inactivating protein from Gypsophila elegans: purification, enzymatic characterization and subcellular localization, *Biosci. Biotechnol. Biochem.*, 1997, **61**, 324–331.
51. F. Rajamohan, C. R. Engstrom, T. J. Denton, L. A. Engen, I. Kourinov and F. M. Uckun, High-level expression and purification of biologically active recombinant pokeweed antiviral protein, *Protein Expression Purif.*, 1999, **16**, 359–368.
52. I. V. Kurinov, D. E. Myers, J. D. Irvin and F. M. Uckun, X-ray crystallographic analysis of the tructural basis for the interactions of pokeweed antiviral protein with its active site inhibitor and ribosomal RNA substrate analogs, *Protein Sci.*, 1999, **8**, 1765–1772.
53. C. A. Vater, L. M. Bartle, J. D. Leszyk, J. M. Lambert and V. S. Goldmacher, Ricin A chain can be chemically cross-linked to the mammalian ribosomal proteins L9 and L10e, *J. Biol. Chem.*, 1995, **270**, 12933–12940.
54. P. Wang and N. E. Tumer, Pokeweed antiviral protein cleaves double-stranded supercoiled DNA using the same active site required to depurinate rRNA, *Nucleic Acids Res.*, 1999, **15**, 1900–1905.
55. T. K. Amukele and V. L. Schramm, Ricin A-chain substrate specificity in RNA, DNA and hybrid stem-loop structures, *Biochemistry*, 2004, **43**, 4913–4922.
56. H. L. Green and W. R. Lane, *Particulate Clouds: Dusts, Smokers and Mists*, Van Nostrand, Princeton, NJ, 1957, p. 425.
57. U.S. Department of the Army, *NATO Handbook on the Medical Aspects of NBC Defensive Operations AMedP-6*, 1973, p. 443.
58. M. G. Kortepeter and G. W. Parker, Potential biological weapons threats, *Emerging Infect. Dis.*, 1999, **5**, 523–527.
59. U.S. Department of the Army, *US Army Activity in the U.S. Biological Warfare Programs*, 1977, p. 68.
60. The Avalon Project, *Laws of war: Laws and customs of war on land (Hague II)*, Yale Law School, 2014, www.yale.edu/lawweb/avalon/lawofwar/hague02.htm, retrieved 23 December.
61. R. R. Williams, *Final Report on Ricin*, Report #OM347.4, Offensive Chemical Research Division, Bureau of Mines, American University Experimental Station, Washington, DC, 1918.
62. R. C. Gupta, *Handbook of Toxicology of Chemical Warfare Agents*, Academic Press, London, 2009, p. 1168.
63. Office of Scientific Research and Development, *Summary Technical Report of Division 9, Volume I: Chemical Warfare Agents and Related Chemical Problems (Parts III-IV)*, National Defense Research Committee, Office of Scientific Research and Development, Washington, DC, 1946, p. 424.

64. D. T. Parker, A. C. Parker and C. K. Ramachandran, *Joint CB Technical Data Source Book, Volume VI: Toxin Agents*, U.S. Army Dugway Proving Ground document no. DPG/JCP-96/007, 1996.
65. R. L. LeTourneau, *Light high explosive bomb for dispersing toxic and insecticidal aerosols*, U.S. Patent 3,207,071, 21 September 1965.
66. L. J. Schep, W. A. Temple, G. A. Butt and M. D. Beasley, Ricin as a weapon of mass terror – separating fact from fiction, *Environ. Int.*, 2009, **35**, 1267–1271.
67. D. V. Bates, B. R. Fish, T. F. Hatch, T. T. Mercer and P. E. Morrow, Deposition and retention models for internal dosimetry of the human respiratory tract, Task group on lung dynamics, *Health Phys.*, 1966, **12**, 173–207.
68. D. A. Edwards, J. Hanes, G. Caponetti, J. Hrkach, A. Ben-Jebria, M. L. Eskew, J. Mintzes, D. Deaver, N. Lotan and R. Langer, Large porous particles for pulmonary drug delivery, *Science*, 1997, **276**, 1868–1871.
69. H. Fischer, I. Polikarpov and A. F. Craievich, Average protein density is a molecular weight – dependent function, *Protein Sci.*, 2004, **13**, 2825–2828.
70. W. S. Winters and D. R. Chenoweth, *Modeling Dispersion of Chemical-Biological Agents in Three Dimensional Living Spaces (Sandia Report SAND2002–8009)*, Sandia National Laboratories, 2002, p. 19.
71. S. M. Burrows, T. Butler, P. Jöckel, H. Tost, A. Kerkweg, U. Pöschl and M. G. Lawrence, Bacteria in the global atmosphere – Part 2: Modeling of emissions and transport between different ecosystems, *Atmos. Chem. Phys.*, 2009, **9**, 9281–9297.
72. D. M. Wilkinson, S. Koumoutsaris, E. A. Mitchell and I. Bey, Modelling the effect of size in the aerial dispersal of microorganisms, *J. Biogeogr.*, 2012, **39**, 89–97.
73. U.S. Army Medical Institute of Infectious Diseases, *Medical Management of Biological Casualties Handbook*, Division of Medicine, U.S. Army Medical Institute of Infectious Diseases, 7th edn, 2011, p. 269.
74. S. L. Wiener, Strategies for the prevention of a successful biological warfare aerosol attack, *Mil. Med.*, 1996, **161**, 251–256.
75. H. Bigalke and A. Rummel, Medical aspects of toxin weapons, *Toxicology*, 2005, **214**, 210–220.
76. E. M. Eitzen, The use of biological weapons, in *The Textbook of Military Medicine, Part 1: Medical Aspects of Chemical and Biological Weapons*, ed. R. Zajtchuk and R. F. Bellamy, Borden Institute, Washington, 1997, pp. 473–450.
77. C. J. Roy, M. Hale, J. M. Hartings, L. Pitt and S. Duniho, Impact of inhalation exposure modality and particle size on the respiratory deposition of ricin in BALB/c mice, *Inhalation Toxicol.*, 2003, **15**, 619–638.
78. H. A. Druett, D. W. Henderson, L. Packman and S. Peacock, Studies on respiratory infection: I. The influence of particle size on respiratory infection with anthrax spores, *J. Hyg.*, 1953, **51**, 359–371.

79. H. A. Druett, J. M. Robinson, D. W. Henderson, L. Packman and S. Peacock, Studies on respiratory infection: II. The influence of aerosol particle size on respiratory infection guinea-pig with Pasturella pestis, *J. Hyg.*, 1956, **54**, 37–48.
80. H. A. Druett, D. W. Henderson and S. Peacock, Studies on respiratory infection: III. Experiments with Brucella suis, *J. Hyg.*, 1956, **54**, 49–57.
81. R. J. Goodlow and F. A. Leonard, Variability and infectivity of microorganisms in experimental airborne infection, *Bacteriol. Rev.*, 1961, **25**, 182–187.
82. W. C. Day and R. F. Berendt, Experimental tularemia in *Macaca mulatta*: Relationship of aerosol particles to size to the infectivity of airborne Pasteurella tularensis, *Infect. Immun.*, 1972, **7**, 77–82.
83. J. A. Doebler, N. D. Wiltshire, T. W. Mayer, J. E. Estep, R. B. Moeller, R. K. Traub, C. A. Broomfield, C. A. Calamaio, W. L. Thompson and M. L. Pitt, The distribution of [125I] ricin in mice following aerosol inhalation exposure, *Toxicology*, 1995, **98**, 137–149.
84. H. Salem and S. A. Katz, *Inhalation Toxicology*, CRC Press, Boca Raton, FL, 3rd edn, 2014, p. 623.
85. B. Ma and K. R. Lutchen, CFD simulation of aerosol deposition in an anatomically based human large-medium airway model, *Ann. Biomed. Eng.*, 2009, **37**, 271–285.
86. International Commission on Radiation Protection, *The Metabolism of Compounds of Plutonium and Other Actinides (ICRP publication 19)*, Pergamon Press, Oxford, 1972, p. 6.
87. E. R. Weibel, *Morphometry of the Human Lung*, Springer Verlag, Berlin, 1963, p. 151.
88. J. D. Brain and P. A. Valberg, Models of lung retention based in ICRP task group reports, *Arch. Environ. Health*, 1974, **28**, 1–11.
89. H. D. Landahl, Particle removal by the respiratory system: Note on the removal of airborne particulates by the human respiratory tract with particular reference to the role of diffusion, *Bull. Math. Biophys.*, 1963, **25**, 29–39.
90. L. Silverman, C. E. Billings, M. W. First and R. Dennis, Air flow measurements on human subjects with and without respiratory resistance at several work rates, *Arch. Ind. Hyg. Occup. Med.*, 1951, **3**, 461–478.
91. B. O. Stuart, Selection of animal models for evaluation of inhalation hazards in man, in *Air Pollution and the Lung: Proceedings of the Twentieth Annual OHOLO Biological Conference*, ed E. F. Aharonson, A. Ben-David and M. A. Klingberg, John Wiley and Sons, New York, 1975, pp. 268–288.
92. L. Silverman and C. E. Billings, Pattern of air flow in the respiratory tract, in *Particles and Vapours*, ed. C. N. Davies, Pergamon Press, London, 1961, pp. 473–450.
93. R. B. Schlesinger, Comparative deposition of inhaled aerosols in experimental animals and humans: A review, *J. Toxicol. Environ. Health*, 1985, **15**, 197–214.

94. R. O. McClellan and R. F. Henderson, *Concepts in Inhalation Toxicology*, Taylor and Francis, Washington, 2nd edn 1995, p. 648.
95. R. Phalen, *Inhalation Studies: Foundations and Techniques*, Informa Healthcare USA, Inc., New York, 2nd edn, 2009, p. 288.
96. B. O. Stuart, Deposition and clearance of inhaled particles, *Environ. Health Perspect.*, 1984, **55**, 369–390.
97. S. Olsnes, The history of ricin, abrin and related toxins, *Toxicon*, 2004, **44**, 361–370.
98. Department of the Army Navy and the Air Force and Commandant of the Marine Corps, *Field Manual: Treatment of Biological Warfare Agent Causalities*, Army FM 8–284, Navy NAVMED P-5042, Air Force AFMAN(I) 44–156, Marine Corps MCRP 4–11, 1C, 2000, pp. 4.7–4.9.
99. J. Hewetson, V. Rivera, P. Limley, L. Pitt and D. Creasia, A formalized toxoid for protection of mice from inhaled ricin, *Vaccine Res.*, 1996, **4**, 179–187.
100. M. A. Poli, V. R. Rivera, M. L. Pitt and P. Vogel, Aerosolized specific antibody protects mice from lung injury associated with aerosolized ricin exposure, *Toxicon*, 1996, **39**, 1037–1044.
101. C. L. Wilhelmsen and M. L. Pitt, Lesions of acute inhaled lethal ricin intoxication in rhesus monkeys, *Vet. Pathol.*, 1996, **33**, 296–302.
102. C. Yan, W. L. Rill, R. Malli, J. Hewetson, H. Naseem, R. Tammariello and M. Kende, Intranasal stimulation of long-lasting immunity against aerosol challenge with ricin toxoid vaccine encapsulated in polymeric microspheres, *Vaccine*, 1996, **14**, 1031–1038.
103. R. W. Wannemacher and J. B. Anderson, Aerosol procedures, animal toxicity and therapy, in *Inhalation Toxicology*, ed. H. Salem and S. A. Katz, CRC Press, Boca Raton, FL, 2nd edn, 2006, pp. 973–982.
104. M. Derenzini, E. Bonetti, V. Marionozzi and F. Stirpe, Toxic effects of ricin: Studies on the pathogenesis of liver lesions, *Virchows Arch. B*, 1976, **20**, 15–28.
105. G. D. Griffiths, P. Rice, A. C. Allenby, S. C. Bailey and D. G. Upshall, Inhalation toxicology and histopathology of ricin and abrin toxins, *Inhalation Toxicol.*, 1995, **7**, 269–288.
106. G. D. Griffiths, G. J. Phillips and J. Holley, Inhalation toxicology of ricin preparations: Animal models, prophylactic and therapeutic approaches to protection, *Inhalation Toxicol.*, 2007, **19**, 873–887.
107. R. F. Brown and D. E. White, Ultrastructure of rat lung following inhalation of ricin aerosol, *Int. J. Exp. Pathol.*, 1997, **78**, 267–276.
108. G. D. Griffiths, C. D. Lindsay, A. C. Allenby, S. C. Bailey, J. W. Scawin, P. Rice and D. G. Upshall, Protection against inhalation toxicity of ricin and abrin by immunization, *Hum. Exp. Toxicol.*, 1995, **14**, 155–164.
109. A. M. Soler-Rodríguez, M. A. Ghetie, N. Oppenheimer-Marks, J. W. Uhr and E. S. Vitetta, Ricin A-chain and ricin A-chain immunotoxins rapidly damage human endothelial cells: implications for vascular leak syndrome, *Exp. Cell Res.*, 1993, **206**, 227–234.

110. V. Korchevia, J. Wong, M. Lindaurer, D. B. Jacoby, M. S. Iordanov and B. Magun, Role of apoptotic signaling pathways in regulation of inflammatory responses to ricin in primary murine macrophages, *Mol. Immunol.*, 2007, **44**, 2761–2771.
111. J. Wong, V. Korcheva, D. B. Jacoby and B. Magnun, Intrapulmonary delivery of ricin at high dosage triggers a systemic inflammatory response and glomerular damage, *Am. Path.*, 2007, **170**, 1497–1510.
112. J. Wong, V. Korcheva, D. B. Jacoby and B. Magnun, Proinflammatory responses of human airway cells to ricin involve stress-activated protein kinases and NF-kappa B, *Am. J. Physiol.: Lung Cell. Mol. Physiol.*, 2007, **293**, L1385–L1394.
113. C. J. Roy, K. Song, S. K. Sivasubramani, D. J. Gardner and S. H. Pincus, Animal models of ricin toxicosis, *Top. Microbiol. Immunol.*, 2012, **357**, 243–257.
114. M. Bhaskaran, P. J. Didier, S. K. Sivasubramani, L. A. Doyle, J. Holley and C. J. Roy, Pathology of lethal and sublethal doses of aerosolized ricin in rhesus macaques, *Toxicol. Pathol.*, 2014, **42**, 573–581.
115. B. B. Moore and C. M. Hogaboam, Murine models pulmonary fibrosis, *Am. J. Physiol.: Lung Cell. Mol. Physiol.*, 2008, **294**, L152–L160.
116. S. M. Bradberry, K. J. Dickers, P. Rice, G. D. Griffiths and J. A. Vale, Ricin poisoning, *Toxicol. Rev.*, 2003, **22**, 65–70.
117. S. Olsnes and A. Phil, Different biological properties of 2 constituent peptide chains of ricin, a toxic protein inhibiting protein synthesis, *Biochemistry*, 1973, **2**, 3121–3126.
118. O. Fodstad, S. Olsnes and A. Pihl, Toxicity, distribution and elimination of the cancerostatic lectins abrin and ricin after parenteral injection into mice, *Br. J. Cancer*, 1976, **34**, 418–425.
119. J. Audi, M. Belson, M. Patel and J. Osterloh, Ricin poisoning: A comprehensive review, *J. Am. Med. Assoc.*, 2005, **294**, 2342–2351.
120. J. E. Smallshaw, J. A. Richardson, S. Pincus, J. Schindler and E. S. Vitetta, Preclinical toxicology and efficacy testing of RiVax, a recombinant protein against ricin, *Vaccine*, 2005, **23**, 4775–4784.
121. J. E. Smallshaw, J. A. Richardson and E. S. Vitetta, RiVax, a recombinant ricin subunit vaccine, protects mice against ricin delivered by gavage or aerosol, *Vaccine*, 2007, **25**, 7459–7469.
122. E. S. Vitetta, J. E. Smallshaw, E. Coleman, H. Jafri, C. Foster, R. Munford and J. Schindler, A pilot clinical trial of a recombinant ricin vaccine in normal humans, *Proc. Natl. Acad. Sci. U. S. A.*, 2010, **103**, 2268–2273.
123. J. E. Smallshaw and E. S. Vitetta, A lyophilized formulation of RiVax, a recombinant ricin subunit vaccine, retains immunogenicity, *Vaccine*, 2010, **28**, 2428–2435.
124. Soligenix, Inc., *Soligenix Announces positive survival results with heat stable ricin toxin vaccine: All vaccinated animals exposed to a lethal aerosol of ricin toxin survive*, Soligenix, Inc., 2015, http://www.soligenix.com/news.aspx?titleId=456, retrieved 7 February.

125. M. F. Sanner, A. J. Olson and J. C. Spehner, Reduced surface: An efficient way to compute molecular surfaces, *Biopolymers*, 1996, **38**, 305–320.
126. E. F. Pettersen, T. D. Goddard, C. C. Huang, G. S. Couch, D. M. Greenblatt, E. C. Meng and T. E. Ferrin, UCSF Chimera – a visualization system for exploratory research and analysis, *J. Comput. Chem.*, 2004, **13**, 1605–1612.
127. J. H. Carra, C. A. McHugh, S. Mulligan, L. M. Machiesky, A. S. Soares and C. B. Millard, Fragment-based identification of determinants of conformational and spectroscopic change at the ricin active site, *BMC Struct. Biol.*, 2007, **7**, 72.
128. C. P. Yu and C. K. Diu, Total and regional deposition of inhaled aerosols in humans, *J. Aerosol Sci.*, 1983, **14**, 599–609.

CHAPTER 6

Bioaerosols in the Environment: Populations, Measurement and Processes

JOSHUA L. SANTARPIA*[a]

[a]Sandia National Laboratories, Albuquerque, NM, USA
*E-mail: jsantar@sandia.gov

6.1 Introduction

Biological particles in the atmosphere include bacteria,[1] fungal spores and hyphae,[2] pollens, algae, proteins, viruses and fragments of fungi, pollens or plants.[3,4] These biological aerosols are generally considered a sub-set of organic carbon (OC) aerosols. Bioaerosols may affect the weather and the global radiative balance because they absorb and emit radiation and can act as cloud-condensation nuclei[5] and ice-condensation nuclei.[6] Many plants, fungi and microorganisms form bioaerosols as a method of spreading to other locations.[7] Some components of bioaerosols have been shown to cause infectious diseases and allergies in humans and animals and also diseases in plants.[8] Throughout history, some bacteria and viruses have been used and contemplated for use as biowarfare agents to cause human disease and agricultural and economic damage. This chapter focuses on the measurement and characterization of the sub-set of biological aerosol that contain bacteria, virus and fungi and exist primarily in the respirable range.

Issues in Toxicology No. 25
Aerobiology: The Toxicology of Airborne Pathogens and Toxins
Edited by Harry Salem and Sidney Katz

Published by the Royal Society of Chemistry, www.rsc.org

6.2 Ambient Bioaerosol Measurement

Characterization of ambient biological particles can be accomplished by a number of techniques, but these techniques can be broadly grouped into two primary categories: characterization of collected particles and *in situ* measurement. Although characterizing collected particles is more time consuming and requires multiple steps in measurement, which can lead to errors, the techniques that can be employed offer extremely detailed information. On the other hand, *in situ* techniques can offer rapid, size-resolved identification of likely biological particles. The amount of information that can be gleaned from these measurements varies significantly by technique. Both *in situ* and collected particle analysis offer different and often complementary views of biological aerosol populations.

6.2.1 Characterization of Collected Particles

6.2.1.1 Collection Techniques

In order to use many analysis techniques on biological aerosols, particles must first be collected from the air. Although many techniques exist, some are better suited for different types of downstream analysis than others. Filtration, impactors, impingers, cyclone collectors and electrostatic precipitators are all routinely used for the collection of biological particles. Detailed descriptions of all of these collectors were given by Hinds,[9] so the discussion here will focus on the compatibility between collection method and the types of bioaerosol analysis. Table 6.1 presents a general guide to the applicability of collection approaches to downstream biological assays.

Filtration is among the most basic types of aerosol collection and also among the most widely used. Depending on the filter type, this method could be used for a wide range of downstream analysis techniques; however, the forces involved in the collection of particles by filtration can often be too harsh to recover live viruses or many vegetative bacteria. Between collection and analysis, air passing through the filter and over the biological particle can desiccate the collected organisms, which may affect viability in many microorganisms and viruses. Depending on the size of the filter, the pressure drop across the filter and the strength of the pump used for sampling filtration techniques can sample air at up to several hundred liters per minute. Such a large range in sample volume makes this method appropriate for applications intended to measure human exposure (usually less than 20 L min^{-1}) using large-volume air sampling to collect biological particles from ambient air. The type of filter selected will affect both the type of downstream analysis that can be performed and the flow rate that can be sampled. Glass-fiber filters are a good choice when the viability of the collected particles is not a concern. These filters and the filter holder can be sterilized and cleaned to remove a large fraction of contaminating or residual nucleic acids and proteins, making them ideal for downstream molecular analysis. Glass-fiber filters will also partially disintegrate in most water-based buffers, ensuring

Table 6.1 A guide to the selection of aerosol collection methods for appropriate downstream biological analysis.[a]

Collection mode	Type	Culture	PCR	Immunoassay	Mass spectrometry	Microscopy	Sequencing
Filter	Glass-fiber	s, f	s, b, f, v	s, b, f, v	s, b, f, v		s, b, f, v
	PTFE		s, b, f, v	vs, b, f, v	s, b, f, v	s, b, f, v	s, b, f, v
	Gelatin	s, b, f, v	s, b, f, v	s, b, f, v			s, b, f, v
Impactor	Solid substrate			s, b, f, v	s, b, f, v	s, b, f, v	s, b, f, v
	Agar	s, b, f	s, b, f, v	s, b, f, v			s, b, f, v
Impinger/WWC		s, b, f, v	s, b, f, v	s, b, f, v	s, b, f, v[b]		s, b, f, v[b]
Electrostatic precipitator		s, b, f, v	s, b, f, v	s, b, f, v	s, b, f, v	s, b, f, v	s, b, f, v

[a]Key: s = spores; b = bacteria; f = fungus; v = virus. WWC, wetted-wall cyclone.
[b]Buffer must be carefully chosen so as not to interfere with the analysis method.

that most of the collected particulate can be recovered for analysis. Gelatin filters are another common choice for bioaerosol sampling, because they provide a softer surface for collection, intended to reduce the stress on collected organisms;[10] however, desiccation after collection can still be an issue, as manufactures generally recommend collection times of no longer than 30 min. These filters also dissolve in most biological buffers, making them well suited for analysis by culture.[10] Various types of polytetrafluoroethylene (PTFE) (Teflon) and metal filters with controlled pore sizes are also used for sampling bioaerosols. Burton *et al.*[11] showed that PTFE filters perform well for the collection of small bacterial and viral aerosols for downstream molecular analysis. These substrates are often desirable because a large fraction of the particulates will remain on or near the filter surface, making it available for observation directly on the filter surface (such as by microscopy)[12] where the physical morphology of the particle, as it was collected, is desirable.

A number of bioaerosol collection techniques use impaction for collection. A major advantage of impaction techniques for bioaerosol collection is the ability to size fractionate particles using multiple impactor stages.[9] Bioaerosols of different sizes may have different effects on the living things with which they interact or on their environment. For instance, biological particles in the respirable range (typically 1–10 μm) may be implicated in pneumonic disease.[13,14] Infectious particles larger than 10 μm may be involved in diseases originating in the mucous membranes and may have a very different presentation from the same disease when inhaled more deeply into the respiratory tract.[13,14] Pollens and other bioaerosol associated with allergies are typically larger than 10 μm. Their association with allergies is likely related to their size, which is primarily filtered by the nasal airway.[9] Therefore, the ability to separate the particles by size prior to detailed analysis may provide additional information into the roles that bioaerosols play in a variety of circumstances. In general, owing to the forces of collection, impactors will not produce viable samples unless they are collected either into nutrient agar or into liquids.[9] A simple example of impaction into agar is the slit-to-agar sampler. In this device, an agar plate is rotated 360°, over a fixed sampling period, under the inlet nozzle, which is slightly smaller than the radius of the agar plate.[9] By rotating the plate during collection, the number of colonies collected over an arc of the plate correspond to a specific time period of sampling. After sampling, the plate is incubated and then colonies are counted in sections. This allows the culturable fraction of biological particles to be measured as a function of time. This method can be very useful for determining the time during which a bioaerosol "event" occurred or for looking at fluctuations in bioaerosol concentration. An Andersen cascade impactor (Copley Scientific, Nottingham, UK) is an example of an impactor that can size fractionate particles while preserving the viability of the biological fraction. Like the slit-to-agar sampler, it is designed to have particles impact onto nutrient agar, although the Anderson impactor uses multiple jets rather than a slit. In practice, the nutrient agar can be replaced with a buffer solution to allow the sample to be collected without encouraging growth, while still retaining viability. Any multijet impactor may be used to collect biological particles if molecular assays, such as polymerase chain reaction (PCR), immunoassay or sequencing, are to be performed on the collected particles, provided that the particles are recovered from the impactor surface with buffer and processes compatible with those downstream assays.

Impingers and wetted-wall cyclones are frequently used to collect bioaerosols into liquid. Both technologies are commonly used to preserve the viability of collected bioaerosols. The most common impingers are all-glass impingers and SKC biosamplers. Both sample at flow rates representative of normal human lavage rates (12.5 L min^{-1}). They are useful for assessing human exposure because of their direct comparability with human inhalation rates,[9] but are limited in environmental settings where bioaerosol concentrations are low. This limitation is due to both the flow rate limitation and the limited time over which these samplers can be run. This time limit is primarily due to evaporation of the collection fluid during sampling.[10] Wetted-wall cyclones are capable of collecting particles into fluid at much higher flow rates. One commercially available system is the Smart Air Sampler System (SASS) 2300 from Research International (Seattle, WA, USA). This system has a high sampling rate (265 L min^{-1}), which provides a balance between preserving aerosol viability and collecting aerosol from a large volume of air. These systems also have advantages over viable impactors because the collected aerosol can be processed using a number of downstream molecular and viable assays. For example, Hietala *et al.*[15] used the SASS to collect Newcastle disease virus particles in commercial poultry flocks by real-time reverse transcriptase PCR (RRT-PCR). This is particularly important for viral aerosols, where collection into agar media has little value.

Electrostatic precipitators (ESPs) are also used in bioaerosol collection. Although only a limited number of studies have demonstrated collection of biological particles with this technique, ESPs are widely used to collect aerosol particles for other applications. Typically, particles collected by an ESP are initially adjusted to a uniform charge distribution using any number of techniques (alpha source, corona discharge, *etc.*) to facilitate downstream collection in the electric field. Several studies have shown that at least some biological particles carry a preferential charge that can be used to facilitate collection.[16,17] Since the forces imparted on a particle do not depend on the sampling flow rate, particles can be collected at high flow rates without directly impacting the velocity of particles impacting the collection surface. The velocity of particles impacting the surface can be controlled through the electric field strength and/or particle charge to within the limits defined by the size of the collection surface. Further, particles can be collected into a small area, facilitating collection by a small liquid volume.[17,18] Although ESPs have seen limited use for bioaerosol collection, it is reasonable to assume that ESP bioaerosol collectors could be suitable for the collection of viable bioaerosols and for providing enough concentration to allow more detailed downstream molecular analysis.[18]

6.2.1.2 Analysis of Collected Bioaerosols

As indicated in the previous section, there are many options for the analysis of bioaerosols after collection. Among the most straightforward is culturing of the collected sample. Culturing is the oldest method used to study the biological fraction of an aerosol, but it is severely limited in its scope. The limitations of culture in environmental microbiology, including the study of bioaerosols, have been discussed by numerous authors.[19–21] It is estimated that only a small fraction (1–2[20]%) of all bacteria in the environment can be cultivated effectively in the laboratory. Even the use of multiple broadband and selective media types is insufficient to overcome these limitations.[21] The above is true of bacteria and fungi, but is even more complicated when considering viral particles. Any attempt to examine ambient infectious viral aerosols using cultivation must be very narrowly focused on a specific virus or small sub-set of viruses, and even then it may be unsuccessful. Viruses are often extremely host specific and the variety of viral hosts is very likely as broad as all life itself. It is therefore effectively impossible to cultivate the breadth of environmental virions that might be found in an air sample. However, if one is narrowly focused on identifying an infectious virus of a specific type from an air sample, cultivating for cytopathic effect or plaque assays on a lawn of host cells can be an effective means of isolating live virus from an environment.[22]

Analysis of bioaerosol proteins offers another means of quantifying both the amount and type of biological material present in an air sample. Three types of analysis are found frequently in the literature: bulk protein analysis,[23] soft ionization mass spectrometry (MS) such as matrix-assisted laser desorption/ionization (MALDI),[24] and identification of specific proteins through antibody response.[25] Each of these three techniques provides different types of information that can be useful in different types of analysis. For example, bulk protein analysis can identify the proteinaceous fraction of the aerosol collected, which is useful in exposure studies aimed at quantifying exposure to biological material.[26] Soft ionization MS can be useful in identifying the presence of specific proteins where the mass of protein or its typical fragments after ionization are known.[27] Immunoassays are far more specific to individual proteins, and therefore the species that produce them, than MS techniques can be. However, antibody production is laborious and requires the ability to cultivate the species of interest in the laboratory and in the animal that will produce the antibodies (typically rabbits, mice and goats). These techniques are useful for viruses, and also bacteria, molds and eukaryotes, but in order to move beyond just understanding the quantity of proteinaceous material in the sample, the species of interest must have been identified and characterized previously.

Analysis of bioaerosol nucleic acids offers a potentially broader look at bioaerosol content,[28] but has its own unique set of limitations. PCR has been used for decades to detect the presence of specific segments of nucleic acids (NAs; both DNA and RNA). These NA segments can be chosen to be specific down to the species or sub-species level,[29] be more generally representative of a group of organisms[30,31] or detect the presence of a gene or family of genes.[30,31] Characterization of bioaerosols by PCR is limited to only those organisms for which there are assays, but PCR is very useful for detecting the presence of specific organisms, or organisms carrying a specific gene, in a sample. PCR requires very little sample (typically as little as a few copies of target NA) and is very specific to the selected segment of NA. It is also possible to multiplex PCR reactions to expand the number of NA segments that can be detected in each reaction. However, multiplex reactions require extensive optimization to maintain sensitivity and the more primer sets are used, the more difficult is the optimization and less sensitive the reaction.[32] Since the identification of organisms by PCR does not require cultivation, it is possible to identify bacteria, virus, fungi and fragments of these and other cell debris, provided that the nucleic acids remain intact.

A common target for PCR reactions in bacteria is the 16S ribosomal RNA region. This has both conserved and hypervariable regions, allowing a general 16S primer set to be built that will recognize all bacterial 16S genes or primers can be built for specific species or sub-species. In addition to traditional single and multiplex PCR reactions, microarrays have been used to identify thousands of bacteria by their 16S genes.[33] In addition, the analogous eukaryotic gene, the 18S rRNA, is often used in phylogenetic studies on fungi and yeasts,[26] and also other eukaryotes.[26,34]

Fundamentally, all approaches involving PCR are biased towards identifying organisms already known to exist. In order to examine the complete diversity in a sample, unbiased methods must be employed. Next-generation sequencing (NGS) offers a number of methods that can help quantify all, or very broad classes, of living things in an air sample. Like PCR, NGS can target 16S and 18S rRNA for the identification of a broad range of prokaryotic and eukaryotic organisms. However, viruses cannot be characterized in this way, as they do not possess these genes. In order to identify viral NAs in a sample, a broader approach is needed. Whole-genome sequencing (WGS) has been used in a number of studies to investigate the entire genomic content of environmental samples.[35] To date, WGS has only rarely been applied to aerosol samples,[36] and unfortunately provides incomplete answers. Genome coverage in many environmental samples, particularly aerosol samples, is often extremely low, leading to only fragments of a genome that can be determined. Further, since so few environmental microorganisms and viruses have been identified and characterized, most sequence fragments from these samples cannot be found in any established genome library (*e.g.* GenBank). Therefore, we are left with considerable information about the NAs in a sample, but only a limited amount of useful information about the details of the microbial communities in those samples. Studies relying on 16S and 18S can characterize the known bacteria present in a sample and identify unknown rRNA sequences, allowing the "discovery" of unknown species[37] with some indication of its genetic origin. WGS, by comparison, can provide much more detail about the gene content, and also relationships to known species, and can identify novel viral sequences.

6.2.2 *In situ* Measurement of Bioaerosols

In situ measurement of biological aerosol particles typically relies on spectroscopy or spectrometry of aerosol particles, where particles of biological origin are expected to have features or chemical make-up that distinguish them from all other particles. A variety of approaches have been developed to exploit these differences, each with its limitations. Mass spectrometry of single particles,[38] IR and Raman spectroscopy[39] of collected samples and fluorescence of single particles[40] are the methods that have primarily featured in the literature. None of these techniques provides the kind of detailed biological information that NA or protein analysis provide. However, they have the distinct advantage of being rapid: spectral properties can be measured and recorded in less than 1 s, compared with minutes or hours for NA or protein analysis. Spectroscopic methods are also much less specific, in that they only provide an indication of whether or not the particle is likely of biological origin. Look-up libraries of mass, fluorescence and atomic spectra can extend this to an indication of the type of biological aerosol (mold, virus, pollen, spore, bacteria, *etc.*),[38,41] but the answers are not as conclusive as those given by biochemical analysis.

Fluorescence-based bioaerosol measurement systems have been developed by several groups[40–54] and they are by far the most prolific *in situ* tools for bioaerosol measurement. Fluorescence measurements of atmospheric aerosols have been made at locations on several continents.[4,41,43,55–60] The commercial availability of some fluorescence detection systems[61] facilitates these measurements.

Two commercial instruments dominate the bioaerosol fluorescence literature: the ultraviolet aerosol particle sizer (UVAPS) and the wideband integrated bioaerosol sensor (WIBS). The UVAPS was first described by Hairston *et al.*,[44] and combines the aerodynamic sizing of the TSI (Shoreview, MN, USA) aerodynamic particle sizer with UV laser-induced fluorescence measurements of single particles excited at 355 nm. It provides size-resolved autofluorescence measurements and has been used extensively in both laboratory[62] and field measurements.[4,63] Despite its widespread use, the UVAPS has been discontinued by TSI. The WIBS was developed by the University of Hertfordshire, the Ministry of Defence and the Defence Science and Technologies Laboratory (DSTL) in the UK as a low-cost, laser-induced fluorescence (LIF) instrument suitable for real-time detection and measurement of bioaerosols.[46] The WIBS uses two filtered xenon flashlamps as a UV light source, centered at approximately 280 and 370 nm, and measures fluorescent emissions between approximately 310–400 and 420–650 nm. Droplet Measurement Technologies (Boulder, CO, USA) obtained the license for the manufacture of the WIBS-4 in 2012 and produced the first commercial version of the instrument, the WIBS-4A, in 2013. The current commercial version has modified the original WIBS-4 design, introducing an improved aerosol jet and significant changes to the electronics and optical system that reduced the variability in aerosol sizing and fluorescent response and improved the overall system signal-to-noise ratio.

To date, the WIBS has been used in several studies to measure ambient bioaerosols. Gabey *et al.*[58] reported measurements of fluorescent particles in a tropical forest and coined the term fluorescent biological aerosol particles (FBAPs) as particles exhibiting significant (above instrument background) fluorescence in tryptophan-like (280 nm excitation; 310–400 nm emission) and NADH-like (370 nm excitation; 420–600 nm emission) excitation–emission channels, which is now common in WIBS analysis. Several studies from the University of Manchester, UK, reported fluxes of aerosols including FBAPs from a tropical forest site in Borneo and compared fluorescent aerosol properties observed in Borneo with observations made in the city of Manchester, UK.[59,64,65] Several additional studies of FBAPs have been carried out in Colorado at both mountain sites[66] and at a forested site.[67–69] Huffman *et al.*[68] also reported a correlation between concentrations of ice nuclei, fluorescent particles and precipitation. Wright *et al.*[70] found a similar correlation between ice nuclei concentrations and relative humidity and ambient aerosols have also been measured using the WIBS-4A in Raleigh, NC, USA.

Fluorescence measurements of ambient aerosols have also been made using a number of additional research instruments that are not commercialized or have limited availability. The single particle fluorescence spectrometer has been used to measure high-resolution spectrographs of fluorescence from single aerosol particles at multiple wavelengths (266 and 365 nm).[56,71] The most recent designs of this instrument also include geometric measurements of particle size.[72,73] Using spectra, particularly from multiple wavelengths, allows the SPFS to discriminate between mold, humic acid and bacteria.[74] Because detailed spectra can be measured for individual particles, this technique is also particularly useful for the detailed study of chemistry that changes bioaerosol fluorescence.[72]

Bioaerosol fluorescence is measured after excitation by light in three specific wavelength ranges (~405, ~360 and ~265 nm). Each of these wavelengths excites a different type of biological molecule, the presence of which is an indicator that the particle is of biological origin. Flavin molecules are excited at 405 nm. Flavin molecules such as riboflavin are produced by bacteria and fungi, and also by many other organisms, and are found in most biological tissues. Like most biologically fluorescent compounds, riboflavin owes its fluorescence to the aromatic rings in its structure. At 355–365 nm, nicotinamide adenine dinucleotides (NADH and NAD^+) are the primary biologically fluorescent compounds. These compounds are also metabolic products with carbon ring structures. Finally, between 260 and 280 nm, the primary fluorescent compounds are amino acids, primarily tryptophan and tyrosine, which also owe their fluorescence to their ring structures. It should be noted that since these are amino acids found in proteins and not metabolic products, they can be found in living organisms, and also in viruses and prions.

Mass spectrometric measurements of ambient biological particles have also been made, although the scope of these measurements is much more limited. MALDI-MS has been used for the analysis of collected aerosol samples,[27] but *in situ* analysis is limited to single-particle mass spectrometry. The literature indicates that only a single group has performed these analyses routinely, using the bioaerosol mass spectrometer (BAMS)[38]/single-particle aerosol mass spectrometer (SPAMS).[75] This technique measures the ion spectra in each particle and can differentiate biological from nonbiological aerosols by matching individual particle spectra to previously measured spectra in a database. Although published results indicate that this technique can distinguish between closely related bacterial species,[38] it is important to note that this differentiation was performed under laboratory conditions and controlled bacterial growth conditions. In the ambient environment, it is unlikely that such close differentiation would be possible. However, it should be noted that all biological material is composed of similar ions (*e.g.* mass peaks associated with carbohydrates and amino acids and nitrogen-containing peaks[76]) and so the general particle-by-particle differentiation between biological and nonbiological particles should be feasible based solely on the ion spectra.

6.3 Bioaerosol Populations and Sources

Studies of ambient bacterial aerosols have sought to provide a better understanding of the diversity, variability and factors that control the biological aerosol population. In principle, all past studies fall into one of three primary categories: studies of viable/culturable bacteria and fungi,[77] studies of phylogenetic diversity using molecular methods such as clone sequencing[78,79] or microarrays[80] and spectroscopic characterization of biological aerosol.[40,50,51,75,81] Several common observations have arisen from these studies. Bacteria can be transported to geographically distant locations from their sources.[7,82] Meteorological conditions can affect both the observable biological aerosol concentration[83] and its diversity.[80,84] Diverse bacteria and/or their genomic fingerprints have been found in air samples.[80] Actinobacteria, Firmicutes and Proteobacteria are ubiquitous in aerosol samples collected from diverse locations.[77,80,84,85] Gram-positive species dominate the culturable fraction of bioaerosols, whereas Gram-negative species are the dominant fraction of the nucleic acid, probably because Gram-negative organisms tend to be more fragile and are less likely to survive aerosol collection or even exist as an aerosol.

Many studies have sought to explain the observed fluctuations in bioaerosols through changes in the source of these aerosols, one of the first being by Lighthart and Shaffer.[83] Viable bacteria were observed with a temporal resolution of 2 min and diurnal variations in the bacterial concentration were observed that appeared to be associated with the local meteorology and climatology associated with land–sea breezes in the Willamette River Valley in coastal Oregon. Shaffer and Lighthart[77] examined fluctuations in viable microbes with a similar technique at urban, rural, forested and coastal sites, showing similar diurnal trends. Gram-positive bacteria, primarily *Bacillus*, accounted for >70% of the observations. Brodie *et al.*[80] used 16S rRNA targeted microarrays to map urban bacterial diversity in San Antonio and Austin, TX, over 17 weeks. This study also indicated that some bacteria appeared to be more sensitive to meteorological conditions than others. They also associated specific taxa with potential sources, such as *Arcobacter* and *Helicobacter*, as a potential indicator of fecal contamination from wastewater treatment facilities. Spectroscopic techniques, using UV-LIF to characterize individual aerosol particles as biological or nonbiological,[40,50,86] have also been used to examine fluctuations in bioaerosol populations. Pöschl *et al.*[63] and Huffman *et al.*[87] used UV-LIF, aerosol mass spectrometry and many other techniques to investigate the roles of biogenic rainforest aerosols, including biological aerosols, as cloud condensation nuclei. Gabey *et al.*[66] also examined biological aerosol populations above and below the forest canopy. Their results further demonstrated a diurnal pattern to bioaerosol concentrations, similarly to previous studies. Further, they observed transient increases in biological aerosol below the canopy, which they attributed to fungal spore release, which correlated with increases in relative humidity. Santarpia *et al.*[86] were the first to suggest that aerosol processes may also play a role in the observed fluctuations in bioaerosol populations. They used 405 nm excitation to monitor bioaerosol populations over several days and correlated them with co-located meteorological measurements at 2 s time intervals. They observed correlations between relative humidity and optically observed bioaerosol concentrations over time scales short enough that they imply that the uptake of water by biological particles changes the observability of those particles by the instrument.

A large fraction of the bioaerosol literature focuses on the characterization of ambient, naturally occurring bioaerosols and the role of these particles in human health and in atmospheric processes such as cloud condensation nuclei and ice nuclei.[88] Anthropogenic bioaerosols, biological particles that are produced through human activity, have also been studied extensively. One of the most important aspects of these is biological weapons; however, biological particles produced from more mundane human activities, such as waste treatment and agriculture, also produce bioaerosols of concern.

Bioaerosol populations in urban areas are studied because of the potential impact they could have on the populations of those urban areas and also as the background "noise" in attempts to detect intentional releases of biological particles. Boreson *et al.*[26] found that the biological loading in particulate matter (as measured by the total protein concentration) collected at an urban site on the fringe of a desert was higher than that from the desert itself. This suggests that urban activities were creating an additional source of biological material in the aerosol. Their analysis of the 18S rDNA signatures from these samples indicates that the eukaryotic populations in these two geographically close sites were significantly different. A study of rural and urban sites in Pakistan indicated that significant concentrations of bioaerosol existed in both rural and urban sites, but that the sources of these populations are likely different, with different implications for the human population.[89] In that study, the highest outdoor bioaerosol concentration was observed at the urban site, with different types of culturable bioaerosol observed between the two rural and one urban site. In general, although the total viable bioaerosol concentrations were lower at the outdoor rural sites than at the urban site, the rural sites had higher fungal and Gram negative bacteria than the urban site. The authors hypothesized that different sources accounted for the different population types. They suggested that poor sanitation at all sites probably accounted for the high concentrations observed, but that agricultural activities likely accounted for the observed difference in bioaerosol types between the urban and rural sites.

The treatment of both human and agricultural waste has been researched extensively for its contribution to biological aerosol populations. Cronholm[90] demonstrated that live aerosolized pathogenic bacteria could be recovered 930 m downwind from a sewage treatment plant. Since then, other waste management activities such as the treatment of agricultural fields with solid waste and wastewater and urban composting have been studied for the potential for hazardous bioaerosol generation that could be transported downwind. Paez-Rubio *et al.*[91] determined that high winds (above 1.9 m s^{-1} with an average of 4.5 m s^{-1}) aerosolized microorganisms during flood irrigation. This conclusion was supported by both culturing and molecular analysis. Moreover, their molecular analysis showed a high degree of similarity between the microbes from the wastewater source and those collected in the air, indicating that high winds increase the exposure risk of workers and nearby populations to the inhalation of wastewater microbes during flood irrigation. Taha *et al.*[92] and Pankhurst *et al.*[93] investigated small-scale urban compost sites in the UK as potential bioaerosol sources. Both studies found elevated levels of bacterial and fungal particles in the vicinity of composting. Taha *et al.*[92] suggested, based on modeling, that bioaerosol concentrations may remain elevated within 100 m of the compost site. These studies indicate that the treatment of anthropogenic waste may be a significant local source of bioaerosols in both urban and rural environments.

Other agricultural activities, in particular animal husbandry such as at swine farms, have been the focus of research on local and regional air quality and disease transmission. Martin *et al.*[94] examined aerosol properties around swine facilities in Ames, IA. Those measurements indicated that the facility studied caused an increase in airborne particulate matter and that a large fraction of the mass was organic carbon. Primary biological particles are generally defined to be a sub-set of organic carbon aerosols, but Martin *et al.*[94] did not extend their analysis to bioaerosols. Agranovski *et al.*[95] used a combination of the UVAPS and impingers to study the bioaerosol in an Australian swine facility. This study found both culturable and biologically fluorescent, respirable aerosols in the facility. They found 12 times more biologically fluorescent particles inside the facility than outside. Other studies have examined the bioaerosol populations around swine barns. Kristiansen *et al.*[96] and Boissy *et al.*[36] studied the composition of bacterial communities in swine barns. Both studies indicated that bacteria previously identified from the pig intestinal tract could be found in particulate matter. Firmicutes were the dominant group identified in both studies, but the dominant genus differed between the studies. *Clostridium* was the dominant genus identified by Boissy *et al.*[36] whereas *Streptococcus* dominated in the analysis of Kristiansen *et al.*[96] Despite the differences, which may have been due to differences in the analysis methods (16S sequencing by Kristiansen *et al.*[96] and shotgun pyrosequencing by Boissy *et al.*[36]), the studies are in general agreement about the predominance of pig intestinal flora in the airborne bacteria in these environments. Recent studies have also implicated aerosol generated by infected swine as a probable route of transmission of influenza A virus to other species. For example, Corzo *et al.*[97] were able to detect three influenza serotypes (H1N1, H1N2 and H2N3) in air samples collected downwind of pig farms where the same serotypes had been identified from the pigs themselves. Influenza identified by PCR and positive air samples were collected from as far away as 2.1 km, whereas infectious virus was only isolated from air samples in and around the barns themselves.

In addition to local sources of bioaerosol, remote sources can also make significant contributions to the bioaerosol population in both urban and rural areas. One of the first compelling arguments for the long-range transmission of bioaerosols was made by Bovallius *et al.*[82] to explain a red snowfall event in Sweden. A combination of microbial analysis and calculation of back-trajectories was used to determine the probable source of the airmass carrying the bacteria. More recently, several groups have investigated the transport of microbes with desert dust and their impacts on distant regions.[7,98,99] Griffin *et al.*[99] demonstrated a significant correlation between satellite-observable dust storms originating in North African deserts in both the northern Caribbean[98] and in a site in Turkey[99] and an increase in local culturable bacteria. Their analysis of the bacteria collected in those studies indicated a variety of potentially pathogenic (human and otherwise) organisms. Prospero *et al.*[7] independently measured increases in the bacterial and fungal load in Barbados during periods when the regional airmass was influenced by North African dust storms. A combination of minerals and opportunistic pathogens carried in African dust deposited in the Caribbean has long been hypothesized to be responsible for the decline in coral reefs in the affected parts of the ocean.[100] African and Asian desert dust storms have been demonstrated to transport bioaerosols, leading to the spread of pathogens and allergens from these regions across the globe.[101] These studies have demonstrated extremely long-range transport of bioaerosols under extreme meteorological conditions, but bioaerosols may also be transported from distant sources even under normal meteorological conditions. Santarpia *et al.*[21] used 16S microarray analysis of air samples collected in Laurel, MD, and back-trajectory analysis to look at the potential origins of that airmass. These analyses indicated oceanic bacterial signatures in some samples where back-trajectory calculations indicated that the airmass passed close to the ocean surface hundreds of miles from the collection point.

In addition to the general public health concern, the study of bioaerosols is often focused on the defense against aerosolized biological weapons. Throughout history there have been many instances where biological agents have been used to spread death and disease among rival forces. As far back of the sixth century BC, there are reports that indicate that the Assyrians used *Claviceps purpurea*, a fungus, to poison the wells of their enemies. In 1346 CE, Mongolian forces catapulted corpses infected with bubonic plague, caused by the bacterium *Yersinia pestis*, over the walls of a city fortress in Kaffa in an attempt to spread the disease amongst the rival city inhabitants. In the late 1700s, the British plotted to spread smallpox amongst Native American tribes by supplying them with contaminated cloth materials.[102] In 1942, the United States began to research the use of biological materials for active offensive warfare and ways of mitigating the effects of biological materials for defensive purposes. In 1969, under the Nixon administration, the United States renounced the use of offensive biological weapons and in 1972 the United States, Great Britain and the Soviet Union signed the Biological and Toxin Weapons Convention, in an attempt to stop the development and use of biological weapons, which was enacted in 1975. However, in 1979 in Sverdlovsk, Russia, *Bacillus anthracis* spores, the causative agent of anthrax, were released from a biological weapons research facility and killed 64 people, notifying the world of Russia's continued research into biological weaponry and indicating that the threat was, and still is, very imminent in modern times. The anthrax letter attacks in 2001, in which the dissemination of a few grams of *Bacillus anthracis* spores in mailed envelopes resulted in 22 infections and five deaths, heightened Americans' sense of vulnerability to biological weapons attacks. In 2004, envelopes containing ricin toxin, a protein extracted from castor beans, were mailed to the office of the US Senate majority leader and processed through a mail sorting facility.[102]

The most effective method of biological weaponization is *via* infectious, respirable aerosols.[102] Biological weapons do not cause an immediate reaction like chemical weapons and can sometimes go undetected for days owing to the incubation period of the particular pathogen. This incubation period can lead to a delay in responsive action which can lead to further spread of the infection amongst targeted populations.[103] It is commonly accepted that to be most effective, aerosolized bioweapons should be in the size range 1–10 μm, since these particles are most effective at entering into the respiratory system and causing infection.[9,10,102]

Detection of biological agents prior to the onset of disease is desirable in order to provide the highest level of defense against biological weapons. The same fundamental technologies used to measure and characterize naturally occurring biological particles may also be used to detect biological weapons. The challenges are that all the pathogenic organisms used for biological weapons are also found in nature, and these pathogens are chemically and physically very similar to the wide range of largely innocuous biological particles found in the atmosphere. In practice, then, it may be desirable to combine the benefits of rapid, *in situ* measurements (such as UV-LIF) and more specific techniques (such as PCR or immunoassay). Optical techniques, such as UV-LIF, have been commercialized and used for early warning systems or to "trigger" more specific analyses, since they can make measurements in real-time, providing measurements at frequencies <1 Hz, and can provide rapid detection of biological particles.[104,105] Since these systems cannot discriminate between pathogenic and innocuous biological particles, molecular techniques, such as PCR and immunoassays, are used to analyze biological particles of concern. However, UV-LIF and other rapid biological detection schemes can be prone to false-positive results, since the ambient population of biological particles is constantly fluctuating. Depending on how the system is being employed, a false-positive result can lead to something as innocuous as an alarm or the unnecessary use of more specific technology to vet the detection, or a false positive could trigger unnecessary use of personal protective gear or the evacuation of a facility. Commonly, repeated false positives in a detection scheme lead to complacency and may cause users to ignore true-positive events.[103–106]

In order to mitigate false positives from natural fluctuations of biological particles, algorithms that attempt to account for the natural variability in biological aerosols are used. To do this, measurements are made of the relevant biological background and detection algorithms are developed and adjusted to be insensitive to ambient fluctuations in the biological aerosol populations.[107] Studies such as those presented in the section are often used to develop these algorithms. Despite the care that has been taken in the development of these algorithms, natural changes in the ambient aerosol population can still greatly affect the operation of a biological detector. Therefore, it is important to characterize system performance under conditions that accurately represent the natural background. Ideally, sensor testing would occur at the site of intended use and under conditions that are relevant to sensor operation.[106] However, transportation and logistical costs associated with testing on-site are prohibitive. Recent advances in biological aerosol sensor testing[62] have sought to improve the quality of field sensors by more accurately representing natural fluctuations in the ambient aerosol in a laboratory environment.

6.4 Factors Affecting Bioaerosol Measurement

In order to interpret environmental bioaerosol measurements properly, it is critical to understand the factors that may affect those measurements. However, there has been only limited investigation into a fraction of the potential factors involved. In particular, the effect that atmospheric processing has on the properties of bioaerosols and how that impacts measurement by the previously described methods has only recently been described. Open air factor (OAF)[108] is a term originally coined to describe the unexplainable loss of viability of certain organisms when released into the environment. Further research indicated that OAF may be caused by the interaction of atmospheric oxidants, such as ozone, with terpenes and other trace organic species that produce low vapor pressure products, which condense onto the biological particles and may be toxic to the affected organisms.[108] These are the same processes that form secondary organic aerosols (SOAs),[109] a component of photochemical smog. Aside from the initial work, only a small amount of published work is available on OAF. Bailey *et al.*[110] looked at the effects of OAF and ozone alone on aerosolized *Micrococcus luteus* and found OAF to be more effective at killing aerosolized bacteria than ozone alone.

The mechanisms responsible for inactivating viruses and bacteria (primarily protein and nucleic acid damage) are also likely to impact their detection and measurement. Pan *et al.*[57] first described the impact of ozone and water vapor on the UV-LIF spectra of aerosolized *Bacillus thuringiensis*. The fluorescence spectra of these particles indicated that the emission peak most commonly associated with tryptophan decreased in intensity and showed a slight blue shift after exposure to ozone. Further studies, reported by Santarpia *et al.*,[86] showed marked decreases in the intensity of fluorescence excited at 263 nm of MS2 bacteriophage, particularly in the presence of water vapor. Further studies have developed a better understanding of the role of water vapor in this interaction. Pan *et al.*[57] used particles composed of an 8-mer peptide to investigate the chemical process responsible for the reduction in fluorescence observed in previous studies. Using spectra from single particles excited at both 266 and 351 nm, they observed decreases in 266 nm fluorescence coincident with increases in 351 nm fluorescence when these particles were exposed to both ozone and high relative humidity (RH). This is consistent with an ozonolysis–hydrolysis process that would produce *N*-formylkynurenine (NFK) and kynurenine (KU) from tryptophan, changing the tryptophan fluorescence, which is excited at 266 nm, to that of NFK and KU, which is excited at 351[111] nm. Ratnesar-Shumate *et al.*[73] further investigated the effects of ozone and water vapor on both MS2 bacteriophage and *Bacillus thuringiensis*. This study was consistent with the peptide studies, indicating a decrease in tryptophan fluorescence with exposure to high ozone and RH, but did not observe the increase in 351 nm fluorescence observed in Pan *et al.* This may have been due to the existing 351 nm fluorescence in the particles masking the changes produced by the ozonolysis–hydrolysis reaction. In addition, because they attempted to use Q-PCR to normalize the physical loss during those experiments, they observed changes in the nucleic acid signature used to detect the *B. thuringiensis*. They hypothesized that the change in PCR signal was due to the ozonolysis of extracellular DNA[112] known to be present in the spore coat of most laboratory-produced *Bacillus* spores.[113] Aside from those of biological origin, there are many other molecules that may be found in the atmosphere,[114] in addition to the possibility of generating additional fluorophores through reactions with biological molecules, ozone and unsaturated hydrocarbons[70,115] that may confuse measurements of biological fluorescence. Ultraviolet radiation from sunlight has the potential to degrade fluorescence at every excitation wavelength. Studies using outdoor chambers at Adelphi, MD, in 2012 showed a distinct correlation between UV radiation degradation in the fluorescence of *B. thuringiensis* and MS2 bacteriophage at both 266 and 365 nm.[116] No significant studies have been carried out on the degradation of 405 nm fluorescence, but the processes that have been shown to affect the molecules at shorter wavelengths are also likely to affect the flavin molecules that fluoresce at 405 nm. The changes in fluorescence that occur during exposure to atmospheric conditions, and also nonbiological molecules that fluoresce at these wavelengths negatively impact studies that attempt to use them to characterize biological particles in the atmosphere.

Mass spectrometry has also been used to investigate the interaction of *Bacillus* spores with ozone. Small acid-soluble proteins (SASPs) of *Bacillus globigii* (*Bg*) spores exposed to ozone were compared with those of unexposed spores using MALDI-MS. Figure 6.1 shows spectra from unexposed control spores and ozone-treated spores. These spectra show the development of secondary peaks near the SASP peaks that indicate partial oxidation of these proteins.

In addition to oxidants such as ozone, free radical species such as the hydroxyl radical may also affect the fate of airborne microorganisms and biological material.[117] Further, many of the intrinsic properties of a biological aerosol are poorly understood. Very limited research has been carried out on the hygroscopicity of biological aerosols.[118,119] The hygroscopicity of a biological aerosol may play a large role in its fate due to changes in particle size, uptake into clouds and susceptibility to certain chemical processes. Santarpia *et al.*[86] demonstrated that changes in local RH were correlated strongly with the detection of biological particles using 405 nm excitation over very short time scales. This indicates that the uptake of water by these particles may have played a role in the increase in measurable biological particles.

It is obvious that the measurement of biological particles using fluorescence can be dramatically affected by common atmospheric processes. The impact that these processes have on other types of biological measurement techniques described in this chapter has not been studied in detail, but may be inferred from some of the data. Because the ozonolysis–hydrolysis process described destroys tryptophan, it may also change the immunological properties of proteins used for detection and therefore the ability of immunoassays to detect them. The mass spectrometric data shown indicate that the protein itself may remain intact, but more study is necessary to determine if this is commonly true, since the RH in those experiments was not varied or tightly controlled, and if hydrolysis was incomplete the peptide bonds may have remained intact. Further, the inactivation of MS2 bacteriophage after exposure to ozone and water vapor[73] indicates that protein damage due to these processes is sufficient to prevent viral recognition or replication. Ratnesar-Shumate *et al.*[73] demonstrated that ozonolysis of DNA can dramatically affect PCR assays when extracellular DNA is present and makes a significant contribution to the detectable DNA in a particle. Notably, despite the protein damage implied by the inactivation of the MS2, the PCR assay did not seem to be affected. This implies that despite damage to the capsid protein, the ozone was unable to penetrate and damage the DNA. It is well known that UV light can cause DNA damage,[120] but the effect that this damage may have on detection by PCR has not been demonstrated. The outdoor data presented by Pan[116] seem to indicate that fluorescence at both 365 and 266 nm can be degraded in the ambient atmosphere. The environmental factors that were present during the exposures were ozone, RH and solar radiation, but the

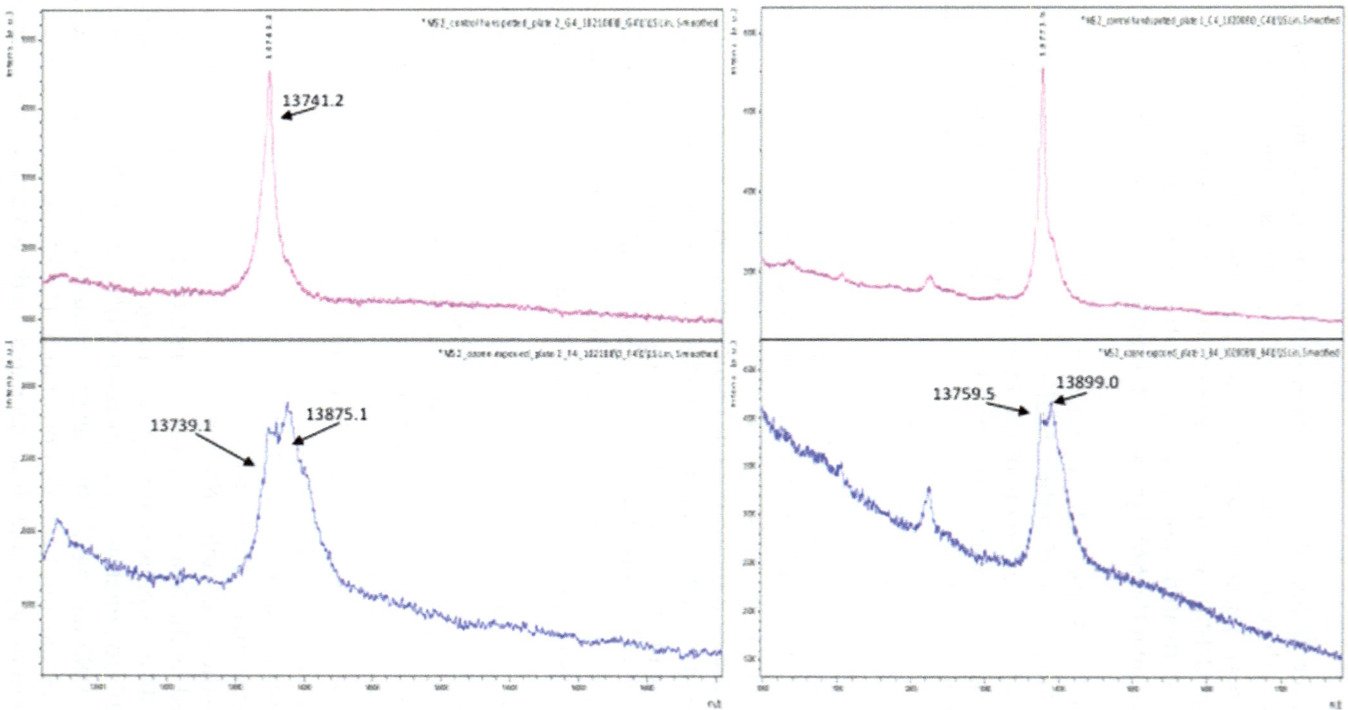

Figure 6.1 MALDI-MS observations of coat protein of ozone-exposed MS2 phage. An MS2 bacteriophage suspension was deposited on slides and exposed to between 0.2 and 0.3 ppm of ozone over a 2 h (bottom left) and a 6 h (bottom right) period. The top panels show measurements of unexposed slides, where the protein peak shown corresponds to the capsid protein of MS2. The enhanced secondary peak in the ozone-treated trace (bottom panels) indicates oxidation of the protein and demonstrates that the protein mass appears to increase with time of exposure. Courtesy of Plamen Demirev and Miquel Antionne.

relative contributions of these factors to the observed degradation in fluorescence has not yet been shown. In summary, common environmental factors have been shown to impact the fluorescence of biological aerosols at both 266 and 365 nm excitation wavelengths. The detection of bioaerosols at 405 nm excitation has been shown to be changed by high RH. Ozone, RH and solar radiation have all been shown to damage DNA, but ozone and RH have only been shown to impact extracellular DNA. The fluorescence degradation at 266 nm indicates the destruction of tryptophan and implies that tryptophan-containing proteins may be damaged and their properties changed. The extent to which these changes in properties impact the immunogenic properties of the protein, and therefore that type of detection, will likely be very protein dependent, given that the position and frequency of tryptophan in the protein sequence will determine the location(s) and extent of the damage caused. It is unclear what effect these factors would have on sequencing technology. Since sequencing generally requires breaking DNA strands into smaller fragments, damage due to environmental factors may not impact this technology significantly. That said, extended exposure has the possibility of extensive damage to nucleic acids and may affect sequencing chemistries.

6.5 Future Directions

The limitations in the study of ambient bioaerosol populations are shared by all studies of environmental microbiology. In addition, the study of ambient bioaerosols includes the need to understand the population fluctuations over very short time scales. This need for rapid data collection has led to the adoption of many optical techniques (such as fluorescence and IR spectroscopy) to make rapid measurements of bioaerosols. The development of this technology, has largely been driven by the need to detect aerosolized biological weapons rapidly. A significant investment has been made by the US and UK governments to develop this technology in the last 20 years, particularly after the 2001 anthrax attacks (Amerithrax). The technology developed during this time has led to major advances in our understanding of environmental microorganisms. High time resolution studies of bioaerosol populations with instruments such as the SPFS, UV-APS and WIBS provide new insights into the fluctuations of ambient bioaerosols and into the roles that these particles play in human health, precipitation and atmospheric chemistry. Despite these advances, these techniques have inherent limitations. Several studies have shown that bioaerosol fluorescence is unstable in the ambient atmosphere and that some nonbiological particles may fluoresce like bioaerosols. Further, intrinsic fluorescence may indicate that something has a biological origin, but does not conclusively define its role in the atmosphere. More specific techniques, such as PCR, immunoassays and even sequencing, give insight into the genes that microorganisms carry and therefore their potential roles in the environment. These techniques are slow, by comparison, so do not provide the type of resolution necessary to understand many atmospheric processes.

Hybrid studies and approaches are necessary both to capture temporal changes in bioaerosols and to determine what organisms are involved. The ideal technique would be one that analyses the genetic composition of every biological particle. This may seem far afield, but much of the component technology required is already available. Single-cell sequencing technology[121] is becoming more mature, but it requires lengthy sample preparation and sequencing itself is still slow. There would be a number of difficulties in the developing DNA isolation, amplification and preparation, but in principle none of these would be impossible on a single particle. Therefore, at the moment, the main limitations are the speed of the chemical processes and analyses, neither of which are insurmountable problems. In the meantime, studies that incorporate multiple, complementary techniques are needed to leverage both the specificity of molecular biology and the speed of optical techniques. Further, additional study is needed in understanding the environmental factors that affect these measurements, so that measurement techniques may be improved and data collected may be put into proper context.

References

1. R. M. Bowers, A. P. Sullivan, E. K. Costello, J. L. Collett Jr., R. Knight and N. Fierer, Sources of Bacteria in Outdoor Air across Cities in the Midwestern United States, *Appl. Environ. Microbiol.*, 2011, 77(**18**), 6350–6356.
2. W. Elbert, P. E. Taylor, M. O. Andreae and U. Poschl, Contribution of fungi to primary biogenic aerosols in the atmosphere: wet and dry discharged spores, carbohydrates, and inorganic ions, *Atmos. Chem. Phys.*, 2007, 7, 4569–4588.
3. *Atmospheric Microbial Aerosols*, ed. B. Lighthart and A. J. Mohr, Chapman and Hall, New York, 1994.
4. J. A. Huffman, B. Treutlein and U. Pöschl, Fluorescent biological aerosol particle concentrations and size distributions measured with an Ultraviolet Aerodynamic Particle Sizer (UVAPS) in Central Europe, *Atmos. Chem. Phys.*, 2010, **10**, 3215–3233, DOI: 10.5194/acp-10-3215-2010.
5. J. M. Sun and P. A. Ariya, Atmospheric organic and bio-aerosols as cloud condensation nuclei (CCN): A review, *Atmos. Environ.*, 2006, **40**(**5**), 795–820.
6. R. M. Bowers, C. L. Lauber, C. Wiedinmyer, M. Hamady, A. G. Hallar, R. Fall, R. Knight and N. Fierer, Characterization of Airborne Microbial Communities at a High-Elevation Site and Their Potential To Act as Atmospheric Ice Nuclei, *Appl. Environ. Microbiol.*, 2009, **75**(**15**), 5121–5130.
7. J. M. Prospero, E. Blades, G. Mathison and R. Naidu, Interhemispheric transport of viable fungi and bacteria from Africa to the Caribbean with soil dust, *Aerobiologia*, 2005, **21**, 1–19.
8. T. Ichinose, S. Yoshida, K. Hiyoshi, K. Sadakane, H. Takano, M. Nishikawa, I. Mori, R. Yanagisawa, H. Kawazato, A. Yasuda and T. Shibamoto,

The Effects of Microbial Materials Adhered to Asian Sand Dust on Allergic Lung Inflammation, *Arch. Environ. Contam. Toxicol.*, 2008, **55**, 348–357.
9. W. Hinds, *Aerosol Technology: Properties, Behavior and Measurement of Airborne Particles*, Wiley-Interscience Publications, 2nd edn, 1999.
10. C. Cox and C. Wathes, *Bioaerosols Handbook*, CRC, 1995.
11. N. C. Burton, S. A. Grinshpun and T. Reponen, Physical Collection Efficiency of Filter Materials for Bacteria and Viruses, *Ann. Occup. Hyg.*, 2007, **51**(**2**), 143–151.
12. P. Baron and K. Willike, *Aerosol Measurement: Principles, Techniques and Application*, Wiley-Interscience Publications, 2nd edn, 2001.
13. R. Thomas, C. Davies, A. Nunez, S. Hibbs, H. Flick-Smith, L. Eastaugh, S. Smither, A. Gates, P. Oyston, T. Atkins and S. Eley, Influence of particle size on the pathology and efficacy of vaccination in a murine model of inhalational anthrax, *J. Med. Microbiol.*, 2010, **59**(**12**), 1415–1427, DOI: 10.1099/jmm.0.024117-0.
14. R. J. Thomas, C. Davies, A. Nunez, S. Hibbs, L. Eastaugh, S. Harding, J. Jordan, K. Barnes, P. Oyston and S. Eley, Particle-size dependent effects in the Balb/c murine model of inhalational melioidosis, *Front. Cell. Infect. Microbiol.*, 2012, 23, DOI: 10.3389/fcimb.2012.00101.
15. S. K. Hietala, P. J. Hullinger, B. M. Crossley, H. Kinde and A. A. Ardans, Environmental air sampling to detect exotic Newcastle disease virus in two California commercial poultry flocks, *J. Vet. Diagn. Invest.*, 2005, **17**, 198–200.
16. G. Mainelis, K. Willeke, P. Baron, T. Reponen, S. A. Grinshpun, R. L. Górny and S. Trakumas, Electrical charges on airborne microorganisms, *J. Aerosol Sci.*, 2001, **32**(**9**), 1087–1110.
17. M. Yao and G. Mainelis, Utilization of natural electrical charges on airborne microorganisms for their collection by electrostatic means, *J. Aerosol Sci.*, 2006, **37**(**4**), 513–527.
18. G. Mainelis, A. Adhikari, K. Willeke, S.-A. Lee, T. Reponen and S. A. Grinshpun, Collection of airborne microorganisms by a new electrostatic precipitator, *J. Aerosol Sci.*, 2002, **33**(**10**), 1417–1432.
19. M. Wagner, R. Amann, H. Lemmer and K.-H. Schleifer, Probing Activated Sludge with Oligonucleotides Specific for Proteobacteria: Inadequacy of Culture-Dependent Methods for Describing Microbial Community Structure, *Appl. Environ. Microbiol.*, 1993, 1520–1525.
20. R. Sharma, R. Ranjan, R. K. Kapardar and A. Grover, Unculturable bacterial diversity: an untapped resource, *Curr. Sci.*, 2005, **89**(**1**), 72–77.
21. J. L. Santarpia, D. Cunningham, J. Gilberry, S. Kim, E. E. Smith, S. Ratnesar-Shumate and J. Quizon, Transport and characterization of ambient biological aerosol near Laurel, MD, *Biogeosci. Discuss.*, 2010, 7, 6725–6747.
22. W. G. Lindsley, F. M. Blachere, R. E. Thewlis, A. Vishnu, K. A. Davis, G. Cao, J. E. Palmer, K. E. Clark, M. A. Fisher, R. Khakoo and D. H. Beezhold, Mesurements of Airborne Influenza Virus in Aerosol PAticles from Human Coughs, *PLoS One*, 2010, **5**(**11**), e15100.

23. T. O. Womiloju, J. D. Miller, P. M. Mayer and J. R. Brook, Methods to determine the biological composition of particulate matter collected from outdoor air, *Atmos. Environ.*, 2003, **37**, 4335–4344.
24. C. Fenselau and P. A. Demirev, Characterization of intact microorganisms by MALDI mass spectrometry, *Mass Spectrom. Rev.*, 2001, **20**(**4**), 157–171.
25. B. J. Hindson, M. T. McBride, A. J. Makarewicz, B. D. Henderer, U. S. Setlur, S. M. Smith, D. M. Gutierrez, T. R. Metz, S. L. Nasarabadi, K. S. Venkateswaran, S. W. Farrow, B. W. Colston, Jr. and J. M. Dzenitis, Autonomous Detection of Aerosolized Biological Agents by Multiplexed Immunoassay with Polymerase Chain Reaction Confirmation, *Anal. Chem.*, 2005, **77**(**1**), 284–289.
26. J. Boreson, A. M. Dillner and J. Peccia, Correlating bioaerosol load with PM2.5 and PM10cf concentrations: a comparison between natural desert and urban-fringe aerosols, *Atmos. Environ.*, 2004, **38**, 6029–6041.
27. P. A. Demirev, Y. P. Ho, V. Ryzhov and C. Fenselau, Microorganism identification by mass spectrometry and protein database searches, *Anal. Chem.*, 1999, **71**(**14**), 2732–2738.
28. J. Peccia and M. Hernandez, Incorporating polymerase chain reaction-based identification, population characterization, and quantification of microorganisms into aerosol science: A review, *Atmos. Environ.*, 2006, **40**, 3941–3961.
29. Y. T. Hua and R. C. Tong, Use of the polymerase chain reaction for specific detection of type A, D and E enterotoxigenic Staphylococcus aureus in foods, *Appl. Microbiol. Biotechnol.*, 1992, **37**, 685–690.
30. B. A. Hales, C. Edwards, D. A. Ritchie, G. Hall, R. W. Pickup and J. R. Saunders, Isolation and identification of methanogen-specific DNA from blanket bog peat by PCR amplification and sequence analysis, *Appl. Environ. Microbiol.*, 1996, **62**(**2**), 668–675.
31. T. Bacchetti De Gregoris, N. Aldred, A. S. Clare and J. G. Burgess, Improvement of phylum- and class-specific primers for real-time PCR quantification of bacterial taxa, *J. Microbiol. Methods*, 2011, **86**(**3**), 351–356.
32. E. M. Elnifro, A. M. Ashshi, R. J. Cooper and P. E. Klapper, Multiplex PCR: Optimization and Application in Diagnostic Virology, *Clin. Microbiol. Rev.*, 2000, **13**(**4**), 559–570.
33. E. L. Brodie, T. Z. DeSantis, D. C. Joyner, S. M. Baek, J. T. Larsen, G. L. Andersen, T. C. Hazen, P. M. Richardson, D. J. Herman, T. K. Tokunaga, J. M. Wan and M. K. Firestone, Application of a high-density oligonucleotide microarray approach to study bacterial population dynamics during uranium reduction and reoxidation, *Appl. Environ. Microbiol.*, 2006, **72**, 6288–6298.
34. L. A. Amaral-Zettler, E. A. McCliment, H. W. Ducklow and S. M. Huse, A method for studying protistan diversity using massively parallel sequencing of V9 hypervariable regions of small-subunit ribosomal RNA genes, *PLoS One*, 2009, **4**, e6372.

35. C. V. Venter, K. Remington, J. F. Heidelberg, A. L. Halpern, D. Rusch, J. A. Eisen, D. Wu, I. Paulsen, K. E. Nelson, D. E. Fouts, S. Levy, A. H. Knap, M. W. Lomas, K. Nealson, O. White, J. Peterson, J. Hoffman, R. Parsons, H. Baden-Tillson, C. Pfannkoch, Y.-H. Rogers and H. O. Smith, Environmental Genome Shotgun Sequencing of the Sargasso Sea, *Science*, 2004, **304**(**5667**), 66–74.
36. R. J. Boissy, D. J. Romberger, W. A. Roughead, L. Weissenburger-Moser, J. A. Poole and T. D. LeVan, Shotgun Pyrosequencing Metagenomic Analyses of Dusts from Swine Confinement and Grain Facilities, *PLoS One*, 2014, **9**(**4**), e95578.
37. T. C. Hazen, *et al.*, Deep-Sea Oil Plume Enriches Indigenous Oil-Degrading Bacteria, *Science*, 2010, **330**(**6001**), 204–208.
38. D. P. Fergenson, M. E. Pitesky, H. J. Tobias, P. T. Steele, G. A. Czerwieniec, S. C. Russell, C. Lebrilla, J. Horn, K. Coffee, A. Srivastava, S. P. Pillai, M.-T. P. Shih, H. L. Hall, A. J. Ramponi, J. T. Chang, R. G. Langlois, P. L. Estacio, R. T. Hadley, M. Frank and E. Gard, *Anal. Chem.*, 2004, **76**, 373–378.
39. A. Tripathi, R. E. Jabbour, J. A. Guicheteau, S. D. Christesen, D. K. Emge, A. W. Fountain, J. R. Bottiger, E. D. Emmons and A. P. Snyder, Bioaerosol Analysis with Raman Chemical Imaging Microspectroscopy, *Anal. Chem.*, 2009, **81**, 6981–6990.
40. R. G. Pinnick, S. C. Hill, P. Nachman, J. D. Pendleton, G. L. Fernandez, M. W. Mayo and J. G. Bruno, Fluorescence Particle Counter for Detecting Airborne Bacteria and other Biological Particles, *Aerosol Sci. Technol.*, 1995, **23**, 653–664.
41. Y. L. Pan, S. C. Hill, R. G. Pinnick, H. Huang, J. R. Bottiger and R. K. Chang, Fluorescence spectra of atmospheric aerosol particles measured using one or two excitation wavelengths: Comparison of classification schemes employing different emission and scattering results, *Opt. Express*, 2010, **18**(**12**), 12436–12457.
42. S. C. Hill, R. G. Pinnick, S. Niles, Y. L. Pan, S. Holler, R. K. Change, J. Bottigoer, B. T. Chen, C. S. Orr and G. Feather, Real-Time Measurement of Fluorescence Spectra from Single Airborne Biological Particles, *Field Anal. Chem. Technol.*, 1999, **3**(**4–5**), 221–239.
43. J. Y.-W. Ho, December 23, 1997 (Filed March 19, 1996). Fluorescent biological particle detection system, United States patent 5,701,012.
44. P. P. Hairston, J. Ho and F. R. Quant, Design of an instrument for real-time detection of bioaerosols using simultaneous measurement of particle aerodynamic size and intrinsic fluorescence, *J. Aerosol Sci.*, 1997, **28**, 471–482.
45. M. Seaver, J. D. Eversole, J. J. Hardgrove, W. K. Cary, Jr. and D. C. Roselle, Size and Fluorescence Measurements for Field Detection of Biological Aerosols, *Aerosol Sci. Technol.*, 1999, **30**, 174–185.
46. P. H. Kaye, W. R. Stanley, E. Hirst, E. V. Foot, K. L. Baxter and S. J. Barrington, Single particle multichannel bio-aerosol fluorescence sensor, *Opt. Express*, 2005, **13**, 3583–3593.

47. A. Manninen, M. Putkiranta, J. Saarela, A. Rostedt, T. Sorvajarvi, J. Toivonen, M. Marjamäki, J. Keskinen and J. R. Hernberg, Fluorescence cross sections of bioaerosols and suspended biological agents, *Appl. Opt.*, 2009, **48**, 4320–4328.
48. Y. L. Pan, J. Hartings, R. G. Pinnick, S. C. Hill and J. Halverson, *et al.*, Single-particle fluorescence spectrometer for ambient aerosols, *Aerosol Sci. Technol.*, 2003, **37**, 628–639.
49. Y. L. Pan, R. G. Pinnick, S. C. Hill and R. K. Chang, Particle-fluorescence spectrometer for real-time measurements of atmospheric organic carbon and biological aerosol, *Environ. Sci. Technol.*, 2009, **43**, 429–434.
50. F. L. Reyes, T. H. Jeys, N. R. Newbury, C. A. Primmerman, G. S. Rowe and A. Scanchez, Bio-aerosol fluorescence sensor, *Field Anal. Chem. Technol.*, 1999, **3**, 240–248.
51. V. Sivaprakasam, A. Huston, C. Scotto and J. D. Eversole, Multiple UV wavelength excitation of bioaerosols, *Opt. Express*, 2004, **12**(**19**), 4457–4466.
52. V. Sivaprakasam, H.-B. Lin, A. L. Huston and J. D. Eversole, Spectral characterization of biological aerosol particles using two-wavelength excited laser-induced fluorescence and elastic scattering measurements, *Opt. Express*, 2011, **19**(7), 6191–6208.
53. D. P. Greenwood, T. H. Jeys, B. Johnson, J. M. Richardson and M. P. Shatz, Optical Techniques for Detecting and Identifying Biological-Warfare Agents, *Proc. IEEE*, 2009, **97**(**6**), 971–989.
54. K. Mitsumoto, K. Yabusaki, K. Kobayashi and H. Aoyagi, Development of a novel real-time pollen-sorting counter using species-specific pollen autofluorescence, *Aerobiologia*, 2010, **26**, 99–111.
55. A. Birenzvige, J. Eversole, M. Seaver, S. Francesconi, E. Valdes and H. Kulaga, Aerosol Characteristics in a Subway Environment, *Aerosol Sci. Technol.*, 2003, **37**, 210–220.
56. R. G. Pinnick, S. C. Hill, Y. L. Pan and R. K. Chang, Fluorescence Spectra of Atmospheric Aerosol at Adelphi, Maryland, USA: Measurement and Classifcation of Single Particles Containing Organic Carbon, *Atmos. Environ.*, 2004, **38**, 1657–1672.
57. Y. L. Pan, S. C. Hill, R. G. Pinnick, J. L. Santarpia, N. Baker, B. Alvarez, S. Ratnesar-Shumate, B. Cottrell and L. McKee, Fluorescence spectra of bioaerosols exposed to ozone in a laboratory reaction chamber to simulate atmospheric processing, *Proc. SPIE*, 2011, **801804**, 801804–801807.
58. A. M. Gabey, M. W. Gallagher, J. Whitehead, J. Dorsey, P. H. Kaye and W. R. Stanley, Measurements and comparison of primary biological aerosol above and below a tropical forest canopy using a dual-channel fluorescence aerosol spectrometer, *Atmos. Chem. Phys.*, 2010, **10**, 4453–4466.
59. A. M. Gabey, W. R. Stanley, M. W. Gallagher and P. H. Kaye, The fluorescence properties of aerosol larger than 0.8μm in urban and tropical rainforest locations, *Atmos. Chem. Phys.*, 2011, **11**, 5491–5504.
60. Y. L. Pan, R. G. Pinnick, S. C. Hill, J. M. Rosen and R. K. Chang, Single-Particle Laser-Induced-Fluorescence Spectra of Biological and

Other Organic-Carbon Aerosols in the Atmosphere: Measurements at New Haven, Connecticut, and Las Cruces, New Mexico, *J. Geophys. Res.*, 2007, **112**, D24S19.

61. R. DeFreez, LIF bio-aerosol threat triggers: then and now, *Proc. SPIE*, 2009, **7484**, 74840H.
62. S. Ratnesar-Shumate, M. L. Wagner, C. Kerechanin, G. House, K. M. Brinkley, C. Bare, N. A. Baker, R. Quizon, J. Quizon, A. Proescher, E. Van Gieson and J. L. Santarpia, Improved Method for the Evaluation of Real-Time Biological Aerosol Detection Technologies, *Aerosol Sci. Technol.*, 2011, **45**(**5**), 635–644.
63. U. Pöschl, S. T. Martin, B. Sinha, Q. Chen, S. S. Gunthe, J. A. Huffman, S. Borrmann, D. K. Farmer, R. M. Garland, G. Helas, J. L. Jimenez, S. M. King, A. Manzi, E. Mikhailov, T. Pauliquevis, M. D. Petters, A. J. Prenni, P. Roldin, D. Rose, J. Schneider, H. Su, S. R. Zorn, P. Artaxo and M. O. Andreae, Rainforest Aerosols as Biogenic Nuclei of Clouds and Precipitation in the Amazon, *Science*, 2010, **329**(**5998**), 1513–1516.
64. J. D. Whitehead, *et al.*, Aerosol fluxes and dynamics within and above a tropical rainforest in South-East Asia, *Atmos. Chem. Phys.*, 2010, **10**, 9369–9382, DOI: 10.5194/acp-10–9369–2010.
65. W. R. Stanley, P. H. Kaye, V. E. Foot, S. J. Barrington, M. Gallagher and A. Gabey, Continuous bioaerosol monitoring in a tropical environment using a UV fluorescence particle spectrometer, *Atmos. Sci. Lett.*, 2011, **12**, 195–199, DOI: 10.1002/asl.310.
66. A. M. Gabey, M. Vaitilingom, E. Freney, J. Boulon, K. Sellegri, M. W. Gallagher, I. P. Crawford, N. H. Robinson, W. R. Stanley and P. H. Kaye, Observations of fluorescent and biological aerosol at a high-altitude site in central France, *Atmos. Chem. Phys.*, 2013, **13**, 7415–7428, DOI: 10.5194/acp-13–7415–2013.
67. N. H. Robinson, J. D. Allan, J. A. Huffman, P. H. Kaye, V. E. Foot and M. Gallagher, Cluster analysis of WIBS single-particle bioaerosol data, *Atmos. Meas. Tech.*, 2013, **6**, 337–347, DOI: 10.5194/amt-6–337–2013.
68. J. A. Huffman, A. J. Prenni, P. J. DeMott, C. Pöhlker, R. H. Mason, N. H. Robinson, J. Fröhlich-Nowoisky, Y. Tobo, V. R. Després, E. Garcia, D. J. Gochis, E. Harris, I. Müller-Germann, C. Ruzene, B. Schmer, B. Sinha, D. A. Day, M. O. Andreae, J. L. Jimenez, M. Gallagher, S. M. Kreidenweis, A. K. Bertram and U. Pöschl, High concentrations of biological aerosol particles and ice nuclei during and after rain, *Atmos. Chem. Phys.*, 2013, **13**, 6151–6164, DOI: 10.5194/acp-13–6151–2013.
69. I. Crawford, N. H. Robinson, M. J. Flynn, V. E. Foot, M. W. Gallagher, J. A. Huffman, W. R. Stanley and P. H. Kaye, Characterisation of bioaerosol emissions from a Colorado pine forest: results from the BEACHON-RoMBAS experiment, *Atmos. Chem. Phys.*, 2014, **14**, 8559–8578, DOI: 10.5194/acp-14–8559–2014.
70. T. P. Wright, J. D. Hader, G. R. McMeeking and M. D. Petters, High Relative Humidity as a Trigger for Widespread Release of Ice Nuclei, *Aerosol Sci. Technol.*, 2014, **48**, i–v.

71. Y.-L. Pan, J. D. Eversole, P. H. Kaye, V. Foot, R. P. Pinnick, S. C. Hill, M. W. Mayo, J. R. Bottiger, A. Huston, V. Sivaprakasam and R. K. Chang, Bio-aerosol fluorescence: Detecting and characterising bio-aerosols via UV light-induced fluorescence spectroscopy, in *Optics of Biological Particles*, ed. A. Hoekstra, V. Maltsev and G. Videen, Springer, Netherlands, 2007, pp. 63–164.
72. Y.-L. Pan, J. L. Santarpia, S. Ratnesar-Shumate, E. Corson, J. Eshbaugh, S. C. Hill, C. C. Williamson, M. Coleman, C. Bare and S. Kinahan, Effects of ozone and relative humidity on fluorescence spectra of octapeptide bioaerosol particles, *J. Quant. Spectrosc. Radiat. Transfer*, 2014, **133**, 538–550.
73. S. Ratnesar-Shumate, Y. L. Pan, S. C. Hill, S. Kinahan, E. Corson, J. Eshbaugh and J. L. Santarpia, Fluorescence spectra and biological activity of aerosolized bacillus spores and MS2 bacteriophage exposed to ozone at different relative humidities in a rotating drum, *J. Quant. Spectrosc. Radiat. Transfer*, 2015, **153**, 13–28, DOI: 10.1016/j.jqsrt.2014.10.003.
74. S. C. Hill, R. G. Pinnick, S. Niles, N. F. Fell, Y. L. Pan, J. Bottiger, B. V. Bronk, S. Holler and R. K. Chang, Fluorescence from air-borne microparticles: dependence on size, concentration of fluorophores, and illumination intensity, *Appl. Opt.*, 2001, **40**, 3005–3013.
75. P. T. Steele, G. R. Farquar, A. N. Martin, K. R. Coffee, V. J. Riot, S. I. Martin, D. P. Fergenson, E. E. Gard and M. Frank, Autonomous, broad-spectrum detection of hazardous aerosols in seconds, *Anal. Chem.*, 2008, **80**, 4583–4589.
76. Q. Chen, *et al.*, Mass spectral characterization of submicron biogenic organic particles in the Amazon Basin, *Geophys. Res. Lett.*, 2009, **36**, L20806, DOI: 10.1029/2009GL039880.
77. B. T. Shaffer and B. Lighthart, Survey of Culturable Airborne Bacteria at Four Diverse Locations in Oregon: Urban, Rural, Forest and Coastal, *Microb. Ecol.*, 1997, **34**, 167–177.
78. N.-P. Hua, F. Kobayashi, Y. Iwasaka, G.-Y. Shi and T. Naganuma, Detailed identification o desert-oriented bacteria carried by Asian dust storms to Japan, *Aerobiologia*, 2007, **23**, 291–298.
79. B. Neheme, V. Letourneau, R. J. Forster, M. Veillette and C. Duchaine, Culture-independent approach of the bacterial bioaerosol diversity in the standard swine confinement buildings, and assessment of the seasonal effect, *Environ. Microbiol.*, 2008, **10**, 665–675.
80. E. L. Brodie, T. Z. DeSantis, J. P. Moberg Parker, I. X. Zubietta, Y. M. Piceno and G. L. Andersen, Urban aerosols harbor diverse and dynamic bacterial populations, *Proc. Natl. Acad. Sci. U. S. A.*, 2007, **104**, 299–304.
81. W. B. Lee, J. Y. Wu, Y. I. Lee and J. Sneddon, Recent applications of laser-induced breakdown spectrometry: A review of material approaches, *Appl. Spectrosc. Rev.*, 2004, **39**, 27–97.
82. A. Bovallius, B. Bucht, R. Roffey and P. Anas, Long range air transmission of bacteria, *Environ. Microbiol.*, 1978, **35**(**6**), 1231–1232.

83. B. Lighthart and B. T. Shaffer, Airborne Bacteria in the Atmospheric Surface Layer: Temporal Distribution above a Grass Seed Field, *Appl. Environ. Microbiol.*, 1995, **61**, 1492–1496.
84. N. Fierer, Z. Liu, M. Rodriguez-Hernandez, R. Knight, M. Henn and M. T. Hernandez, Short-term temporal variability in airborne bacterial and fungal populations, *Appl. Environ. Microbiol.*, 2008, **74**, 200–207.
85. R. L. Mancinelli and W. A. Shulls, Airborne bacteria in an urban environment, *Appl. Environ. Microbiol.*, 1978, **35**, 1095–1101.
86. J. L. Santarpia, S. Ratnesar-Shumate, J. U. Gilberry and J. J. Quizon, Relationship Between Biologically Fluorescent Aerosol and Local Meteorological Conditions, *Aerosol Sci. Technol.*, 2013, **47**(**6**), 655–661.
87. J. A. Huffman, B. Sinha, R. M. Garland, A. Snee-Pollmann, S. S. Gunthe, P. Artaxo, S. T. Martin, M. O. Andreae and U. Poschl, Size distributions and temporal variations of biological aerosol particles in the Amazon rainforest characterized by microscopy and real-time UV-APS fluorescence techniques during AMAZE-08, *Atmos. Chem. Phys.*, 2012, **12**, 11997–12019.
88. V. R. Despres, J. A. Huffman and S. H. Burrows, *et al.*, Primary biological aerosol particles in the atmosphere: a review, *Tellus*, 2012, **64**, 1–44.
89. Z. A. Nasir, I. Colbeck, S. Sultan and S. Ahmed, Bioaerosols in residential micro-environments in low income countries: A case study from Pakistan, *Environ. Pollut.*, 2012, **168**, 15–22.
90. L. S. Cronholm, Potential health hazards from microbial aerosols in densely populated urban regions, *Appl. Environ. Microbiol.*, 1980, **39**, 6–12.
91. T. Paez-Rubio, E. Viau, S. Romero-Hernandez and J. Peccia, Source Bioaerosol Concentration and rRNA Gene-Based Identification of Microorganisms Aerosolized at a Flood Irrigation Wastewater Reuse Site, *Appl. Environ. Microbiol.*, 2005, **71**(**2**), 804–810.
92. M. P. M. Taha, G. H. Drew, P. J. Longhurst, R. Smith and S. J. T. Pollard, Bioaerosol releases from compost facilities: Evaluating passive and active source terms at a green waste facility for improved risk assessments, *Atmos. Environ.*, 2006, **40**, 1159–1169.
93. L. J. Pankhurst, U. Akeel, C. Hewson, I. Maduka, P. Pham, J. Saragossi, J. Taylor and K. M. Lai, Understanding and mitigating the challenge of bioaerosol emissions from urban community composting, *Atmos. Environ.*, 2011, **45**, 85–93.
94. R. S. Martin, P. J. Silva, K. Moore, M. Erupe and V. S. Doshi, Particle composition and size distributions in and around a deep-pit swine operation, Ames, IA, *J. Atmos. Chem.*, 2008, **59**, 135–150.
95. V. Agranovski, Z. Ristovski, M. Hargreaves, P. J. Blackall and L. Morawska, Performance evaluation of the UVAPS: influence of physiological age of airborne bacteria and bacterial stress, *J. Aerosol Sci.*, 2004, **34**, 1711–1727.

96. A. Kristiansen, A. M. Saunders, A. A. Hansen, P. H. Nielsen and J. L. Nielsen, Community structure of bacteria and fungi in aerosols of a pig confinement building, *FEMS Microbiol. Ecol.*, 2012, **80**, 390–401.
97. C. A. Corzo, M. Culhane, S. Dee, R. B. Morrison and M. Torremorell, Airborne Detection and Quantification of Swine Influenza A Virus in Air Samples Collected Inside, Outside and Downwind from Swine Barns, *PLoS One*, 2013, **8**(**8**), e71444.
98. D. W. Griffin, C. A. Kellogg, V. H. Garrison, J. T. Lisle, T. C. Borden and E. A. Shinn, Atmospheric microbiology in the northern Caribbean during African dust events, *Aerobiologia*, 2003, **19**, 143–157.
99. D. W. Griffin, N. Kubilay, M. Kocak, M. A. Gray, T. C. Borden and E. A. Shinn, Airborne desert dust and aeromicrobiology over the Turkish Mediterranean coastline, *Atmos. Environ.*, 2007, **41**, 4050–4062.
100. E. A. Shinn, G. W. Smith, J. M. Propero, P. Betzer, M. L. Hayes, V. Garrison and R. T. Barber, African dust and the demise of Caribbean coral reefs, *Geophys. Res. Lett.*, 2000, **27**(**19**), 3029–3032.
101. C. A. Kellogg and D. W. Griffin, Aerobiology and the global transport of desert dust, *Trends Ecol. Evol.*, 2006, **21**(**11**), 638–644.
102. Eric A. Croddy and J. J. Wirtz, *Weapons of Mass Destruction*, ABC-CLIO, Santa Barabara, CA, 2005.
103. D. Ivnitski, D. Morrison and D. J. O'Neil, Critical Elements of Biological Sensor Technology for Deployment in an Environmental Network System, in *Defense Against Bioterrorism: Detection Technologies, Implementation Strategies and Commercial Opportunities*, ed. D. Morrison, F. Milanovich, D. Ivnitski and Th. R. Austin, Springer, Netherlands, 2005, pp. 207–220.
104. A. Samuels, J. L. Santarpia, J. R. Bottiger, S. Hunter and E. W. Stuebing, Test Methodology Development for Biological Agent Detection Systems, *Proc. SPIE*, 2006, **2006**, 6738.
105. L. A. Vanderberg, Detection of biological agents: Looking for bugs in all the wrong places, *Appl. Spectrosc.*, 2000, **54**(**11**), 376A.
106. J. Carrano, T. Jeys, D. Cousins, J. Eversole, J. Gillespie, D. Healy, N. Licata, W. Loerop, M. O'Keefe, A. Samuels, J. Schultz, M. Walter, N. Wong, B. Billotte, M. Munley, E. Reich and J. Roots, *Chemical and Biological Sensor Standards Study*, Defence Advanced Research Project Agency, 2005.
107. M. Jamriska, T. C. DuBois and A. Skvortsov, Statistical characterisation of bio-aerosol background in an urban environment, *Atmos. Environ.*, 2012, **54**(**439**), 448.
108. F. A. Dark and T. Nash, Comparative toxicity of various ozonised olefins to bacteria suspended in air, *J. Hyg. Camb.*, 1970, **68**, 245–252.
109. S. John and S. Pandis, *Atmospheric Chemistry and Physics: From Air Pollution to Climate Change*, Willey-Intersciences Publications, New Jersey, 1998.
110. R. Bailey, L. Fielding, A. Young and C. Griffith, Effect of Ozone and Open Air Factor against Aerosolized *Micrococcus luteus*, *J. Food Prot.*, 2007, **70**(**12**), 2769–2773.

111. A. V. Ignatenko, Use of the method of tryptophan fluorescence to characterize disruptions of the structure of ozonized proteins, *J. Appl. Spectrosc.*, 1988, **49**(**1**), 691–695.
112. F. Cataldo, DNA degradation with ozone, *Int. J. Biol. Macromol.*, 2006, **38**(**3–5**), 248–254.
113. M. Johns, L. Harrington, R. W. Titball and D. L. Leslie, Improved methods for the detection of *Bacillus anthracis* spores by the polymerase chain reaction, *Lett Appl Microbiol.*, 1994, **18**(**4**), 236–238.
114. C. Pohlker, J. A. Huffman and U. Poschl, Autofluorescence of atmospheric bioaerosols – fluorescent biomolecules and potential interferences, *Atmos. Meas. Tech. Discuss.*, 2011, **4**, 5857–5933.
115. D. L. Bones, D. K. Henricksen, S. A. Mang, M. Gonsior, A. P. Bateman, T. B. Nguyen, W. J. Cooper and S. A. Nizkorodov, Appearance of strong absorbers and fluorophores in limonene-O_3 secondary organic aerosol due to NH^+_4-mediated chemical aging over long time scales, *J. Geophys. Res.*, 2010, **115**, 1–14.
116. Y. L. Pan, Detection and characterization of biological and other organic-carbon aerosol particles in atmosphere using fluorescence, *J. Quant. Spectrosc. Radiat. Transfer*, 2015, **150**, 12–35.
117. C. S. Cox, *The Aerobiological Pathway of Microorganisms*, John Wiley and Sons, Chichester, 1987.
118. G. O. Rubel, Measurement of Water Vapor Sorption by Single Biological Aerosols, *Aerosol Sci. Technol.*, 1997, **27**(**4**), 481–490.
119. J. B. Bateman, C. L. Stevens, W. B. Mercer and E. L. Carstensen, Relative humidity and the killing of bacteria: the variation of cellular water content with external relative humidity or osmolality, *J. Gen. Microbiol.*, 1962, **29**, 207–219.
120. R. P. Sinha and D.-P. Hader, UV-induced DNA damage and repair: a review, *Phochem. Photobiol. Sci.*, 2002, **1**, 225–236.
121. H. Shintaku, H. Nishikii, L. A. Marshall, H. Kotera and J. G. Santiago, On-Chip Separation and Analysis of RNA and DNA from Single Cells, *Anal. Chem.*, 2014, **86**(**4**), 1953–1957, DOI: 10.1021/ac4040218.

CHAPTER 7

Bacillus anthracis: *An Aerobiological Threat*[†]

SANJIV R. SHAH*[a], STEPHEN A. MORSE[b], MICHAEL W. CALFEE[c], AND SHAWN P. RYAN[c]

[a]US Environmental Protection Agency, Office of Research and Development, National Homeland Security Research Center, Washington, DC, USA;
[b]Centers for Disease Control and Prevention, Atlanta, GA, USA;
[c]US Environmental Protection Agency, Office of Research and Development, National Homeland Security Research Center, Research Triangle Park, NC, USA
*E-mail: shah.sanjiv@epa.gov

7.1 Introduction

According to the "miasma theory," diseases such as cholera, plague and chlamydia were caused by "bad air" or "poisonous air."[1–3] Although this theory was eventually proven to be wrong, it eventually became known that certain diseases could be spread through the air due to the aerosolization of their respective causative microbial agents. Anthrax, the first disease directly

[†]This chapter has been subjected to the US Environmental Protection Agency's review through its Office of Research and Development and has been approved for publication. Note that approval does not signify that the contents necessarily reflect the views of the Agency. Mention of trade names, products or services does not convey official EPA approval, endorsement or recommendation.

Issues in Toxicology No. 25
Aerobiology: The Toxicology of Airborne Pathogens and Toxins
Edited by Harry Salem and Sidney Katz

Published by the Royal Society of Chemistry, www.rsc.org

linked to a microorganism and to the life cycle of its etiological agent, *Bacillus anthracis*, confirmed the "germ theory" of disease. *B. anthracis* is a spore-forming bacterium whose spores are highly resistant to a variety of environmental conditions, such as heat, cold, ultraviolet and ionizing radiation, pressure and chemical agents. Both spores and vegetative cells of *B. anthracis* are the infectious forms and can cause the disease anthrax in mammals, including humans. There are four clinical forms of human anthrax based on the portal by which spores or cells enter the host, namely inhalational, cutaneous, gastrointestinal and injectional. Inhalational anthrax is the most serious form of the disease.[4,5] The high mortality rate, ease of aerosolization and resistance of its spores to environmental factors have made *B. anthracis* an attractive biological weapon. During the era of offensive bioweapons development, advances were made in the weaponization and dissemination of *B. anthracis* spores. However, most nations signed the 1972 Biological Weapons Convention (BWC) treaty[6] and ceased their offensive research and development programs. As of January 2015, 173 countries were BWC members.[7] However, because of the threat that *B. anthracis* spores could be used as a weapon by terrorist groups or rogue nations, many countries have developed defensive programs to protect their soldiers and civilians. These programs and the 2001 anthrax bioterrorism incident in the United States have increased awareness of anthrax worldwide. As a result, unprecedented advances have been made in anthrax diagnostics, anthrax vaccine development, detection of *B. anthracis* spores in environmental matrices, decontamination technology and strategies for *B. anthracis* contamination and anthrax incident waste management.

7.2 Historical Aspects

Anthrax, primarily a disease of grazing herbivorous mammals such as cattle, sheep, goats, horses and bison, can be transmitted sporadically to humans.[8] In fact, all mammals are susceptible to anthrax. The disease has probably been recognized since biblical times. Perhaps the first reference to anthrax is in the Bible, where the fifth plague was most likely anthrax.[9] In the 5th century BCE, the Greek physician Hippocrates coined the term "anthrax" (from *anthrakis* = coal in Greek) for a disease characterized by the black color of the skin lesions that arise during the course of cutaneous infections.[9]

The identification of the etiological agent of anthrax holds a very special place in the history of medicine and microbiology. It was the first time that an infectious disease could be attributed to a specific microorganism.[9] Although numerous scientists and physicians played significant roles in this discovery, two individuals stand out for their landmark discoveries. The first was Robert Koch, an unknown physician scientist at the time, who was the first to demonstrate that (i) a rod-shaped microorganism was consistently present in blood and tissue of animals with anthrax, (ii) this microorganism developed spores under conditions of nutrient starvation, (iii) these spores could transform into rod-like bacilli under nutrient-rich conditions, (iv) this rod-shaped microorganism could be cultured in pure form and (v) the cultured microorganism, in the form of either rod-shaped bacilli or spores, caused anthrax in experimentally infected animals.[10] The discovery that this bacterium formed spores also greatly contributed to our understanding of the epidemiology of anthrax. Koch's work led to the birth of medical microbiology in the late 19th century along with the establishment of what was to become known as "Koch's Postulates," *i.e.* the reproducible identification of the agent in diseased tissue, *in vitro* culture of the responsible agent in pure form and transfer of the pure culture material to elicit disease in experimental animals.[11]

The second individual was Louis Pasteur, a well-established scientist known as the father of microbiology. Over the years, his personal interests had shifted from fermentation to focus increasingly on infectious diseases. Pasteur demonstrated unequivocally that it was the bacterium itself that was the cause of anthrax, not some accompanying toxic substance. He also performed one of the most famous experiments in public, the anthrax vaccine trial.[12] This vaccine, based on the use of an attenuated strain (*i.e.* Pasteur strain) was very effective and its worldwide use has been credited with reducing the burden of anthrax in livestock.[11]

Anthrax is a zoonotic infectious disease. When an animal dies from anthrax, the infected blood and body fluids enter the soil from the natural orifices of the dead animal. The bacteria, which are in the vegetative stage in blood, form spores upon exposure to air and proper environmental conditions. These spores are extremely resistant to environmental conditions and can remain dormant in the soil for decades or cause new infections among susceptible animals that graze in the field.[13] Generally, humans become infected when they come into contact with spore-contaminated animal products. For epidemiological purposes, it was formerly convenient to divide human anthrax into industrial and nonindustrial anthrax, reflecting the relationship between occupation and exposure.[14] Nonindustrial anthrax referred to its occurrence in farmers and veterinarians who had contact with infected animals and butchers who had contact with contaminated meat; industrial anthrax referred to its occurrence in persons employed in the processing of bones, hides, wool and other products from animals with anthrax. Today, we refer to human anthrax by the route that spores enter a susceptible human host, *e.g.* cutaneous, gastrointestinal, injectional or inhalational.

In Victorian England, anthrax became known as woolsorters' disease because it occurred in mill workers sorting wool in enclosed factory spaces where spores could be inhaled. However, it was not until 1879 that woolsorters' disease was shown to be a form (inhalational) of anthrax.[15] After a large number of cases of woolsorters' disease occurred in the mills in and around Bradford, England, the name Bradford disease became synonymous with woolsorters' disease. This eventually led to institution of the Bradford rules, which in 1897 became law and led to a decrease in the incidence of inhalational anthrax among sorters. However, the actual elimination of inhalational anthrax as an industrial hazard in England followed passage of the

Anthrax Prevention Act in 1919. This law required that all dangerous wool and animal hair products entering England undergo decontamination with formaldehyde in a government-run facility.[16]

The history of anthrax also has a dark side in which *B. anthracis* was misused for unethical purposes.[17,18] Between 1932 and 1945, the Japanese Army's infamous Unit 731, which was based in Manchuria, experimentally infected prisoners with *B. anthracis*.[11] Also during World War II, the British Army conducted anthrax bioweapons testing on Gruinard Island in Scotland and prepared large quantities of "cattle cake" contaminated with anthrax spores for dispersal by aircraft over German agricultural areas; fortunately, this weapon was not used. In 1979 in Sverdlovsk in the former Soviet Union, an anthrax outbreak was documented with at least 77 cases of human disease, 66 of whom died from inhalational anthrax.[4] This was subsequently found to be the result of an accidental release of finely milled and electrostatically neutral spores of *B. anthracis* from a bioweapons research and production facility.[19] This event served to demonstrate that inhalation of anthrax spores can cause inhalational disease in the absence of conventional occupational exposure to contaminated animal products and supported the view that anthrax is a potent weapon for biological warfare and bioterrorism.[4] In 1993, the Aum Shinrikyo, an apocalyptic religious cult based in Japan, released aerosolized spores of *B. anthracis* from the roof top of a building in Kameido in the Tokyo metropolitan area. No casualties occurred as a result of this release because the strain they used was an attenuated vaccine strain.[20,21] Finally, in the aftermath of the terror attacks of 11 September 2001, at least five letters containing small quantities of spores of *B. anthracis* were sent to media companies and political figures along the east coast of the United States.[22,23] These attacks resulted in 22 confirmed anthrax cases, of which 11 were inhalational anthrax (five of whom died) and 11 cases of cutaneous anthrax; more than 30 000 individuals received prophylactic treatment and numerous public and private buildings had to be decontaminated.

7.3 *Bacillus anthracis*

B. anthracis is a facultative, Gram-positive, nonmotile, spore-forming bacterium. It belongs to the *B. cereus* group, which consists of six very closely related species: *B. cereus*, *B. anthracis*, *B. thuringiensis*, *B. mycoides*, *B. pseudomycoides* and *B. weihenstephanensis*. The classification of these organisms remains unclear. Some studies have suggested that *B. cereus sensu stricto*, *B. anthracis* and *B. thuringiensis* should be considered as members of a single species designated *B. cereus sensu lato* based in part on a comparison of their 16S rRNA sequences, which exhibit less than 1% divergence.[24–27] However, other studies have been able to discriminate genetically between *B. anthracis*, *B. cereus* and *B. thuringiensis*.[28–31] Thus, at the moment there is no clear consensus as to whether these bacteria should be considered as separate species or variants of a single species.[32]

B. anthracis appears to be a recently emerged clone within the *B. cereus sensu lato* group of highly successful soil microorganisms.[29,33–36] The primary virulence factors produced by *B. anthracis*, an antiphagocytic poly-D-glutamic acid capsule and three toxin components (protective antigen, edema factor and lethal factor), are due to the presence of two plasmids, pXO2 and pXO1, respectively.[32] Much of the disease and host specificity in this group can be attributed to the presence or absence of these plasmids.[37,38] Although the genomes of *B. anthracis*, *B. cereus* and *B. thuringiensis* are very similar, there are some significant differences that can affect the phenotypic characteristics of these species.[32] For example, PlcR is a pleiotropic regulator that putatively belongs to the PlcR regulon on the basis of an upstream PlcR DNA-binding sequence, which controls the expression of more than 100 genes in *B. cereus*; many of these genes are associated with virulence.[39,40] All strains of *B. anthracis* have the same genes but they also have a nonsense mutation in the *plcR* gene, which results in an inactive PlcR protein. Thus, none of these genes are expressed in *B. anthracis*,[41] which, in turn, results in some of the phenotypic characteristics that are used to differentiate *B. anthracis*, *B. cereus* and *B. thuringiensis*.[32] For example, *B. anthracis* is nonmotile and nonhemolytic when grown on sheep blood agar whereas both *B. cereus* and *B. thuringiensis* are motile and hemolytic when grown on this medium. Additionally, only *B. anthracis* is lysed by the species-specific γ-phage and produces a poly-D-glutamic acid capsule. *B. thuringiensis* produces a parasporal body, which is the only established characteristic that distinguishes this species from *B. cereus*.

7.3.1 *Bacillus anthracis* Spores

The discovery that *B. anthracis* can form a dormant, highly resistant cell type called a spore was a seminal event in the history of microbiology.[10] Koch not only described the process by which spores were formed (sporulation) but also demonstrated that the spore was the infectious form responsible for the disease anthrax.[10] Vegetative cells were only infectious if inoculated directly into the bloodstream of an experimental animal whereas spores were infectious regardless of the route of inoculation.[10] Carrera *et al.* determined the sizes of spores from seven strains of *B. anthracis* and observed that there were two groups based on length.[42] One group (three strains) had mean lengths between 1.23 and 1.26 μm whereas the other group (four strains) had mean lengths between 1.49 and 1.67 μm ($p < 0.001$); the mean spore diameters (0.81–0.86 μm) were similar for all strains. The mean volumes of the *B. anthracis* spores (range 0.420–0.642 μm^3) are almost twice that of surrogates commonly used in aerodynamic studies (0.160 and 0.273 μm^3 for *B. subtilis* and *B. atrophaeus*, respectively). In addition, spores of *B. subtilis* and *B. atrophaeus* are denser than spores of *B. anthracis*, suggesting that they may not behave in the same way aerodynamically.[43] *B. anthracis* spores are highly resistant to a variety of environmental conditions, such as heat, cold, ultraviolet and ionizing radiation, pressure and chemical agents.[32] *B. anthracis* spores have been shown to persist for long periods (at least 68 years) in the environment.[44,45]

All *Bacillus* spp. spores share a common structure.[46,47] The innermost compartment is called the core and contains the spore chromosome. The core is first surrounded by an inner membrane and then a shell of specialized peptidoglycan, referred to as the cortex. Proteins in the core and cortex maintain spore dormancy and protect the DNA against heat and other assaults.[48,49] The inner membrane helps protect the spore DNA and is the site of receptors that sense germinants and, as a result, initiate germination. Surrounding the cortex are the spore coat and exosporium. The coat is present in spores of all *Bacillus* species examined so far. The spore coat of *B. anthracis* (and also *B. cereus* and *B. thuringiensis*) is thinner than the spore coat of other *Bacillus* species and the multiple layers in the coat, while present, are difficult to distinguish by electron microscopy.[46,50] The outermost spore structure in *B. anthracis* is the exosporium, which is separated from the coat by a region called the interspace.[42] An exosporium is not present on spores of all *Bacillus* species. It is found in members of the *B. cereus* group, but is absent from many others, including *B. subtilis* and *B. licheniformis*. The exosporium of *B. anthracis* possesses hair-like projections, called the nap, that extend outward. The exosporium is most likely involved in spore adhesion to environmental matrices and also host cells. Mutations affecting exosporium proteins have been described. For example, *bclA* mutants lack the protein BclA and the exosporium of spores from these mutants lacks the hair-like projections or nap.[51] BclA is not required for disease in several animal models; however, it may still be involved in natural infection as BclA has been shown to facilitate binding to macrophages and to the cell surface integrin CD11b/CD18.[52,53] In the moist state, anthrax spores carry a negative charge and are hydrophobic.[54]

7.3.2 Ecology of *Bacillus anthracis*

More than a century ago, Pasteur made seminal observations on the role of soil in anthrax. He noted that «... les régions dans l'Aveyron, où nous trouvons la maladie du charbon, ont un sol argilo-calcaire, et les pièces où le charbon est inconnu ont un sol schisteux ou granit.» ("... the regions in the Aveyron, where we find anthrax, have a calcareous clay soil and the parts where anthrax is unknown have a schistose or granite soil.").[55] Van Ness observed that in the Western Hemisphere, vegetative cells and spores of *B. anthracis* were eliminated in acid soils but multiplication was favored by either calcareous soils or alkaline groundwater.[56] Based upon his observations, he proposed the existence of "incubator areas" for *B. anthracis* that are characterized by (1) depressions in which water has remained sufficiently long to devitalize or kill grass, (2) desiccated ephemeral watercourses or (3) hillside seep areas.[56] These observations were subsequently confirmed for the Eastern Hemisphere by Kasi'anenko *et al.*[57] The calcium level in the soil appears to be important for spore survival.[58,59] However, despite releases of up to 7 × 10^9 vegetative cells at sites of anthrax carcasses, fewer than 5 × 10^4 spores are detected after 60 months.[60] Hence additional factors are of importance for spore viability and continued transmission of anthrax. For example, De Vos noted that 100–250 spores parenterally administered consistently killed kudu (African antelope) in the Kruger National Park; however, the oral ID_{50} with the same strain was approximately 1.5 × 10^7 spores.[61] A similar discrepancy between parenteral and oral infectious doses was observed with unstressed sheep, cattle and horses.[62]

The classical view of the *B. anthracis* lifecycle has the vegetative form sporulating after host death and the spores remaining dormant until encountering its next host.[14,63] This view is supported by evidence such as (1) the durability and resistance properties of spores,[64,65] (2) poor survival of vegetative cells in soil and water under laboratory conditions[60] and (3) long-term survival of spores at sites where animals died of anthrax.[66–68] In this scenario, *B. anthracis* is considered an obligate pathogen because it is thought to multiply almost exclusively in the animal host.[14] However, more recent studies suggest that the lifecycle is far more complicated than previously thought. Saile and Koehler employed a grass plant–soil model system and showed that spores of *B. anthracis* could germinate on and around roots, growing as long filaments whereas spores inoculated into soil without plants remained as heat-resistant spores.[69] Thus, *B. anthracis* can actually survive as a saprophyte outside the host. Furthermore, they were able to demonstrate transfer of an antibiotic resistance plasmid between two *B. anthracis* strains in this model system and suggested that horizontal gene transfer in the rhizosphere of grass plants may play a role in the evolution of the *B. cereus* group species.

B. cereus is capable of forming biofilms, which may represent an important mechanism for survival in an agricultural setting.[70] Lee *et al.* demonstrated that *B. anthracis* is also able to form biofilms, which, like other organisms, consist of cells that are resistant to antibiotics.[71] The role of biofilms in the pathogenesis of anthrax or survival of vegetative cells in the environment remains to be determined. There have been additional studies that have suggested that anthrax spores may germinate in the soil. Dey *et al.* recently developed an *Acanthamoeba castellanii* infection model that supported spore germination and intracellular growth of *B. anthracis* under conditions that simulate standing water or moist environments in which this bacterium persists.[72] Soil-dwelling amoebae, such as *A. castellanii*, may have a role in the amplification of spores in soil. Spores ingested by *A. castellanii* undergo germination and outgrowth within the amoebal phagosome, the vesicle enclosing the spores, followed by multiplication of the vegetative bacterial cells. Following the death of the amoeba, the vegetative cells are released into the extracellular environment where they sporulate, resulting in a 50-fold increase in the number of spores.

Schuch and Fischetti showed that survival of *B. anthracis* in the soil was the result of a complex interplay involving bacteriophages and earthworms.[73] A dying animal can release up to 1 × 10^9 bacteria per milliliter of blood into the surrounding soil where those bacteria can interact with bacteriophages.

Ashelford *et al.* demonstrated that there are about 1.5×10^8 bacteriophages per gram of soil, which is equivalent to about 4% of the total number of bacteria.[74] Some of these bacteriophages can specifically infect *B. anthracis* by a process in which the phage DNA becomes incorporated into the genome of the host, creating a lysogenic state.[73] The added viral DNA encodes sigma factors that switch on certain genes in *B. anthracis* and in so doing provides new abilities that enhance survival of *B. anthracis* cells and allows them to colonize an intermediate host – the earthworm.[73] Lysogeny with some bacteriophages was able to speed the transition from growing cells to spores when conditions such as low nutrient conditions or low temperatures (24 instead of 30 °C) were unfavorable for growth whereas other phages triggered the formation of biofilms under the same conditions.[73] Both types of phages were critical to the colonization of the earthworm gut by *B. anthracis* where, once established, the bacteria were able to survive for at least 6 months without harming the invertebrates.[73] The difference between strains with and without phages was so great that Schuch and Fischetti proposed that phage infections are actually necessary for the survival of *B. anthracis* in soil.[73]

7.3.3 *B. anthracis* Virulence Factors

As mentioned previously, fully virulent strains of *B. anthracis* possess two large plasmids designated pXO1 (182 kb) and pXO2 (96 kb), which encode the primary virulence factors, toxins and capsule, respectively. Multiple strains of *B. anthracis* have been sequenced and their complete chromosomal and plasmid DNA sequences are available on public databases. The base composition or guanine–cytosine content (33%) of these plasmids is similar to that of the *B. anthracis* chromosome.[63] The plasmids are not required for growth and strains lacking one or both plasmids are not uncommon. Strains can be cured of plasmids by growth at 43 °C or the presence of antibiotics.[63] Plasmid pXO2 is easily and spontaneously lost; $pX02^-$ strains have been found in environmental samples and among fully virulent strains.[75] However, curing of pX01 is rare. Attempts to attenuate the virulence of anthrax strains for vaccine purposes also led to plasmid curing. Two widely available attenuated vaccine strains lack plasmids: the Sterne strain, which is $pXO1^+$ $pXO2^-$ and the Pasteur strain, which is $pXO1^-$ $pXO2^+$.

Plasmid pXO1 carries the toxin genes (*pagA*, *lef* and *cya*), a regulatory gene (*atxA*), a resolvase, a transposase and a three-gene germination operon (*gerX*).[76] The gene *pagA* encodes the 83 kDa protective antigen (PA), *lef* encodes the 90 kDa lethal factor (LF) and *cya* encodes the 89 kDa edema factor (EF). Functional anthrax toxins require the combination of PA with LF for lethal toxin (LT) and PA with EF for edema toxin (ET). PA has been shown to bind to two cell surface receptors, the tumor endothelial marker 8 (TEM8) and the capillary morphogenesis protein 2 (CMG2). These receptors are widely expressed on many cell types, including immune cells.[77,78] PA binds to TEM8 and CMG2, which then interact with host low-density lipoprotein (LDL)-receptor-related protein 6 (LRP6).[79] After PA binds to the host cell receptor, the 20 kDa C-terminal region is proteolytically cleaved by furin and the truncated PA spontaneously oligomerizes to form heptamers, which then bind three EF and/or LF molecules.[77] The resulting toxin complex is phagocytized and, after acidification of the phagosome, the PA oligomer inserts into the membrane and mediates the translocation of the bound LF and/or EF into the cytoplasm of the host cell.[80] LF is a zinc-dependent metalloprotease that has been shown to cleave host cell mitogen-activated protein kinase kinases (MAPKK) in the cell cytosol.[81,82] Brittingham *et al.*[83] demonstrated that cleavage of MAPKK by LT interrupts host cell signal transduction, consequently inhibiting the expression of some cytokines. EF is a calmodulin-dependent adenylate cyclase,[84,85] which creates a gradient of cyclic AMP with a high concentration in the perinuclear area.[86] EF also disrupts host cell responses.[84,87] The *gerX* operon codes for three proteins, GerXA, GerXB and GerXC. It is postulated that these proteins form a receptor, which specifically detects germinant within the host.[63] Deletion of this operon affects spore germination in macrophages. The *atxA* gene regulates the expression of genes on pXO1 and pXO2.

The majority of the *in vitro* and *in vivo* work describing the effects of anthrax toxins has been performed with purified toxins. Limited information is available on the effects of anthrax toxins in the context of an infection.[88] Loving *et al.*[88] demonstrated that LT was required for outgrowth of bacilli in the draining lymph nodes and subsequent progression to disseminated disease in the mouse aerosol challenge model of infection.

The pXO2 plasmid carries *capB*, *capC*, *capA* and *dep*, which are known to encode the synthesis and degradation of a capsule consisting of a linear polymer of poly-D-glutamic acid.[89] The function of the capsule in disease pathogenesis has not been completely elucidated but it is hypothesized that it prevents phagocytosis of vegetative cells by macrophages and thus evades the host immune defenses and promotes septicemia.[89,90] The capsule is only very weakly immunogenic.[11]

7.4 Clinical Forms of Anthrax

There are several clinical forms of human anthrax that can be differentiated by the mode of entry of the infectious spore.

7.4.1 Cutaneous Anthrax

About 95% of reported anthrax cases are cutaneous and most occur in Africa, Asia and Eastern Europe where preventive measures such as animal and worker vaccination are limited.[91] Disease typically results when spores are introduced subcutaneously through a cut or abrasion in the skin.[23,92] The primary skin lesion is usually a painless pruritic papule that appears 3–5 days after infection. The lesion forms a vesicle that undergoes central necrosis, leaving a characteristic black eschar surrounded by edema.[92] The eschar dries and sloughs over the next 1–2 weeks.[93] Histological examination of anthrax skin lesions reveals necrosis and massive edema with lymphocytic infiltrates. There is no liquefaction or abscess formation but focal points of hemorrhage are evident, with some thrombosis.[94] Mortality of untreated cutaneous anthrax due to dissemination has been reported to be as high as 20%.[95] With appropriate treatment, death due to cutaneous anthrax is uncommon.

7.4.2 Gastrointestinal Anthrax

Gastrointestinal (GI) anthrax occurs by eating food contaminated with spores of *B. anthracis* (most often contaminated meat).[13] GI anthrax occurs in oropharyngeal or lower GI forms. The oropharyngeal form is associated with the development of an oral or esophageal ulcer followed by regional lymphadenopathy, edema and sepsis.[96,97] The lower GI form is characterized by primary intestinal lesions occurring predominantly in the terminal ileum or cecum.[98] Patients with lower GI anthrax present initially with nausea, vomiting and malaise, which then progresses rapidly to bloody diarrhea and acute abdomen or sepsis.[98] Ulceration of the GI mucosa may result in hematemesis. The lower GI tract demonstrates mucosal necrosis, extensive edema and mesenteric lymphadenitis.[99,100] *B. anthracis* cells can be seen in the mucosal, submucosal and lymphatic tissues along with inflammatory infiltrates.[100] Morbidity is due to blood loss, fluid and electrolyte imbalances and subsequent shock. Death results from intestinal perforation or toxemia. The mortality rate for untreated GI anthrax can approach 100%.[101] The last confirmed case of GI anthrax in the United States occurred in 2009 in a person who likely swallowed aerosolized anthrax spores released from a contaminated animal-hide drum.[99] It is possible that some cases of oropharyngeal anthrax are also due to the inhalation of aerosolized spores.

7.4.3 Inhalational Anthrax

Inhalational anthrax occurs when spores of *B. anthracis* are inhaled and reach the lower respiratory tract.[102] The particle size of the inhaled spore(s) is important and will be discussed below. In the lower respiratory tract, alveolar macrophages phagocytose and transport the spores to the hilar and mediastinal lymph nodes, where they germinate and produce toxins.[23] The clinical presentation of inhalational anthrax consists of two phases. The prodromal phase consisting of flu-like symptoms with cough, fever and fatigue that lasts from hours to a few days and then may briefly resolve.[5] The fulminant stage, which follows, is characterized by rising fever, dyspnea, diaphoresis and shock. Hemorrhagic meningitis may complicate inhalational anthrax, resulting in patients presenting with meningismus, delirium and obtundation. In patients presenting with advanced disease, cyanosis and hypotension progress rapidly and death can occur within hours.[23] Patients with progressive inhalational anthrax are typically found to have mediastinal adenopathy and hemorrhagic pleural effusions that can be detected on chest radiography.[103–105] Pathological examination usually shows a focus of necrotizing hemorrhagic pneumonitis at the portal of infection, with hemorrhage and necrosis of peribronchial and mediastinal lymph nodes. Regardless of antibiotic or antiserum therapy or other medical intervention, 97% of patients who progressed to the fulminant phase died.[103] The mortality rate of patients in the 2001 US outbreak, although lower than in earlier cases, was still high (45%).[106] Earlier antibiotic treatment and the use of pleural drainage or toxin inhibitors may decrease mortality.[103,106]

7.4.4 Injectional Anthrax

Injectional anthrax is a recently identified form of anthrax. It was first observed in India in 1988 following the injection of a contaminated antibiotic[107] and in 2000 in a heroin user.[108] There were no further reports until 2009, when two hospitalized injection drug users from Glasgow, UK, were found to have blood cultures positive for *B. anthracis*.[109] By the end of this outbreak in December 2010, there were 47 confirmed injectional anthrax cases with 13 deaths in Scotland and five cases with four deaths in England.[110] There were also two cases with one death reported from Germany.[111] Ten of these cases have been described and are notably distinct from cutaneous anthrax in that these soft tissue infections were not associated with black eschar formation.[98]

In spite of the information presented above, the disease anthrax appears to be relatively rare among humans. However, factors such as laboratory capacity, uniform case definitions and reporting may underestimate the true burden of this disease. The data obtained from infected individuals provide only a fragmentary view of the dynamics of infection and human host–pathogen interactions. For that reason, animal models have been developed to study the pathogenesis of anthrax, particularly the inhalational form.[112] Much of what we know about inhalational anthrax comes from studies employing animal models.

7.5 Aerobiology and Inhalation Exposure to *B. anthracis* Spores

Aerosol dispersal is the most likely method for the dissemination of spores of *B. anthracis* by a bioterrorist.[113] Factors that relate to outcome after inhalation of spores include the particle size and virulence of the strain, deposition within the respiratory tract and the host response.

7.5.1 Particle Size

Bioaerosols are defined as "a collection of particles suspended on a column of air derived from or incorporating material of biological origin."[114] Particle size dictates where aerosolized spores deposit in the respiratory tract; once deposited, the potential to cause disease is influenced by tissue tropism, clearance kinetics and the host immunological response (see Table 7.1). Individual spore-containing aerosol particles may vary widely in size from 1 µm to several millimeters depending on the production method, the method of aerosol

Table 7.1 Influence of aerosol particle size on the respiratory lethal dose values for spores of *B. anthracis* in a guinea pig. Data from ref. 112 and 115

Particle size/μm	Cfu[c]
1.0–3.0[a]	$(0.23–3.4) \times 10^5$
4.0–6.5[b]	2.21×10^5
8.5–13.0[b]	$(0.7–5.7) \times 10^6$

[a]Particles generated with a Collison nebulizer.
[b]Particles generated with a spinning top aerosol generator.
[c]Cfu, colony-forming units.

generation and attachment to larger particles.[114] It is difficult to predict the particle size that an individual may inhale from a bioaerosol because any source consists of a particle size distribution that changes with time and distance due to the local climate (*e.g.* humidity, temperature, turbulent activity, ventilation) and removal from the air column. The final inhaled particle size is dependent on a number of factors, including the solid organic content of the initial particle (including the spore) and location of an individual relative to the aerosol source. One can generalize that the closer an individual is to an aerosol source, the greater the likelihood of large particles being inhaled.

No studies exist that investigated the effects of particle size on respiratory infection with *B. anthracis* in humans.[114] However, studies using animal models have been conducted using methods that can differentially deposit spores in the lower respiratory tract (LRT) and upper respiratory tract (URT) within small or large particle aerosols.[102,115] A number of animal models have been used to study various aspects of inhalational anthrax, including the virulence of the strain and the effectiveness of vaccines and therapeutics. To date, mice, Guinea pigs, rabbits and nonhuman primates are the principal animal species that have been used.[112,116] Nonhuman primates (*e.g. Macaca mulatta*, *Macaca fascicularis* and *Chlorocebus aethiops*) consistently develop the full range of classic lesions of human inhalational anthrax, including meningitis, lymphadenitis, splenitis, mediastinitis, pneumonia, vasculitis, and hemorrhage, congestion and edema in multiple tissues.[116] However, cost, availability and ethical issues limit research with nonhuman primates.[112]

The respiratory tract is complex, comprising a collection of specialized organs, tissues and cells ranging from the nares to the alveoli.[117] Additional tissues may also be exposed during inhalation of a bioaerosol (see Table 7.2). The aerodynamic diameter of inhaled particles determines where within the respiratory tract spores incorporated within the particles deposit and interact with host tissues. The deposition of spores in the respiratory tract occurs through the physical processes of impaction, sedimentation and diffusion.[118,119] These three processes are governed by the physicochemical characteristics of the particle (*i.e.* size, shape, charge, density, hygroscopicity, solubility, chemical reactivity) and by host factors, such as respiratory route (nasal *versus* mouth breathing), respiratory rate and tidal volume and respiratory tract anatomy.[112,119] Impaction is the dominant mechanism governing the deposition of spore-containing particles with an aerodynamic diameter >1–2 μm. Impaction typically occurs in the nasopharyngeal and

Table 7.2 Tissues that may represent the primary site of deposition, clearance and infection during exposure to a bioaerosol. Modified from Thomas.[114]

Anatomic region	Tissue(s)	Epithelium
Nasopharyngeal/ oropharyngeal	Nares to nasopharynx	Ciliated
	Olfactory epithelium	Sensory
	Mouth to oropharynx	Nonciliated
	Tonsils, adenoids[a]	Lymphoepithelial
Tracheobronchial	Larynx	Ciliated/nonciliated
	Trachea	Ciliated
	Bronchi to bronchioles	Ciliated
Pulmonary	Respiratory bronchioles to alveoli	Nonciliated
Ocular	Ocular conjunctiva	Nonciliated
	Nasolacrimal duct	Nonciliated
Gastrointestinal	Esophageal, stomach and intestinal epithelium	Nonciliated
	Peyer's patches	Lymphoepithelial

[a]Nasal-associated lymphoid tissue (NALT) represents the rodent equivalent.

Table 7.3 Effect of aerosol particle size and deposition site on infection kinetics and pathology of inhalational anthrax in a mouse model of infection. Data from ref. 122 and modified from ref. 114.[a]

	Particle size	
Characteristic	1 μm	12 μm
Deposition site	Lungs	Nasal mucosa
MLD/cfu	2.43×10^3	7.65×10^3
MTTD/h	101.6 ± 10.4	161.0 ± 16.1
Mortality/%	78	56
Pathology	BALT infection	Nasal ulceration
	Mediastinal lymph nodes	NALT infection
	Splenitis	Cervical adenitis
	Septicemia	Splenitis
		Primary gastritis
		Peyer's patches
		Mesenteric lymph nodes

[a]Abbreviations: MLD = minimum lethal dose; MTTD = minimum time to death; BALT = broncho-pulmonary-associated lymphoid tissue; NALT = nasopharyngeal-associated lymphoid tissue.

tracheobronchial regions, particularly at airway branching points where the bulk air flow is faster and more turbulent. Sedimentation is an important deposition mechanism for larger particles (>8 μm) where they are deposited in a size-dependent manner from the nasal passages to the bronchioles. Diffusion deep into the lung tissue depositing small particles (0.1–2.0 μm) in the alveoli is an important deposition mechanism; however, sedimentation and electrostatic effects also play important roles. Results from animal models of inhalational anthrax demonstrated that the deposition site can markedly influence infection kinetics and pathology (see Table 7.3). The majority of inhaled spores, which do not germinate in the lung and may persist for up to 2 months, are the basis for the duration of post-exposure prophylaxis.[120,121]

Inhaled spores can be cleared from the respiratory tract. Clearance kinetics are important in determining the dose of deposited spores within the respiratory tract and ultimately systemically. The nose effectively filters foreign particles that enter the nasal cavity in a manner dependent on particle size and air flow rate.[114] Once spores have been deposited, the speed of nasal clearance depends on the deposition location in the nasopharynx. In healthy humans, clearance from the ciliated anterior region is more rapid than from the nonciliated posterior region.[123] Clearance mechanisms include entrapment of spores in mucus within the nasal cavity and tracheobronchial region with subsequent mucociliary transport towards the oral cavity (along with sneezing, swallowing, blowing and coughing actions to expel the material), where spores may be swallowed. Alternatively, they may cause disease locally within the URT (see Table 7.4). In the alveolar region, alveolar macrophages phagocytize spores and destroy them locally or clear them by migrating to the conducting airways for mucociliary transport or to the lymphatics at the bronchioalveolar junction.

In summary, spores are deposited within the URT or LRT dependent on particle size. Once deposited, they may undergo one of several fates, including (1) removal by a clearance mechanism, (2) production of local infection in the URT and (3) phagocytosis by alveolar or dendritic macrophages and, if not killed, migration of the macrophage to local lymph nodes where spores germinate and establish an infection. The steps by which an infection is established are discussed below using information obtained from animal models.

Table 7.4 Upper respiratory symptoms in humans with anthrax.[96,103,124–126]

	Percentage of patients with symptom/%		
Symptom	Nasopharyngeal	Oropharyngeal	Laryngeal
Fever	83	97	50
Tonsillitis	—	63	—
Pharyngitis	33	72	—
Nasal/sinus complaints	83	3	0
Malaise/fatigue	33	—	25
Headache	17	6	0
Cervical/submandibular lymphadenitis	50	100	—
Abdominal pain	17	25	
Vomiting/nausea	0	3	25
Cough	0	6	0
Dysphagia	17	—	0
Diarrhea	—	3	—
Hemoptysis/dyspnea	17	39	75
Anorexia	17	—	—
Depression	17	—	0
Mortality	50	40	50

7.5.2 Crossing the Respiratory Epithelium

Once deposited in the respiratory tract, the next crucial step in the infection process is the crossing of the epithelial barrier by the noninvasive spore. Studies have demonstrated that the spores are phagocytized by alveolar macrophages and/or dendritic cells in the alveolar space.[127,128] Migration of host cells having phagocytized the spores to the mediastinal/tracheobronchial lymph nodes is then thought to occur with subsequent germination, bacterial multiplication and expression of virulence factors (*i.e.* toxins and capsule) in the lymph nodes.[112] The involvement of the lung epithelial cells in the capture and crossing of the epithelial barrier has also been proposed,[129,130] but not substantiated.

Once the spores have crossed the epithelial barrier, they disseminate *via* the lymphatic route to the draining lymph nodes. *B. anthracis* does not multiply primarily in the lung parenchyma and therefore does not cause a primary pneumonia. Infection initiates in the draining lymph nodes. Involvement of the mediastinal lymph nodes has been described as a prominent feature of inhalational anthrax. However, not all patients with inhalational anthrax have mediastinal widening.[131] Animal studies using bioluminescence imaging have indicated that other lymph nodes draining the respiratory tract can also be involved in the first stages of dissemination.[131–134] Transport of the spores most probably involves macrophages and dendritic cells.

7.5.3 Hematogenous Dissemination

After multiplication in the lymph nodes, *B. anthracis* exits in the efferent lymphatics and reaches the blood compartment *via* the main lymphatic ducts (thoracic duct and right lymphatic duct). In animal models, an observed increase in bioluminescent signal indicates that the encapsulated bacilli subsequently colonize the spleen where they multiply extensively and produce toxins.[135] The spleen plays a central role as a reservoir because when the bacterial load becomes too high, *B. anthracis* will exit the spleen and return to the lung by the hematogenous route. Chains (4–8 cells) of encapsulated bacteria can become trapped in the alveolar and pleural capillaries.[135] Unrestricted multiplication of the bacteria in the alveolar and pleural capillaries followed by lesions in the lung parenchyma and severe alteration of gas exchanges lead to secondary pneumonia and death if treatment is delayed.

7.6 Laboratory Diagnosis

In the absence of a recognized outbreak or known cases, the initial nonspecific and flu-like symptoms of anthrax can be misleading to a healthcare provider and result in a delay in diagnosis. Owing to the short incubation period and rapid progression of the disease, accurate and timely diagnosis of anthrax, especially the inhalational form, is critical for successful treatment. A preliminary diagnosis based on clinical signs and early symptoms must be quickly confirmed by specific laboratory tests. Anthrax can be diagnosed by performing bacteriological, serological, immunological and molecular tests on a variety of clinical samples such as nasal swabs, skin lesion exudates, blood, tissues, cerebrospinal fluid, pleural fluid, pericardial fluid, sputum and feces.[98,101,106,136] Presumptive identification is based on standard bacteriological tests, including Gram staining of the clinical samples. *B. anthracis* is a Gram-positive, encapsulated, rod-shaped bacterium that forms central or terminal endospores. On blood agar medium plates, *B. anthracis* forms non-hemolytic, opaque, white to gray colored, raised and undulated colonies (2–5 mm diameter). Comma-shaped projections may arise from the colony edge, producing the "Medusa-head" colony. On nutrient agar plates containing 0.8 sodium bicarbonate, virulent *B. anthracis* usually forms its typical, heavily encapsulated colonies when incubated in the presence of 5% carbon dioxide. *B. anthracis* does not grow on MacConkey agar medium. Rapid confirmation from colonies or cell suspensions can be achieved using real-time polymerase chain reaction (PCR) assays or immunoassays to detect the presence of specific toxin genes or antigenic markers, respectively. Other confirmatory tests include γ-phage sensitivity and direct fluorescent-antibody staining of the capsule. Serological diagnostic tests are useful only in acute and convalescent phases of the disease. Inclusion of antibiotic susceptibility testing in laboratory diagnosis can help guide appropriate treatment of patients.

7.7 Prevention and Treatment

7.7.1 Prevention and Treatment in Animals

Vaccination is the most effective way to protect animals from anthrax. Since the development of the first effective anthrax animal vaccine *via* attenuation of a strain of *B. anthracis* by Louis Pasteur,[63] live or inactivated vaccines have been widely used to immunize animals.[5,137–140] In known endemic areas, annual vaccination of grazing animals should be done 2–4 weeks before the expected seasonal outbreaks. Anthrax can be effectively controlled by vaccination, rapid detection, quarantine and treatment of affected animals and incineration, burning or deep burial of dead animals following appropriate procedures. Rapid identification and isolation of source(s) of infection such as contaminated feed, pastures, water and carcasses of dead animals can reduce further spread of the disease in unexposed animals. Animals should be moved to uncontaminated areas. When possible, proper decontamination of contaminated feed, pastures, water, soil and other materials should be promptly performed. To prevent further spread of the diseases to other animals or humans and to safeguard their own health, animal owners and veterinary professionals should follow proper hygienic and sanitary practices when handling infected and dead animals or carcasses and contaminated materials. Although a majority of naturally occurring *B. anthracis* strains are susceptible to several antibiotics,[141–145] penicillin is widely used to treat anthrax in animals. However, the proper dosage and duration of treatment should be followed to prevent relapse and potentially to prevent β-lactamase induction leading to penicillin-resistant *B. anthracis*. Oxytetracycline is also a highly effective antibiotic and should be used in adequate doses to treat anthrax in animals, especially, in areas where penicillin-resistant strains of *B. anthracis* are found. As a prophylactic measure, either penicillin or oxytetracycline should be given to asymptomatic but potentially exposed animals. Additionally, vaccination of these animals 7–10 days after antibiotic treatment can further reduce the number of deaths.

7.7.2 Prevention and Treatment in Humans

The majority of human anthrax cases have occurred through animal to human transmission. In enzootic areas, the occurrence and spread of anthrax in humans can be controlled by implementation of proper procedures for the livestock and livestock-product industries and strict restrictions and regulations for the importation of animal products. Additionally, appropriate management of infections in livestock, proper disposal of dead animals and thorough decontamination of contaminated materials can reduce transmission to humans. Although person-to-person transmission of anthrax is exceptionally rare, medical practitioners and hospital staff should use standard infection control precautions and proper personal protective equipment such as an appropriate mask, apron and gloves while taking care of anthrax patients and also when handling waste. Since *B. anthracis* is a spore-forming pathogen, additional protective measures are required while handling and disposing of clothing and other material contaminated with blood or body fluids from a patient with anthrax. All waste generated should be discarded as clinical waste and incinerated.

Vaccination is the most effective preventive measure, especially for veterinarians, animal handlers and ranchers in areas where anthrax is enzootic. Vaccination is also highly recommended for laboratory personnel conducting research on virulent *B. anthracis* strains or handling and testing clinical or environmental samples during an anthrax outbreak.

The first effective anthrax vaccines using live, attenuated cultures of *B. anthracis* were developed by Greenfield in 1880[146] and by Pasteur in 1881.[147] Live attenuated vaccines were also developed in the former Soviet Union[138] and are still used in some countries today.[148] The United States and the United Kingdom developed PA-based acellular vaccines.[98,149,150] The US vaccine, which became known as anthrax vaccine absorbed or AVA, is currently prepared by precipitation of PA from cell-free culture filtrate of a toxigenic, nonencapsulated strain of *B. anthracis* V770-NP1-R with aluminum hydroxide.[98,149,150] The production cultures are grown in a chemically defined protein-free medium. The final product contains benzethonium chloride and formaldehyde as preservatives. The UK vaccine, known as anthrax vaccine precipitated (AVP), is an alum-precipitated PA preparation containing small amounts of LF from culture supernatants of *B. anthracis* Sterne strain.[98,149–151] AVA is the only licensed human anthrax vaccine in the United States. It has been recommended for both pre- and post-exposure

Table 7.5 Recommended pre- and post-exposure vaccination schedules for anthrax vaccine adsorbed[152]

Type of prophylaxis	Schedule	Route	Dose/mL
Pre-exposure	5 doses (0 weeks, 4 weeks, 6 months, 12 months and 18 months). Annual booster to maintain immunity	Intramuscular	0.5
Post-exposure[a]	3 doses (0, 2 and 4 weeks)[b,c]	Subcutaneous	0.5

[a]For previously unvaccinated persons.
[b]In conjunction with 60 day antimicrobial post-exposure prophylaxis.
[c]Administered under an Investigational New Drug (IND) protocol or an Emergency Use Authorization (EUA).

prophylaxis[152] (see Table 7.5); however, the number of doses and the route of administration differ. Because AVA is not licensed for post-exposure use, the vaccine will likely be made available either through an Investigational New Drug (IND) protocol or Emergency Use Authorization (EUA) during a public health emergency.[153] The types of populations for which pre-event or post-exposure use of anthrax vaccine is recommended[152] are shown in Table 7.6. The use of AVA in children aged 0–17 years is not contraindicated in a post-event setting that poses a high risk for exposure to aerosolized spores. During such an event, public health authorities will determine whether to offer vaccine to children under the existing IND protocol.

Asymptomatic individuals with suspected exposure to *B. anthracis* spores can also be protected by antibiotic treatment.[101] During anthrax epizootics and epidemics and during an intentional or accidental release of *B. anthracis* spores, contingent upon the availability of an appropriate acellular vaccine, combined administration of vaccine and antibiotic to exposed individuals can reduce the number of human cases (see Tables 7.5 and 7.7). Similarly, as a post-exposure prophylactic measure, such a combined treatment for asymptomatic but exposed humans can also help reduce the number of human cases. Antimicrobial therapy should be administered as soon as possible. Ideally, the first dose of vaccine should be administered within 10 days. A 6 week or longer course of appropriate doses of ciprofloxacin or doxycycline is usually recommended depending on the suspected or measured extent of spore contamination. Also, owing to a longer retention of spores in lungs,[4,101,120,121] in cases of suspected and known aerosol exposure, longer treatment with antibiotic may be required.[101,120,154]

Several antibiotics have been approved by the US Food and Drug Administration (FDA) for the prevention and treatment of anthrax.[156] Recommended antibiotic treatment for adults is presented in Table 7.7. Penicillin, doxycycline and ciprofloxacin were approved before the 2001 anthrax attack in the United States. Levofloxacin was approved for adults in 2004[157] and for children in 2008.[158] Several other antibiotics were also approved by the FDA.[159] Owing to the ease of availability in many parts of the world, penicillin has been widely used to treat patients with active anthrax. However, some

Table 7.6 Recommendations for use of anthrax vaccine adsorbed, by type of population. Recommended pre- and post-exposure vaccination schedules for anthrax vaccine adsorbed[152]

Population	Pre-event[a]	Post-exposure prophylaxis (PEP)[b]
General public	Not recommended	Recommended
Special populations in the general public		
Pregnant women	Not recommended	Recommended
Breastfeeding women	Not recommended	Recommended
Children (aged <18 years)	Not recommended	Determined on an event-by-event basis
Medical professionals	Not recommended	Recommended
Populations at risk for occupational exposure		
Persons who handle animals or animal products	Not routinely recommended[c]	Recommended
Persons who perform certain types of laboratory work	Recommended[d]	Based on pre-event vaccination status
Persons who work in postal facilities	Not recommended	Recommended
Military personnel	As recommended by the department of defense	As recommended by the department of defense
Persons involved in environmental investigations or remediation efforts	Recommended	Based on pre-event vaccination status
Persons involved in emergency response activities[e]	Not routinely recommended; may be offered on a voluntary basis under the direction of a comprehensive occupational health and safety program	Recommended

[a]Five 0.5 mL doses administered intramuscularly at 0 weeks, 4 weeks, 6 months, 12 months and 18 months. Annual boosters are required to maintain immunity.
[b]Three 0.5 mL doses administered subcutaneously at 0, 2 and 4 weeks after exposure to aerosolized *B. anthracis* spores for persons who have not completed the pre-event vaccination schedule.
[c]Recommended only if handling potentially infected animals in research settings or in areas with a high incidence of enzootic anthrax or when standards and restrictions are insufficient to prevent exposure to *B. anthracis* spores.
[d]Laboratorians who work (1) with high concentrations or pure cultures of *B. anthracis* spores, (2) with environmental samples associated with anthrax investigations or (3) in spore-contaminated areas or other settings with exposure to aerosolized *B. anthracis* spores. Laboratorians who do not work in these settings are not recommended for pre-event vaccine.
[e]Persons involved in emergency response activities might include persons who work in police departments, fire departments, hazardous material units and the National Guard, and also other government responders. These persons might perform site investigations, respond to suspicious substance reports (also known as white powder incidents) and perform other related activities, such as evacuation procedures or other activities critical to the maintenance of infrastructure.

Table 7.7 Recommended treatment for anthrax in adults[155]

Category	Recommended treatment[a]	Duration
Oral post-exposure prophylaxis	For all strains regardless of penicillin susceptibility or if susceptibility is unknown *Ciprofloxacin*, 500 mg every 12 h OR *Doxycycline*, 100 mg every 12 h OR Levofloxacin, 750 mg every 24 h OR Moxifloxacin, 400 mg every 24 h OR Clindamycin, 600 mg every 8 h OR Alternatives for penicillin-susceptible strains Amoxicillin, 1 g every 8 h OR Penicillin VK, 500 mg every 6 h	60 days
Intravenous treatment for systemic anthrax[b] with possible or confirmed meningitis	Bactericidal agent (fluoroquinolone) *Ciprofloxacin*, 400 mg every 8 h OR Levofloxacin, 750 mg every 24 h OR Moxifloxacin, 400 mg every 24 h PLUS Bactericidal agent (β-lactam). For all strains, regardless of penicillin susceptibility or if susceptibility is unknown *Meropenem*, 2 g every 8 h OR Imipenem, 1 g every 6 h OR Doripenem, 500 mg every 8 h OR Alternatives for penicillin-susceptible strains Penicillin G, 4 million units every 4 h OR Ampicillin, 3 g every 6 h PLUS Protein synthesis inhibitor *Linezolid*, 600 mg every 12 h OR Clindamycin, 900 mg every 8 h OR Rifampin, 600 mg every 12 h OR Chloramphenicol, 1 g every 6–8 h	≥2–3 weeks until clinical criteria for stability are met Patients exposed to aerosolized spores will require prophylaxis to complete an antimicrobial drug course of 60 days from onset of disease

(continued)

Table 7.7 (*continued*)

Category	Recommended treatment[a]	Duration
Intravenous treatment for systemic anthrax[b] when meningitis has been excluded	For all strains, regardless of penicillin susceptibility or if susceptibility is unknown *Ciprofloxacin*, 400 every 8 h OR Levofloxacin, 750 mg every 24 h OR Moxifloxacin, 400 mg every 24 h OR Meropenem, 2 g every 8 h OR Imipenem, 1 g every 6 h OR Doripenem, 500 mg every 8 h OR Vancomycin, 60 mg kg^{-1} day^{-1} intravenous divided every 8 h (maintain serum trough concentrations of 15–20 μg mL^{-1}) OR Alternatives for penicillin-susceptible strains Penicillin G, 4 million units every 4 h OR Ampicillin, 3 g every 6 h PLUS Protein synthesis inhibitor *Clindamycin*, 900 mg every 8 h OR *Linezolid*, 600 mg every 12 h OR Doxycycline, 200 mg initially, then 100 mg every 12 h OR Rifampin, 600 mg every 12 h	2 weeks Until clinical criteria for stability are met Patients exposed to aerosolized spores will require prophylaxis to complete an antimicrobial drug course of 60 days from onset of illness
Oral treatment for cutaneous anthrax without systemic involvement	For all strains, regardless of penicillin susceptibility or if susceptibility is unknown *Ciprofloxacin*, 500 mg every 12 h OR *Doxycycline*, 100 mg every 12 h OR *Levofloxacin*, 750 mg every 24 h OR *Moxifloxacin*, 400 mg every 24 h OR Clindamycin, 600 mg every 8 h OR Alternatives for penicillin-susceptible strains Amoxicillin, 1 g every 8 h OR Penicillin VK, 500 mg every 6 h	60 days For bioterrorism-related cases; 7–10 days for naturally acquired cases

[a]Preferred drugs are in bold type. Alternative drugs are listed in order of preference for treatment for patients who cannot take first-line treatment or if first-line treatment is unavailable.
[b]Systemic anthrax includes anthrax meningitis, inhalation, injection and gastrointestinal anthrax, and cutaneous anthrax with systemic involvement, extensive edema or lesions of the head or neck.

B. anthracis strains are resistant to penicillin.[143,160–162] Therefore, to avoid treatment failure due to a penicillin-resistant *B. anthracis* strain, inclusion of doxycycline or ciprofloxacin as a combined therapy is recommended.[161,163,164] In high-risk situations, the same antibiotic regimen is recommended for children under 18 years of age and for pregnant or nursing women.[23,155,165–167] Some *B. anthracis* isolates have also been found to be resistant to several antibiotics, including cefuroxime, cefotaxime, ceftazidime, aztreonam and the sulfamethoxazole–trimethoprim combination.[141,168–170]

Unless there is an anthrax outbreak or known aerosol release of *B. anthracis* spores, prompt treatment of inhalational anthrax can be difficult owing to initial nonspecific and flu-like symptoms.[23,161] However, with the first suspicion of inhalational anthrax, intravenous administration of ciprofloxacin or doxycycline is currently recommended.[98,155,167] Since ciprofloxacin is bactericidal and doxycycline is bacteriostatic, treatment with the former is preferred. A 60 day course of antibiotic treatment is usually recommended; however, a switch from intravenous to oral administration can be made after the patient's condition improves and when it is clinically appropriate.[155,167] Clindamycin can also be used to treat the patients because it inhibits anthrax toxin synthesis.[155,167] In patients with severe symptoms, including meningitis, combined therapy including at least one other antibiotic that has been shown to penetrate the central nervous system (CNS) effectively is recommended,[98,155,167] such as ampicillin, rifampicin, vancomycin and meropenem. Also, corticosteroid therapy may be considered for anthrax patients with meningitis.[148,171,172]

For mild and naturally acquired cases of the cutaneous form of anthrax, patients can be treated orally with ciprofloxacin or doxycycline for 7–10 days.[23,161,167] Upon determination that the infecting strain of *B. anthracis* is susceptible to amoxicillin, ciprofloxacin or doxycycline may be replaced with amoxicillin. Amoxicillin can also be used when the patient cannot be given or tolerate fluoroquinolones or tetracycline. For bioterrorism-related cases, treatment with ciprofloxacin or doxycycline should continue for 60 days presuming that the patient may also have inhaled spores.[23,155,167] Patients with a severe form of cutaneous anthrax involving systemic spread and/or extensive edema or lesions on the head or neck should be treated with the regimen recommended above for inhalational anthrax.[23,155,167]

Gastrointestinal anthrax should be treated in the same way as inhalational anthrax.[23,155,167] For suspected injectional anthrax, intravenous administration of ciprofloxacin and clindamycin along with other antibiotics such as penicillin, flucloxacillin and metronidazole is recommended.[98,173] For the control of soft tissue infection in injectional anthrax patients, surgery may be required to remove the tissue.[98,110,149,173,174]

For post-exposure prophylaxis, the Centers for Disease Control and Prevention (CDC) has recommended a 60 day oral regimen of ciprofloxacin or doxycycline as first-line agents.[155,167] Since the AVA is not a live vaccine, it can be used for post-exposure prophylaxis in combination with the antibiotics.[98,149,155,167]

There are many limitations and problems associated with the current antibiotic and vaccine recommendations.[175,176] One of the major concerns is the generation of a multidrug-resistant (MDR) *B. anthracis* strain *via* either a natural route or by intentional genetic manipulation. For some anthrax patients, antibiotic treatment may be contraindicated. It is also possible that a 60 day antibiotic treatment may make the patients less compliant. Owing to the short incubation period, the rapid onset of severe symptoms and progression of inhalational anthrax, current multidose vaccine regimens alone will not provide protection to exposed or symptomatic individuals. Most importantly, in the absence of an early diagnosis resulting in delay in appropriate and aggressive antibiotic treatment, patients may die due to the toxins produced by replicating as well as dead bacteria. Therefore, protective antibody-based passive immunization is considered to be an attractive treatment approach. During an intentional anthrax incident, passive immunization may be the most effective countermeasure.[177] Currently, raxibacumab and anthrax immunoglobulin intravenous (AIGIV) antitoxins are available from the CDC's Strategic National Stockpile. The antitoxins have been shown to inhibit binding of PA to cell receptors and translocation of the LT and ET into cells.[155] Raxibacumab is the first recombinant human monoclonal antitoxin antibody that is effective against anthrax toxins.[155,178,179] Raxiacumab is approved by the FDA for the treatment or prevention of anthrax in adults and children.[166] Antitoxin-based treatment can also be highly effective during an intentional anthrax attack involving a genetically engineered strain, which can evade the protective effect of currently available vaccines and antibiotics.

7.8 Anthrax as a Biological Weapon

Owing to its virulence characteristics, a short incubation period and rapid onset of severe symptoms, *B. anthracis* has been one of the most feared biological weapons.[170] *B. anthracis* can be easily obtained from epizootic areas. The nonfastidious nutritional requirements and relatively simple growth conditions[169,170] make it is relatively easy and inexpensive to cultivate in large quantities. It can be grown in growth media under conditions that promote high levels of spore formation. Since the spores are 1–2 μm in size,[101,169,170] a large amount of spores can be easily stored. Owing to their small size, they are relatively easy to aerosolize and disseminate.[170] During the offensive bioweapons development era in the 20th century, significant and troublesome advances were made in the techniques for weaponization and dissemination of *B. anthracis* spores. During this era, *B. anthracis* spores were tested and used as a bioweapon to spread anthrax.[17,18] During World War I, Germany used *B. anthracis* as a weapon against livestock and draft animals. Japan conducted field trials with *B. anthracis* as a biological weapon in Manchuria during World War II.[11] Other countries, including the USA, the UK, the former Soviet Union and Iraq, conducted anthrax weapons research.[17,18] To prevent further advancements in the weaponization of *B. anthracis* spores and other biological threat agents, most nations signed the 1972 Biological Weapons Convention (BWC) treaty and ceased offensive research and development.[6] As of January 2015, 173 countries were BWC members.[7] On the bioterrorism front, Aum Shinrikyo, a religious group in Japan, sprayed a liquid suspension of *B. anthracis* as an aerosolized biological weapon near Tokyo, in 1993.[20,180] However, the attack was unsuccessful owing to the use of a nonencapsulated vaccine strain.[20,21] Owing to the potential use of *B. anthracis* spores as a weapon, many countries have been investing in biodefense programs to protect their soldiers and civilians.

7.9 Amerithrax

Anthrax in humans is extremely rare in the United States. Prior to October 2001, the last case of inhalational anthrax there occurred in 1976, when a home craftsman died after working with imported yarn contaminated with spores of *B. anthracis*.[181] The anthrax attacks of 2001, also known as Amerithrax, were an unprecedented incident involving the dissemination of *B. anthracis* spores *via* envelopes and using the US Postal Service mail system.[23] Spore-laden letters were mailed to several news organizations and the offices of two US Senators. A total of 22 cases of bioterrorism-related anthrax occurred as a result. There were nine cases among individuals at targeted locations, nine cases among postal service employees, two cases of individuals who handled cross-contaminated mail and two cases of individuals with an undetermined source of exposure. An additional case of cutaneous anthrax in a laboratory worker handling environmental samples suspected of containing *B. anthracis* was diagnosed later. Out of the 22 cases, there were 11 cases of inhalational anthrax which resulted in five deaths and 11 cases of cutaneous anthrax.[106] More than 30 000 individuals received post-exposure prophylaxis and numerous public and private buildings had to be decontaminated. The total cost of the response to this incident was estimated to exceed $1 billion. Decontamination of various buildings and postal facilities accounted for a significant part of the cost.[182] Although effective decontamination of contaminated areas and facilities was achieved, serious weaknesses were recognized in terms of timeliness and cost-effectiveness. Additionally, the response efforts highlighted the serious challenges concerning the timely, cost-effective and adequate sampling for and detection of *B. anthracis* spores in contaminated indoor and outdoor environments after a wide-area anthrax attack.

The 2001 anthrax incidents and subsequent response efforts led to heightened awareness of potential bioterrorist attacks and increased funding for preparedness to respond to such attacks. Since 2001, the United States has spent tens of billions of dollars on biosecurity infrastructure and biodefense research programs.[183] As a result, significant progress has been made to prepare for, protect against, respond to and recover from future bioterror attacks. Establishment of the Public Health Emergency Medical Countermeasures Enterprise (PHEMCE), including the Strategic National Stockpile[184–186] and the Integrated Consortium of Laboratory Networks (ICLN),[187,188] is among the major active efforts and accomplishments. The PHEMCE is the coordinating body for the US federal agencies charged with protecting the civilian population from adverse health impacts from both intentional and naturally emerging threats through the development and use of medical countermeasures. These countermeasures include medicines, devices or other medical interventions that can lessen the harmful effects of these threats. The ICLN is a coordinated and operational system of laboratory networks of US government departments and agencies that provide timely, high-quality and interpretable results for early detection and effective consequence management of acts of terrorism and other events requiring an integrated laboratory response.

7.10 Detection of *B. anthracis* Spores in Environmental Samples

Deposition of aerosolized spores of *B. anthracis* resulting from natural, intentional or unintentional incidents can contaminate both indoor and outdoor environments. Detection of spores in indoor and outdoor environments is achieved through the analysis of diverse types of environmental samples. Environmental samples include air, surface, water, wastewater, vegetation and soil.[188] A significant source of samples may be urban materials, such as building materials, both exterior (*e.g.* concrete, granite, brick and glass) and interior (*e.g.* glass, steel, drywall, carpet, laminate and ceiling tiles).[188] Spores can be collected on a variety of sampling materials, including heating, ventilation and air-conditioning (HVAC) system filters, swabs, wipes, 3M Sponge-Stick sampling tool (3M, St Paul, MN, USA), vacuum socks and vacuum filters. During the first few weeks to months following a wide-area anthrax incident, thousands of samples will need to be collected and analyzed for various purposes such as determining how far the spores have spread, who might have been exposed and how many indoor and outdoor facilities and materials may need to be decontaminated. Samples may arise not only from contamination resulting from the incident, but also from the multitude of activities surrounding the response to the incident, including decontamination.[189]

Analysis of environmental samples for detection of *B. anthracis* spores is still a significant challenge.[190] Environmental samples will likely contain materials that may interfere with the analytical method used. Therefore, sample processing is a critical step before analysis. Sample processing itself is a serious bottleneck for high-throughput sample analysis during a wide-area event. Sample processing may include recovery of *B. anthracis* spores from the sample itself (*e.g.* powder, water and food), sampling buffer or sampling tool followed by concentration or dilution of the suspension of recovered spores. Depending upon the analytical method used, extraction and purification of nucleic acid may be required. Thus, spores must be disrupted prior to extraction. Furthermore, analytical methods need to be highly sensitive and specific. Sample processing and analysis will require the availability of large quantities of reagents, equipment and supplies; therefore, the detection methods should also be cost-effective.

There are several analytical approaches for the detection and identification of *B. anthracis* in environmental samples.[190,191] Culture-, real-time PCR- and immunoassay-based methods are the most commonly used approaches. Traditional microbiological culture is regarded as the "gold standard" for the detection and identification of *B. anthracis.* A selective medium such as polymyxin–lysozyme EDTA–thallous acetate (PLET) agar is often used since there is a high likelihood that spores of other *Bacillus* species are present in environmental samples.[192] This culture method, however, requires a relatively long incubation period (1–2 days); other microbiological, biochemical or molecular confirmatory tests taking an additional 1–2 days are required for confirmatory identification. Therefore, culture-based sample analysis can be very challenging if thousands of samples need to be analyzed. Fluorogenic end-point or real-time PCR, a nucleic acid-based analytical method, is widely used for environmental sample analysis. For the correct detection of a fully virulent strain, PCR assays targeting multiple *B. anthracis*-specific gene targets are used. Usually, the gene targets include at least one virulence gene on each virulence plasmid (*i.e.* pXO1 and pXO2) and one unique chromosomal marker. Such a strategy not only helps to differentiate a virulent *B. anthracis* strain from an avirulent one, but also distinguishes *B. anthracis* from other *Bacillus* species, especially *B. thuringiensis* and *B. cereus* that are genetically very close.[26,193] Although PCR assays can be very sensitive, specific and rapid, they cannot confirm the viability of spores. The rapid viability PCR (RV-PCR) method, however, can detect and identify viable *B. anthracis* spores in environmental samples.[194] The RV-PCR method integrates high-throughput sample processing, rapid enrichment culture and sensitive, specific and fast real-time PCR analysis. The PCR assays are performed before and after incubation of the processed sample in liquid growth media. A difference in the resultant PCR cycle threshold (C_T) values indicates the presence (C_T difference) or absence (no C_T difference) of viable spores of *B. anthracis* in a sample. The method can provide results in 16 h with a limit of detection similar to or lower than that of the culture-based method.[195] Immunoassays targeting the surface antigens present on the spores are also used to detect and identify *B. anthracis* spores in environmental samples. Enzyme-linked immunosorbent assay (ELISA) and electrochemiluminescence (ECL) immunoassays are more commonly used. The ECL assays are more sensitive than ELISA.[196]

7.11 Decontamination

As a direct result of the 2001 Amerithrax letters, at least 23 buildings were contaminated and a further 263 were suspected to be contaminated.[197] Decontamination efforts for some of the most highly contaminated facilities were not completed until late 2003 and the total recovery costs were estimated at upwards of US$1 billion.[197] The time and cost required to remediate these facilities highlight the complexity of the decontamination process following contamination incidents.[182] Remediation of the Senator Daschle suite of the Hart Senate Office Building (HSOB)[198] (3000 ft^2) following the release of ~2 g of *B. anthracis* spores from an envelope cost a reported $27 million over a period of more than 3 months.[199,200] The Capitol Hill remediation efforts represented the first of several major cleanups following the 2001 incident and were atypical in that considerable resources and funding were devoted to reopening the buildings safely as soon as possible. Other contaminated facilities were remediated at a lower overall cost, but required much more time to complete. Facilities that required major cleanup efforts included the Joseph Curseen Jr and Thomas Morris Jr Processing and Distribution Center (PDC) (formerly the Brentwood PDC) in Washington, DC (700 000 ft^2),[201–203] the Trenton PDC in Hamilton, NJ (317 000 ft^2),[199,204] the Department of State's Diplomatic Pouch and Mail Facility (SA-32) in Sterling, VA (70 000 ft^2)[205,206] and the American Media Incorporated (AMI) building in Boca Raton, FL (70 000 ft^2).[199]

Decontamination can be defined as "the irreversible inactivation of infectious agents such that an area or object is rendered safe."[207] Disinfection is the process of eliminating many or all microorganisms on inanimate objects, with the exception of bacterial spores.[208] Sterilization is the complete destruction or removal of all microbial life, including bacterial spores.[209] Cleaning typically involves chemical (often detergents), manual and mechanical means, to remove visible soil or grime from surfaces.[208] Decontamination of surfaces includes the use of cleaning, disinfection or sterilization products or processes to remove or reduce surface-bound microorganisms. Although a simple concept, decontamination is typically considered on a relatively small scale and successful decontamination of whole building interiors or vast outdoor areas will likely require diverse and innovative decontamination methods and strategies.

At the time of the 2001 incidents, no decontamination products (sporicides) were registered by the US Environmental Protection Agency (EPA) under the Federal Insecticide, Fungicide and Rodenticide Act (FIFRA) for proven use against *B. anthracis* spores.[210,211] Further, most EPA-registered sporicides were registered only for hard, precleaned, nonporous surfaces and the use of sporicidal decontaminants on the scale required for full-facility decontamination was unprecedented. Nonetheless, using on-site trials[212] and vendor-supplied efficacy data, sporicidal products and decontamination procedures were selected. The affected facilities were ultimately decontaminated using sporicidal liquids and fumigants, in combination with removal and off-site treatment/disposal of some equipment, materials and waste items.[197] Waste management (handling, transport and disposal) costs and the willingness of facilities to accept waste potentially contaminated with *B. anthracis* spores are major issues that have to be addressed during a biological remediation.[213–215]

In the United States, pesticides (which includes those chemicals used to kill microorganisms) must be registered through the EPA under the FIFRA. Prior to registration, data demonstrating the efficacy of the product against the particular class of organism targeted, using approved test methodologies, must be presented to EPA for review. For sporicidal claims, approved test methodologies include the AOAC Method 966.04 – Sporicidal Activity of Disinfectants,[216] ASTM E 2197-02 – Quantitative Disk Carrier Method (QCT-2),[217] and AOAC Method 2008.05 – Quantitative Three Step Method (TSM).[217] The use of these methods has demonstrated the ability to evaluate repeatably the antimicrobial potential of test products; however, extrapolation to product performance outside the clinical setting is problematic.[218] Standardized laboratory test methods can fail to predict accurately a decontaminant's efficacy on complex surfaces (*e.g.* building interiors or outdoor surfaces) or challenging environments (cold weather, *etc.*). Since the 2001 incident, numerous studies have been undertaken to understand the challenges of decontamination under real-world conditions.

For registration of sporicidal decontaminants under FIFRA, demonstration of a >6 log inactivation when challenged with 6–7 log of *Clostridium difficile* or *B. subtilis* spores, using an approved test method, is required.[217] For FIFRA sporicidal registration against spores of *B. anthracis*, these efficacy tests must be conducted with *B. anthracis* spores or a suitable surrogate (*e.g. B. subtilis*).[219] The use of nonpathogenic surrogates for efficacy tests is desirable to reduce risk and cost; however, the surrogate must exhibit equivalent or slightly higher resistance to a particular inactivation mechanism than the agent it models. The 6 log reduction milestone is based on the concept of a 10^{-6} sterility assurance level (SAL) adhered to by the medical device industry,[220,221] which indicates the probability of contaminants occurring after sterilization. The 6 log reduction benchmark has also been used in decontamination-focused research endeavors and is a useful way to compare the effectiveness across trials, studies, methods and decontaminant technologies.

Outside the laboratory, a different set of success criteria must be utilized since the concentration of contaminants prior to decontamination may be unknown and/or highly variable spatially. Further, the risk of exposure or disease from any remaining viable microorganism is difficult to predict.[222] Accordingly, during the Amerithrax remediation, decontamination success was defined as "no growth" from any environmental sample where decontamination techniques were used.[223] Any remaining contamination was deemed to pose a hazard to building occupants and was therefore necessary to be removed or killed. To date, all *B. anthracis* remediations in the United States have adopted the "no growth" clearance benchmark.[222]

In a 2003 review of publications documenting *B. anthracis* spore inactivation, several key issues were noted regarding translation of then current data to real-world remediation activities.[207] First, cross-study comparisons of results were not appropriate since each study utilized unique test methods. Second, most studies were conducted on surrogate organisms, without adequate side-by-side comparison with the agent of interest (*B. anthracis*). Third, few laboratory studies evaluated complex, porous, difficult to decontaminate material types, commonly encountered outside hospitals (*e.g.* carpet, wallboard, soil). Lastly, a holistic evaluation of decontamination strategies, on a building-scale, had yet to be tested.

In the years since the 2003 review, numerous advancements in our knowledge of *B. anthracis* spore inactivation have been made. Initially, bench-scale studies were conducted to identify liquids and gases with adequate sporicidal efficacy when challenged with spore-contaminated building materials expected to be encountered during facility decontaminations.[224–233] Significant investments were also made into the investigation and refinement of standardized methods used to evaluate decontamination efficacy.[218,234–236] These studies were instrumental in determining which technologies demonstrated sufficient potential for further evaluation or enhanced our understanding of the limitations of laboratory-based decontaminant evaluations. For example, it was shown that inoculation method (liquid *versus* aerosol methods for contaminating test surfaces) can have significant impacts on decontamination efficacy for some material types.[237] Also, decontamination of environmental surfaces, using real-world application methods, can be more challenging than decontamination of standardized materials used in some tube-based efficacy tests. Biological indicators also often overestimate the efficacy of many fumigant-based decontamination approaches.[238–240]

Efficient waste management can significantly reduce costs and speed recovery following a biological incident. The amount of waste generated during a remediation can be significantly affected by the decontamination approach selected. The amount of waste generated and the types of materials comprising the waste drastically affect the cost of waste management.[213] Early investigations on the destruction of spores in waste materials were also conducted. These studies focused on heat-based treatment methods.[241–245] Subsequently, studies have evaluated chemical-based "expedient" approaches for waste decontamination[214] and the efficient management of waste streams.[246]

Most recently, the focus of many studies has been on the operational aspects of conducting decontaminations in facilities and outdoor areas. For example, one study evaluated the efficacy of several spray-based sporicide delivery systems when used to decontaminate concrete and wood and/or grimed surfaces.[247–250] Decontamination of environments that pose significant challenges due to organic content, such as soil,[251] or access and material challenges, such as HVAC ducts,[252–254] has also been evaluated. Many fumigants are simply unsuited for decontamination of particular material types. For example, decontamination of bare concrete or galvanized metals is challenging for hydrogen peroxide-based technologies.[255,256] Understanding the impacts of various fumigant chemicals on sensitive electronics and valuable items (*e.g.* valuable artwork or historic documents) is imperative, as collateral damage can be significant during large-scale decontamination operations. Recent studies have determined that ethylene oxide, hydrogen peroxide vapor and methyl bromide are less damaging to many electronic components than gaseous chlorine dioxide, when used under conditions (temperature, humidity, fumigant concentration, exposure durations) efficacious against *B. anthracis* spores.[257] Hot, humid air may be effective at killing *B. anthracis* spores in airplanes or expensive vehicles without destruction of vital electronic components.[258,259] However, significant technical challenges remain in the ability to deploy this approach beyond the laboratory scale.

Lastly, the ability to scale up technologies to full-facility or wide-area implementation is currently the focus of numerous research efforts. A field-scale chlorine dioxide generator was evaluated for its ability to achieve target concentrations within a large structure.[260] Similarly, three different decontamination approaches (chlorine dioxide fumigation, vaporous hydrogen peroxide fumigation and a liquid-based pH-adjusted bleach procedure) were evaluated in an 8000 ft^2 facility during a multi-agency field-scale test.[213] Safe and effective means for capturing decontamination chemicals during and following use are also of importance.[261] Effective means for capture of chlorine dioxide gas and methyl bromide have been studied.[261,262]

Several decision support tools have been developed to support distinct phases of the overall remediation process. Sandia National Laboratories has developed the Prioritization Analysis Toolset for All-Hazards (PATH) and the Analyzer for Wide-Area Restoration Effectiveness (AWARE). PATH assists decision-makers with prioritization across multiple systems and critical infrastructure. Impacted areas overlaid onto detailed area maps and associations between locale assets (*e.g.* roadways necessary for assess to a facility or one critical infrastructure asset dependent upon another for functionality) are utilized to assist in prioritization for remediation activities. Coupled with AWARE, the prototype software provides analysis of remediation resource allocations and provides estimates of remediation timelines and cost. PATH/AWARE enables "what-if" scenarios to be run and analyzed, allowing for capability gaps analyses to be conducted.

Several sampling support tools have been developed that assist with developing sampling plans and data management. The Visual Sampling Plan (VSP) was developed by Pacific Northwest National Laboratories to assist with the development of sampling plans based on statistical sampling theory.[263] Sandia National Laboratories developed the Building Restoration Operations Optimization Model (BROOM) software system to assist with the collection, visualization and analysis of environmental sampling data.[264] More recent efforts by Sandia National Laboratories have been in the development of the Site Exploitation System for Situational Awareness (SESSA) toolkit that allows for three-dimensional mapping and visualization of areas, including visualization of sampling locations and results.[265] Johns Hopkins University Applied Physics Laboratory is currently developing a Tactical Dynamic Operational Guided Sampling (TacDOGS) Tool, which incorporates the tracking of sample collection information and analysis of results through the use of map layers, and also assists in the decision-making for sample collection planning and asset coordination.[266] Sandia National Laboratories, in collaboration with the EPA and Department of Homeland Security (DHS), have developed a Single Building Decontamination Tool (DeconST) that provides a comprehensive framework for considering decontamination options.[266] DeconST incorporates information on decontamination methods' effectiveness and cost and site-specific details to assess decontamination options.

The assessment includes information on sampling requirements, waste generation and overall time and cost for implementation. DeconST is readily updatable to include the latest information on decontamination options. The EPA has also developed Incident Waste (I-WASTE), a web-based planning and decision-making tool that provides information on waste characterization, treatment and disposal options.[267] Each of these tools may be useful in assisting decision-makers with the many decisions that are required during remediation planning and implementation. However, at present, there is not a single comprehensive decision support tool or toolset.

From an operational perspective, numerous challenges remain in our ability to decontaminate following a large-scale *B. anthracis* contamination incident.[247,268] In particular, deployment of decontamination and sampling personnel and equipment on a scale required following contamination of a large metropolitan area would be difficult. The current inventory of commercially available generators needed for volumetric (fumigant) decontamination approaches could be inadequate in a large-scale incident due to the number of facilities needing remediation. Incident Commanders would therefore need to make difficult yet strategic decisions, to deploy decontamination equipment, resources and personnel based on need, effectiveness, cost, time and availability. Development of effective, expedient or self-help methods for reducing contaminant concentrations and thus exposure risk could enhance our ability to respond to and recover from such incidents. Such methods, when used in areas outside heavily contaminated areas, would likely help speed recovery and minimize exposure to the agent. In the absence of sufficient laboratory capacity to analyze samples for every suspected contaminated surface, waste or area, clearance and reoccupancy decisions may need to rely more heavily on other evidence such as decontamination conditions. Management of unprecedented volumes of waste items would also be required following a large metropolitan incident. With limited sampling and analysis prior to reoccupancy, those citizens and businesses adjacent to, but outside, the known highly contaminated zone would likely generate significant quantities of waste as many items will likely be discarded due to fear of contamination. This, coupled with the large quantities of waste generated during decontamination and demolition operations within the hot zone, would undoubtedly overwhelm our ability to manage waste (waste transport, treatment, sampling and disposal) effectively. The development of effective and efficient on-site waste management approaches, including developments in on-site waste treatment and sampling methods, can help to reduce the overall waste disposal burden.

Acknowledgements

The authors thank Dr Staci Kane (Lawrence Livermore National Laboratory), Dr Laura Rose (CDC), Dr Eugene Rice (EPA) and Dr Michele Burgess (EPA) for the chapter content review and constructive comments and suggestions. The authors also thank Ms Marti Sinclair (Alion Science and Technology for SRA) for editorial assistance with this chapter.

References

1. J. M. Last, *A Dictionary of Public Health*, Oxford University Press, New York, 2007.
2. A. Karlen, *Man and Microbes: Disease and Plagues in History and Modern Times*, Touchstone Book, Simon & Schuster, New York, 1995.
3. N. Kokayeff, Dying to be Discovered: Miasma vs. Germ Theory, *ESSAI*, 2012, **10**, 1–3.
4. M. Meselson, J. Guillemin, M. Hughjones, A. Langmuir, I. Popova, A. Shelokov and O. Yampolskaya, The Sverdlovsk Anthrax Outbreak of 1979, *Science*, 1994, **266**, 1202–1208.
5. P. S. Brachman, Inhalation Anthrax, *Ann. N. Y. Acad. Sci.*, 1980, **353**, 83–93.
6. U.N., *Convention on the Prohibition of the Development, Production and Stockpiling of Bacteriological (Biological) and Toxin Weapons and on Their Destruction*, 2014, http://disarmament.un.org/treaties/t/bwc.
7. U.N., *The Latest Information on the Biological Weapons Convention, Efforts of States Parties, the Work of the Chair and Activities Undertaken by the Implementation Support Unit*, 2015, http://www.unog.ch/bwc/news.
8. G. Stembach, The History of Anthrax, *J. Emerg. Med.*, 2003, **24**, 463–467.
9. M. Schwartz, Dr. Jekyll and Mr. Hyde: A Short History of Anthrax, *Mol. Aspects Med.*, 2009, **30**, 347–355.
10. R. Koch, The Etiology of Anthrax, Based on the Life History of Bacillus anthracis, *Beitr. Biol. Pflanz.*, 1876, **2**, 277–310.
11. M. J. Hudson, W. Beyer, R. Bohm, A. Fasanella, G. Garofolo, R. Golinski, P. L. Goossens, U. Hahn, B. Hallis, A. King, M. Mock, C. Montecucco, A. Ozin, F. Tonello and S. H. Kaufmann, Bacillus anthracis: Balancing Innocent Research with Dual-Use Potential, *Int. J. Med. Microbiol.*, 2008, **298**, 345–364.
12. L. Pasteur, C.-E. Chamberlain and E. Roux, Compte rendu sommaire des experiences faites à Pouille-le-Fort, près Melun, sur la vaccination charbonneuse, *C. R. Acad. Sci.*, 1881, **92**, 1378–1383.
13. A. K. Goel, Anthrax: A Disease of Biowarfare and Public Health Importance, *World J. Clin. Cases*, 2015, **3**, 20–33.
14. P. C. B. Turnbull, *Introduction: Anthrax History, Disease and Ecology*, ed. T. M. Koehler, Springer-Verlag, Berlin, 2002, pp. 1–19.
15. J. H. Bell, On Anthrax and Anthracaemia in Wool-Sorters, Heifers and Sheep, *Br. Med. J.*, 1880, **2**, 656–657.
16. F. M. Laforce, Woolsorters' Disease in England, *Bull. N. Y. Acad. Med.*, 1978, **54**, 956–963.
17. E. Eitzen and E. Takafuji, Historical Overview of Biological Warfare, in *Medical Aspects of Chemical and Biological Warfare*, ed. F. Sidell, E. Takafuji and D. Franz, Borden Institute, Washington, DC, 1997, pp. 415–423.
18. G. W. Christopher, T. J. Cieslak, J. A. Pavlin and E. M. Eitzen, Jr., Biological Warfare: a Historical Perspective, in *Biological Weapons: Limiting the Threat*, ed. J. Lederburg, MIT Press, Cambridge, MA, 1999, pp. 17–35.

19. D. M. Missiakas and O. Schneewind, Bacillus anthracis and the Pathogenesis of Anthrax, in *Biological Weapons Defense: Infectious Diseases and Counterterrorism*, ed. L.E. Lindler and G. W. Korch, Humana Press, Inc., Totowa, N.J., 2005, pp. 79–97.
20. A. F. Kaufmann and P. S. Keim, The Kameido Anthrax Incident: A Microbial Forensics Case, in *Microbial Forensics*, ed. B. Budowle, S. E. Schutzer, R. G. Breeze, P. S. Keim and S. A. Morse, Elsevier Inc., San Diego, 2nd edn, 2011, pp. 5–13.
21. H. Takahashi, P. Keim, A. F. Kaufmann, C. Keys, K. L. Smith, K. Taniguchi, S. Inouye and T. Kurata, Bacillus anthracis Bioterrorism Incident, Kameido, Tokyo, 1993, *Emerging Infect. Dis.*, 2004, **10**, 117–120.
22. L. M. Bush, B. H. Abrams, A. Beall and C. C. Johnson, Index Case of Fatal Inhalational Anthrax due to Bioterrorism in the United States, *N. Engl. J. Med.*, 2001, **345**, 1607–1610.
23. T. V. Inglesby, T. O'Toole, D. A. Henderson, J. G. Bartlett, M. S. Ascher, E. Eitzen, A. M. Friedlander, J. Gerberding, J. Hauer, J. Hughes, J. McDade, M. T. Osterholm, G. Parker, T. M. Perl, P. K. Russell, K. Tonat and B. Working Group on Civilian, Anthrax as a Biological Weapon, 2002: Updated Recommendations for Management, *JAMA, J. Am. Med. Assoc.*, 2002, **287**, 2236–2252.
24. D. Daffonchio, A. Cherif and S. Borin, Homoduplex and Heteroduplex Polymorphisms of the Amplified Ribosomal 16S-23S Internal Transcribed Spacers Describe Genetic Relationships in the "Bacillus cereus Group", *Appl. Environ. Microbiol.*, 2000, **66**, 5460–5468.
25. E. Helgason, D. A. Caugant, I. Olsen and A. B. Kolsto, Genetic Structure of Population of Bacillus cereus and B. thuringiensis Isolates Associated with Periodontitis and Other Human Infections, *J. Clin. Microbiol.*, 2000, **38**, 1615–1622.
26. E. Helgason, O. A. Okstad, D. A. Caugant, H. A. Johansen, A. Fouet, M. Mock, I. Hegna and A. B. Kolsto, Bacillus anthracis, Bacillus cereus and Bacillus thuringiensis-One Species on the Basis of Genetic Evidence, *Appl. Environ. Microbiol.*, 2000, **66**, 2627–2630.
27. S. G. Bavykin, Y. P. Lysov, V. Zakhariev, J. J. Kelly, J. Jackman, D. A. Stahl and A. Cherni, Use of 16S rRNA, 23S rRNA and gyrB Gene Sequence Analysis to Determine Phylogenetic Relationships of Bacillus cereus Group Microorganisms, *J. Clin. Microbiol.*, 2004, **42**, 3711–3730.
28. L. J. Harrell, G. L. Andersen and K. H. Wilson, Genetic Variability of Bacillus anthracis and Related Species, *J. Clin. Microbiol.*, 1995, **33**, 1847–1850.
29. P. Keim, A. Kalif, J. Schupp, K. Hill, S. E. Travis, K. Richmond, D. M. Adair, M. Hugh-Jones, C. R. Kuske and P. Jackson, Molecular Evolution and Diversity in Bacillus anthracis as Detected by Amplified Fragment Length Polymorphism Markers, *J. Bacteriol.*, 1997, **179**, 818–824.
30. Y. H. Chang, Y. H. Shangkuan, H. C. Lin and H. W. Liu, PCR Assay of the groEL Gene for Detection and Differentiation of Bacillus cereus Group Cells, *Appl. Environ. Microbiol.*, 2003, **69**, 4502–4510.

31. L. Radnedge, P. G. Agron, K. K. Hill, P. J. Jackson, L. O. Ticknor, P. Keim and G. L. Andersen, Genome Differences that Distinguish Bacillus anthracis from Bacillus cereus and Bacillus thuringiensis, *Appl. Environ. Microbiol.*, 2003, **69**, 2755–2764.
32. G. T. Vilas-Boas, A. P. Peruca and O. M. Arantes, Biology and Taxonomy of Bacillus cereus, Bacillus anthracis and Bacillus thuringiensis, *Can. J. Microbiol.*, 2007, **53**, 673–687.
33. G. B. Jensen, B. M. Hansen, J. Eilenberg and J. Mahillon, The Hidden Lifestyles of Bacillus cereus and Relatives, *Environ. Microbiol.*, 2003, **5**, 631–640.
34. P. Keim, L. B. Price, A. M. Klevytska, K. L. Smith, J. M. Schupp, R. Okinaka, P. J. Jackson and M. E. Hugh-Jones, Multiple-Locus Variable-Number Tandem Repeat Analysis Reveals Genetic Relationships Within Bacillus anthracis, *J. Bacteriol.*, 2000, **182**, 2928–2936.
35. P. Keim, M. N. Van Ert, T. Pearson, A. J. Vogler, L. Y. Huynh and D. M. Wagner, Anthrax Molecular Epidemiology and Forensics: Using the Appropriate Marker for Different Evolutionary Scales, *Infect., Genet. Evol.*, 2004, **4**, 205–213.
36. S. Yamada, E. Ohashi, N. Agata and K. Venkateswaran, Cloning and Nucleotide Sequence Analysis of gyrB of Bacillus cereus, B. thuringiensis, B. mycoides and B. anthracis and Their Application to the Detection of B. cereus in Rice, *Appl. Environ. Microbiol.*, 1999, **65**, 1483–1490.
37. D. A. Rasko, M. R. Altherr, C. S. Han and J. Ravel, Genomics of the Bacillus cereus Group of Organisms, *FEMS Microbiol. Rev.*, 2005, **29**, 303–329.
38. D. A. Rasko, M. J. Rosovitz, O. A. Okstad, D. E. Fouts, L. Jiang, R. Z. Cer, A. B. Kolsto, S. R. Gill and J. Ravel, Complete Sequence Analysis of Novel Plasmids from Emetic and Periodontal Bacillus cereus Isolates Reveals a Common Evolutionary History Among the B. cereus-Group Plasmids, Including Bacillus anthracis pXO1, *J. Bacteriol.*, 2007, **189**, 52–64.
39. N. Ivanova, A. Sorokin, I. Anderson, N. Galleron, B. Candelon, V. Kapatral, A. Bhattacharyya, G. Reznik, N. Mikhailova, A. Lapidus, L. Chu, M. Mazur, E. Goltsman, N. Larsen, M. D'Souza, T. Walunas, Y. Grechkin, G. Pusch, R. Haselkorn, M. Fonstein, S. D. Ehrlich, R. Overbeek and N. Kyrpides, Genome Sequence of Bacillus cereus and Comparative Analysis with Bacillus anthracis, *Nature*, 2003, **423**, 87–91.
40. T. D. Read, S. N. Peterson, N. Tourasse, L. W. Baillie, I. T. Paulsen, K. E. Nelson, H. Tettelin, D. E. Fouts, J. A. Eisen, S. R. Gill, E. K. Holtzapple, O. A. Okstad, E. Helgason, J. Rilstone, M. Wu, J. F. Kolonay, M. J. Beanan, R. J. Dodson, L. M. Brinkac, M. Gwinn, R. T. DeBoy, R. Madpu, S. C. Daugherty, A. S. Durkin, D. H. Haft, W. C. Nelson, J. D. Peterson, M. Pop, H. M. Khouri, D. Radune, J. L. Benton, Y. Mahamoud, L. Jiang, I. R. Hance, J. F. Weidman, K. J. Berry, R. D. Plaut, A. M. Wolf, K. L. Watkins, W. C. Nierman, A. Hazen, R. Cline, C. Redmond, J. E. Thwaite, O. White, S. L. Salzberg, B. Thomason, A. M. Friedlander, T. M. Koehler, P. C. Hanna, A. B. Kolsto and C. M. Fraser, The Genome Sequence of Bacillus anthracis Ames and Comparison to Closely Related Bacteria, *Nature*, 2003, **423**, 81–86.

41. L. Slamti, S. Perchat, M. Gominet, G. Vilas-Boas, A. Fouet, M. Mock, V. Sanchis, J. Chaufaux, M. Gohar and D. Lereclus, Distinct Mutations in PlcR Explain Why Some Strains of the Bacillus cereus Group are Nonhemolytic, *J. Bacteriol.*, 2004, **186**, 3531–3538.
42. M. Carrera, R. O. Zandomeni, J. Fitzgibbon and J. L. Sagripanti, Difference Between the Spore Sizes of Bacillus anthracis and Other Bacillus Species, *J. Appl. Microbiol.*, 2007, **102**, 303–312.
43. M. Carrera, R. O. Zandomeni and J. L. Sagripanti, Wet and Dry Density of Bacillus anthracis and Other Bacillus Species, *J. Appl. Microbiol.*, 2008, **105**, 68–77.
44. M. J. Kennedy, S. L. Reader and L. M. Swierczynski, Preservation Records of Micro-organisms: Evidence of the Tenacity of Life, *Microbiology*, 1994, **140(Pt 10)**, 2513–2529.
45. C. Redmond, M. J. Pearce, R. J. Manchee and B. P. Berdal, Deadly Relic of the Great War, *Nature*, 1998, **393**, 747–748.
46. A. I. Aronson and P. Fitz-James, Structure and Morphogenesis of the Bacterial Spore Coat, *Bacteriol. Rev.*, 1976, **40**, 360–402.
47. S. C. Holt and E. R. Leadbetter, Comparative Ultrastructure of Selected Aerobic Spore-Forming Bacteria: A Freeze-Etching Study, *Bacteriol. Rev.*, 1969, **33**, 346–378.
48. P. Setlow, Small, Acid-Soluble Spore Proteins of Bacillus species: Structure, Synthesis, Genetics, Function and Degradation, *Annu. Rev. Microbiol.*, 1988, **42**, 319–338.
49. P. Setlow, Spores of Bacillus subtilis: Their Resistance to and Killing by Radiation, Heat and Chemicals, *J. Appl. Microbiol.*, 2006, **101**, 514–525.
50. R. Giorno, J. Bozue, C. Cote, T. Wenzel, K. S. Moody, M. Mallozzi, M. Ryan, R. Wang, R. Zielke, J. R. Maddock, A. Friedlander, S. Welkos and A. Driks, Morphogenesis of the Bacillus anthracis Spore, *J. Bacteriol.*, 2007, **189**, 691–705.
51. T. N. Brahmbhatt, B. K. Janes, E. S. Stibitz, S. C. Darnell, P. Sanz, S. B. Rasmussen and A. D. O'Brien, Bacillus anthracis Exosporium Protein BclA Affects Spore Germination, Interaction with Extracellular Matrix Proteins and Hydrophobicity, *Infect. Immun.*, 2007, **75**, 5233–5239.
52. J. Bozue, K. L. Moody, C. K. Cote, B. G. Stiles, A. M. Friedlander, S. L. Welkos and M. L. Hale, Bacillus anthracis Spores of the bclA Mutant Exhibit Increased Adherence to Epithelial Cells, Fibroblasts and Endothelial Cells but Not to Macrophages, *Infect. Immun.*, 2007, **75**, 4498–4505.
53. C. R. Oliva, M. K. Swiecki, C. E. Griguer, M. W. Lisanby, D. C. Bullard, C. L. Turnbough, Jr. and J. F. Kearney, The integrin Mac-1 (CR3) Mediates Internalization and Directs Bacillus anthracis Spores into Professional Phagocytes, *Proc. Natl. Acad. Sci. U. S. A.*, 2008, **105**, 1261–1266.
54. M. Hugh-Jones and J. Blackburn, The Ecology of Bacillus anthracis, *Mol. Aspects Med.*, 2009, **30**, 356–367.
55. L. Pasteur, C. Chamberland and E. Roux, Sur l'étiologie du charbon, *C. R. Seances Acad. Sci.*, 1880, **91**, 86–94.

56. G. B. Van Ness, Ecology of Anthrax, *Science*, 1971, **172**, 1303–1307.
57. A. M. Kasi'anenko, O. M. Verner, A. M. Zaviriukha and K. M. Siniak, Mapping Method in the Epidemiological Study of Infections and Infestations. II. Soil Characteristics in Locations Permanently at Risk for Anthrax, *Zh. Mikrobiol., Epidemiol. Immunobiol.*, 1984, **61**, 90–95.
58. K. L. Smith, V. De Vos, H. B. Bryden, M. E. Hugh-Jones, A. Klevytska, L. B. Price, P. Keim and D. T. Scholl, Meso-Scale Ecology of Anthrax in Southern Africa: A Pilot Study of Diversity and Clustering, *J. Appl. Microbiol.*, 1999, **87**, 204–207.
59. C. G. Himsworth, The Danger of Lime Use in Agricultural Anthrax Disinfection Procedures: The Potential Role of Calcium in the Preservation of Anthrax Spores, *Can. Vet. J.*, 2008, **49**, 1208–1210.
60. P. M. Lindeque and P. C. Turnbull, Ecology and Epidemiology of Anthrax in the Etosha National Park, Namibia, *Onderstepoort J. Vet. Res.*, 1994, **61**, 71–83.
61. V. De Vos, The Ecology of Anthrax in the Kruger National Park, South Africa, *Salisbury Med. Bull.*, 1990, **68S**, 19–23.
62. P. C. Turnbull, M. Diekmann, J. W. Kilian, W. Versfeld, V. De Vos, L. Arntzen, K. Wolter, P. Bartels and A. Kotze, Naturally Acquired Antibodies to Bacillus anthracis Protective Antigen in Vultures of Southern Africa, *Onderstepoort J. Vet. Res.*, 2008, **75**, 95–102.
63. M. Mock and A. Fouet, Anthrax, *Annu. Rev. Microbiol.*, 2001, **55**, 647–671.
64. W. L. Nicholson, N. Munakata, G. Horneck, H. J. Melosh and P. Setlow, Resistance of Bacillus Endospores to Extreme Terrestrial and Extraterrestrial Environments, Microbiol, *Mol. Biol. Rev.*, 2000, **64**, 548–572.
65. D. C. Dragon and R. P. Rennie, The Ecology of Anthrax Spores: Tough but Not Invincible, *Can. Vet. J.*, 1995, **36**, 295–301.
66. P. C. B. Turnbull, J. E. Bowen and J. W. Mann, Stubborn Contamination with Anthrax Spores, *Environ. Health*, 1996, **106**, 171–173.
67. D. C. Dragon, D. E. Bader, J. Mitchell and N. Woollen, Natural Dissemination of Bacillus anthracis Spores in Northern Canada, *Appl. Environ. Microbiol.*, 2005, **71**, 1610–1615.
68. P. C. B. Turnbull, J. M. Hofmeyr, A. M. T. McGetrick and B. A. Oppenheim, Isolation of Bacillus anthracis, the Agent of Anthrax, in the Etosha National Park. A Comparison of Various Media and Methods, *Madoqua*, 1986, **14**, 321–331.
69. E. Saile and T. M. Koehler, Bacillus anthracis Multiplication, Persistence and Genetic Exchange in the Rhizosphere of Grass Plants, *Appl. Environ. Microbiol.*, 2006, 72, 3168–3174.
70. S. Auger, E. Krin, S. Aymerich and M. Gohar, Autoinducer 2 Affects Biofilm Formation by Bacillus cereus, *Appl. Environ. Microbiol.*, 2006, **72**, 937–941.
71. K. Lee, J. W. Costerton, J. Ravel, R. K. Auerbach, D. M. Wagner, P. Keim and J. G. Leid, Phenotypic and Functional Characterization of Bacillus anthracis Biofilms, *Microbiology*, 2007, **153**, 1693–1701.

72. R. Dey, P. S. Hoffman and I. J. Glomski, Germination and Amplification of Anthrax Spores by Soil-Dwelling Amoebas, *Appl. Environ. Microbiol.*, 2012, **78**, 8075–8081.
73. R. Schuch and V. A. Fischetti, The Secret Life of the Anthrax Agent Bacillus anthracis: Bacteriophage-Mediated Ecological Adaptations, *PloS One*, 2009, **4**, e6532.
74. K. E. Ashelford, M. J. Day and J. C. Fry, Elevated Abundance of Bacteriophage Infecting Bacteria in Soil, *Appl. Environ. Microbiol.*, 2003, **69**, 285–289.
75. P. C. Turnbull, R. A. Hutson, M. J. Ward, M. N. Jones, C. P. Quinn, N. J. Finnie, C. J. Duggleby, J. M. Kramer and J. Melling, Bacillus anthracis but Not Always Anthrax, *J. Appl. Bacteriol.*, 1992, **72**, 21–28.
76. R. T. Okinaka, K. Cloud, O. Hampton, A. R. Hoffmaster, K. K. Hill, P. Keim, T. M. Koehler, G. Lamke, S. Kumano, J. Mahillon, D. Manter, Y. Martinez, D. Ricke, R. Svensson and P. J. Jackson, Sequence and Organization of pXO1, the Large Bacillus anthracis Plasmid Harboring the Anthrax Toxin Genes, *J. Bacteriol.*, 1999, **181**, 6509–6515.
77. R. J. Collier and J. A. Young, Anthrax Toxin, *Annu. Rev. Cell Dev. Biol.*, 2003, **19**, 45–70.
78. H. M. Scobie and J. A. Young, Interactions Between Anthrax Toxin Receptors and Protective Antigen, *Curr. Opin. Microbiol.*, 2005, **8**, 106–112.
79. W. Wei, Q. Lu, G. J. Chaudry, S. H. Leppla and S. N. Cohen, The LDL Receptor-Related Protein LRP6 Mediates Internalization and Lethality of Anthrax Toxin, *Cell*, 2006, **124**, 1141–1154.
80. J. A. Young and R. J. Collier, Anthrax Toxin: Receptor Binding, Internalization, Pore Formation and Translocation, *Annu. Rev. Biochem.*, 2007, **76**, 243–265.
81. N. S. Duesbery, C. P. Webb, S. H. Leppla, V. M. Gordon, K. R. Klimpel, T. D. Copeland, N. G. Ahn, M. K. Oskarsson, K. Fukasawa, K. D. Paull and G. F. Vande Woude, Proteolytic Inactivation of MAP-Kinase-Kinase by Anthrax Lethal Factor, *Science*, 1998, **280**, 734–737.
82. G. Vitale, R. Pellizzari, C. Recchi, G. Napolitani, M. Mock and C. Montecucco, Anthrax Lethal Factor Cleaves the N-Terminus of MAPKKs and Induces Tyrosine/Threonine Phosphorylation of MAPKs in Cultured Macrophages, *Biochem. Biophys. Res. Commun.*, 1998, **248**, 706–711.
83. K. C. Brittingham, G. Ruthel, R. G. Panchal, C. L. Fuller, W. J. Ribot, T. A. Hoover, H. A. Young, A. O. Anderson and S. Bavari, Dendritic Cells Endocytose Bacillus anthracis Spores: Implications for Anthrax Pathogenesis, *J. Immunol.*, 2005, **174**, 5545–5552.
84. D. L. Hoover, A. M. Friedlander, L. C. Rogers, I. K. Yoon, R. L. Warren and A. S. Cross, Anthrax Edema Toxin Differentially Regulates Lipopolysaccharide-Induced Monocyte Production of Tumor Necrosis Factor Alpha and Interleukin-6 by Increasing Intracellular Cyclic AMP, *Infect. Immun.*, 1994, **62**, 4432–4439.

85. S. H. Leppla, Anthrax Toxin Edema Factor: A Bacterial Adenylate Cyclase that Increases Cyclic AMP Concentrations of Eukaryotic Cells, *Proc. Natl. Acad. Sci. U. S. A.*, 1982, **79**, 3162–3166.
86. F. Dal Molin, F. Tonello, D. Ladant, I. Zornetta, I. Zamparo, G. Di Benedetto, M. Zaccolo and C. Montecucco, Cell Entry and cAMP Imaging of Anthrax Edema Toxin, *EMBO J.*, 2006, **25**, 5405–5413.
87. J. N. Tournier, A. Quesnel-Hellmann, J. Mathieu, C. Montecucco, W. J. Tang, M. Mock, D. R. Vidal and P. L. Goossens, Anthrax Edema Toxin Cooperates with Lethal Toxin to Impair Cytokine Secretion During Infection of Dendritic Cells, *J. Immunol.*, 2005, **174**, 4934–4941.
88. C. L. Loving, T. Khurana, M. Osorio, G. M. Lee, V. K. Kelly, S. Stibitz and T. J. Merkel, Role of Anthrax Toxins in Dissemination, Disease Progression and Induction of Protective adaptive Immunity in the Mouse Aerosol Challenge Model, *Infect. Immun.*, 2009, **77**, 255–265.
89. M. Drysdale, S. Heninger, J. Hutt, Y. Chen, C. R. Lyons and T. M. Koehler, Capsule Synthesis by Bacillus anthracis is Required for Dissemination in Murine Inhalation Anthrax, *EMBO J.*, 2005, **24**, 221–227.
90. A. Scorpio, D. J. Chabot, W. A. Day, K. D. O'Brien, N. J. Vietri, Y. Itoh, M. Mohamadzadeh and A. M. Friedlander, Poly-Gamma-Glutamate Capsule-Degrading Enzyme Treatment Enhances Phagocytosis and Killing of Encapsulated Bacillus anthracis, *Antimicrob. Agents Chemother.*, 2007, **51**, 215–222.
91. S. Shafazand, R. Doyle, S. Ruoss, A. Weinacker and T. A. Raffin, Inhalational Anthrax: Epidemiology, Diagnosis and Management, *Chest*, 1999, **116**, 1369–1376.
92. M. Doganay, G. Metan and E. Alp, A Review of Cutaneous Anthrax and its Outcome, *J. Infect. Public Health*, 2010, **3**, 98–105.
93. M. T. Kutluk, G. Secmeer, G. Kanra, A. Celiker and H. Aksoyek, Cutaneous Anthrax, *Cutis*, 1987, **40**, 117–118.
94. E. Mallon and P. H. McKee, Extraordinary Case Report: Cutaneous Anthrax, *Am. J. Dermatopathol.*, 1997, **19**, 79–82.
95. W. D. Tutrone, N. S. Scheinfeld and J. M. Weinberg, Cutaneous Anthrax: A Concise Review, *Cutis*, 2002, **69**, 27–33.
96. T. Sirisanthana, N. Navachareon, P. Tharavichitkul, V. Sirisanthana and A. E. Brown, Outbreak of Oral-Oropharyngeal Anthrax: An Unusual Manifestation of Human Infection with Bacillus anthracis, *Am. J. Trop. Med. Hyg.*, 1984, **33**, 144–150.
97. T. Sirisanthana and A. E. Brown, Anthrax of the Gastrointestinal Tract, *Emerging Infect. Dis.*, 2002, **8**, 649–651.
98. C. W. Hicks, D. A. Sweeney, X. Cui, Y. Li and P. Q. Eichacker, An Overview of Anthrax Infection Including the Recently Identified Form of Disease in Injection Drug Users, *Intensive Care Med.*, 2012, **38**, 1092–1104.
99. M. S. Klempner, E. A. Talbot, S. I. Lee, S. Zaki and M. J. Ferraro, Case Records of the Massachusetts General Hospital. Case 25-2010. A 24-Year-Old Woman with Abdominal Pain and Shock, *N. Engl. J. Med.*, 2010, **363**, 766–777.

100. D. R. Nalin, B. Sultana, R. Sahunja, A. K. Islam, M. A. Rahim, M. Islam, B. S. Costa, N. Mawla and W. B. Greenough, 3rd, Survival of a Patient with Intestinal Anthrax, *Am. J. Med.*, 1977, **62**, 130–132.
101. T. C. Dixon, M. Meselson, J. Guillemin and P. C. Hanna, Anthrax, *N. Engl. J. Med.*, 1999, **341**, 815–826.
102. H. A. Druett, D. W. Henderson, L. Packman and S. Peacock, Studies on Respiratory Infection. I. The Influence of Particle Size on Respiratory Infection with Anthrax Spores, *J. Hyg.*, 1953, **51**, 359–371.
103. J. E. Holty, D. M. Bravata, H. Liu, R. A. Olshen, K. M. McDonald and D. K. Owens, Systematic Review: A Century of Inhalational Anthrax Cases from 1900 to 2005, *Ann. Intern. Med.*, 2006, **144**, 270–280.
104. K. Vessal, J. Yeganehdoust, W. Dutz and E. Kohout, Radiological Changes in Inhalation Anthrax. A Report of Radiological and Pathological Correlation in Two Cases, *Clin. Radiol.*, 1975, **26**, 471–474.
105. B. J. Wood, B. DeFranco, M. Ripple, M. Topiel, C. Chiriboga, V. Mani, K. Barry, D. Fowler, H. Masur and L. Borio, Inhalational Anthrax: Radiologic and Pathologic Findings in Two Cases, *AJR, Am. J. Roentgenol.*, 2003, **181**, 1071–1078.
106. D. B. Jernigan, P. L. Raghunathan, B. P. Bell, R. Brechner, E. A. Bresnitz, J. C. Butler, M. Cetron, M. Cohen, T. Doyle, M. Fischer, C. Greene, K. S. Griffith, J. Guarner, J. L. Hadler, J. A. Hayslett, R. Meyer, L. R. Petersen, M. Phillips, R. Pinner, T. Popovic, C. P. Quinn, J. Reefhuis, D. Reissman, N. Rosenstein, A. Schuchat, W. J. Shieh, L. Siegal, D. L. Swerdlow, F. C. Tenover, M. Traeger, J. W. Ward, I. Weisfuse, S. Wiersma, K. Yeskey, S. Zaki, D. A. Ashford, B. A. Perkins, S. Ostroff, J. Hughes, D. Fleming, J. P. Koplan, J. L. Gerberding and the National Anthrax Epidemiologic Investigation Team, Investigation of Bioterrorism-Related Anthrax, United States, 2001: Epidemiologic Findings, *Emerging Infect. Dis.*, 2002, **8**, 1019–1028.
107. M. K. Lalitha, V. Anandi, N. Walter, J. O. Devadatta and B. M. Pulimood, Primary Anthrax Presenting as an Injection "Abscess", *Indian J. Pathol. Microbiol.*, 1988, **31**, 254–256.
108. S. H. Ringertz, E. A. Hoiby, M. Jensenius, J. Maehlen, D. A. Caugant, A. Myklebust and K. Fossum, Injectional Anthrax in a Heroin Skin-Popper, *Lancet*, 2000, **356**, 1574–1575.
109. C. N. Ramsay, A. Stirling, J. Smith, G. Hawkins, T. Brooks, J. Hood, G. Penrice, L. M. Browning, S. Ahmed, G. G. C. NHS and Scottish National Outbreak Control Teams, An Outbreak of Infection with Bacillus anthracis in Injecting Drug Users in Scotland, *Euro. Surveill.*, 2010, **15**, 19465.
110. M. G. Booth, J. Hood, T. J. Brooks and A. Hart, Anthrax Infection in Drug Users, *Lancet*, 2010, **375**, 1345–1346.
111. D. Radun, H. Bernard, M. Altmann, I. Schoneberg, V. Bochat, U. van Treeck, R. M. Rippe, R. Grunow, M. Elschner, W. Biederbick and G. Krause, Preliminary Case Report of Fatal Anthrax in an Injecting Drug User in North-Rhine-Westphalia, Germany, December 2009, *Euro. Surveill.*, 2010, **15**, 19464.

112. P. L. Goossens, Animal Models of Human Anthrax: The Quest for the Holy Grail, *Mol. Aspects Med.*, 2009, **30**, 467–480.
113. M. G. Kortepeter, T. J. Cieslak and E. M. Eitzen, Bioterrorism, *J. Environ. Health*, 2001, **63**, 21–24.
114. R. J. Thomas, Particle Size and Pathogenicity in the Respiratory Tract, *Virulence*, 2013, **4**, 847–858.
115. L. D. Fothergill, Biological Warfare: Nature and Consequences, *Tex. Med.*, 1964, **60**, 8–14.
116. N. A. Twenhafel, Pathology of Inhalational Anthrax Animal Models, *Vet. Pathol.*, 2010, **47**, 819–830.
117. C. S. Wang, Inhaled Particles, *Interface Sci. Technol.*, 2005, **5**, 1–187.
118. R. B. Schlesinger, Comparative Deposition of Inhaled Aerosols in Experimental-Animals and Humans – A Review, *J. Toxicol. Environ. Health*, 1985, **15**, 197–214.
119. C. J. Roy, D. S. Reed and J. A. Hutt, Aerobiology and Inhalation Exposure to Biological Select Agents and Toxins, *Vet. Pathol.*, 2010, **47**, 779–789.
120. A. M. Friedlander, S. L. Welkos, M. L. Pitt, J. W. Ezzell, P. L. Worsham, K. J. Rose, B. E. Ivins, J. R. Lowe, G. B. Howe, P. Mikesell and W. B. Lawrence, Postexposure Prophylaxis Against Experimental Inhalation Anthrax, *J. Infect. Dis.*, 1993, **167**, 1239–1243.
121. D. W. Henderson, S. Peacock and F. C. Belton, Observations on the Prophylaxis of Experimental Pulmonary Anthrax in the Monkey, *J. Hyg.*, 1956, **54**, 28–36.
122. R. Thomas, C. Davies, A. Nunez, S. Hibbs, H. Flick-Smith, L. Eastaugh, S. Smither, A. Gates, P. Oyston, T. Atkins and S. Eley, Influence of Particle Size on the Pathology and Efficacy of Vaccination in a Murine Model of Inhalational Anthrax, *J. Med. Microbiol.*, 2010, **59**, 1415–1427.
123. E. Puchelle, F. Aug, J. M. Zahm and A. Bertrand, Comparison of Nasal and Bronchial Mucociliary Clearance in Young Non-Smokers, *Clin. Sci.*, 1982, **62**, 13–16.
124. M. Doganay, A. Almac and R. Hanagasi, Primary Throat Anthrax. A Report of Six Cases, *Scand. J. Infect. Dis.*, 1986, **18**, 415–419.
125. N. Navacharoen, T. Sirisanthana, W. Navacharoen and K. Ruckphaopunt, Oropharyngeal Anthrax, *J. Laryngol. Otol.*, 1985, **99**, 1293–1295.
126. H. Ozdemir, K. Demirdag, T. Ozturk and E. Kocakoc, Anthrax of the Gastrointestinal Tract and Oropharynx: CT Findings, *Emerg. Radiol.*, 2010, **17**, 161–164.
127. A. Cleret, A. Quesnel-Hellmann, A. Vallon-Eberhard, B. Verrier, S. Jung, D. Vidal, J. Mathieu and J. N. Tournier, Lung Dendritic Cells Rapidly Mediate Anthrax Spore Entry Through the Pulmonary Route, *J. Immunol.*, 2007, **178**, 7994–8001.
128. J. M. Ross, The Pathogenesis of Anthrax Following the Administration of Spores by the Respiratory Route, *J. Pathol. Bacteriol.*, 1957, **73**, 485–494.
129. B. H. Russell, Q. Liu, S. A. Jenkins, M. J. Tuvim, B. F. Dickey and Y. Xu, In Vivo Demonstration and Quantification of Intracellular Bacillus anthracis in Lung Epithelial Cells, *Infect. Immun.*, 2008, **76**, 3975–3983.

130. B. H. Russell, R. Vasan, D. R. Keene, T. M. Koehler and Y. Xu, Potential Dissemination of Bacillus anthracis Utilizing Human Lung Epithelial Cells, *Cell. Microbiol.*, 2008, **10**, 945–957.
131. J. A. Jernigan, D. S. Stephens, D. A. Ashford, C. Omenaca, M. S. Topiel, M. Galbraith, M. Tapper, T. L. Fisk, S. Zaki, T. Popovic, R. F. Meyer, C. P. Quinn, S. A. Harper, S. K. Fridkin, J. J. Sejvar, C. W. Shepard, M. McConnell, J. Guarner, W. J. Shieh, J. M. Malecki, J. L. Gerberding, J. M. Hughes, B. A. Perkins and Anthrax Bioterrorism Investigation Team, Bioterrorism-Related Inhalational Anthrax: The First 10 Cases Reported in the United States, *Emerging Infect. Dis.*, 2001, **7**, 933–944.
132. I. J. Glomski, J. P. Corre, M. Mock and P. L. Goossens, Noncapsulated Toxinogenic Bacillus anthracis Presents a Specific Growth and Dissemination Pattern in Naive and Protective Antigen-Immune Mice, *Infect. Immun.*, 2007, **75**, 4754–4761.
133. I. J. Glomski, A. Piris-Gimenez, M. Huerre, M. Mock and P. L. Goossens, Primary Involvement of Pharynx and Peyer's Patch in Inhalational and Intestinal Anthrax, *PLoS Pathog.*, 2007, **3**, e76.
134. C. L. Loving, M. Kennett, G. M. Lee, V. K. Grippe and T. J. Merkel, Murine Aerosol Challenge Model of Anthrax, *Infect. Immun.*, 2007, **75**, 2689–2698.
135. P. Sanz, L. D. Teel, F. Alem, H. M. Carvalho, S. C. Darnell and A. D. O'Brien, Detection of Bacillus anthracis Spore Germination In Vivo by Bioluminescence Imaging, *Infect. Immun.*, 2008, **76**, 1036–1047.
136. A. A. Adalja, E. Toner and T. V. Inglesby, Clinical Management of Potential Bioterrorism-Related Conditions, *N. Engl. J. Med.*, 2015, **372**, 954–962.
137. P. C. Turnbull, Anthrax Vaccines: Past, Present and Future, *Vaccine*, 1991, **9**, 533–539.
138. E. N. Shlyakhov and E. Rubinstein, Human Live Anthrax Vaccine in the Former USSR, *Vaccine*, 1994, **12**, 727–730.
139. P. C. B. Turnbull, *Guidelines for the Surveillance and Control of Anthrax in Humans and Animals*, World Health Organization, 3rd edn, 1998, http://www.who.int/csr/resources/publications/anthrax/whoemczdi986text.pdf.
140. S. S. Lewerin, M. Elvander, T. Westermark, L. N. Hartzell, A. K. Norstrom, S. Ehrs, R. Knutsson, S. Englund, A. C. Andersson, M. Granberg, S. Backman, P. Wikstrom and K. Sandstedt, Anthrax Outbreak in a Swedish Beef Cattle Herd–1st Case in 27 Years: Case Report, *Acta Vet. Scand.*, 2010, **52**, 1–8.
141. M. Doganay and N. Aydin, Antimicrobial Susceptibility of Bacillus anthracis, *Scand. J. Infect. Dis.*, 1991, **23**, 333–335.
142. M. Z. Bakici, N. Elaldi, M. Bakir, I. Dokmetas, M. Erandac and M. Turan, Antimicrobial Susceptibility of Bacillus anthracis in an Endemic Area, *Scand. J. Infect. Dis.*, 2002, **34**, 564–566.
143. J. D. Cavallo, F. Ramisse, M. Girardet, J. Vaissaire, M. Mock and E. Hernandez, Antibiotic Susceptibilities of 96 Isolates of Bacillus anthracis Isolated in France Between 1994 and 2000, *Antimicrob. Agents Chemother.*, 2002, **46**, 2307–2309.

144. P. C. Turnbull, N. M. Sirianni, C. I. LeBron, M. N. Samaan, F. N. Sutton, A. E. Reyes and L. F. Peruski, Jr., MICs of Selected Antibiotics for Bacillus anthracis, Bacillus cereus, Bacillus thuringiensis and Bacillus mycoides from a Range of Clinical and Environmental Sources as Determined by the Etest, *J. Clin. Microbiol.*, 2004, **42**, 3626–3634.
145. V. A. Luna, D. S. King, J. Gulledge, A. C. Cannons, P. T. Amuso and J. Cattani, Susceptibility of Bacillus anthracis, Bacillus cereus, Bacillus mycoides, Bacillus pseudomycoides and Bacillus thuringiensis to 24 antimicrobials using Sensititre Automated Microbroth Dilution and Etest Agar Gradient Diffusion Methods, *J. Antimicrob. Chemother.*, 2007, **60**, 555–567.
146. W. D. Tigertt, Anthrax. William Smith Greenfield, M.D., F.R.C.P., Professor Superintendent, the Brown Animal Sanatory Institution (1878–81). Concerning the Priority due to Him for the Production of the First Vaccine Against Anthrax, *J. Hyg.*, 1980, **85**, 415–420.
147. L. Pasteur, De l'atténuation des virus et de leur retour à la virulence, *C. R. Acad. Sci., Ser. III*, 1881, **92**, 429–435.
148. M. Nass, Anthrax Vaccine. Model of a Response to the Biologic Warfare Threat, *Infect. Dis. Clin.*, 1999, **13**, 187–208.
149. D. A. Sweeney, C. W. Hicks, X. Cui, Y. Li and P. Q. Eichacker, Anthrax Infection, *Am. J. Respir. Crit. Care Med.*, 2011, **184**, 1333–1341.
150. J. M. Beierlein and A. C. Anderson, New Developments in Vaccines, Inhibitors of Anthrax Toxins and Antibiotic Therapeutics for Bacillus anthracis, *Curr. Med. Chem.*, 2011, **18**, 5083–5094.
151. L. Baillie, T. Townend, N. Walker, U. Eriksson and D. Williamson, Characterization of the Human Immune Response to the UK Anthrax Vaccine, *FEMS Immunol. Med. Microbiol.*, 2004, **42**, 267–270.
152. J. G. Wright, C. P Quinn, S. Shadomy and N. Messonnier, Use of Anthrax Vaccine in the United States, Recommendations of the Advisory Committee on Immunization Practices (ACIP), 2009, *Morb. Mort. Weekly Report*, 2010, **59**, 1–30.
153. S. L. Nightingale, J. M. Prasher and S. Simonson, Emergency Use Authorization (EUA) to Enable Use of Needed Products in Civilian and Military Emergencies, United States, *Emerging Infect. Dis.*, 2007, **13**, 1046–1051.
154. J. Vancurik, Causes of the Failure of Antibiotic Prophylaxis of Inhalation Anthrax and Clearance of the Spores from the Lungs, *Folia Microbiol.*, 1966, **11**, 459–464.
155. K. A. Hendricks, M. E. Wright, S. V. Shadomy, J. S. Bradley, M. G. Morrow, A. T. Pavia, E. Rubinstein, J. E. Holty, N. E. Messonnier, T. L. Smith, N. Pesik, T. A. Treadwell, W. A. Bower and the Workgroup on Anthrax Clinical Guidelinesl, Centers for Disease Control and Prevention Expert Panel Meetings on Prevention and Treatment of Anthrax in Adults, *Emerging Infect. Dis.*, 2014, **20**(2), e130687.
156. A. A. Adalja, Anthrax: Has the Clinical Milieu Changed Since 2001?, *J. Community Hosp. Intern. Med. Perspect.*, 2012, **2**, 18191.
157. FDA, *Approval Letter for Post-Exposure Prophylaxis in Adults (Levofloxacin)*, 2004, http://www.fda.gov/downloads/Drugs/EmergencyPreparedness/BioterrorismandDrugPreparedness/UCM133680.pdf.

158. FDA, *Approval Letter for Post-Exposure Prophylaxis in Pediatric Patients (Levofloxacin)*, 2008, http://www.accessdata.fda.gov/drugsatfda_docs/summary_review/2008/020634se5-047, 021721se5-015_SUMR.pdf.
159. C. Stroud, K. Viswanathan, T. Powell and R. R. Bass, *Prepositioning Antibiotics for Anthrax*, Institute of Medicine, The National Academies Press, Washington, DC, 2011.
160. M. K. Lalitha and M. K. Thomas, Penicillin Resistance in Bacillus anthracis, *Lancet*, 1997, **349**, 1522.
161. J. G. Bartlett, T. V. Inglesby, Jr. and L. Borio, Management of Anthrax, *Clin. Infect. Dis.*, 2002, **35**, 851–858.
162. N. Bradaric and V. Punda-Polic, Cutaneous Anthrax due to Penicillin-Resistant Bacillus anthracis Transmitted by an Insect Bite, *Lancet*, 1992, **340**, 306–307.
163. I. Brook, The Prophylaxis and Treatment of Anthrax, *Int. J. Antimicrob. Agents*, 2002, **20**, 320–325.
164. G. Ressel, CDC Updates Interim Guidelines for Anthrax Exposure Management and Antimicrobial Therapy, *Am. Fam. Physician*, 2001, **64**, 1901–1902, 1904.
165. D. Meaney-Delman, M. E. Zotti, A. A. Creanga, L. K. Misegades, E. Wako, T. A. Treadwell, N. E. Messonnier, D. J. Jamieson and Workgroup on Anthrax in Pregnant, Postpartum, Women, Special Considerations for Prophylaxis for and Treatment of Anthrax in Pregnant and Postpartum Women, *Emerging Infect. Dis.*, 2014, **20**(**2**), e130611.
166. J. S. Bradley, G. Peacock, S. E. Krug, W. A. Bower, A. C. Cohn, D. Meaney-Delman, A. T. Pavia, FAAP, FIDSA and AAP Committee on Infectious Diseases and Disaster Preparedness Advisory Council, Pediatric Anthrax Clinical Management: Executive Summary, *Pediatrics*, 2014, **133**, 940–942.
167. E. J. Stern, K. B. Uhde, S. V. Shadomy and N. Messonnier, Conference Report on Public Health and Clinical Guidelines for Anthrax, *Emerging Infect. Dis.*, 2008, **14**, DOI: 10.3201/eid1404.070969.
168. M. W. Odendaal, P. M. Pieterson, V. de Vos and A. D. Botha, The Antibiotic Sensitivity Patterns of Bacillus anthracis Isolated from the Kruger National Park, *Onderstepoort J. Vet. Res.*, 1991, **58**, 17–19.
169. M. N. Swartz, Recognition and Management of Anthrax–an Update, *N. Engl. J. Med.*, 2001, **345**, 1621–1626.
170. T. V. Inglesby, D. A. Henderson, J. G. Bartlett, M. S. Ascher, E. Eitzen, A. M. Friedlander, J. Hauer, J. McDade, M. T. Osterholm, T. O'Toole, G. Parker, T. M. Perl, P. K. Russell and K. Tonat, Anthrax as a Biological Weapon: Medical and Public Health Management. Working Group on Civilian Biodefense, *JAMA, J. Am. Med. Assoc.*, 1999, **281**, 1735–1745.
171. J. J. Sejvar, F. C. Tenover and D. S. Stephens, Management of Anthrax Meningitis, *Lancet Infect. Dis.*, 2005, **5**, 287–295.
172. M. A. Borg and W. C. Tieck, Jr., Responding to Anthrax, *J. Dent. Technol.*, 2001, **18**, 12–14, 28.

173. T. Berger, M. Kassirer and A. A. Aran, Injectional Anthrax – New Presentation of an Old Disease, *Euro. Surveill.*, 2014, **19**(**32**), 20877.
174. B. J. Parcell, A. D. Wilmshurst, A. J. France, L. Motta, T. Brooks and W. J. Olver, Injection Anthrax Causing Compartment Syndrome and Necrotising Fasciitis, *J. Clin. Pathol.*, 2011, **64**, 95–96.
175. Z. Chen, M. Moayeri and R. Purcell, Monoclonal Antibody Therapies Against Anthrax, *Toxins*, 2011, **3**, 1004–1019.
176. A. Schneemann and M. Manchester, Anti-Toxin Antibodies in Prophylaxis and Treatment of Inhalation Anthrax, *Future Microbiol.*, 2009, **4**, 35–43.
177. A. Casadevall, Passive Antibody Administration (Immediate Immunity) as a Specific Defense Against Biological Weapons, *Emerging Infect. Dis.*, 2002, **8**, 833–841.
178. C. E. Kummerfeldt, Raxibacumab: Potential Role in the Treatment of Inhalational Anthrax, *Infect. Drug Resist.*, 2014, **7**, 101–109.
179. T. S. Migone, G. M. Subramanian, J. Zhong, L. M. Healey, A. Corey, M. Devalaraja, L. Lo, S. Ullrich, J. Zimmerman, A. Chen, M. Lewis, G. Meister, K. Gillum, D. Sanford, J. Mott and S. D. Bolmer, Raxibacumab for the Treatment of Inhalational Anthrax, *N. Engl. J. Med.*, 2009, **361**, 135–144.
180. K. B. Olson, Aum Shinrikyo: Once and Future Threat?, *Emerging Infect. Dis.*, 1999, **5**, 513–516.
181. S. C. Suffin, W. H. Carnes and A. F. Kaufmann, Inhalation Anthrax in a Home Craftsman, *Hum. Pathol.*, 1978, **9**, 594–597.
182. K. Schmitt and N. A. Zacchia, Total Decontamination Cost of the Anthrax Letter Attacks, *Biosecur. Bioterrorism*, 2012, **10**, 98–107.
183. T. K. Sell and M. Watson, Federal Agency Biodefense Funding, FY2013-FY2014, *Biosecur. Bioterrorism*, 2013, **11**, 196–216.
184. U.S. Department of Health & Human Services, *Public Health Emergencies*, http://www.phe.gov/preparedness/Pages/default.aspx.
185. U.S. Department of Health & Human Services, *PHEMCE, Public Health Emergency Medical Countermeasures Enterprise*, http://www.phe.gov/Preparedness/mcm/phemce/Pages/strategy.aspx.
186. CDC, *Emergency Preparedness and Response*, http://emergency.cdc.gov/bioterrorism/prep.asp.
187. ICLN, *Integrated Consortium of laboratory Networks*, https://www.icln.org/.
188. M. Magnuson, H. Ernst, J. Griggs, S. Fitz-James, L. Mapp, M. Mullins, T. Nichols, S. Shah, T. Smith and E. Hedrick, Analysis of Environmental Contamination Resulting from Catastrophic Incidents: Part 1. Building and Sustaining Capacity in Laboratory Networks, *Environ. Int.*, 2014, **72**, 83–89.
189. M. Magnuson, R. Campisano, J. Griggs, S. Fitz-James, K. Hall, L. Mapp, M. Mullins, T. Nichols, S. Shah, E. Silvestri, T. Smith, S. Willison and H. Ernst, Analysis of Environmental Contamination Resulting from Catastrophic Incidents: Part 2. Building Laboratory Capability by Selecting and Developing Analytical Methodologies, *Environ. Int.*, 2014, **72**, 90–97.

190. L. M. Irenge and J. L. Gala, Rapid Detection Methods for Bacillus anthracis in Environmental Samples: A Review, *Appl. Microbiol. Biotechnol.*, 2012, **93**, 1411–1422.
191. S. S. Rao, K. V. Mohan and C. D. Atreya, Detection Technologies for Bacillus anthracis: Prospects and Challenges, *J. Microbiol. Methods*, 2010, **82**, 1–10.
192. R. F. Knisely, Selective Medium for Bacillus anthracis, *J. Bacteriol.*, 1966, **92**, 784–786.
193. A. R. Hoffmaster, J. Ravel, D. A. Rasko, G. D. Chapman, M. D. Chute, C. K. Marston, B. K. De, C. T. Sacchi, C. Fitzgerald, L. W. Mayer, M. C. Maiden, F. G. Priest, M. Barker, L. Jiang, R. Z. Cer, J. Rilstone, S. N. Peterson, R. S. Weyant, D. R. Galloway, T. D. Read, T. Popovic and C. M. Fraser, Identification of Anthrax Toxin Genes in a Bacillus cereus Associated with an Illness Resembling Inhalation Anthrax, *Proc. Natl. Acad. Sci. U. S. A.*, 2004, **101**, 8449–8454.
194. S. E. Letant, G. A. Murphy, T. M. Alfaro, J. R. Avila, S. R. Kane, E. Raber, T. M. Bunt and S. R. Shah, Rapid-Viability PCR Method for Detection of Live, Virulent Bacillus anthracis in Environmental Samples, *Appl. Environ. Microbiol.*, 2011, **77**, 6570–6578.
195. S. Kane, S. Shah, S. Létant, G. Murphy, T. Alfaro, J. Avila, E. Salazar, M. Mullins and T. Nichols, Operational Evaluation of the Rapid Viability PCR Method for PostDecontamination Clearance Sampling, *J. Bioterrorism Biodef.*, 2013, **S3**, 016.
196. N. Morel, H. Volland, J. Dano, P. Lamourette, P. Sylvestre, M. Mock and C. Creminon, Fast and Sensitive Detection of Bacillus anthracis Spores by Immunoassay, *Appl. Environ. Microbiol.*, 2012, **78**, 6491–6498.
197. R. J. Sharp and A. G. Roberts, Anthrax: The Challenges for Decontamination, *J. Chem. Technol. Biotechnol.*, 2006, **81**, 1612–1625.
198. H. Kennedy, *New York Daily News*, New York, 2001.
199. D. A. Canter, Remediating Anthrax-Contaminated Sites: Learning from the Past to Protect the Future, *Chem. Health Saf.*, 2005, **12**, 13–19.
200. U.S. Government Accountability Office, *Capitol Hill Anthrax Incident: EPA's Cleanup Was Successful; Opportunities Exist to Enhance Contract Oversight, GAO-03-686*, U.S. Government Accountability Office, Washington, DC, 2003.
201. W. T. Sanderson, R. R. Stoddard, A. S. Echt, C. A. Piacitelli, D. Kim, J. Horan, M. M. Davies, R. E. McCleery, P. Muller, T. M. Schnorr, E. M. Ward and T. R. Hales, Bacillus anthracis Contamination and Inhalational Anthrax in a Mail Processing and Distribution Center, *J. Appl. Microbiol.*, 2004, **96**, 1048–1056.
202. P. K. Dewan, A. M. Fry, K. Laserson, B. C. Tierney, C. P. Quinn, J. A. Hayslett, L. N. Broyles, A. Shane, K. L. Winthrop, I. Walks, L. Siegel, T. Hales, V. A. Semenova, S. Romero-Steiner, C. Elie, R. Khabbaz, A. S. Khan, R. A. Hajjeh and A. Schuchat, Inhalational Anthrax Outbreak Among Postal Workers, Washington, DC, 2001, *Emerging Infect. Dis.*, 2002, **8**, 1066–1072.

203. C. Strohm, Washington Postal Facility Anthrax-Free, but Psychological Specter Looms, in *GovernmentExecutive.com*, 2003. http://www.govexec.com/dailyfed/1103/111203c1.htm.
204. D. J. Valiante, D. P. Schill, E. A. Bresnitz, G. A. Burr and K. R. Mead, Responding to a Bioterrorist Attack: Environmental Investigation of Anthrax in New Jersey, *Appl. Occup. Environ. Hyg.*, 2003, **18**, 780–785.
205. B. Coffin and T. Sgroi, Anthrax Clean-up at the State Department, *Risk Management Magazine*, 2003, **50**, 40–43. http://cf.rims.org/Magazine/PrintTemplate.cfm?AID=2210.
206. P. N. Price, K. Hamachi, J. McWilliams and M. D. Sohn, *Anthrax Sampling and Decontamination: Technology Trade-Offs*, Lawrence Berkeley National Laboratory, Environmental Energy Technologies Division, Berkeley, CA, 2008, https://eaei.lbl.gov/sites/all/files/lbnl-1519e_2.pdf.
207. E. A. S. Whitney, M. E. Beatty, T. H. Taylor, R. Weyant, J. Sobel, M. J. Arduino and D. A. Ashford, Inactivation of Bacillus anthracis Spores, *Emerging Infect. Dis.*, 2003, **9**, 623–627.
208. W. Rutala, *Disinfection, Sterilization and Antisepsis: Principles, Practices, Current Issues and New Research*, Association for Professionals in Infection Control and Epidemiology, Washington, DC, 2007.
209. W. A. N. Dorland, *Dorland's Illustrated Medical Dictionary (Standard Version)*, ed. W. E. Dorland, W.B. Saunders & Co, Philadelphia, 29th edn, 2000.
210. G. B. Martin, *Practical Experiences with Technologies for Decontamination of B. anthracis in Large Buildings*, AWMA/EPA Indoor Air Quality Problems and Engineering Solutions Specialty Conference and Exhibition, Research Triangle Park, N.C., 2003, http://cfpub.epa.gov/si/si_public_record_Report.cfm?dirEntryID=119258.
211. D. A. Canter, *Remediating Sites with Anthrax Contamination: Building on Experience*, AWMA/EPA Indoor Air Quality Problems and Engineering Solutions Specialty Conference and Exhibition, Research Triangle Park, N.C., July 21–23, 2003.
212. R. P. Schaudies and D. A. Robinson, *Analysis of Chlorine Dioxide Remediation of Washington, DC Bacillus anthracis Contamination*, EPA/903/R-03/013, U.S. Environmental Protection Agency, Washington, DC, 2003.
213. U.S. Environmental Protection Agency, *Bio-Response Operational Testing and Evaluation (BOTE) Project – Phase 1: Decontamination Assessment*. EPA/600/R-13/168, U.S. Environmental Protection Agency, Washington, DC, 2013, http://cfpub.epa.gov/si/si_public_file_download.cfm?p_download_id=517087.
214. U.S. Environmental Protection Agency, *Expedient Approaches for the Management of Wastes Generated from Biological Decontamination Operations in an Indoor Environment – Evaluation of Waste Sampling and Decontamination Procedures*, EPA/600/R-14/262, U.S. Environmental Protection Agency, Washington DC, 2014, http://cfpub.epa.gov/si/si_public_file_download.cfm?p_download_id=521653.

215. A. M. Lesperance, S. Stein, J. F. Upton and C. Toomey, Challenges in Disposing of anthrax Waste, *Biosecur. Bioterrorism*, 2011, **9**, 310–314.
216. AOAC International, *Official Methods of Analysis*, AOAC International, Gaithersburg, MD, 18th edn, 2006.
217. U.S. Environmental Protection Agency, *Product Performance Test Guidelines – OPPTS 810.2100: Sterilants – Efficacy Data Recommendations*, EPA/712/C-07/056, U.S. Environmental Protection Agency, Washington, DC, 2007, http://www.regulations.gov/#!documentDetail;D=EPA-HQ-OPPT-2009–0150–0020.
218. S. F. Tomasino, R. M. Pines, M. P. Cottrill and M. A. Hamilton, Determining the Efficacy of Liquid Sporicides Against Spores of Bacillus subtilis on a Hard Nonporous Surface Using the Quantitative Three Step Method: Collaborative Study, *J. AOAC Int.*, 2008, **91**, 833–852.
219. US Environmental Protection Agency, *Guidance on Test Methods for Demonstrating the Efficacy of Antimicrobial Products for Inactivating Bacillus anthracis Spores on Environmental Surfaces*, Presented in Part at the FIFRA Scientific Advisory Panel Meeting, Arlington, VA, 2007, U.S. Environmental Protection Agency, Washington, DC, 2007, http://archive.epa.gov/scipoly/sap/meetings/web/pdf/agenda-4.pdf.
220. M. Favero, *Dininfection, sterilization and antisepsis in health care*, ed. W. Rutala, Association for Professionals in Infection Control and Epidemiology, Washington, DC, 1998, pp. 119–132.
221. International Organization for Standardization, Geneva, 2006, vol. ISO/TS 11139:2006.
222. D. A. Canter, Addressing Residual Risk Issues at Anthrax Cleanups: How Clean is Safe?, *J. Toxicol. Environ. Health*, 2005, **68**, 1017–1032.
223. U.S. Department of Homeland Security and U. S. Environmental Protection Agency, *Draft Planning Guidance for Recovery Following Biological Incidents*, Washington, DC, 2009, http://www.frwa.net/uploads/4/2/3/5/42359811/guidanceforrecoveryfollowingbioincident092509.pdf.
224. J. V. Rogers, Y. W. Choi, W. R. Richter, D. C. Rudnicki, D. W. Joseph, C. L. K. Sabourin, M. L. Taylor and J. C. S. Chang, Formaldehyde Gas Inactivation of Bacillus anthracis, Bacillus subtilis and Geobacillus stearothermophilus Spores on Indoor Surface Materials, *J. Appl. Microbiol.*, 2007, **103**, 1104–1112.
225. U.S. Environmental Protection Agency, *CDG Research Corp. Bench-Scale Chlorine Dioxide Gas: Solid Generator*, EPA/600/R-11/199, U.S. Environmental Protection Agency, Washington, DC, 2004, http://cfpub.epa.gov/si/si_public_file_download.cfm?p_download_id=504529.
226. U.S. Environmental Protection Agency, *BIOQUELL, Inc. CLARUS C Hydrogen Peroxide Gas Generator*, EPA/600/R-11/200, U.S. Environmental Protection Agency, Washington, DC, 2004, http://cfpub.epa.gov/si/si_public_file_download.cfm?p_download_id=504534.
227. U.S. Environmental Protection Agency, *CERTEK, Inc. 1414RH Formaldehyde Generator/Neutralizer*, EPA/600/R-04/150, 2004, U.S. Environmental

Protection Agency, Washington, DC, 2004, http://cfpub.epa.gov/si/si_public_record_Report.cfm?dirEntryID=87659.

228. U.S. Environmental Protection Agency, *Evaluation of Spray-Applied Sporicidal Decontamination Technologies*, EPA/600/R-06/146, U.S. Environmental Protection Agency, Washington, DC, 2006, http://cfpub.epa.gov/si/si_public_record_report.cfm?subject=Homeland%20Security%20Research&dirEntryId=160628.
229. U.S. Environmental Protection Agency, *Evaluation of Liquid and Foam Technologies for the Decontamination of B. anthracis and B. subtilis Spores on Building and Outdoor Materials*, EPA 600/R-09/150, U.S. Environmental Protection Agency, Washington, DC, 2009, http://cfpub.epa.gov/si/si_public_record_report.cfm?dirEntryId=212543.
230. J. P. Wood, M. W. Calfee, M. Clayton, N. Griffin-Gatchalian and A. Touati, Optimizing Acidified Bleach Solutions to Improve Sporicidal Efficacy on Building Materials, *Lett. Appl. Microbiol.*, 2011, **53**, 668–672.
231. U.S. Environmental Protection Agency, *Systematic Investigation of Liquid and Fumigant Decontamination Efficacy Against Biological Agents Deposited on Test Coupons of Common Indoor Materials*, EPA/600/R-11/076, U.S. Environmental Protection Agency, Washington, DC, 2011, http://cfpub.epa.gov/si/si_public_file_download.cfm?p_download_id=502793.
232. J. P. Wood, Y. W. Choi, J. V. Rogers, T. J. Kelly, K. B. Riggs and Z. J. Willenberg, Efficacy of Liquid Spray Decontaminants for Inactivation of Bacillus anthracis Spores on Building and Outdoor Materials, *J. Appl. Microbiol.*, 2011, **110**, 1262–1273.
233. M. W. Calfee, Y. Choi, J. Rogers, T. Kelly, Z. Willenberg and K. Riggs, Lab-Scale Assessment to Support Remediation of Outdoor Surfaces Contaminated with Bacillus anthracis Spores, *J. Bioterrorism Biodef.*, 2011, **2**, 110.
234. S. F. Tomasino, V. K. Rastogi, L. Wallace, L. S. Smith, M. A. Hamilton and R. M. Pines, Use of Alternative Carrier Materials in AOAC Official Method(SM) 2008.05, Efficacy of Liquid Sporicides Against Spores of Bacillus subtilis on a Hard, Nonporous Surface, Quantitative Three-Step Method, *J. AOAC Int.*, 2010, **93**, 259–276.
235. V. K. Rastogi, L. Wallace, L. S. Smith, S. P. Ryan and B. Martin, Quantitative Method To Determine Sporicidal Decontamination of Building Surfaces by Gaseous Fumigants and Issues Related to Laboratory-Scale Studies, *Appl. Environ. Microbiol.*, 2009, **75**, 3688–3694.
236. U.S. Environmental Protection Agency, *Determining the Efficacy of Liquids and Fumigants in Systematic Decontamination Studies for Bacillus anthracis Using Multiple Test Methods*, EPA/600/R-10/088, U.S. Environmental Protection Agency, Washington, DC, 2010, http://cfpub.epa.gov/si/si_public_file_download.cfm?p_download_id=518018.
237. S. P. Ryan, S. D. Lee, M. W. Calfee, J. P. Wood, S. McDonald, M. Clayton, N. Griffin-Gatchalian, A. Touati, L. Smith and M. Nysewander, Effect of Inoculation Method on the Determination of Decontamination Efficacy Against Bacillus spores, *World J. Microbiol. Biotechnol.*, 2014, **30**, 2609–2623.

238. V. K. Rastogi, S. P. Ryan, L. Wallace, L. S. Smith, S. S. Shah and G. B. Martin, Systemic Evaluation of the Efficacy of Chlorine Dioxide in Decontamination of Building Interior Surfaces Contaminated with Anthrax Spores, *Appl. Environ. Microbiol.*, 2010, **76**, 3343–3351.
239. S. Ryan, V. K. Rastogi, L. Wallace, G. B. Martin, L. S. Smith, S. S. Shah and P. Clark, *A study on the Use of Biological Indicators in Determining the Efficacy of the Decontamination of Building Materials Contaminated with Bacillus anthracis*, National Conference on Environmental Sampling and Detection for Bio Threat Agents, Brooklyn, NY, October 25–27, 2006.
240. U.S. Environmental Protection Agency, *Decontamination Process Indicators: Biological Indicators*, EPA/600/R-14/239, U.S. Environmental Protection Agency, Washington DC, 2014, http://cfpub.epa.gov/si/si_public_file_download.cfm?p_download_id=521475.
241. U.S. Environmental Protection Agency, *Destruction of Spores on Building Decontamination Residue in a Commercial Autoclave*, EPA/600/R-05/081, U.S. Environmental Protection Agency, Washington, DC, 2005, http://nepis.epa.gov/Exe/ZyPDF.cgi?Dockey=90150O00.PDF.
242. P. Lemieux, *Air and Waste Management Association Annual Meeting*, AWMA, Minneapolis, MN, June 21–24, 2005, http://cfpub.epa.gov/si/si_public_record_report.cfm?dirEntryId=95801&fed_org_id=1253&address=nhsrc/si/&view=desc&sortBy=pubDateYear&showCriteria=1&count=25&searchall=%27indoor%20outdoor%20decontamination%27.
243. P. Lemieux, R. Sieber, A. Osborne and A. Woodard, Destruction of Spores on Building Decontamination Residue in a Commercial Autoclave, *Appl. Environ. Microbiol.*, 2006, **72**, 7687–7693.
244. J. P. Wood, P. Lemieux, D. Betancourt, P. Kariher and N. G. Gatchalian, Dry Thermal Resistance of Bacillus anthracis (Sterne) Spores and Spores of Other Bacillus Species: Implications for Biological Agent Destruction via Waste Incineration, *J. Appl. Microbiol.*, 2010, **109**, 99–106.
245. J. A. Tufts and J. A. Rosati, Thermal Inactivation of Bacillus anthracis Surrogate Spores in a Bench-Scale Enclosed Landfill Gas Flare, *J. Air Waste Manage. Assoc.*, 2012, **62**, 151–159.
246. P. Lemieux, S. Thorneloe, K. Nickel and M. Rogers, *Air and Waste Management Association Annual Meeting*, AWMA, New Orleans, LA, 2006, http://cfpub.epa.gov/si/si_public_record_report.cfm?dirEntryId=151404&fed_org_id=1253&address=nhsrc/si/&view=desc&sortBy=pubDateYear&showCriteria=1&count=25&searchall=%27indoor%20outdoor%20decontamination%27.
247. M. W. Calfee, S. P. Ryan, J. P. Wood, L. Mickelsen, C. Kempter, L. Miller, M. Colby, A. Touati, M. Clayton, N. Griffin-Gatchalian, S. McDonald and R. Delafield, Laboratory Evaluation of Large-Scale Decontamination Approaches, *J. Appl. Microbiol.*, 2012, **112**, 874–882.

248. U.S. Environmental Protection Agency, *Effectiveness of Physical and Chemical Cleaning and Disinfection Methods for Removing, Reducing or Inactivating Agricultural Biological Threat Agents*, EPA/600/R-11/092, U.S. Environmental Protection Agency, Washington, DC, 2011.
249. U.S. Environmental Protection Agency, *Assessment of Liquid and Physical Decontamination Methods for Environmental Surfaces Contaminated with Bacterial Spores: Evaluation of Spray Method Parameters and Impact of Surface Grime*, EPA/600/S-13/064, U.S. Environmental Protection Agency, Washington, DC, 2012, http://cfpub.epa.gov/si/si_public_record_report.cfm?address=nhsrc/si/&dirEntryId=254661.
250. U.S. Environmental Protection Agency, *Assessment of Liquid and Physical Decontamination Methods for Environmental Surfaces Contaminated with Bacterial Spores: Development and Evaluation of the Decontamination Procedural Steps*, EPA/600/R-12/025, U.S. Environmental Protection Agency, Washington, DC, 2012, http://cfpub.epa.gov/si/si_public_record_report.cfm?address=nhsrc/&dirEntryId=241144.
251. U.S. Environmental Protection Agency, *Inactivation of Bacillus anthracis Spores in Soil Matrices with Chlorine Dioxide Gas*, EPA/600/R-12/517, U.S. Environmental Protection Agency, Washington DC, 2012, http://cfpub.epa.gov/si/si_public_file_download.cfm?p_download_id=519063.
252. U.S. Environmental Protection Agency, *Evaluation of Chlorine Dioxide Gas and Peracetic Acid Fog for the Decontamination of a Mock Heating, Ventilation and Air Conditioning Duct System*, EPA/600/R-14/014, U.S. Environmental Protection Agency, Washington, DC, 2014, http://cfpub.epa.gov/si/si_public_file_download.cfm?p_download_id=517856.
253. K. M. Meyer, M. W. Calfee, J. P. Wood, L. Mickelsen, B. Attwood, M. Clayton, A. Touati and R. Delafield, Fumigation of a Lab-Scale HVAC System with Hydrogen Peroxide for Decontamination Following a Biological Contamination Incident, *J. Appl. Microbiol.*, 2014, **116**, 533–541.
254. U.S. Environmental Protection Agency, *Evaluation of Hydrogen Peroxide Fumigation for HVAC Decontamination*, EPA/600/R/12/586, U.S. Environmental Protection Agency, Washington, DC, 2012, http://cfpub.epa.gov/si/si_public_file_download.cfm?p_download_id=507840.
255. U.S. Environmental Protection Agency, *Material Demand Studies: Materials Sorption of Vaporized Hydrogen Peroxide*, EPA/600/R-10/002, U.S. Environmental Protection Agency, Washington, DC, 2010, http://cfpub.epa.gov/si/si_public_record_report.cfm?subject=Homeland%20Security%20Research&dirEntryId=216864.
256. U.S. Environmental Protection Agency, *Material Demand Studies: Interaction of Chlorine Dioxide Gas with Building Materials*, EPA/600/R-08/091, U.S. Environmental Protection Agency, Washington, DC, 2008, http://cfpub.epa.gov/si/si_public_record_report.cfm?subject=Homeland%20Security%20Research&dirEntryId=198943.

257. U.S. Environmental Protection Agency, *Compatibility of Material and Electronic Equipment with Ethylene Oxide Fumigation*, EPA/600/R-14/399, U.S. Environmental Protection Agency, Washington, DC, 2014, http://cfpub.epa.gov/si/si_public_file_download.cfm?p_download_id=520698.
258. T. L. Buhr, A. A. Young, Z. A. Minter, C. M. Wells, D. C. McPherson, C. L. Hooban, C. A. Johnson, E. J. Prokop and J. R. Crigler, Test Method Development to Evaluate Hot, Humid Air Decontamination of Materials Contaminated with Bacillus anthracis Sterne and B. thuringiensis Al Hakam Spores, *J. Appl. Microbiol.*, 2012, **113**, 1037–1051.
259. E. J. Prokop, J. R. Crigler, C. M. Wells, A. A. Young and T. L. Buhr, Response Surface Modeling for Hot, Humid Air Decontamination of Materials Contaminated with Bacillus anthracis Sterne and Bacillus thuringiensis Al Hakam Spores, *AMB Express*, 2014, **4**, 21.
260. J. P. Wood and G. Blair Martin, Development and Field Testing of a Mobile Chlorine Dioxide Generation System for the Decontamination of Buildings Contaminated with Bacillus anthracis, *J. Hazard. Mater.*, 2009, **164**, 1460–1467.
261. J. P. Wood, S. P. Ryan, E. G. Snyder, S. D. Serre, A. Touati and M. J. Clayton, Adsorption of Chlorine Dioxide Gas on Activated Carbons, *J. Air Waste Manage. Assoc.*, 2010, **60**, 898–906.
262. J. P. Wood, M. J. Clayton, T. McArthur, S. D. Serre, L. Mickelsen and A. Touati, Capture of Methyl Bromide Emissions with Activated Carbon Following the Fumigation of a Small Building Contaminated with a Bacillus anthracis Spore Simulant, *J. Air Waste Manage. Assoc.*, 2015, **65**, 145–153.
263. L. L. Nuffer, N. L. Hassig, L. H. Sego, B. A. Pulsipher, J. E. Wilson and B. Matzke, *Validation of Statistical Sampling Algorithms in Visual Sample Plan (VSP)-Summary Report*, Report# PNNL-18253, Pacific Northwest National Laboratory, Richland, Washington, 2009.
264. R. G. Knowlton, B. J. Melton, P. D. Finley and J. L. Ramsey, *User's Manual for the BROOM Decision Support System*, Report# SAND2012–8052, Sandia National Laboratory, Albuquerque, New Mexico, 2012, http://prod.sandia.gov/techlib/access-control.cgi/2012/128052.pdf.
265. R. G. Knowlton, B. J. Melton and R. J. Anderson, *Phase 1 Development Report for the SESSA Toolkit*, Report# SAND2014–17383, Sandia National Laboratory, Albuquerque, New Mexico and Livermore, California, 2014, http://prod.sandia.gov/techlib/access-control.cgi/2014/1417383.pdf.
266. U. S. Environmental Protection Agency, *Report on the 2013 U.S. Environmental Protection Agency (EPA) International Decontamination Research and Development Conference*, EPA/600/R-14/210, U.S. Environmental Protection Agency, Office of Research and Development, Washington, DC, 2014, http://cfpub.epa.gov/si/si_public_file_download.cfm?p_download_id=520120.

267. U. S. Environmental Protection Agency, *Incident Waste Decision Support Tool (I-WASTE), v.6.3*, EPA/600/R-13/160, U.S. Environmental Protection Agency, Office of Research and Development, Washington, DC, 2013, http://cfpub.epa.gov/si/si_public_file_download.cfm?p_download_id=515891.
268. C. G. Campbell, R. D. Kirvel, A. H. Love, C. G. Bailey, R. Miles, J. Schweickert, M. Sutton and E. Raber, Decontamination After a Release of B. anthracis Spores, *Biosecur. Bioterrorism*, 2012, **10**, 108–122.

CHAPTER 8

Detection of Airborne Pathogens and Toxins

POLONCA TREBŠE[a], OLGA MALEV[b], AND SIDNEY A. KATZ*[c]

[a]Faculty of Health Sciences, University of Ljubljana, 1000 Ljubljana, Slovenia; [b]Clinical Research Department, Children's Hospital Srebrnjak, 10000 Zagreb, Croatia; [c]Chemistry Faculty, Rutgers University, 450 Kingston Drive, Cherry Hill, NJ 08034, USA
*E-mail: skatz@camden.rutgers.edu

8.1 Introduction

Classically, the detection and characterization of pathogens and toxins in the atmosphere was accomplished by the collection of aerosol particulate matter, concentration of the microbiological components and incubation on an agar medium followed by observations of growth, morphology, stain uptake and reactions with biochemical reagents. These procedures allowed public health agencies to assess the microbiological components of the ambient air and to recommend countermeasures for avoiding exposures to airborne pathogens and toxins and also for prescribing treatments to combat the infections they cause. This classical approach has been refined by incorporating spectrochemical and chromatographic modules into offline monitoring systems for airborne pathogens and toxins and by utilizing genomic, proteomic and immunochemical techniques to enhance the online, automated identification of these pathogens and toxins.

Issues in Toxicology No. 25
Aerobiology: The Toxicology of Airborne Pathogens and Toxins
Edited by Harry Salem and Sidney Katz

Published by the Royal Society of Chemistry, www.rsc.org

8.2 Sample Collection

While clinical signs and symptoms are often characteristic manifestations of infections with airborne pathogens and intoxications with airborne toxins, the processes and procedures for their detection usually begin with the collection of an appropriate sample. This sample must be representative of the atmosphere under investigation, free of contamination and sufficiently stable to ensure that the results of its evaluation accurately reflect the composition of the atmosphere under investigation. Very often this means that the sample must be collected and manipulated under antiseptic or sterile conditions. The inclusion of replicate positive and negative controls is an important aspect of the quality assurance/quality control (QA/QC) protocols for the determination of airborne pathogens and toxins. A variety of sample collection procedures have been reported, ranging from settling under the influence of gravity to forced air filtration, impaction or impingement. Jensen and Schafer[1] described the general principles of bioaerosol collection and the theoretical aspects of sampling bioaerosols were discussed by Cox and Wathes[2] and Yang and Heinsohn.[3]

Durand *et al.*[4] addressed the question of airborne fungal spore and bacterial cell viability during long-term sampling. Six samplers were placed side-by-side at two locations in each of three composting facilities. Air samples were collected for 2 h on 0.8 μm polycarbonate membrane filters in polystyrene cassettes at a flow rate of 2 L min^{-1}. After this time, two of the cassettes at each location were capped and high-efficiency particulate adsorption (HEPA) filters were attached to the intakes of the remaining four cassettes so that additional clean air could be drawn through these filters for a further 2 h without depositing additional biomaterials. After 2 h, two of the remaining four cassettes were capped and additional clean air was drawn through the remaining two filters for a further 5 h. It was expected that all six of the cassettes from a given location would contain the same amounts of viable fungal and bacteriological material if there were no effects from the prolonged exposures to air. Analysis of the results showed that this was the case.

In order to collect data for guidance in the establishment of indoor air quality standards, Qudiesat *et al.*[5] measured the quality and the quantity of airborne pathogens in indoor air at two selected Jordanian hospitals in Zarqa city. Using a microbiological air sampler based on the Andersen impaction principle, 500 L air samples were collected from intensive care units, operating rooms, neonatal wards, patient rooms and main entrances at both hospitals. Three air samples were taken at each location for three different periods (10:00–12:00, 14:00–16:00 and 19:00–21:00 h). The sampler was operated at a flow rate of 100 L min^{-1} for 5 min (total volume 500 L for each sample and site) to avoid drying of the agar surface and overloading of the collection plate. The air sampler was set up at a height representative of the normal human breathing zone, 1.5 m above floor level. The sampler was cleaned by swabbing with 70% ethanol between measurements. Nine bacterial genera and four fungal genera were isolated and identified from indoor air in the hospitals in Zarqa city. The type of hospital was a significant factor influencing the number of indoor air microorganisms because of differences in the age of the hospital buildings, number of beds and disinfection strategies. Bacterial concentrations were more sensitive to the type of unit inside the hospital than were the fungal concentrations. Limitations on visitation times resulted in increases in the number of people in the hospital building for short periods and consequently increased the airborne microbial concentrations during visitation times. The type of hospital, the type of room and the time of sampling were three factors that affected the indoor airborne microbial concentrations. Qudiesat *et al.*[5] concluded that more studies were required to establish a surveillance base on the Jordanian healthcare facilities and to investigate the antibiotic-resistance patterns of the airborne microorganisms they isolated. They recommended the implementation of more stringent, frequent and comprehensive disinfection procedures in all healthcare facilities in Jordan.

Verreault *et al.*[6] reviewed the sampling techniques for airborne viruses. The operation of solid and liquid impactors, filters and electrostatic precipitators was described and the details of numerous virus aerosol sampling studies reported for the 70 year period from 1936 to 2007 were tabulated. It was concluded that although the techniques have been improved greatly over the years and were definitely better equipped at the time of the review to tackle the important health issue of airborne viruses, the lack of standardization must be addressed, as it limited the development of general recommendations for sampling of airborne viruses. Given the wide range of aerodynamic properties of airborne viruses, which can be from nanometers to micrometers in size, the issue of standardization was of the utmost importance. The detection of viruses in air samples depended on the type of aerosol and the sampling and analytical methodologies. Studies to date had rarely included quantitative analyses of total viral load. Although culture was often used to determine virus concentrations, most sampling methods affected viral infectivity, making culture inadequate for calculating the true concentrations of infectious airborne viruses. Although filters caused more damage to viruses than other methods, they are more efficient for determining viral loads in aerosols. Lastly, viruses were an astonishingly diverse group of microorganisms and this diversity had to be taken into consideration to select the most appropriate sampling devices. It was concluded that the consideration of a representative sample containing nanoparticles together with larger airborne particles was important whatever techniques or recommendations were proposed for sampling virus aerosols. Future studies would contribute to predicting better the potential risk of infection by airborne viruses.

To overcome difficulties encountered in correlating airborne levels of bacteria and fungi with epidemiological data, the ASTM[7] designed a standard practice for sampling airborne microorganisms at municipal solid-waste processing facilities and also at other locations. This was done to facilitate the formation of a database that would aid in the assessment of airborne microorganisms in public and occupational health issues. It was intended that the use of this practice would improve sampling precision and thereby facilitate comparisons between sampling operations. This practice applied only to sampling airborne bacteria and fungi, not viruses, which was significantly more difficult. Reliable methods of sampling viruses were not yet available.

An earlier edition of the current ASTM standard practice was recommended by the US Environmental Protection Agency (EPA).[8] This earlier standard allowed the determination of the concentrations of selected airborne bacteria and fungi using both liquid impinger and multistage impactor samplers. It contained recommendations for selecting the sampling locations, the number of samples to be determined, the types of microorganisms to be sampled, sample preservation and the intervals between sample collection and analysis. *Aspergillus flavus*, *A. fumigatus* and *A. niger*, were among the fungi identified and quantified using this standard practice. This standard practice was also applied to bacteria such as *Staphylococcus aureus*, *Klebsiella* species, *Mycobacterium* species and other members of the order Actinomycetales.

The Andersen N-6 single-stage sampler or its equivalent is specified in NIOSH Method 0800[9] for collecting indoor air samples for the identification of thermophilic actinomycetes. For the identification of airborne fungi and mesophilic bacteria, this standard specifies collection of the sample with the Andersen two-stage cascade impactor or its equivalent. The Andersen impactor is specified in NIOSH Method 0801[10] for collecting air samples for the identification of Gram-positive and Gram-negative bacteria and *Mycobacterium* species by gas chromatography of characteristic fatty acid methyl esters (see Section 8.6). In NIOSH Method 0900,[11] samples for the identification of airborne *Mycobacterium tuberculosis* are collected with PTFE filters.

Corzo *et al.*[12] investigated the role that infectious aerosols play in the transmission of swine influenza A virus, paying particular to attention to detecting the virus in air samples collected in swine farms and downwind from them and to providing an estimate of quantities of viral load found in swine farm aerosols under field conditions. Air samples were collected using a liquid cyclonic collector capable of processing 400 L min^{-1} of air. A 10 mL volume of an Eagle's minimum essential medium supplemented with 4% bovine serum albumin was added to the collection vessel of the liquid cyclonic collector. The cyclonic collector was run for 30 min, allowing airborne particles to be mixed with the collection medium. Once air sampling was completed, a sterile syringe was used to recover the sample and place it in a sterile plastic vial. Air samples were then stored on ice until they were transported to the laboratory for examination. The device was cleaned and disinfected by spraying alkyldimethylbenzylammonium chloride on the turbine and the collection vessel. These two surfaces were then sprayed with water to remove remaining disinfectant and dried with paper towels. Upon arrival at the farm and confirming the presence of clinical respiratory disease, the first set of samples was collected inside the barn. Fifteen 30 min air samples were collected by simultaneously placing four or five cyclonic collectors equally distributed throughout the barn. Cyclonic collectors were placed 1.5 m above the floor and 1 m below the ceiling and secured to a feed line using rubber bungee cords. Pigs did not have direct access to the air collection devices. Immediately after collecting inside air samples, samples of the air exhausted to the outside of the barn were collected. Fifteen 30 min samples were collected by placing the cyclonic collectors as close as possible to the draft of air exhausted from the pig barn. Cyclonic collectors were placed either on the ground when samples were collected from exhaust manure pit fans or were hung from a tripod when samples were collected from an external wall exhaust fan. Fifteen air samples were collected inside the barn, 30 samples at the exhaust location and 60 samples downwind, giving a total of 105 air samples per farm. Downwind samples were collected after completing collection of exhaust samples and collection started at the location closest to the farm and ended at the furthest location. Exterior samples were collected for two consecutive days, first at dusk and into the night and at dawn into the morning to increase the chances of virus detection. Samples were collected between 0.9 and 2.1 km downwind from the infected pig population. Between 13 and 100% of the samples collected inside the barns tested real-time reverse transcriptase polymerase chain reaction (PCR) positive with an average load of 3.20×10^5 influenza A virus RNA copies m^{-3} air (see Section 8.10). In exhaust air, positive samples also ranged between 13 and 100% with an average viral load of 1.79×10^4 RNA copies m^{-3} air. Influenza virus RNA was detected in air samples collected between 1.5 and 2.1 km away from the farms. Here the viral concentrations were significantly lower, 4.65×10^3 RNA copies m^{-3}. H1N1, H1N2 and H3N2 subtypes were detected in the air samples and the hemagglutinin gene sequences identified in the swine samples matched those in aerosols, providing evidence that the viruses detected in the aerosols originated from the pigs in the farms under study. Overall the results indicated that pigs can be a source of swine influenza A virus infectious aerosols and that these aerosols can be exhausted from pig barns and be transported downwind. The results from this study provided evidence of the risk of aerosol transmission in pigs under field conditions.

Harper *et al.*[13] determined the prevalence, types and quantity of airborne microorganisms in the small animal clinic of a veterinary teaching hospital to assess variations in microbial counts at different times and locations within the facility. An SKC Standard Biostage Impactor was used for bioaerosol sampling. The impactor was mounted on a 5 ft tall stand and connected to a sampling pump with flexible tubing. The pump was set to the flow specifications of 28.3 L min^{-1}. The impactor allowed air to be drawn directly on to blood agar plates. Sampling times of 2 and 4 min were used for each location to ensure that total aerobic plate counts per cubic meter were obtained for at least one sampling time without overgrowth on the plate. The impactor was steam sterilized before each sampling day. It was sterilized with 99% ethanol between locations. A total of 319 samples were collected and immediately taken to the laboratory for further processing.

For their investigations comparing the diversities of fungal and bacterial communities in indoor environments, Adams *et al.*[14] passively collected airborne particulate matter by gravitational settling for 1 month in Petri dishes suspended from the ceilings of university housing units. Analysis of the samples collected by this simple technique showed that human intervention had a greater influence on the diversity of the bacterial community than it had on that of the fungal community.

Adell *et al.*[15] compared gravitational settling, impingement and impaction for the collection of samples prior to the determination of airborne *Salmonella* spp. in poultry barns. Petri dishes containing Brilliant Green agar (Kauffmann medium) were exposed to the air in the barns for 24 h prior to incubation and colony counting. A six-stage Andersen impactor with Brilliant Green agar plates was used to collect air samples for the determination of *Salmonella*. It was operated for 90 s at a flow rate of 28.3 L min^{-1} prior to incubating the plates and counting the colonies. An Ace Glass impinger containing 20 mL of buffered peptone water was operated for 15 min at a flow rate of 12.5 L min^{-1} to collect samples for the determination of *Salmonella*. Based on culture-dependent detection, no cultivable *Salmonella* spp. was recovered from samples collected by liquid impingement. Cultivable *Salmonella* spp. was recovered from samples collected by gravitational settling and by impaction. The authors suggested that the difference in the results among techniques could be explained by considering the differences related to the sampling technique used: "...(i) the use of forced air and direct impaction on agar or not; (ii) the cut-off diameter of each sampling device (*i.e.* the size of sampled particles); and (iii) their detection limits."

Earlier, Fahlgren *et al.*[16] compared impingement, impaction and filtration with PTFE membranes for the collection of airborne bacteria at a mountainous location in Norway and at a seaside location in Sweden. Microbial samples were collected in parallel at both locations. Bacteria belonging to the Bacteroidetes and Proteobacteria phyla were dominate at both locations with *Sphingomonas* sp. and *Pantoea* sp. being the most common genera. The authors noted, "Considering the completely different collection principles of the sampling devices, *i.e.* liquid impinging, solid surface impaction and filtering, we find the results encouraging since the different sampling approaches yielded comparable results regarding the overall composition of the airborne bacterial community."

Tseng and Li[17] compared the collection efficiencies of an impactor, an impinger, a gelatin filter and a Nuclepore filter for virus-containing aerosols. Four different bacteriophages were used as surrogates for the mammalian viruses. The results showed the collection efficiencies were significantly affected by the morphology of the virus particles. For hydrophilic viruses, the collection efficiencies of the Andersen impactor, impinger and gelatin filter were 10 times higher than that of the Nuclepore filter. For hydrophilic viruses, the collection efficiencies of all four samplers were 10–100 times higher than for hydrophobic viruses. The infectivity of the virus in the collected samples was also evaluated for an Ace Glass impinger. The results showed that the

Table 8.1 Some samplers for airborne pathogens.

Principle	Example	Collection medium	Size cut-off/μm	Air flow rate/L min^{-1}
Filtration	Sartorius MD8	Membrane	0.8	41.7–133.3
Filtration	Gravikon PM4	Membrane	0.8	66.7
Filtration	SASS 3100	"Electret" media	1	50–310
Impaction	One-stage Andersen	Agar plate	<2	28.3
Impaction	Six-stage Andersen	Agar plate	0.65	28.3
Impaction	OXOID	Agar plate	0.8	100
Impingement	Ace Glass impinger	Saline solution	0.31	12.5
Impingement	SKC impinger	Saline solution	—	12.5
Impingement	Coriolis	Liquid	0.5	100–300

viruses retained more infectivity when the samples were refrigerated during storage than when stored at room temperature. Therefore, even when refrigerated, airborne virus samples collected using an impinger should be processed as soon as possible to avoid loss of virus infectivity.

Lednicky and Loeb[18] compared the efficiencies of a liquid impinger and a personal cascade impactor for the collection of an airborne H3N2 influenza virus. Viable virus was recovered from the liquid collection medium of the impinger and from some but not all of the of the PTFE impactor filters. It was uncertain if the difference was due to a difference in collection efficiency or to the possible desiccation of the virus on the PTFE filter by the flowing air stream.

Apparently, there is no universally applicable sampling procedure for airborne pathogens and toxins at present. Some of the more common approaches to sample collection are listed in Table 8.1.

8.3 Sample Preparation

The isolation and characterization of the microorganisms in the bioaerosol often begins with incubation of the sample. Collection by gravitational settling and by impaction can deposit the sample directly on an agar medium. Aliquots of samples collected in liquid impingers can be streaked across agar media and samples collected by filtration can be washed from the filters and treated as samples collected in liquid impingers. Contamination control is an important aspect of sample preparation.

Casein soy peptone agar and tryptic soy agar are "broad-spectrum media" for the collection and incubation of bacteria. The Brilliant Green agar mentioned above[15] is selective for incubation and colony counting of *Salmonella* spp. (except for *S. typhi* and *S. paratyphi*) while inhibiting the growth of Gram-positive and Gram-negative bacteria. Some other "selective media" are Eosin Methylene Blue agar for the incubation and colony counting of

Gram-negative bacteria and phenylethyl alcohol agar for inhibiting the growth of Gram-negative bacteria. Malt extract agar and Rose Bengal agar are "broad-spectrum media" for the collection and incubation of fungi. The incubation temperatures range from 20 to 25 °C for fungi, from 25 to 30 °C for environmental bacteria, from 35 to 37 °C for bacteria from human sources and from 50 to 56 °C for thermophilic bacteria. After incubation, the numbers of cultivable microorganisms in the samples are estimated from visual observations of the individual colonies and the concentrations of microorganisms are reported as the number of colony-forming units per unit volume of air, cfu L^{-1}. This value rarely reflects the numbers and types of microorganisms in bioaerosols because (1) some microorganisms are damaged during collection and hence unable to form colonies, (2) some microorganisms do not form colonies under the conditions of incubations and/or (3) some microorganisms are masked by the presence of others in the samples.

Microbial growth is dependent on the nutrient medium and the incubation conditions. DeMiguel *et al.*[19] demonstrated that CITA medium (blood agar supplemented with antimicrobial agents) was superior to both Farrell medium and Thayer-Martin medium for inhibiting most contaminant microorganisms while allowing the growth of *Brucella* species for the isolation of *Brucella* species from field samples. Vicente *et al.*[20] reported that CITA medium and Farrell medium inhibited the growth of contaminant microorganisms better than did *Brucella* agar (enzymatic digested casein and animal tissue, yeast extract, dextrose, salt and sodium bisulfite in agar). Ifeanyi *et al.*[21] compared, blood agar, mannitol–salt agar, MacConkey's agar, nutrient agar and Salmonella–Shigella agar for total colony formation and for comparative growth efficiency of *Escherichia coli*, *Salmonella*, *Shigella* and *Staphylococcus aureus* from a single sample. They found blood agar > mannitol–salt agar > Salmonella–Shigella agar > MacConkey's agar > nutrient agar in terms of total colony formation. MacConkey's agar was highly selective in favoring the growth of *E. coli* while inhibiting the growth of *Salmonella*, *Shigella* and *S. aureus*. Blood agar promoted the growth of *Shigella* in terms of comparative growth efficiency. The selectivities of some media are summarized in Table 8.2. The details of the compositions and selectivities of these and other culture media can be found in the *Merck Manual*[22] and in the *Difco & BBL Manual*.[23]

8.4 Identification Based on Macroscopic Methods

The colony morphology for given bacterial species is fairly consistent. Observations of physical characteristics of the colonies can be useful in their identification. Growth or lack of growth on various media is another characteristic for a given bacterial species that can be used for its identification.

In addition to the growth characteristics on the media described above and listed in Table 8.2, the size, shape, edge, height, color and surface of the colony often provide preliminary information on its identity. Sizes of the

Table 8.2 Selectivities of some culture media.

Medium	Selectivity agent	Microbes isolated
Blood agar	5% sheep's blood	Almost all bacteria; differential for hemolytic organisms
Chocolate agar	1% Hemoglobin and supplements	Most fastidious pathogens, such as *Neisseria* and *Hemophilus*
Eosin methylene blue agar	Lactose, eosin, methylene blue	Gram-negative enteric bacilli
Hektoen enteric agar	Lactose, sucrose, bile salts, $Fe(NH_4)(SO_4)_2$, $Na_2S_2O_3$, Bromothymol Blue, fuchsin	*Salmonella* and *Shigella* species enteric pathogens
Mannitol–salt agar	7.5% NaCl, mannitol	Staphylococci and micrococci
MacConkey's agar	Lactose, bile salts, neutral red, crystal violet	Gram-negative enteric bacilli
Phenylethyl alcohol agar	Phenylethyl alcohol	Gram-positive bacteria

bacterial colonies have been reported to range from tiny pin-points (punctiform), through small and medium to as large as 10 mm, and their shapes have been classified as round, irregular, filamentous or rhizoid with edges (or margins) described as smooth without projections (or entire), wavy (or undulate), lobed (or lobate), filamentous (or filiform) or rhizoid. Flat, raised, umbonate, crateriform, convex and pulvinate are descriptive of the height (or elevation) of the bacterial colonies on the medium. Observations on the pigmentation of the colonies often include other relevant optical characteristics such as opacity, cloudiness, translucency and iridescence. In addition, the colonies have surface characteristic textures (dry, brittle, moist, mucoid, viscid, butyrous) and appearances (smooth, glistening, dull, rough, wrinkled, veined).

Some examples when grown on tryptic soy agar are:

- *Escherichia coli* form shiny, mucoid colonies with entire margins.
- *Salmonella choleraesuis* form shiny colonies with entire margins and convex elevations.
- *Enterobacter aerogenes* form shiny colonies with entire margins and convex elevations.
- *Staphylococcus epidermidis* form circular, punctiform, convex colonies with entire margins. They appear whitish or the color of the agar.
- *Staphylococcus aureus* form circular, punctiform, golden brown, convex colonies with entire margins.
- *Micrococcus luteus* form circular, punctiform, bright yellow, convex colonies with entire margins.
- *Rhodospirillum rubrum* form circular, punctiform, red convex colonies with entire margins.

8.5 Identification Based on Microscopic Methods

At the cellular level, the size and shape are characteristic for given bacterial species, as are the presence or absence of flagellae, a capsule or endospores. These characteristics can be useful in identification. To enhance observation of these characteristics, they are highlighted with dyes or stains such as methylene blue, crystal violet or safranin O. The use of differential stains makes it possible to distinguish among the bacterial structures or species. For example, Gram stain is able differentiate between bacteria having cell walls composed of relatively thick peptidoglycan layers and teichoic acids and those with thin peptidoglycan layers and outer membranes consisting of lipoproteins, lipopolysaccharides and phospholipids. The former retain the dye whereas the latter do not, hence the former are Gram-positive and the latter are Gram-negative, for examples, *Listeria* is Gram-positive whereas *Legionella* stains Gram-negative.

Branched-chain fatty acids are common in many Gram-positive bacteria, whereas Gram-negative bacteria are composed of predominately straight-chain fatty acids. The presence of lipopolysaccharide in Gram-negative bacteria gives rise to the presence of hydroxy fatty acids in those genera. Thus, the presence of 10:0 3–OH, 12:0 3–OH, and/or 14:0 3–OH fatty acids indicates that the organism is Gram-negative, and conversely, the absence of the lipopolysaccharide and hydroxy fatty acids indicates that the organism is Gram-positive. More than 300 fatty acids and related compounds are found in bacteria. The information contained in these compounds is both in the qualitative differences (usually at genus level) and quantitative differences (commonly at species level). The MIDI Sherlock Microbial Identification System is a fully automated gas chromatographic analytical system that identifies over 2000 microbial species based on their unique cellular fatty acid profiles (see Section 8.6).

8.6 Identification Based on GC-FAME

Gas chromatography (GC) is a versatile technique for the separation, identification and quantification of volatile compounds. The gas chromatography of fatty acid methyl esters (GC-FAME) has been reviewed by Shantha and Napolitano[24] and by Gutnikov.[25]

Kunitsky *et al.*[26] stated, "The MIDI Sherlock® Microbial Identification System is a fully automated gas chromatographic analytical system that identifies over 2000 microbial species based on their unique cellular fatty acid profiles. The Sherlock System has many features that are important for pharmaceutical industry microbiology QC. The combination of accuracy of identification, high throughput, low cost per sample, extensive environmental libraries, compliance with 21 CFR Part 11, data export for trend analysis, reliable Agilent hardware and fully automated strain tracking ability is unique. Because no subjective tests are required, the naming is highly objective and reproducible." This system requires the bacteria be cultured, the fatty acids be extracted and saponified with sodium hydroxide–methanol solution, esterified with hydrochloric acid–methanol solution and extracted into hexane–methyl *tert*-butyl ether prior to GC analysis. The Sherlock software automates all analytical operations and uses a sophisticated pattern recognition algorithm to match the unknown FAME profile to the stored library entries for identification.

Rajan *et al.*[27] compared GC-FAME analysis and 16S rDNA sequencing PCR analysis for the identification of bacteria (see Section 8.10). The two bacteria were pure cultured onto trypticase soya broth agar medium. The fatty acids were extracted and methylated to form the methyl esters, which were analyzed by GC using the MIDI Sherlock software library. DNA was extracted from the bacterial culture for PCR analysis and the purified products were taken for cycle sequencing. The bioinformatics analysis was performed using BLAST (Basic Local Alignment Search Tool; http://ncbi.nlm.nih.gov/blast) and the microorganisms were identified using online databases. GC-FAME analysis identified one sample to be *Citrobacter freundii* with a similarity index of 0.71. The genotypic analysis also gave similar results. According to BLAST, the organism was identified as *C. freundii* with a 97% identity match. In the case of the other sample, GC-FAME analysis identified the organism to be *Pseudomonas aeruginosa* with a similarity index of 0.92. Sequencing also identified the organism with BLAST to be *P. aeruginosa* showing a 98% identity match.

Four methods were used to isolate and characterize bacteria from air samples collected by Wilson *et al.*[28] Bacteria were isolated and initially characterized using Gram stain, malachite green spore staining, microscopy and the BIOLOG Micr olog Bacterial Identification System (see Section 8.8). Twenty-one bacteria from eight separate filters were selected as potential human pathogens based on this preliminary BIOLOG analysis. Identification of the bacterial isolates was achieved using (1) GC-FAME, (2) 16S rRNA gene sequence analysis (see Section 8.10), (3) the BIOLOG Micrology System and (4) repetitive extragenic palindromic (REP)–PCR analysis (see Section 8.10). Overall, FAME, 16S rRNA gene sequence analysis and the BIOLOG Micrology System yielded complementary results. Only one of the isolates could not be identified by 16S rRNA gene sequence analysis. FAME was unable to identify four of the isolates. The BIOLOG Micrology System also failed to identify four of the isolates including three of those that were not identified by FAME. The majority of isolates (18/21) were identified as species of the genus *Bacillus*. Three isolates were classified within the *Bacillus cereus senso lato* group, which includes *B. anthracis*, *B. thuringiensis* and *B. cereus* strains. One isolate was identified as a *Staphylococcus* sp., most closely related to species (*e.g. S. kloosii*, *S. warneri*) that are commonly associated with human or animal skin, but can also act as an opportunistic pathogen. Another isolate was tentatively identified as *Tsukamurella inchonensis*, a known respiratory pathogen and resistant to the 10 antibiotics tested, including vancomycin.

The BIOLOG Micrology System (see Section 8.8) and GC-FAME were among the tools employed by Ross *et al.*[29] in their investigations of bacteria closely related to *Pseudomonas brassicacearum* and *P. thivervalensis*. Cluster analyses of BIOLOG and GC-FAME data revealed distinct groups for the isolates. Most isolates (82%) fell within two distinct BIOLOG groups, with one group further divided into two sub-groups containing a similar number of isolates in each. Two distinct GC-FAME groups accounted for 73% of all isolates tested. When BIOLOG and GC-FAME clusters were compared with each other, it was found that 81% of the isolates analyzed remained within the same group. Comparison of amplified rDNA restrictive analysis (ARDRA) profiles with BIOLOG and GC-FAME clusters revealed that isolates from each ARDRA group fell into distinct BIOLOG or GC-FAME clusters (91 and 85%, respectively), suggesting that the results from the three separate analyses were consistent with each other in assessing phonetic relationships among bacterial isolates.

Mass spectrometry coupled with gas chromatography or high-performance liquid chromatography has been applied to the detection of pathogens and toxins, and some of these applications are described below (see Section 8.7).

8.7 Identification Based on Spectrochemical Methods

Mass spectrometry (MS) is a powerful tool for the identification and quantification of organic compounds. Ho and Reddy[30] reviewed some applications of MS to the identifications of pathogens. Identification is usually based on protein profiling. These proteomic technologies include matrix-assisted laser desorption/ionization time-of-flight mass spectrometry (MALDI-TOFMS), electrospray ionization mass spectrometry (ESI-MS) and surface-enhanced laser desorption/ionization mass spectrometry (SELDI-MS) in addition to the hyphenated techniques gas chromatography-mass spectrometry (GC-MS) and high-performance liquid chromatography-mass spectrometry (HPLC-MS). For MALDI-TOFMS, the sample to be analyzed is mixed with an organic matrix and deposited on the sample plate as a small spot. The mixture is ionized by a laser beam. The resulting ions, along with molecular fragments, leak into the vacuum chamber of the mass spectrometer. Under the influence of a high voltage, the lighter ions travel faster than, and consequently arrive at the end of the vacuum chamber before, the heavier ions. The arrival times of the ions are recorded and the resulting display of the number of ions arriving at a given time as a function of their arrival time or mass-to-charge ratio (m/z) is the mass spectrum. For ESI-MS, a solution containing the sample is sprayed through a fine capillary under the influence of a high-voltage source to give multiply charged ions. The ions are desolvated as they leak into the mass spectrometer, where they travel at rates determined by the mass-to-charge ratio. The multiple charges allow heavy ions to be measured. Like MALDI, the mass spectrum obtained by ESI represents the number of ions arriving at a given time as a function of mass-to-charge ratio. MALDI-TOFMS and SELDI-MS are applicable to bioaerosol samples collected on agar plates by impaction, to samples pure cultured on agar media and to samples resolved by one- or two-dimensional sodium dodecyl sulfate polyacrylamide gel electrophoresis (SDS-PAGE). Samples collected in liquid impingers or pure cultured in liquid media can be subjected to the chromatographic analysis by ESI. Some applications of MS to the identification and quantification of airborne pathogens and toxins are summarized below.

Tobias *et al.*[31] evaluated single-particle laser desorption/ionization time-of-flight MS using bioaerosol mass spectrometry (BAMS) for the rapid detection of individual airborne, micron-sized *Mycobacterium tuberculosis* H37Ra particles as single cells or a small number of clumped cells. The BAMS mass spectral signatures for aerosolized *M. tuberculosis* H37Ra particles were found to be distinct from *M. smegmatis*, *Bacillus atrophaeus* and *B. cereus* particles, using a characteristic biomarker. This is the first time that a potentially unique biomarker was measured in *M. tuberculosis* H37Ra on a single-cell level. In addition, *M. tuberculosis* H37Ra and *M. smegmatis* were aerosolized into a bioaerosol chamber and were sampled and analyzed using BAMS, an aerodynamic particle sizer, a viable Andersen six-stage sampler and filter cassette samplers that permitted direct counts of cells. In a background-free environment, BAMS was able to sample and detect *M. tuberculosis* H37Ra at airborne concentrations of >1 *M. tuberculosis* H37Ra-containing particles per liter of air in 20 min as determined by direct counts of filter cassette-sampled particles and concentrations of >40 *M. tuberculosis* H37Ra cfu per liter of air in 1 min as determined by using viable Andersen six-stage samplers. This is a first step towards the development of a rapid, stand-alone airborne *M. tuberculosis* particle detector for the direct detection of *M. tuberculosis* bioaerosols generated by an infectious patient. Additional instrumental development is currently under way to make BAMS useful in realistic environmental and respiratory particle backgrounds expected in tuberculosis diagnostic scenarios.

The resolution of complex mixtures by chromatographic techniques coupled with the identification of the separated fractions by mass spectrometric techniques has been applied to the identification and quantification of airborne pathogens and toxins.

Hines *et al.*[32] measured endotoxin levels in air and in dust samples collected during four commercial aircraft flights. Samples were analyzed for endotoxin biological activity using the Limulus Amebocyte Lysate (LAL) assay and 3-hydroxy fatty acids of carbon chain lengths C10:0–C18:0 were determined in dust by GC-ion trap tandem MS. The geometric mean and the geometric standard deviation for endotoxin levels in air were 1.5 and 1.9 EU m^{-3}, respectively, for $n = 28$. However, significant differences were found by flight and within aircraft type. Mean endotoxin levels were significantly higher in carpet dust than in they were in seat dust (140 ± 81 *versus* 51 ± 25 EU mg^{-1} dust for $n = 32$ each, $p < 0.001$). Airborne endotoxin levels were not significantly related to either carpet or seat dust endotoxin levels. Mean 3–OH fatty acid levels were significantly higher in carpet dust than in seat dust for the C10:2, C12:0 and C14:0 fatty acids ($p < 0.001$ for each), whereas the mean level of C16:0 fatty acids was significantly higher in seat dust than in carpet dust ($p < 0.01$). The carpet dust endotoxin was significantly, but moderately, correlated with 3-OH–C12:0 and 3-OH–C14:0 fatty acid levels (Pearson $r = 0.52$ and 0.48, respectively), whereas correlation of seat dust endotoxin with individual 3-OH fatty acids depended on the test statistic used. Mean endotoxin potency was significantly higher for carpet dust than for seat dust (6.3 ± 3.0 *versus* 3.0 ± 1.4 EU pmol^{-1} lipopolysaccharide, $p < 0.0001$). Mean endotoxin levels in the air and dust of commercial aircraft cabins were generally higher than mean levels reported in homes and office buildings. These results suggested that the exposure route and the dust source are important considerations when relating endotoxin exposures.

Bloom *et al.*[33] employed both GC-MS and HPLC-MS to demonstrate the presence of mycotoxins produced by *Aspergillus* spp. and *Stachybotrys* spp. in buildings with either ongoing dampness or a history of water damage. Verrucarol and trichodermol, hydrolysis products of macrocyclic trichothecenes (including satratoxins) and trichodermin, produced by *Stachybotrys chartarum*, were analyzed by GC-tandem MS and sterigmatocystin produced by *Aspergillus versicolor*, satratoxin G and satratoxin H were analyzed by HPLC-tandem MS. These mycotoxins were found in 45 of 62 samples studied, in three of eight settled dust samples and in five of eight cultures of airborne dust samples. This work had application to the direct detection of the highly toxic sterigmatocystin and macrocyclic trichothecene mycotoxins in indoor environments for assessing potential health impacts.

Alam *et al.*[34] reported an application of tandem MALDI-TOF-TOFMS for the detection of airborne protein toxins, such as the botulinum neurotoxins, the *Clostridium perfringens* epsilon toxin, staphylococcal enterotoxin B, shiga toxin and ricin. They considered these toxins to be threat agents for bioterrorism and biological warfare likely to cause casualties *via* the inhalation route. The methodology they developed demonstrated the ability to identify the toxins in blind spiked samples and a novel directed search approach using a list of unique peptides was used to identify toxins in a complex protein mixture. The bioinformatic analysis provided high confidence for detecting these toxins. The use of tandem MS data with peptide sequence information increased the specificity of the method. Tandem MS combined with protein database searching was used to detect the toxin. *C. perfringens* epsilon toxin was present in aerosols at concentrations of 0.2 ppb.

Dybwad *et al.*[35] were also concerned with the detection of airborne biological threat agents. Their focus was on the performance criteria of the detector, its operational environment and improvement of its performance. Of specific concern were the background levels of bacteria in subway stations because they are enclosed public environments and could be regarded as potential targets for incidents involving biological threat agents. In their study, the airborne bacterial community at the Nationaltheatret metro station in Oslo was characterized in terms of concentration level, diversity and virulence- and survival-associated properties. Airborne particulate matter was collected with an SASS 3100 high-volume air sampler and a MALDI-TOFMS-based isolate screening procedure was used to characterize the microorganisms. The daytime level of airborne bacteria at this station was higher than the night-time and outdoor levels and the relative bacterial spore number was higher in outdoor air than at the station. The bacterial content, particle concentration and size distribution were stable within each environment throughout the study. The majority of the airborne bacteria belonged to the genera *Bacillus*, *Micrococcus* and *Staphylococcus*. A total of 37 different genera were identified in the air. These results suggest that anthropogenic sources were the major contributors to airborne bacteria at this metro station and that such airborne communities could harbor virulence- and survival-associated properties of potential relevance for biological detection and surveillance, and also for public health. The findings also contributed to the development of realistic testing and evaluation schemes for biological detection–surveillance systems by providing information that could be used to mimic real-life operational airborne environments in controlled aerosol test chambers.

Malebo and Shale[36] employed MALDI-TOFMS to investigate the spread of airborne microbial contaminants from heating ventilation air conditioning (HVAC) systems and the consequent exposure of indoor surfaces and building occupants. In this study, samples were collected from the surfaces of the ventilation grills at a public hospital using swabs. In general, bacterial counts were above 12 cfu cm^{-2} while fungal counts were below 5 cfu cm^{-2}, ranging from clean (<5) to very contaminated (>10). Microbial contaminants were identified and characterized using the MALDI-TOFMS and scanning electron microscopy. Identified genera included *Bacillus*, *Arthrobacter*, *Rhodotorula* and *Penicillium*. The microorganisms identified in this study using MALDI-TOFMS have no reported clinical implications. However, some are potential pathogens due to the presence of immune-compromised patients in the hospital settings. The probable aerosolization of these microorganisms from ventilation systems could result in microorganisms settling on hospital surfaces and possible contamination of medical equipment. This study also demonstrated the MALDI-TOFMS is a rapid, inexpensive and effective method for characterization of microorganisms isolated from HVAC systems.

Setlhare *et al.*[37] assessed the microbial composition of the air in the kitchen and in selected wards at a typical district hospital in South Africa. Air samples were collected by impaction on agar using an SAS Super 90 air sampler and by gravitational deposition on agar plates. These microbial samples were quantified and identified by MALDI-TOFMS and analytical profile index (API). Microbial counts were found to be higher in the fourth (≤60 cfu m^{-3}) sampling rounds when compared to the first (≥2 cfu m^{-3}), the second (≤30 cfu m^{-3}) and the third (≤15 cfu m^{-3}) sampling rounds. Genera identified included *Bacillus*, *Kocuria*, *Staphylococcus*, *Arthrobacter*, *Candida*, *Aureobasidium*, *Penicillium* and *Phoma*, among others. Because of their ability to cause diseases in humans, especially in those with suppressed host immunity defenses, the presence of these pathogens was of concern. Furthermore, fungal genera identified in this study are also known to cause food spoilage and fungal infections in patients. Results from this study indicate the importance of air quality monitoring in healthcare settings to prevent possible hospital-acquired infections and contamination of hospital surfaces, including food contact surfaces, by airborne contaminants.

Woods *et al.*[38] studied the performance of MALDI-TOFMS for the identification of the *Streptococcus anginosus* group to species level. In total, 116 stored bacteremia isolates identified by conventional methods as belonging to this group were analyzed by MALDI-TOFMS. Partial 16S rRNA gene sequencing, supplemented with sialidase activity testing, was performed on all isolates to provide "gold standard" identification against which to compare MALDI-TOFMS performance. Overall, 100% of the isolates were correctly identified to the genus level and 93.1% to the species level by MALDI-TOFMS. However, only 77.6 and 59.5% of these isolates were correctly identified to the genus to the species level, respectively, by a MALDI-TOFMS direct transfer method alone. Use of a rapid *in situ* extraction method significantly improved the identification rates compared with the direct transfer method ($p < 0.001$). Routine use of this method to reduce the number of time-consuming full extractions required was recommended for identification of this group of bacteria by MALDI-TOFMS in the routine diagnostic laboratory. Only 22% (2/9) of *Streptococcus intermedius* isolates were reliably identified by MALDI-TOFMS to the species level, even after full extraction. MALDI-TOFMS reliably identified *S. anginosus* and *S. constellatus* to the species level but did not reliably identify *S. intermedius*.

In addition to their detection by MS, airborne pathogens and toxins have been detected by Fourier transform infrared (FTIR) spectroscopy. Duygu *et al.*[39] reviewed some of these applications. Fischer *et al.*[40] reported on the use of FTIR spectroscopy for the rapid identification and intra-species characterization of airborne filamentous fungi. Valentine *et al.*[41] were able to distinguish between three *Bacillus* species: *B. thuringiensis israelensis*, *B. cereus* and *B. megaterium*, using FTIR spectroscopy.

8.8 Identification Based on Chemical/Biochemical Methods

The BIOLOG Microlog System mentioned above is based on 96 simultaneous chemical assays that allow the identification of aerobic Gram-negative and Gram-positive bacteria in the same test panel. The system's powerful carbon source utilization technology accurately identifies environmental and pathogenic microorganisms by producing a characteristic substrate utilization pattern or "metabolic fingerprint" from discrete test reactions performed within a 96-well microplate. Culture suspensions are tested with a panel of preselected assays, then incubated, read and compared with extensive databases of environmental organisms, human pathogens, veterinary pathogens and plant pathogens. The scope of the 96 assay reactions, coupled with sophisticated interpretative software, delivers a high level of accuracy that is comparable to that of molecular methods. The 1 min per sample setup is much simpler and faster than some of the gene probes described below.

Chan *et al.*[42] employed the BIOLOG Microlog System, the MIDI Sherlock Microbial Identification System described above and the DuPont Qualicon RiboPrinter Microbial Identification System to determine airborne bacteria in two Hong Kong fast food restaurants (see DuPont Qualicon RiboPrinter in Section 8.10). Quarterly, 1 hour air samples were collected in mineral oil with an SKF BioSampler at a flow rate of 12.5 L min^{-1} during business hours. The airborne particulate matter was recovered from the mineral oil by filtration with 0.45 μm cellulose ester membranes and cultured on tryptic soy agar. Colonies were pure cultured and examined using these three methods of identification, and fifteen genera of the bacteria were identified, *Micrococcus* and *Bacillus* species being the most abundant. The majority of these bacteria were opportunistic pathogens. This study employed chemical, chromatographic and genomic techniques for identifying airborne pathogens.

8.9 Identification Based on Immunoassays

Immunoassays or antibody-based assays are highly sensitive and selective techniques built on chemical and immunological principles. These techniques are adaptable to the detection of any biologically active substance that has the potential to elicit an immune response, including protein toxins that are not amenable to PCR-based methods. Their high selectivity results from the use of antibodies and purified antigens as immunological reagents that are the backbone of every immunoassay system. Antigens include structures such as viruses, toxins, proteins, vegetative cells and endospores that elicit antibody response in organisms. The antibody is chosen based on its ability to bind the target antigen with high affinity while at the same time maintaining a high specificity. Selecting the correct antibody is the most critical aspect of designing immunoassays. Furthermore, the antibody must be tested against an extensive panel of molecules, including compounds closely related to the target and also potential interferents. Immunoassays measure the formation of antibody–antigen complexes by the indicator reaction and are used when an analyte needs to be detected in complex biological systems in a qualitative (positive or negative) or quantitative (amount measurement) manner. Many immunodetection methods use secondary antibodies conjugated to enzymes in order to amplify the signal produced by the indicator reaction *via* the catalytic properties of the enzyme. Common uses include measurement of therapeutic and abused drugs, hormones, specific proteins and cancer antigens and to detect antigens on hepatitis B, *Chlamydia trichomatis*, *Hemophilus*, *Cryptococcus* and *Streptococcus*. Immunoassays use many types of supports and detection system. The detection signal can be radioactive, colorimetric or fluorescent. Widely used variants of immunoassays include radioimmunoassay (RIA), enzyme-linked immunosorbent assay (ELISA), lateral flow immunoassay (LFI), luminescent immunoassay (LIA) and fluorescence immunoassay (FIA).

Well-known immunoassay methods such as LFI and ELISA, although lacking sensitivity, are available commercially and are conventionally used.

Although biosensor-based methods such as time-resolved FIA, chemiluminescence assay, electrochemical immunoassay, surface plasmon resonance sensors, fiber-optic sensors and microfluidic biochips have, in some cases, demonstrated improved sensitivity, they still require further standardization with real samples.

Immunoassays have been used in numerous recent studies and were capable of rapidly detecting and identifying aerosols of multiple pathogens or toxins such as cyanotoxins, molecules associated with groups of microbes and bacterial and plant toxins.

Cheng *et al.*[43] used an ELISA method to detect aerosolized cyanobacterial toxins (microcystins) which are transferred from water to air *via* a bubble-bursting process. Using a commercial ELISA kit, the limit of detection for the most common microcystin variant (MCYST-LR) in the aerosol ranged from 0.0014 to 0.094 ng m^{-3}, depending on the sampler, the sampling flow rate and the sampling tine. However, the ELISA analysis did not differentiate the chemical forms of cyanotoxins. In addition, the sampling and ELISA analysis were also successfully used in a pilot field study to measure microcystin aerosol *in situ*.

In 2006, Haskard[44] published a detailed report on microcystin immunoassays based on the clone AD4G2, which is recommended as providing a full coverage of all microcystins. More recently, Weller[45] comprehensively reviewed all commercially available immunoassay kits and the overall antibody repertoire for microcystins and related toxins.

Noss *et al.*[46] reported the development of inexpensive but highly sensitive enzyme immunoassays (EIAs) to measure dust-associated β-(1,3)-glucans in environments with low exposure levels. The β-glucan EIAs and reagents previously reported by three other laboratories served as the basis for this immunoassay development. Three new sensitive glucan sandwich assays were developed using the combination of one polyclonal (U1) and two monoclonal (P1, B2) anti-β-(1,3)-glucan antibodies. The specificities of the new EIAs were tested with a panel of 13 different glucans. The highest sensitivity and acceptable background reactions were achieved with the antibody combinations U1B2, P1U1 and P1B2. The feasibility of measuring glucans in airborne dust with a recently introduced electronic dust fall collection method was investigated by selecting the most suitable of the three EIAs to measure β-(1,3)-glucan levels in actively collected airborne dust samples. All of these EIAs in combination with electronic dust fall sampling were applicable to airborne dust β-(1,3)-glucan measurements in a home environment or other low-exposure environments.

The immuno-biosensors are alternatives to immunoassays for monitoring antibody–antigen reactions in real time, without the use of labeling. A biosensor comprises a biorecognition molecule (enzyme, antibody, sequence of DNA, peptide or microorganism) immobilized over a signal transducer to give a reagentless analytical device. Over the last five decades, the development of biosensors has attracted considerable interest and the biorecognition molecules provide the biosensor with its selectivity for the target analyte so that the molecule of interest can be detected and identified by the biosensor from a complex matrix. Miniaturized or microfluidic analysis systems, also called micro total analysis systems (μTAS) or lab-on-a-chip (LOC) have gained particular attention, as they offer advantages over other biosensors.

Weingart *et al.*[47] described a microfluidic device, the Inca Bioanalytical System, consisting of two essential components, IncaTrace and IncaSlide, which is a microarray-based system for simultaneous detection of bacterial and plant toxins. This system is a bioanalytical platform that combines the specificity of covalently immobilized capture probes with dedicated instrumentation and immuno-based microarray analytics. The bioanalytical platform consisted of a micro-structured polymer slide, the IncaSlide, serving both as the support for printed arrays and as the incubation chamber. The platform also included an easy-to-operate instrument for simultaneous slide processing at the selectable assay temperature. Streptavidin labeled with cyanine 5 (Cy5) was used for fluorescence detection of toxin–antibody complexes. Upon scanning of processed IncaSlides, fluorescence signal intensities were registered and quantified with commercial instrumentation. The performance of the system was investigated by immunological detection of *Botulinum* neurotoxin type A, *Staphylococcus* enterotoxin B and the plant toxin ricin. Toxins were detectable at levels in the ng mL^{-1} range. The advantages of the described bioanalytical platform are its ability to analyze multiple samples for multiple analytes simultaneously with multiple reaction media fast assay performance and fast assay development and exploration of different assay strategies.

Zhu *et al.*[48] reviewed the principles and practices of immunoassays, paying particular attention to some of the more recent developments in the detection of bacterial toxins. Applications of nanotechnology to bioanalysis were among the attractive innovations they reviewed. Nanomaterials have been utilized directly for signal sensing or indirectly as anchored supports for other labels such as enzymes and fluorescent probes. In this review, the authors cited optical sensing by gold and silver nanoparticles (NPs) bridging antibody–antigen complexes, which allowed the detection of microcystin (MCYST-LR) at concentrations two orders of magnitude lower than those reported by Cheng *et al.*[43] In addition to their use as a signal sensor, gold nanoparticles have been used to support a peptide with specific binding properties for the anthrax protective antigen in an electrochemical ELISA. The limit of detection for this assay was reported to be 3 pM. The combination of magnetic nanoparticles (MNPs) functionalized with antibodies for capture (*e.g.* conductometric immunosensors) and modified AuNPs for detection is widely used for the detection of bacteria.[49,50]

Overall, Zhu *et al.*[48] concluded, "Nanomaterials, such as AuNPs, AgNPs, MNPs, quantum dots, carbon nanomaterials and metal–organic frameworks, have the potential to improve sample separation, target recognition and signal amplification. The integration of these new materials into conventional immunoassays represents a promising, alternative solution to improve analytical performance."

In their market survey on current and future detectors for biological threat agents, Fruchey and Emanuel[51] evaluated over 150 devices in terms of effectiveness, logistics, operations and detection of biological agents. Mostly the devices operated on immunoassay- or PCR-based technology. Very few of the devices that they evaluated were based on chromatography or spectrometry. The guide prepared by Ozanich *et al.*[52] described some of the commercially available biological sampling and detection technologies available for use by first responders. In preparing this guide, the information provided, the ease of use in the field, the cost, the time from sampling to output and the size and weight of the device were considered. Of the devices described, more than twice as many were based on immunoassay technology as PCR technology. Where available, this guide also included information from the peer-reviewed literature.

A good example of integration between immune-based and PCR-based detection is represented by the standalone instrument called the Autonomous Pathogen Detection System (APDS), capable of continuous and fully autonomous monitoring of multiple airborne biological agents.[53,54] The APDS has a virtual impactor and a cyclone sampler in series to concentrate and continuously collect aerosol samples. It performs multiplexed antibody-based tests, immunoassays and automated real-time flow-through nucleic acid assays (PCR). The immunoassay allows the detector to respond to a wide range of airborne pathogens or toxins and also protein toxins, while PCR provides a secondary, confirmatory assay. By utilizing two highly specific biological assays that rely on fundamentally different molecular interactions for detection, the probability of reporting false-positive results is greatly reduced.

8.10 Identification Based on Nucleic Acids

Nucleic acid-based detection methods offer many advantages over traditional methods (*e.g.* microbiological) for the detection of airborne pathogens. Developments in technologies have allowed the recovery of nucleic acids from complex environmental matrices and the use of nucleic acid-based methods for the detection of bioactive agents by identifying their nucleic acids by gene probes or PCR.

For the detection and identification of viruses and bacteria, but not toxins, nucleic acid techniques are exceedingly attractive because they can provide unambiguous identification of a characteristic sequence. These techniques are highly specific because the use of PCR amplification techniques allows the detection of minute amounts of nucleic acids.[55]

The PCR is an *in vitro* technique used to amplify enzymatically, in an exponential manner, minute amounts of a nucleic acid to identifiable levels. The possibilities of this technique have been reviewed by Ogram and Sayler,[56] Peccia and Hernandez,[57] Pusteria *et al.*,[58] Rahman *et al.*[59] and Garibyan and Avashia.[60] Especially the sensitivity of advanced nucleic acid techniques was demonstrated by the detection of a single anthrax spore and single *Francisella tularensis* bacteria that was differentiated from not only non-*Francisella* bacteria but also from *Francisella philomiragia*.[61,62]

In PCR, DNA is extracted from the sample and the DNA target is amplified by repeating three major steps: (1) denaturation of double-stranded DNA into single-stranded DNA; (2) annealing of primers to the complementary single-stranded target sequences; and (3) extension of the primers in the 5′ to 3′ direction by heat-stable DNA polymerase to produce double-stranded molecules of the target DNA. During each cycle, the number of copies of the target sequence doubles, and since newly synthesized copies also serve as templates for following synthesis cycles, the amount of DNA generated increases exponentially. Detection of amplification products can be carried out in a variety of ways, for example, ethidium bromide staining, using gel electrophoresis and examination of the gel using UV light.[63] Southern blot-based hybridization with a specific DNA probe following gel electrophoresis of the PCR products confirms the identity of the PCR products, increasing the sensitivity of the overall assay.[64] The sensitivity of the PCR/DNA probe assays depends on the PCR conditions used, the biological matrix and the post-PCR detection method applied.

Because the method does not require live or intact cells, PCR is a valuable tool for detecting pathogens from clinical specimens, where bacteria die or lyse easily due to inappropriate storage conditions or prior antibiotic treatment. Alternatively, the application of assays that target the *in vitro* amplification and detection of RNA will ensure also the detection of viable organisms.[65]

PCR is widely used in the diagnosis and surveillance of bacterial pathogens because of its high sensitivity and specificity and high throughput capabilities. It provides a complementary tool to the classical phenotype-based methods and often enhances confirmatory results. A number of conventional PCR assays have been developed for the detection and subtyping of bacterial and viral pathogens. Conventional PCR detects products at the endpoint of DNA amplification by visualizing amplicons using gel electrophoresis. In contrast, real-time PCR (qPCR) monitors the accumulation of PCR product in a reaction while it is taking place. qPCR is less time consuming and has higher sensitivity and specificity. It provides the quantitative detection of a PCR product in a fast turnover time in a closed tube, in this way reducing the possible risk of contamination.[64] In addition, the use of qPCR is expanding rapidly because it is easier to perform.

Sometimes the sequence of interest consists of RNA and not DNA. RNA, an alternative to DNA as the target for nucleic acid-based detection methods, represents a labile and quickly degradable molecule, especially once the organism has died. This characteristic makes handling RNA more complicated than handling DNA, but RNA has the advantage of distinguishing viable organisms from nonviable ribosomal RNA (rRNA) and remains the "gold standard" target for bacterial pathogen identification. Gen-Probe Inc. (merged with Hologic in 2012) patented the use of rRNA as a target technology and have exploited this target in a range of assays for bacterial pathogens (http://www.hologic.com/).

All cells contain ribosomes, which consist of ribosomal proteins and rRNAs. The ribosome is composed of two sub-units, one larger than the other. Although there are also associated proteins helping to make up the functional units of the ribosome, the smaller sub-unit is usually coded for by the 16S rRNA gene and the larger sub-unit is usually coded for by the 23S rRNA and 5S rRNA genes in bacteria. The 16S rRNA gene is often used for bacterial classification. 16S rRNA gene sequences are easily obtained from unknown organisms by PCR amplification followed by DNA sequencing.[66] The 16S rRNA gene is relatively short at 1.5 kb (1500 nucleotides), making it faster and cheaper to sequence than other unique bacterial genes. Furthermore, 16S rRNA gene sequences can be classified and the best-match species is given by a database search of the Ribosomal Database Project (RDP), which is a database updated monthly with 16S rRNA gene sequences from the International Nucleotide Sequence Database Collaboration (DDBJ, EMBL-EBI and GenBank).[67] Species-specific probes constructed from portions of the gene coding for rRNA have been used for the rapid identification and quantification of bacterial species in environmental samples.

An integrated air sampling and PCR-based method for detecting airborne fungal spores (*e.g. Penicillium roqueforti*) was described by Williams *et al.*[68] The spores were collected directly in small tubes using a miniature cyclone-type air sampler. Three different types of treated spores were used to produce DNA for PCR tests: (i) untreated spores, (ii) spores disrupted with Ballotini beads to fracture their walls for the release of the contents and (iii) disrupted spores followed by DNA purification. Three *P. roqueforti*-specific assays were also tested: single-step PCR, nested PCR and PCR followed by Southern blotting and probing. These PCR techniques had the potential for precise detection of fungal spores in air samples. It was possible to detect some fungal species by simple addition of unprocessed spores directly to the PCR. However, for sensitive detection, the spores needed additional processing, mainly disruption of the spore walls to release DNA before it is amplified by PCR.

Makino and Cheun[69] reported a rapid detection method for airborne bacterial spores. They examined *Bacillus anthracis* spores by RT-PCR. The biosolids from 100 L of air were trapped on the nitrocellulose filter of the air sampling device and subsequently suspended in phosphate-buffered saline. The suspension was used for qPCR using anthrax-specific primers. It was shown that one *B. anthracis* spore can be detected in a 100 L air sample within 1 h using PCR. The results confirmed that anthrax spores from the atmosphere can be detected rapidly, suggesting that qPCR provides a flexible and powerful tool for this purpose.

Ziros *et al.*[70] reported the development of a method for the detection of airborne viruses using qPCR analysis. In this study, a simple, low-cost method for sampling of airborne viruses using low-melting agarose plates and a conventional microbial air sampling device was developed. The use of low-melting agarose plates permitted the direct extraction of nucleic acids from the captured viruses without the need for any preliminary elution steps. Molecular detection and quantification of airborne viruses was performed using this RT-PCR technique. The method was tested using *Adenovirus* and *Norovirus* G II, as representatives of DNA and RNA viruses, respectively. Moreover, the method has been tested successfully in outdoor experiments, by detecting and quantifying human adenoviruses in the airborne environment of a wastewater treatment plant. The major advantage of using the low-melting agarose plates was that the extraction of the nucleic acids can be performed directly on the low-melting agarose plates, while the eluted nucleic acids are totally free of inhibitory substances. Coupled with qPCR, the entire procedure can be completed in reduced amount of time (less than 3 h).

To trace environmental sample contamination with accurate genetics-based typing, beyond species level, is possible with the use of the DuPont Qualicon RiboPrinter System technology. This system investigates regions encoding not only the 16S rRNA sequence, but also the 5S and 23S sequences, plus the spacer regions and flanking genes on either side. It automates restriction fragment length polymorphism (RFLP) analysis and targets the rRNA-coding region of the bacterial genome. Restriction enzymes, such as *Eco*RI or *Pvu*II, cut bacterial DNA into fragments that are processed to form a characteristic banding pattern or "fingerprint." The system captures an image of the banding pattern and digitizes it as a RiboPrint pattern. This pattern can then be compared with a reference database of patterns from historical samples analyzed previously. The RiboPrinter System technology investigates a much larger portion of the bacterial genome than other bacterial identification systems and allows highly precise differentiation among strains of the same species, even those with the same 16S sequence. For example, a comprehensive characterization of airborne bacterial profiles in indoor environments was performed for the first time by Chan *et al.*[42] using three different bacterial identification systems, including the RiboPrinter System. Their results suggested that the identity of airborne bacteria should also be included in indoor air quality features to ensure there is a safety guideline for the public.

8.11 Conclusion

A variety of sample collection procedures have been reported, ranging from settling under the influence to gravity to forced air filtration, impaction or impingement. None of these approaches is universally applicable to sampling for airborne pathogens and toxins. Although protocols for specific targets have been validated, results obtained with different sampling strategies influenced the apparent composition of the airborne microbiological community. Cell damage by desiccation or by lysis could reduce viability and impact culture-dependent assessments. Consequently, at present it is virtually impossible to complete a comprehensive assay of all microbial species in a single sample.

Traditionally, the determination of airborne pathogens began with culture-dependent techniques, which provided taxonomic information for the identification of microorganisms. Batteries of tests, in addition to macroscopic and microscopic examinations, were employed to classify microorganisms. Very often, the reaction to Gram stain is the first step in the classification process. Differences in the structures and fatty acid compositions of the bacterial cell walls influence the uptake of the Crystal Violet primary stain, allowing bacteria to be separated into the Gram-positive and Gram-negative classification.

These differences in compositions have been used to identify bacteria by GC-FAME. The MIDI Sherlock Microbial Identification System is a fully automated GC analytical system that identifies over 2000 microbial species based on their unique cellular fatty acid profiles. The hyphenated chromatographic techniques GC-MS and HPLC-MS have been applied to the determination of airborne pathogens and toxins.

MALDI-TOFMS, ESI-MS and SELDI-MS are powerful technologies for bacterial identification and characterization. In addition, FTIR spectrometry has been used to classify and identify bacterial samples. Bioinformatics is crucial for successfully relating the spectral characteristics to the identification of the pathogens and toxins.

Bioinformatics plays a similar crucial role in the interpretation when the metabolic fingerprint generated by microorganisms in the BIOLOG Microlog System are recorded and compared with hundreds of identification profiles in the database. The databases are periodically updated to reflect the newest taxonomy and incorporate additional species.

The simplicity and specificity of immunoassays coupled with automated plate reading systems and computerized data analysis led to the development of instrumentation for the rapid and reliable detection of airborne pathogens and toxins. Immunoassay-based technology has been adapted to on-site and laboratory utilization.

The ability of the PCR to amplify microbial nucleic acid exponentially s has made possible the detection of difficult-to-culture, noncultivable and even nonviable specimens. PCR-based technology has been adapted to on-site and laboratory utilization.

Airborne pathogens and toxins pose potentially serious public health threats to members of unprotected populations. To protect public health, identification and quantification of these pathogens and toxins are essential for the prompt initiation of countermeasures and therapeutic regimens. The classical macroscopic and microscopic identification techniques have been complemented with an array of sophisticated offline chromatographic, spectrometric and chemometric methodologies and also online methodologies based on immunoassays and gene probes.

These threats have stimulated continued research on and development of devices for the detection of airborne pathogens and toxins with greater selectivity, sensitivity, speed and ease of operation and at lower cost. A variety of different biosensors have been developed in the past two decades, but there is still a need for further research and future development of miniaturized ("sample-in and answer-out" platforms at the centimeter scale), low-cost or disposable biosensors capable of rapid detection and accurate identification of a wide range of airborne pathogens and toxins.

References

1. P. A. Jensen and M. P. Schafer, *Chapter J, Sampling and Characterization of Bioaerosols, NIOSH Manual of Analytical Methods*, 4th edn, 1994.
2. *Bioaerosols Handbook*, ed. C. S. Cox and C. M. Wathes, CRC Press, Boca Raton, FL, 1995.
3. C. S. Yang and P. A. Heinsohn, *Sampling and Analysis of Indoor Microorganisms*, John Wiley and Sons, Hoboken, NJ, 2007.
4. K. T. H. Durand, M. L. Muilenberg, H. A. Burge and N. S. Seixas, Effect of Sampling Time on the Cultivability of Airborne Fungi and Bacteria Sampled by Filtration, *Ann. Occupat. Hyg.*, 2002, **46**(**1**), 113–118.
5. K. Qudiesat, K. Abu-Elteen, A. Elkarmi, M. Hamad and M. Abussaud, Assessment of Airborne Pathogens in Healthcare Settings, *Afr. J. Microbiol. Res.*, 2009, **3**(**2**), 66–76.
6. D. Verreault, S. Moineau and C. Duchaine, Methods for Sampling of Airborne Viruses, *Microbiol. Mol. Biol. Rev.*, 2008, **72**(**3**), 413.
7. ASTM E884-82(2012), *Standard Practice for Sampling Airborne Microorganisms at Municipal Solid-Waste Processing Facilities*, ASTM International, West Conshohocken, PA, 2012, www.astm.org, DOI: 10.1520/E0884-82R12, accessed 02 November 2014.
8. US EPA, ASTM E 884–82, *Standard Practice for Sampling Airborne Microorganisms at Municipal Solid Waste Processing Facilities*, RTI Report Number 94U-5400-24A (3), July 1993, EPA Contract 66V.
9. NIOSH Method 0800, *Bioaerosol Sampling (Indoor Air), NIOSH Manual of Analytical Methods*, 4th edn, 1994.
10. NIOSH Method 0801, *Aerobic Bacteria by GC – FAME, NIOSH Manual of Analytical Methods*, 4th edn, 1994.
11. NIOSH Method 0900, *Mycobacterium Tuberculosis, Airborne, NIOSH Manual of Analytical Methods*, 4th edn, 1994.
12. C. A. Corzo, M. Culhane, S. Dee, R. B. Morrison and M. Torremorell, Airborne Detection and Quantification of Swine Influenza A Virus in Air Samples Collected Inside, Outside and Downwind from Swine Barns, *PLoS One*, 2013, **8**(**8**), e71444.
13. T. A. M. Harper, S. Bridgewater, L. Brown, P. Pow-Brown, A. Stewart-Johnson and A. A. Adesiyun, Bioaerosol sampling for airborne bacteria in a small animal veterinary teaching hospital, *Infection Ecol. Epidemiol.*, 2013, **3**, 20376.

14. R. I. Adams, M. Miletto, S. E. Lindow, J. W. Taylor and T. D. Burns, Airborne Bacterial Communities in Residences: Similarities and Differences with Fungi, *PLoS One*, 2014, **9**(**3**), e91283.
15. E. Adell, V. Moset, Y. Zhao, A. Jiménez-Belenguer, A. Cerisuelo and M. Cambra-Lopez, Comparative Performance of Three Sampling Techniques to Detect Airborne *Salmonella* Species in Poultry Farms, *Ann. Agric. Environ. Med.*, 2014, **21**(**1**), 15–24. www.aaem.pl, accessed 02 November 2014.
16. C. Fahlgren, G. Bratbak, R.-A. Sandaa, R. Thyrhang and U. Li Zweifel, Diversity of Airborne Bacteria in Samples Collected Using Different Devices for Aerosol Collection, *Aerobiologia*, 2011, **27**(**2**), 107–120.
17. C.-C. Tseng, and C.-S. Li, Efficiencies of Aerosol Samplers for Virus-Containing Aerosols, *J. Aerosol Sci.*, 2005, 36(5–6), 593–607.
18. J. A. Lednicky and J. C. Loeb, Detection and Isolation of Airborne Influenza A H3N2 Virus Using a Sioutas Personal Cascade Impactor Sampler, *Influenza Res. Treatment*, 2013, 656825.
19. M. J. DeMiguel, C. M. Marín, P. M. Muñzo, L. Dieste, M. J. Grilló and J. M. Blasco, Development of a Selective Culture Medium for Primary Isolation of the Main *Brucella* Species, *J. Clin. Microbiol.*, 2011, **49**(**4**), 1458–1463.
20. A. F. Vicente, J. M. A. P. Antunes, G. H. B. Lara and S. R. Mioni, Evaluation of Three Formulations of Culture Media for Isolation of *Brucella spp*. Regarding Their Ability to Inhibit the Growth of Contaminating Organisms, *BioMed Res. Int.*, 2014, 702072.
21. V. O. Ifeanyi, S. C. Nwosu, J. O. Okafor, C. P. Onnegbu, and E. Nwebunnia., Comparative Studies on Five Culture Media for Bacterial Isolation, *Afr. J. Microbiol. Res.*, 2014, 8(36), 3330–3334.
22. *Merck Manual*, Merck Sharp & Dohme Corp, Whitehorse Station, NJ, USA, 2014.
23. *Difco™ & BBL™ Manual*, Becton, Dickenson and Company, Sparks, MD, USA, 2009.
24. N. C. Shantha and G. E. Napolitano, Gas Chromatography of Fatty Acids, *J. Chromatogr.*, 1992, **624**(**1–2**), 37–51.
25. G. Gutnikov, Fatty Acid Profiles of Lipid Samples, *J. Chromatogr. B: Biomed. Appl.*, 1995, **71**(**1–2**), 71–89.
26. C. Kunitsky, G. Osterhout and M. Sasser, Identification of Microorganisms Using Fatty Acid Methyl Ester (FAME) Analysis and the MIDI Sherlock® Microbial Identification System, in *Encyclopedia of Rapid Microbiological Methods, III*, ed. M. Miller, ODA Technical Books, Bethesda, 2006, www.pda.org/books, accessed 25 November 2014.
27. A. Rajan, N. Aruna and S. Kaur, Comparitive Study of FAME and Sequence Analysis for Identification of Bacteria, *Biotechnol. Bioinformat. Bioengin.*, 2011, **1**(**3**), 319–323.
28. C. A. Wilson, R. L. Brigmon, C. Yeager, W. Smith Garriet and S. W. Polson, Genotypic and Phenotypic Characterization of Aerosolized Bacteria Collected From African Dust Events, *J. SC Acad. Sci.*, 2014, **12**(**1**), 5.
29. I. L. Ross, Y. Alami, P. R. Harvey, W. Achouak and M. H. Ryder, Genetic Diversity and Biological Control Activity of Novel Species of Closely

Related Pseudomonads Isolated from Wheat Field Soils in South Australia, *Appl. Environ. Microbiol.*, 2000, **66**(**4**), 1609–1616.
30. Y.-P. Ho and M. Reddy, Identification of Pathogens by Mass Spectrometry, *Clin. Chem.*, 2010, **56**(**4**), 525–536.
31. H. J. Tobias, M. P. Schafer, M. Pitesky, D. P. Fergenson, J. Horn, M. Frank and E. E. Gard, Bioaerosol Mass Spectrometry for Rapid Detection of Individual Airborne *Mycobacterium tuberculosis* H37Ra Particles, *Appl. Environ. Microbiol.*, 2005, **71**(**10**), 6086–6095.
32. C. J. Hines, M. A. Waters, L. Larsson, M. R. Petersen, A. Saraf and D. K. Milton, Characterization of Endotoxin and 3-Hydroxy Fatty Acid Levels in Air and Settled Dust from Commercial Aircraft Cabins, *Indoor Air*, 2003, **13**(**2**), 166–173.
33. E. Bloom, K. Bal, E. Nyman, A. Must and L. Larsson, Mass Spectrometry-Based Strategy for Direct Detection and Quantification of Some Mycotoxins Produced by *Stachybotrys* and *Aspergillus* spp. in Indoor Environments, *Appl. Environ. Microbiol.*, 2007, **73**(**13**), 4211–4217.
34. S. I. Alam, B. Kumar and D. V. Kamboj, Multiplex Detection of Protein Toxins Using MALDI-TOF-TOF Tandem Mass Spectrometry: Application in Unambiguous Toxin Detection from Bioaerosol, *Analyt. Chem.*, 2012, **84**(**23**), 10500–10507.
35. M. Dybwad, P. E. Granum, P. Bruheim and J. M. Blatnya, Characterization of Airborne Bacteria at an Underground Subway Station, *Appl. Environ. Microbiol.*, 2012, **78**(**6**), 1917–1929.
36. N. J. Malebo and K. Shale, MALDI Biotyper Characterization of Microorganisms Colonizing Heating Ventilation Air-Conditioning Systems at a South African Hospital, *Life Sci. J.*, 2013, **10**(**4**), 314–417.
37. G. Setlhare, N. Malebo, K. Shale and R. Lues, Identification of Airborne Microbiota in Selected Areas in a Health-Care Setting in South Africa, *BMC Microbiol.*, 2014, **14**, 100.
38. K. Woods, D. D. Beighton and J. L. Klein, Identification of the Streptococcus nginosus Group by Matrix-Assisted Laser Desorption Ionization Time-of-Flight Mass Spectro-metry, *J. Med. Microbiol.*, 2014, **63**(**9**), 1143–1147.
39. D. Duygu, T. Baykal, D. Acikgoz and K. Yildiz, Fourier Transform Infrared (FT-IR) Spectroscopy for Biological Studies, *G.U. J. Sci.*, 2009, **22**(**3**), 117–121.
40. G. Fischer, S. Braun, R. Thissen and W. Dott, FT-IR Spectroscopy as a Tool for Rapid Identification and Intra-Species Characterization of Airborne Filamentous Fungi, *J. Microbiol. Methods*, 2006, **64**(**1**), 63–77.
41. N. B. Valentine, T. J. Johnson, Y. F. Su and J. Forrester, FTIR Spectroscopy for Bacterial Spore Identification and Classification, *Proc. SPIE*, 2006, **6378**, p. 63780P.
42. P. L. Chan, P. H. Yu, Y. W. Cheng, C. Y. Chan and P. K. Wong, Comprehensive Characterization of Indoor Airborne Bacterial Profile, *J. Environ. Sci.*, 2009, **21**, 1148–1152.

43. Y. S. Cheng, Y. Zhou, C. M. Irvin, B. Kirkpatrick and L. Baker, Characterization of Aerosols Containing Microsystin, *Mar. Drugs*, 2007, **5**, 136–150.
44. C. A. Haskard, *Development of an Elisa for Microcystins, Research Report 27; Cooperative Research Center for Water Quality and Transport*, Salisbury, Australia, 2006, p. 84.
45. M. G. Weller, Immunoassays and Biosensors for the Detection of Cyanobacterial Toxins in Water, *Sensor*, 2013, **13**, 15085–15112.
46. I. Noss, I. M. Wouters, G. Bezemer, N. Metwali, I. Sander, M. Raulf-Heimsoth, D. J. J. Heederik, P. S. Thorne and G. Gert Doekes, β-(1,3)-Glucan Exposure Assessment by Passive Airborne Dust Sampling and New Sensitive Immunoassays, *Appl. Environ. Microbiol.*, 2010, **76**(**4**), 1158–1167.
47. O. G. Weingart, H. Gao, F. Crevoisier, F. Heitger, M. A. Avondet and H. Sigrist, A Bioanalytical Platform for Simultaneous Detection and Quantification of Biological Toxins, *Sensors*, 2012, **12**, 2324–2339.
48. K. Zhu, R. Dietrich, A. Didier, D. Doyscher and E. Märtbauer, Recent Developments in Antibody-Based Assays for the Detection of Bacterial Toxins, *Toxins*, 2014, **6**, 1325–1348.
49. Y. J. Sung, H. J. Suk, T. Li, H. Poo and M. G. Kim, Novel Antibody/Gold Nanoparticle/Magnetic Nanoparticle Nanocomposits for Immunomagnetic Separation and Rapid Colorimetric Detection of *Staphylococcus aureus* in Milk, *Biosens. Bioelectron.*, 2013, **43**, 432–439.
50. A. S. Afonso, B. Perez-Lopez, R. C. Faria, L. H. Mattoao, M. Hernandez-Herrero, A. X. Roig-Sauges, M. Maltez-de Costa and A. Merkoci, Electrochemical Detection of Salmonella Using Gold Nanoparticles, *Biosens. Bioelectron.*, 2013, **40**, 121–126.
51. I. R. Fruchey and P. A. Emanuel., *Market Survey Biological Detectors 2007 Edition*, U.S. Army Edgewood Chemical Biological Center, Aberdeen Proving Ground, MD 21010, 2007.
52. R. M. Ozanich, C. I. Baird, R. A. Bartholomew, H. A. Colburn, T. M. Straub and C. J. Bruckner-Lea, *Biodetection Technologies for First Responders: 2014 Edition*, Pacific Northwest National Laboratory, Richland, WA 99352, March 2014, http://biodetectionrresource.pnnl.gov, accessed 20 December 2014.
53. J. M. Dzenitis, B. J. Hindson, M. T. McBride, A. J. Makarewicz, B. D. Henderer, U. S. Sathyam, S. L. Nasarabadi, S. M. Smith, D. M. Gutierrez, D. M. Metz, D. M. Masquelier, S. B. Brown, A. M. Erler, K. S. Venkateswaran, and B. W. Colston, The Autonomous Pathogen Detection System, *23rd International Symposium on the Separation of Proteins, Peptides, and Polynucleotides*, Delray beach, FL, 10 November 2003, UCRL-PRES 200795.
54. B. Hindson, A. Markarewicz, U. Sethur, B. Henderer, M. McBride and J. Dzenitis, The Autonomous Pathogen Detection System, *Biosens. Bioelectron.*, 2005, **20**(1), 1925–1931.

55. J. J. Gooding, Bioosensor Technology for Detecting Biological Warfare Agents: Recent Progress and Recent Trends, *Anal. Chim. Acta*, 2006, **559**, 137–151.
56. A. V. Ogram and G. S. Sayler, The Use of Gene probes in the Rapid Analysis of Natural Microbial Communities, *J. Ind. Microbiol.*, 1988, **3**, 281–292.
57. J. Peccia and M. Hernandez, Incorporating Polymerase Chain Reaction-Based Identification, Population Characterization, and Quantification of Microorganisms Into Aerosol Science: A Review, *Atmos. Environ.*, 2006, **40**, 3941–3961.
58. N. Pusteria, J. E. Madigan and C. M. Leutenegger, Real-Time Polymerase Chain Reaction: A Novel Molecular Diagnostic Tool for Equine Infectious Diseases, *J. Vet. Intern. Med.*, 2006, **20**, 3–12.
59. M. T. Rahman, M. S. Uddin, R. Sultarm, A. Moue and M. Setu, Polymerase Chain Reaction (PCR): A Short Review, *Anwer Khan Mod. Med. Coll. J.*, 2013, **4**(**1**), 30–38.
60. L. Garibyan and N. Avashia, Polymerase Chair Reaction, *J. Invest. Dermatol.*, 2013, **133**, e6.
61. H. A. Hartley and A. J. Beaumner, Biosensor for the Specific Detection of a Single Viable *B. anthracis* Spore, *J. Analyt. Bioanalyt. Chem.*, 2003, **376**(**3**), 319–327.
62. J. I. Versage, D. D. Severin, M. C. Chu and J. M. Petersen, Development of a Multielement Real-Time TaqMan PCR for Enhanced Detection *Francisella tularensis* in Complex Specimens, *J. Clin. Microbiol.*, 2003, **41**(**12**), 5492–5499.
63. E. van Pelt-Verkuli, A. van Belkum and J. P. Hays, *Analysis of PCR Amplification Products in Principles and Technical Aspects of PCR Amplification*, ed. E. van Pelt-Verkuli, A. van Belkum and J. P. Hays, Springer, 2008, pp. 141–182.
64. B. Glynn, *Rapid Nucleic Acid-Based Diagnostics Methods for the Detection of Bacterial Pathogens in Principles of Bacterial Detection, Biosensors, Recognition Receptors, and Microsystems*, ed. M. Zourob, S. Elwary, and A. Turner, Springer, 2008.
65. J. L. McKillup, L. A. Jaykus and M. Drake, RNA Stability in Heat-Killed and UV Irradiated Enterotoxigenic *Staphylococcus aureus* and *Escherichia coli* 0157117, *Appl. Microbiol.*, 1998, **64**, 4264–4268.
66. J. E. Clarridge, Impact of 16S rRNA Gene Sequence Analysis for Identification of Bacteria on Clinical Microbiology and Infectious Diseases, *Clin. Microbiol. Rev.*, 2004, **17**(**4**), 840–862.
67. J. R. Cole, B. Chai, R. J. Farris, Q. Wang, A. S. Kulam-Syed-Mohideen, D. M. McGarrell, A. M. Bandel, E. Cardenas, G. M. Garrity and J. M. Tiedje, The Ribosomal Database Project (RDP-II): Introducing MyRDP Space and Quality Controlled Public Data, *Nucleic Acid Res.*, 2007, **35**(**Database Issue**), D169–D172.

68. R. H. Williams, E. Ward and H. A. McCartney, Methods for Integrating Air Sampling and DNA Analysis for Detection of Airborne Fungal Spores, *Appl. Environ. Microbiol.*, 2001, **67**, 2453–2459.
69. S.-I. Makino and H.-I. Cheun, Application of the Real Time–PCR for the Detection of Airborne Microbial Pathogens in Reference to the Anthrax Spores, *J. Microboil. Methods*, 2003, **53**(**8**), 141–147.
70. P. G. Ziros, P. A. Kokkinos, E. Legaki and A. Vantarakis, Development of an Optimized Method for the Detection of Airborne Viruses with Real-Time PCR Analysis, *Virol. J.*, 2011, **8**, 369–375.

CHAPTER 9

Aerobiological Aspects of Biological Warfare†

JOSEPH L. CORRIVEAU*[a]

[a]US Army Edgewood Chemical Biological Center, Aberdeen Proving Ground, MD 21010, USA
*E-mail: joseph.l.corriveau.civ@mail.mil

9.1 Introduction

"We're in the cloud," shouted the technician as he jumped out of the trailer containing the laser-based standoff early warning system that was tracking the biological aerosol "cloud." Minutes earlier and a kilometer away, a truck with an agricultural sprayer drove down a road disseminating a cloud of *Bacillus globigii* (BG) spores (*B. globigii* is now known as *B. atrophaeus*). The BG cloud of spores had quickly disappeared from view, but was tracked successfully by the laser-based biological aerosol detector as it drifted away. A sudden change in wind direction had moved the cloud of spores over our position. Curiously, I sniffed at the air. My senses indicated that I was inhaling fresh Dugway desert air. Yet, I knew that the air was filled with respirable

†The views and opinions expressed in this chapter are those of the author and should not be construed as an official Department of the Army position, policy, or decision, unless so designated by other official documentation. Citation of trade names in this chapter does not constitute an official Department of the Army endorsement or approval of the use of such commercial items.

Issues in Toxicology No. 25
Aerobiology: The Toxicology of Airborne Pathogens and Toxins
Edited by Harry Salem and Sidney Katz

Published by the Royal Society of Chemistry, www.rsc.org

BG spores, a biological warfare agent simulant for anthrax because of its particle size and dissemination similarities to *B. anthracis*.

Aerosolized biological warfare agents (particles approximately 1–5 μm in diameter) pose a substantial inhalation risk to soldiers on the battlefield or, in the case of a bioterrorist attack, to public health.[1] The natural process of breathing provides a mechanism for the inhalation of infectious or toxic particles that can become deposited in the alveoli of the lungs. In the absence of a warning of the presence of an odorless and "invisible" biological aerosol threat, unprotected individuals may unknowingly become infected, thereby delaying the application of medical countermeasures until treatment is triggered by the appearance of disease signs and symptoms. And therein lies the *raison d'être* for biological warfare. It is a silent killer, a weapon of mass destruction with the potential to kill or harm large numbers of humans.

The magnitude of a strategic-level aerosolized biological attack on a city was captured by the World Health Organization (WHO) publication entitled *Public Health Response to Biological and Chemical Weapons*.[2] The scenario described in the WHO study is for one or a few aircraft to disseminate 50 kg of dried powder biological agent (representing 6×10^{15} organisms) along a 2 km line at a right-angle to the wind direction moving over the target area under stable atmospheric conditions (likely at night). In a developed city of five million inhabitants, the estimated number of casualties from an attack using *Fransicella tularensis* is 19 000 deaths and 250 000 incapacitated (including deaths). The estimated number of casualties from an attack using *B. anthracis* is 100 000 deaths and 250 000 incapacitated (including deaths).

The Centers for Disease Control and Prevention (CDC) has established a list of potential bioterrorism agents that the United States must be prepared to address.[3] The list of highest priority agents, termed Category A agents, is provided in Table 9.1. Category A agents pose a significant risk to national security because they (1) can be easily disseminated or transmitted person

Table 9.1 CDC list of Category A biological agents. Data from CDC[3]

Biological agent	Disease	Likely method of dissemination[a]
Variola major	Smallpox	Aerosol
Bacillus anthracis	Anthrax	Spores in aerosol
Yersinia pestis	Plague	Aerosol, infected vectors
Clostridium botulinum	Botulism	Aerosol, food, water
Francisella tularensis	Tularemia	Aerosol
Ebola virus	Ebola hemorrhagic fever	Aerosol
Marburg virus	Marburg hemorrhagic fever	Not listed
Lassa virus	Lassa fever	Aerosol
Junin virus	Argentine hemorrhagic fever	Aerosol

[a]Likely method of dissemination as a potential biological warfare agent as assessed in Department of the Army FM 3-11.9.[5]

to person, (2) can cause high mortality and have potential for major public health impact, (3) may cause public panic and social disruption and (4) require special action for public health preparedness. Also provided in Table 9.1 is a United States military assessment as to the likely method of dissemination of biological agents that are included in the CDC Category A list.[5] The likely method of dissemination for every agent is by aerosol.

The 2014 naturally occurring outbreak of Ebola virus disease in West Africa serves as an example of the devastation that can result from a Category A biological agent. There is no scientific evidence that Ebola virus is transmitted person to person by inhalation of virus-containing aerosols, yet as of 24 December 2014 there were 19 497 reported cases of Ebola virus disease with 7588 deaths.[4] One can imagine the profound impact of a biological attack using aerosolized Ebola virus. The United States military assesses that the likely method of dissemination of Ebola virus as a potential biological warfare agent is as an aerosol.[5]

The United States,[6] former Soviet Union[7,8] and to a lesser extent Iraq[9] had substantial biological warfare programs in the 20th century that focused on agents that could be weaponized for aerosol delivery. The United States biological weapons program included *B. anthracis* (the causative agent of anthrax), *F. tularensis* (tularemia), *Brucella suis* (brucellosis), *Coxiella burnetii* (Q fever), Venezuelan equine encephalitis virus, *Staphylococcus* enterotoxin B and botulinum toxin.[1] Ricin derived from castor bean was also investigated by the United States, in addition to plant pathogens such as *Puccinia graminis* (wheat stem rust) and *Magnaporthe grisea* (rice blast).[6] Likely candidate biological agents considered by former Soviet Union experts included *B. anthracis*, *F. tularensis*, *Yersinia pestis* (plague), *C. burnetii*, *Rickettsia prowaseckii* (typhus), *Burkholderia pseudomallei* (melioidosis), variola major (smallpox), equine encephalitis virus, Marburg virus and botulinum toxin.[7,8] Iraq's biological weapons program included *B. anthracis*, *Clostridium perfringens* (gas gangrene), botulinum toxin, ricin, aflatoxin and the plant pathogen *Tilletia* sp. (wheat cover smut).[9] Many of the pathogens developed by the United States, former Soviet Union and Iraq for warfare are included in the CDC Category A agents (Table 9.1). Others are CDC Category B agents, to include *C. burnetii*, *Brucella* sp., alpha viruses such as Venezuelan equine encephalitis virus and ricin.[3] Category B agents are the second highest priority pathogens to the CDC. Category B agents differ from Category A agents in that they are, for example, not as easily disseminated and result in "moderate" *versus* "high" mortality rates.

The United States biological warfare program was renounced and abandoned under President Richard Nixon in 1969. Subsequently, the United States ratified the Biological Weapons Convention (BWC) treaty in 1975. Each State Party to the BWC undertakes to neither develop, produce, stockpile or otherwise acquire or retain biological agents/toxins in quantities that have no justification for prophylactic, protective or other peaceful purposes, nor weapons, equipment or means of delivery designed to use such agents/toxins for hostile purposes or in armed conflict. The former Soviet Union also ratified the BWC treaty in 1975, but there is strong evidence that it continued to develop and produce biological warfare agents, as evidenced by the inhalation anthrax outbreak in Sverdlovsk, Russia (now named Yekaterinburg) in 1979.[10,11] Iraq ratified the BWC treaty in 1991, in the aftermath of the Gulf War.

In this chapter, we examine aerobiological aspects of bacterial, viral, fungal and toxic agents that were developed for aerosol delivery as biological warfare agents. This includes deep dives into the spore-forming bacterial warfare agent (*B. anthracis*) that was included in the United States, former Soviet and Iraqi biological weapons programs; the vegetative bacterial agent (*F. tularensis*) developed in the United States and former Soviet Union; Ebola and Marburg filoviruses investigated by the former Soviet Union; the toxin ricin that was developed by the United States and Iraq (and apparently Syria[12]); and the fungus *Tilletia* sp. that was developed by Iraq to attack wheat crops. However, first we examine some fundamental aspects of biological agents as respirable aerosols.

9.2 Biological Agents As Infectious Aerosols

Biological warfare agent properties include (1) lethality, (2) ease of production, (3) stability as an aerosol and (4) ability to be dispersed as an aerosol.[1] All of these properties are associated directly or indirectly with the size of the particles containing the biological agent. Indeed, a particle's size is of the utmost importance in determining whether it becomes and remains airborne and infectious/toxic. In this section, we focus on two aerobiological aspects of infectious aerosols, namely the optimal particle size for lower respiratory tract deposition and airborne movement.

A starting point in the examination of infectious aerosols as a biological weapon is the natural mechanism for aerosol transmission of infectious pathogens from person to person, which can occur by "droplet" or "airborne" means. Droplet transmission can be defined as the transmission of disease by expelled particles (>5 μm in size) generated by coughing, sneezing or talking that settle on a surface quickly, typically within a few feet of the source.[13] Sneezing can produce up to 40 000 moisture-laden droplets 0.5–12 μm in diameter.[14] These droplets can be expelled at a velocity of 100 m s^{-1}, reaching distances of several meters.[15] Diseases transmitted by droplets include pneumonia, pertussis, diphtheria, influenza, mumps and meningococcal meningitis. Airborne transmission occurs when droplets become desiccated to form what are termed "droplet nuclei" that are <5 μm in size. These particles can remain suspended in the air for longer periods, especially if they are bound to dust particles.[13] Examples of diseases transmitted by airborne transmission include tuberculosis, measles and chickenpox. Droplet nuclei or infectious particles of similar size, notably <1–3 μm, are capable of deep lung deposition in the alveoli.[16]

Synthetic biological warfare agent formulations would be expected to contain particles that are ~1–5 μm in size.[1] Particles of this size are amenable to

Table 9.2 Influence of aerosolized anthrax spore-containing particle size on infectivity in guinea pigs. Adapted from data reported by Druett *et al.*[17]

Particle diameter/ μm	Spores/particle	Particles L^{-1} air × 10^4	Dead animals/total animals
3.6	18	9.2	40/40
8.4[a]	19	7.3	11/40
11.6[a]	18	6.1	10/40

[a]Particles made to size by the addition of dextrin to the spray fluid.

both respiratory tract delivery and aerosol dissemination. Scientific experimentation by Druett *et al.*, using *B. anthracis* spore aerosol exposures on guinea pigs and rhesus macaques, illustrated the importance of particle size in respiratory infection.[17] As shown in Table 9.2, when guinea pigs were exposed to particles of size 3.6, 8.4 and 11.6 μm (with number of spores per particle held constant at ~18), the 3.6 μm particles were far more infective than 8.4 or 11.6 μm particles. The infectivity of the smaller particles was even more pronounced when considering aerosols containing single spores, which would be ~1 μm in size. Indeed, single spores were about 17 times more infective in guinea pigs than aerosols containing 12 μm particles. The same held true in the experiments with the rhesus macaques, in which single spores were about 14 times more infective than 12 μm spore-containing particles. The authors attributed the greater infectivity of the smaller particles (<4 μm) to deposition in the lung, whereas larger particles were retained in the upper respiratory tract.

Finally, biological warfare agents have to be capable of airborne transmission for considerable distances to be considered effective. Movement of particles in the air is determined by Stokes settling law, which essentially determines how quickly a sphere falls under the opposing forces of gravity (downwards) and air friction (upwards). Knight estimated that 10, 20 and 100 μm particles would fall the height of a room in 17 min, 4 min and 10 s, respectively.[18] Particles of 1–3 μm, on the other hand, could remain airborne "indefinitely." He concluded that particles should be ≤10 μm in size to be effective in airborne transmission.

9.3 Anthrax Spores (*Bacillus anthracis*)

Anthrax spores meet the characteristics of a biological warfare agent in large part because of their size. Using transmission electron microscopy, Carrera *et al.* determined that the spores of *B. anthracis* Ames measure ~1.52 μm in length and 0.81 μm in diameter.[19] Three other virulent strains had similar sized spores. Spores this size are optimal for deep lung deposition and for dissemination as an aerosol. Furthermore, *B. anthracis* spores are extremely stable in the environment, capable of remaining viable for years.[20] The combination of virulence, size and environmental persistence makes *B. anthracis* perhaps the biological agent of choice for a bioweaponeer.

Anthrax is primarily a disease of herbivores such as sheep, goats, cattle and other animals.[21] This disease had a significant impact on 19th century agriculture, which was a contributing factor to a number of seminal scientific advancements in the cause and prevention of disease, including Davaine's identification of the pathogenic role of bacteria,[22] the development of Koch's postulates[23] and Pasteur's demonstration of the protective action of vaccination.[24] Animals grazing on anthrax spore-contaminated plants are at risk of gastrointestinal anthrax. Human cases of anthrax have typically been associated with agricultural workers coming into contact with infected animals or industrial workers coming into contact with infected animal products, notably workers in animal hair processing. Cutaneous anthrax accounts for 95% of human anthrax cases, with inhalation anthrax accounting for 5% and the gastrointestinal form 0–5% (gastrointestinal anthrax has not been reported in some parts of the world, *e.g.* the United States).[21]

Two historical events in the United States regarding human inhalation anthrax are worthy of discussion. First, an epidemic of inhalation anthrax occurred in 1957 among workers at a goat hair processing mill in Manchester, NH.[25,26] There were four cases of cutaneous anthrax and five cases of inhalation anthrax, of which four were fatal. Albrink *et al.* concluded that, based on the pathogenesis of the disease and previously reported experimental work, the inhalation anthrax was caused by *B. anthracis* spores on small particles (judged to be ≤5 μm in size) that became deposited in the alveoli of the victim's lungs.[25] The mill in which the epidemic arose was eventually decontaminated and razed to the ground. The second event is the anthrax in the mail bioterrorist attack of 2001. A powdered formulation containing *B. anthracis* Ames strain spores was "disseminated" by four letters that were transported through the United States Postal Service.[27] There were 11 cutaneous and 11 inhalation cases of anthrax. Five of the victims died of inhalation anthrax. An unsettling aspect of this bioterrorist event is that it demonstrated the potential of *B. anthracis* spores as a bioweapon.

In 1979 in Sverdlovsk (now Yekaterinburg), Russia, there was an anthrax epidemic that has been attributed to the airborne release of *B. anthracis* spores from a military biological facility named Compound 19.[10] Meselson *et al.* were able to compile a list of 77 victims of the epidemic. For 66 of them, they were able to obtain residential and workplace addresses. An examination of the geological location of these 66 victims at the time of the onset of the epidemic in concert with local meteorological conditions (northerly wind) for 2 April revealed that the majority of the victims (60 of 66) could be placed in a narrow zone (about 4 km long) extending from Compound 19 to the southern city limit. Further south along the axis of the narrow zone there were six villages with reports of animal anthrax. One of the villages, Abramovo, was 50 km south-southeast of Compound 19. Meselson *et al.*[10] concluded that there was an accidental release of *B. anthracis* spores on 2 April from the military biological facility. This conclusion is supported by Abramova *et al.*, who performed necropsies on 42 of the victims and found pathological evidence for inhalation anthrax.[11] To date, the Russian government has not provided an explanation for what happened in Compound 19 that led to the release of the cloud of infectious *B. anthracis* spores.

9.4 Vegetative Bacteria (*Francisella tularensis*)

Francisella tularensis is a Gram-negative bacterium with a diameter of 0.2–0.7 μm that is the causative agent of tularemia, a disease found globally in mammals and arthropod vectors.[28] It was first described in squirrels in California in 1911. Human infections by *F. tularensis* can occur by ingestion, dermal contact or inhalation. Most natural cases of human inhalation tularemia tend to occur in rural areas. For example, a significant outbreak of tularemia (over 600 cases) occurred in 1966–1967 in Sweden, where farm workers unwittingly generated infectious *F. tularensis* aerosols as they worked, notably with rodent-infested hay.[29] Tularemia is also recognized as one of the top 10 most frequently reported laboratory infections, with the greatest risk coming from infectious aerosols generated by the manipulation of *F. tularensis* cultures.[30]

A number of studies have been performed to determine the infectious dose for inhalation tularemia.[28] Here we examine the work of Day *et al.*, who conducted experiments that demonstrated a relationship between particle size and infectivity of aerosolized *F. tularensis*.[31] Rhesus macaques were exposed to aerosols containing *F. tularensis*-laden particles with median diameters of 2.1, 7.5, 12.5 and 24 μm. All 48 animals exposed to 2.1 or 7.5 μm particles developed infections in their lower respiratory tract and died within 4–8 days post-infection. Of the 45 animals exposed to 12.5 or 24 μm particles, 38 became infected. Many of them had infections in their upper respiratory tract and death occurred in 7–10 days. Seven of them developed severe conjunctivitis and subsequently died 10–21 days post-infection. As shown in Table 9.3, LD_{50} values increased from 14 to 4447 cells as the median diameter of the infectious particles increased from 2.1 to 24 μm. Furthermore, based on extensive experimentation by others regarding the estimated infectious dose of *F. tularensis*, it is estimated that just one cell deposited in the respiratory tract can produce infection in humans.[28] Given the exceptional infectivity of *F. tularensis* particles, which appear to be amenable to aerosol dissemination, it is not surprising that tularemia was of keen interest as a biological warfare agent.

Table 9.3 Respiratory LD_{50} values for *Macaca mulatta* exposed to aerosols of different particles sizes containing *F. tularensis*. Table of data as reported in Day *et al.*[31]

ONMD[a]/μm	Respiratory LD_{50} (cells)	95% confidence limits (cells)	Respiratory ID_{50} (cells)
2.1	14	7–27	<5
7.5	378	50–630	<240
12.5	872	250–1500	<556
24.0	4447	1600–8500	<340

[a]Organism number median diameter is defined by Day *et al.*[31] as the diameter (in μm) of that particle representative of the median number of organisms.

Unlike *B. anthracis*, which has been shown to have been a sought-after biological agent in part because of its ability to form resistant spores, *F. tularensis* is not a spore-forming bacterium. However, the bacterium is known to survive for weeks at low temperatures in water, moist soil, hay and animal carcasses.[29] Other studies have shown that *F. tularemia* under favorable conditions can be moderately stable when disseminated as an aerosol.

Beebe studied the stability of dry and wet disseminated aerosol clouds of *F. tularensis* subjected to simulated solar range UV light at various humidities at 30 °C.[32] The dry *F. tularensis* formulation containing sucrose and skim-milk solids was milled to yield particles with a mass median diameter of about 3.5 μm. Liquid *F. tularensis* cultures were used to generate wet-disseminated aerosols containing particles with a mass median diameter of about 1.5 μm. In general, the dry-disseminated aerosols were found to be more resistant to the action of solar range UV light than the wet-disseminated aerosols. For example, at 25% relative humidity (RH) the decay rate of wet aerosols was 72% min^{-1} compared with a decay rate of 31.7% min^{-1} for dry aerosols. However, in general, wet and dry aerosols had similar decay rates in the absence of light. At RH 25 and 60%, the decay rates were 7.0 and 17.8% min^{-1} for wet aerosols and 10.5 and 18.7% min^{-1} for dry aerosols, respectively.

Finally, an important aspect of the decay of biological aerosols is their ability to maintain infectivity. The effect of aerosol age on *F. tularensis* infectivity was studied by Sawyer *et al.* using rhesus macaques and volunteer soldiers.[33] In this study, wet aerosols containing particles ≤5 μm in size were generated and held in darkness at 24 °C in a 10^6 L spherical static chamber. Under these experimental conditions, the decay rate for the *F. tularensis* was 1–2% min^{-1}. Data collected for the human infectivity of aged aerosols of *F. tularensis* are presented in Table 9.4. The infectivity of the *F. tularensis* aerosols did not appear to diminish after 60 min of aging. However, infectivity was reduced after aging for either 120 or 180 min. Neither the incubation period nor the severity of the illness varied with the age of the aerosol. Worthy of mention, early treatment with streptomycin was used to terminate the illness without complications in all of the volunteer soldiers.

Table 9.4 Infectivity of aged *F. tularensis* aerosols in humans. Table of data as reported in Sawyer *et al.*[33]

Inhaled dose (viable cells)	No. of infected/No. exposed to aerosol aged for			
	30 min	60 min	120 min	180 min
80			0/4	
150	2/4	3/4		
350		4/4		0/4
750		4/4		0/4
7500			8/8	8/8
12 500				3/4
50 000			7/8	10/12

9.5 Filoviruses (Marburg and Ebola Viruses)

Marburg and Ebola viruses are enveloped negative strand RNA filoviruses that cause severe viral hemorrhagic fever disease.[34] Marburg virus was first identified in 1967 after an outbreak of viral hemorrhagic fever in Marburg, Germany. Almost a decade later, in 1976, Ebola virus was identified during outbreaks in Zaire and Sudan. There are four recognized sub-types of Ebola virus, namely the human pathogenic Zaire and Sudan sub-types and also the Reston and Ivory Coast sub-types, which are pathogenic in nonhuman primates (NHPs) but not in humans. Unlike the inhalation anthrax and tularemia diseases discussed above, the filoviruses are not naturally transmitted by aerosol.[34] Transmission appears to be by direct contact with infectious blood or body fluids. However, the filoviruses reportedly were investigated as biological weapons by the former Soviet Union.[8]

The infection of NHPs by wet-disseminated Ebola and Marburg virus preparations has demonstrated that very low doses of these filoviruses are required for aerosol transmission. Johnson *et al.* used rhesus macaques to evaluate the aerosol infectivity of Ebola virus (Zaire type).[35] Virus-containing droplets 0.8–1.2 μm in size at doses as low as 400 plaque-forming units (pfu) caused disease, with death occurring 4–5 days post-infection. Alves *et al.* used cynomolgus macaques to evaluate the aerosol infectivity of Marburg virus.[36] Virus-containing droplets at doses as low as 2–14 pfu caused disease, with death occurring 9 days post-infection. These studies indicate that pathogens for which there is no natural airborne transmission may still hold potential for anthropogenic aerosolization as a biological weapon.

Further evidence that suggests that filoviruses could be considered by a bioweaponeer is their relative stability in the environment. Piercy *et al.* studied the stability of Marburg and Ebola (Zaire) viruses on surfaces and as aerosols.[37] They found that viruses could not be recovered from surfaces (plastic, stainless steel and glass) when experiments were run at room temperature (RH ~55%). However, when the experiments were run at 4 °C they were able to show that the filoviruses could be recovered from plastic and glass surfaces for over a 3 week period. The aerosol studies were carried out using a 40 L rotating Goldberg drum (temperature 22 °C and RH 50–55%) in which an aerosol cloud of 1–3 μm particles was generated. The estimated decay rate determined for both Marburg and Ebola viruses was about 3% min^{-1}.

9.6 Toxins (Ricin)

The history of ricin as a biological weapon can be traced to United States efforts during World War I to examine the potential for ricin to coat shrapnel bullets and for dissemination as "dust clouds."[6] Ricin was to remain a biological agent of interest by other nations over the next century, including Iraq[9] and Syria[12] (in the summer of 2014, as part of its chemical weapons program declaration, Syria declared a ricin chemical weapons production facility to the Organization for the Prohibition on Chemical Weapons). Perhaps one of the reasons for the interest in ricin as a bioweapon is its relative, or perceived, ease of production. Ricin is an extremely potent ribosome-inactivating protein derived from the seeds of the castor bean plant.[38] The seeds for the castor bean plant are readily available because it is a legitimate crop grown throughout the world and it is appreciated as an ornamental plant.

The inhalation toxicity of aerosols containing ricin particles has been determined for a number of animal models, including rhesus macaques.[38,39] Bhaskaran *et al.* reported that the LD_{50} of ricin by the inhalation route was 5.8 μg kg^{-1} body weight.[38] Particles about 1 μm in size were used in the study because they had been shown to be more efficient at reaching the lower regions of the lungs. Here the toxin causes pulmonary edema and necrosis of the alveoli. Previous work using a mouse model by Roy *et al.* had shown that particle size was important in ricin aerosol lethality.[40] An estimated 50% of aerosolized 1 μm particles were deposited in the alveoli, whereas larger particles were far less efficient at reaching the lower regions of the lungs and the subsequent lethality of the toxic aerosol was diminished.

In addition to the extreme inhalation hazard of ricin aerosols within a size range that is amenable to aerosol dissemination, the ricin protein has been shown to be moderately stable under some environmental conditions. The US Environmental Protection Agency (EPA) performed a persistency study of ricin on galvanized metal and painted concrete surfaces under different test conditions: ambient conditions (20 °C and RH 40–70%) and at an elevated temperature of 30 °C (>70% high RH or <40% low RH).[41] Ricin was extracted at specific intervals and bioactivity was measured with a cytotoxicity assay. The bioactivity of ricin persisted on painted concrete surfaces for up to 14 days under all environmental test conditions and on galvanized metal under low RH conditions.

9.7 Fungal Spores (Wheat Cover Smut – *Tilletia* sp.)

Plant pathogens, like human pathogens, have contributed to shaping the course of civilization.[42] In the 1840s, a blight disease of potato crops (caused by the fungus *Phytophthora infestans*) led to the Irish Potato Famine. There were over one million deaths in Ireland and another million Irish emigrated. A century later, in the midst of World War II, brown spot on rice (caused by the fungus *Helminthosporium oryzae*) was the major contributing factor to the Great Bengal Famine of 1943, in which two million people perished.[43] Given the potential for catastrophic loss of major food sources by a nation, it is no wonder that the concept of targeting food supplies with biological weapons was conceived.

As part of its biological weapons declaration to the United Nations, Iraq disclosed that *Tilletia* sp., the causative agent of wheat cover smut, was included in their biological weapons program.[9] Their declared intent was to use wheat cover smut as an "economic" weapon. A 5 ton stockpile of *Tilletia* spores was produced in the late 1980s. This was accomplished by harvesting *Tilletia*-infected spikes from a 5 ha crop of wheat. The stockpile ultimately was destroyed by fire in 1991 after the Gulf War ended.

Fungal spores for former biological warfare agents such as *P. graminis* (wheat stem rust), *M. grisea* (rice blast) and *Tilletia* spores (wheat cover smut) possess aerobiological traits that are amenable to bioweaponeers, such as resistant spores with a size range amenable to dissemination over fields of food crops. In the case of *Tilletia*, when wheat spikes are infected each diseased kernel can contain up to millions of resistant teliospores.[44] These fungal spores are about 15–25 μm in size, can be harvested from an infected field of wheat (as demonstrated by the Iraqis, with a successful spike infection rate of 80%), and can remain viable for five years or longer, notably in a dry environment.

9.8 Future Prospects for Biological Warfare

The aerobiological aspects of biological warfare were determined for a number of weaponized biological agents during the 20th century and examined for many more candidate agents. Two of the more important aerobiological aspects of the agents are size (as related to respirability and aerosol dissemination) and stability. The spore-forming bacterium, vegetative bacterium, filoviruses, biologically derived toxin and fungal spores discussed here are historical examples of biological agents that have an adequate size and stability for use in biological warfare. However, aerobiology is only one aspect of a biological weapons capability. A discussion of the future prospects of biological warfare requires the consideration of technical requirements beyond agent size and stability.

Bioweaponeers must be capable of large-scale agent production and processing, weaponization and delivery of stable virulent/toxic biological agents with adequate aerosol characteristics. The entire process is extremely hazardous and requires robust safety measures. Furthermore, even if all of the technical barriers are overcome, one cannot control the potential for uncontrollable meteorological and environmental conditions that would thwart a biological aerosol attack.

Arguably, only a state actor or highly sophisticated non-state actor could overcome these technical barriers to develop a viable biological weapon capable of widespread harm *via* aerosol dissemination. Indeed, even the Aum Shinrikyo terrorist group was unable to develop and use a biological weapon successfully.[45] This included a number of botched attacks with botulinum toxin and *B. anthracis*. The inability to produce a virulent biological agent appeared to be a major obstacle. In one case, a cult member fell into a fermentation tank for *C. botulinum* production; he nearly drowned but did not show any signs of disease. In another case, cult members inadvertently inhaled *B. anthracis* spores but never became ill.

Despite the technical barriers associated with biological weapons, it is recognized that the astonishing advances in the life sciences could be used for malevolent purposes.[46–48] Petro *et al.* assessed that advances in biosciences could be used for the development of new biological warfare agents that, for example, are resistant to medical countermeasures, have enhanced virulence and aerosol stability and can even be tailored to specific populations.[46]

This harsh reality has not escaped the attention of world leaders. Soon after coming to office, President Obama in 2009 signed the National Strategy for Countering Biological Threats. More recently, Vladimir Putin in 2012 wrote an essay in which he included "genetic"-based weapons in a list of "hi-tech" weapons that will be developed in the future and be comparable in effect to nuclear weapons.[49] There is no doubt that the "genetic" weapons that Putin referred to are biological weapons.

In conclusion, there are three overarching factors that will help to avert a large-scale biological attack in the 21st century. The first two are linked. First, there are a number of technical barriers to overcome to obtain and use a biological weapon capable of large-scale harm. History has shown that it would likely be a dedicated and sophisticated military program with substantial tacit knowledge in biological weapons development that would pose a significant biological warfare capability. Second, the Biological and Toxin Weapons Convention (BWC) has proven to be an impactful international forum to reduce and ultimately eliminate tacit knowledge associated with former biological weapons programs and to reinforce norms against the potential for future misuse of the life sciences. Nations need to continuously advance as a norm the safe and beneficial use of the life sciences and to decry the misuse of science. The overwhelming majority of nations have ratified the BWC treaty: 171 nations as of December 2014. Finally, global health security to counter outbreaks of disease, whether they are of natural or deliberate origin, not only serves to protect and save lives, but also acts as a deterrent. This includes an ability for biological agent recognition (environmental detection, diagnostics and forensics/attribution), personal protection, decontamination and medical countermeasures.

References

1. M. K. Kortepeter and G. W. Parker, Potential Biological Weapons Threats, *Emerging Infect. Dis.*, 1999, **5**(**4**), 523–527.
2. World Health Organization Group of Consultants, *Public Health Response to Biological and Chemical Weapons*, World Health Organization, 1970.
3. Centers for Disease Control and Prevention, Biological and Chemical Terrorism: Strategic Plan for Preparedness and Response, Recommendations of the CDC Strategic Planning Workgroup, *MMWR*, 2000, **49**(**RR-4**), 1–14.
4. World Health Organization, *Ebola Response Roadmap Situational Report*, 24 December 2014.
5. Department of the Army, *Potential Military Chemical/Biological Agents and Compounds*, FM 3–11.9, 2005.
6. J. K. Smart, in *History of Chemical and Biological Warfare: An American Perspective, Medical Aspects of Chemical and Biological Warfare*, ed. F. R. Sidell, E. T. Takafuji and D. R. Franz, TMM Publications, Washington, DC, 1997, ch. 2, pp. 9–86.
7. Defense Intelligence Agency, *Soviet Biological Warfare Threat, DST-1610F-057–86*, U.S. Department of Defense, Washington, DC, 1986.

8. K. Alibek and S. Handelman, *Biohazard*, Random House, New York, NY, 1999.
9. United Nations Monitoring, *Verification and Inspection Commision, Compendium of Iraq's Proscibed Weapons Programmes in the Chemical, Biological and Missile Areas*, June 2007.
10. M. Meselson, J. Guillemin, M. Hugh-Jones, A. Langmuir, I. Popova, A. Shelokov and O. Yampolskaya, The Sverdlovsk Anthrax Outbreak of 1979, *Science*, 1994, **266(5188)**, 1202–1208.
11. F. A. Abramova, L. M. Grinberg, O. V. Yampolskaya and D. H. Walker, Pathology of Inhalation Anthrax in 42 Cases from the Sverdlovsk Outbreak in 1979, *Proc. Natl. Acad. Sci. U. S. A.*, 1993, **90**, 2291–2294.
12. Note by the Director-General, *Progress in the Elimination of the Syrian Chemical Weapons Programme*, Organisation for the Prohibition of Chemical Weapons, 25 July 2014.
13. World Health Organization, *Practical Guidelines for Infection Control in Health Care Facilities, Vol. 41 of SEARO Regional Publication*, Regional Office for Western Pacific, Manila, Philippines, 2005.
14. E. C. Cole and C. E. Cook, Characterization of Infectious Aerosols in Health Care Facilities: An Aid to Effective Engineering Controls and Preventive Strategies, *Am. J. Infect. Control*, 1998, **26**(**4**), 453–464.
15. W. F. Wells, *Airborne Contagion and Air Hygiene*, Harvard University Press, Cambridge, 1955.
16. R. J. Thomas, Particle Size and Pathogenicity in the Respiratory Tract, *Virulence*, 2013, **4**(**8**), 847–858.
17. H. A. Druett, D. W. Henderson, L. Packman and S. Peacock, Studies on Respiratory Infection. I. The Influence of Particle Size on Respiratory Infection with Anthrax Spores, *J. Hyg.*, 1953, **51**(**3**), 359–371.
18. V. Knight, Viruses as Agents of Airborne Contagion, *Ann. N. Y. Acad. Sci.*, 1980, **353**, 147–156.
19. M. Carrera, R. O. Zandomeni, J. Fitzgibbon and J. L. Sagripanti, Difference Between the Spore Sizes of *Bacillus anthracis* and other *Bacillus* species, *J. Appl. Microbiol.*, 2007, **102**(**2**), 303–312.
20. D. C. Dragon and R. P. Rennie, The Ecology of Anthrax Spores: tough but not Invincible, *Can. Vet. J.*, 1995, **36**(**5**), 295–301.
21. J. C. Pile, J. D. Malone, E. M. Eitzen and A. M. Friedlander, Anthrax as a Potential Biological Warfare Agent, *Arch. Intern. Med.*, 1998, **158**(**5**), 429–434.
22. C. Davaine, Recherches sur les Infusoires du Sang dans la Malaidie Connue sous le Nom de Sang de Rate, *Compt. Rend. Acad. de Sci.*, 1863, **57**, 220–223.
23. R. Koch, Untersuchungen uber Bacterien. Die Ätiologie der Milzbrand-Krankheit, Begründet auf die Entwichlungsgeschichte des Bacillus anthracis, *Beitr. Biol. Pflanz.*, 1877, **2**, 277–308.
24. L. Pasteur, De l'atténuation des virus et de leur retour à la virulence, *C. R. Acad. Sci.*, 1881, **92**, 429–435.
25. W. S. Albrink, S. M. Brooks, R. E. Biron and M. Kopel, Human Inhalation Anthrax, *Am. J. Pathol.*, 1960, **36**(**4**), 457–471.

26. P. S. Brachman, A. F. Kaufman and F. G. Dalldorf, Industrial Inhalation Anthrax, *Bacteriol. Rev.*, 1966, **30**(**3**), 646–659.
27. D. A. Rasko, P. L. Worsham, T. G. Abshire, S. T. Stanley, J. D. Bannan, M. R. Wilson, R. J. Langham, R. S. Decker, L. Jiang, T. D. Read, A. M. Phillippy, S. L. Salzberg, M. Pop, M. N. Van Ert, L. J. Kenefic, P. S. Keim, C. M. Fraser-Liggett and J. Ravel, *Bacillus anthracis* Comparative Genome Analysis in Support of the Amerithrax Investigation, *Proc. Natl. Acad. Sci. U. S. A.*, 2011, **108**(**12**), 5027–5032.
28. R. M. Jones, M. Nicas, A. Hubbard, M. D. Sylvester and A. Reingold, The Infectious Dose of *Francisella tularensis* (Tularemia), *Appl. Biosafety*, 2005, **10**(**4**), 227–239.
29. D. T. Dennis, T. V. Inglesby, D. A. Henderson, J. G. Bartlett, M. S. Ascher, E. Eitzen, A. D. Fine, A. M. Friedlander, J. Hauer, M. Layton, S. R. Lillibridge, J. E. McDade, M. T. Osterholm, T. O'Toole, G. Parker, T. M. Perl, P. K. Russell and K. Tonat, Tularemia as a Biological Weapon, *JAMA*, 2001, **285**(**21**), 2763–2773.
30. K. Singh, Laboratory-Acquired Infections, *Clin. Infect. Dis.*, 2009, **49**, 142–147.
31. W. C. Day and R. F. Berendt, Experimental Tularemia in *Macaca mulatte*: Relationship of Aerosol Particle Size to the Infectivity of Airborne *Pasteurella tularensis*, *Infect. Immun.*, 1972, **5**(**1**), 77–82.
32. J. M. Beebe, Stability of Disseminated Aerosols of *Pasteurella tularensis* Subjected to Simulated Solar Radiations at Various Humidities, *J. Bacteriol.*, 1959, **78**(**1**), 18–24.
33. W. D. Sawyer, J. V. Jemski, A. L. Hogge, Jr., H. T. Eigelsbach, E. K. Wolfe, H. G. Dangerfield, W. S. Gochenour Jr. and D. Crozier, Effect of Aerosol Age on the Infectivity of Airborne *Pasteurella tularensis* for *Macaca mulatta* and Man, *J. Bacteriol.*, 1966, **91**(**6**), 2180–2184.
34. E. K. Leffel and D. S. Reed, Marburg and Ebola Viruses as Aerosol Threats, *Biosecur. Bioterror.*, 2004, **2**(**3**), 186–191.
35. E. Johnson, N. Jaax, J. White and P. Jarling, Lethal Experimental Infections of Rhesus Monkeys by Aerosolized Ebola Virus, *Int. J. Exp. Pathol.*, 1995, **76**, 227–236.
36. D. A. Alves, A. R. Glynn, K. E. Steele, M. G. Lackemeyer, N. L. Garza, J. G. Buck, C. Mech and D. S. Reed, Aerosol Exposure to the Angola Strain of Marburg Virus Causes Lethal Viral Hemorrhagic Fever in Cynomolgus Macaques, *Vet. Pathol.*, 2010, **47**(**5**), 831–851.
37. T. J. Piercy, S. J. Smither, J. A. Steward, L. Eastaugh and M. S. Lever, The Survival of Filoviruses in Liquids, on Solid Substrates and in a Dynamic Aerosol, *J. Appl. Microbiol.*, 2010, **109**(**5**), 1531–1539.
38. M. Bhaskaran, P. J. Didier, S. K. Sivasubramani, L. A. Doyle, J. Holley and C. J. Roy, Pathology of Lethal and Sublethal Doses of Aerosolized Ricin in Rhesus Macaques, *Toxicol. Pathol.*, 2014, **42**(**3**), 573–581.
39. C. L. Wilhelmsen and M. L. M. Pitt, Lesions of Acute Inhaled Lethal Ricin Intoxication in Rhesus Monkeys, *Vet. Pathol.*, 1996, **33**, 296–302.

40. C. J. Roy, M. Hale, J. M. Hartings, L. Pitt and S. Duniho, Impact of Inhalation Exposure Modality and Particle Size on the Respiratory Deposition of Ricin in BALB/c Mice, *Inhal. Toxicol.*, 2003, **15**(**6**), 619–638.
41. H. J. Stone, J. V. Rogers, E. J. Fleming, Y. W. Choi, J. D. Waugh, W. R. Richter, M. L. Taylor, K. B. Riggs, Z. J. Willenberg, and R. T. Krile, *Impact of Temperature and Humidity on the Persistence of Vaccinia Virus and Ricin Toxin on Indoor Surfaces*, Environmental Protection Agency, 2006, EPA/600/R-08/002.
42. G. N. Agrios, *Plant Pathology*, Elsevier Academic Press, San Diego, CA, 5th edn, 2005.
43. S. Y. Padmanabhan, The Great Bengal Famine, *Annu. Rev. Phytopathol.*, 1973, **11**, 11–24.
44. A. Halasz, D. Magyar and J. Bobvos, Aerobiological Aspects on Quarantine Risks in Grain Warehouses: A Study on Bunt (*Tilletia* spp.) Dispersal, *Aerobiologia*, 2014, **30**(2), 161–171.
45. R. Danzig, H. Yuki, L. Hough, M. Sageman, R. Kotani and T. Leighton, *Aum Shinrikyo: Insights Into How Terrorists Develop Biological and Chemical Weapons*, Center for a New American Security, 2011.
46. J. B. Petro, T. R. Plasse and J. A. McNulty, Biotechnology: Impact on Biological Warfare and Biodefense, *Biosecur. Bioterror.*, 2003, **1**(**3**), 161–168.
47. J. Warner, J. Ramsbotham, E. Tunia and J. J. Valdes, *Analysis of the Threat of Genetically Modified Organisms for Biological Warfare*, Center for Technology and National Security Policy, National Defense University, 2011.
48. *National Strategy for Countering Weapons of Mass Destruction*, National Security Council, November 2009.
49. V. Putin, *Being Strong: National Security Guarantees for Russia*, Rossiiskaya Gazeta, 20 February 2012, http://archive.premier.gov.ru/eng/events/news/18185, accessed 29 December 2014.

CHAPTER 10

Aerosol Physics for Bioaerosols†

JANA KESAVAN*[a], LUPITA D. MONTOYA[b], AND BETH L. LAUBE[c]

[a]US Army Edgewood Chemical Biological Center, Aberdeen Proving Ground, MD 21010, USA; [b]University of Colorado Boulder, Boulder, CO 80389, USA; [c]The Johns Hopkins University School of Medicine, Baltimore, MD 21287, USA
*E-mail: jana.s.kesavan.civ@mail.mil

10.1 Introduction

Aerosols play a critical role in many industries and processes, including those of manufacturing, pharmaceuticals, nuclear power generation, petrochemicals and tobacco, to name just a few. Aerosols are used either as part of the process itself or produced as a by-product or waste. The US government has an interest in aerosols because of their many chemical, biological, nuclear, radiological and explosive capabilities, which can be used to affect vast geographical areas. Additionally, public health agencies have an interest in aerosols because of their role in the spread of diseases in people, animals and plants. These include a variety of human bacterial diseases such as tuberculosis and inhalation anthrax, viral diseases such as influenza, the common cold, viral pneumonia, measles, mumps and chickenpox, and

†The views and opinions expressed in this chapter are those of the author and should not be construed as an official Department of the Army position, policy, or decision, unless so designated by other official documentation. Citation of trade names in this chapter does not constitute an official Department of the Army endorsement or approval of the use of such commercial items.

Issues in Toxicology No. 25
Aerobiology: The Toxicology of Airborne Pathogens and Toxins
Edited by Harry Salem and Sidney Katz

Published by the Royal Society of Chemistry, www.rsc.org

fungal diseases such as paracoccidioidomycosis and systemic mycoses such as coccidioidomycosis.[1–3] The recent outbreak of Ebola demonstrates how the spread of infectious disease is difficult to contain and has the potential to become a national or even international threat. Infectious biological aerosols may also damage plants such as wheat, which can acquire yellow rust disease through the air.[4] In addition to infectious diseases, airborne particles play a central role in causing noninfectious diseases such as bronchitis, pulmonary emphysema, asthma, lung cancer and nasal cancer. It is imperative that efficient sampling, detection, identification and decontamination procedures be developed and put in place to prevent the cause and spread of airborne disease.

Aerosol particles have a variety of physical, chemical and biological properties that affect their movement and interaction in the environment. Airborne biological particles behave in the same way as other inert particles with similar physical properties. The physical characteristics of airborne particles, such as size, shape and density, will determine how long they remain airborne and their potential to spread disease, and also where they deposit in the respiratory tract to cause diseases. The chemical and biological composition of a particle varies depending on the particle's origin and this will play a role in the resulting health effects. In general, biological particles are responsible for infectious, allergic and toxic diseases, whereas chemical particles are responsible for toxic and irritant diseases. The biological characteristics of the particles (*i.e.* viability, culturability, allergenicity and toxicity) are critical to their disease potential. For example, inhalation of large quantities of nonviable organisms may not cause infectious diseases but may cause allergies, irritation or toxic responses. In comparison, the inhalation of large quantities of toxic materials may cause diseases such as cancer.

Airborne particles are removed from the air over time due to settling, electrostatic deposition and diffusion to local surfaces and walls. In addition to these mechanisms, particles are removed from air under high airflow conditions by impaction and interception. In addition to particle removal from air, bioparticles also lose their biological characteristics such as culturability and viability as they age in air. Temperature, relative humidity and oxygen content of the air are some of the factors that degrade biological properties over time.[5] While these properties are critical to their resultant health effects, particle movement in air is governed strictly by aerosol physics and these basic principles are the subject of this chapter.

10.2 Definition of Aerosols

Aerosols are two-phase systems in which liquid or solid particles are suspended in a surrounding gas-phase medium. In general, the particle size range of interest is 0.001–100 μm in diameter due to the health effects of these particles. Particle behavior in air is influenced by the properties of the particle, the surrounding air and external forces such as gravity and electric fields.

Airborne particles are ubiquitous in our environment due to primary and secondary aerosol formation. Primary aerosols are directly introduced into the atmosphere, whereas secondary aerosols are formed by condensation of gas-phase chemical species or *via* chemical reactions in air. Airborne particles can also be classified based on the activity that produced the aerosol: those created through natural means, such as sea salt aerosols, wind-blown dust, smoke from forest fires and volcanic material, and those created anthropogenically, such as smoke produced from power and manufacturing plants and automobile exhausts.

10.3 Bioaerosols

Bioaerosols commonly found in the environment are living material and/or originate from living organisms. Examples of biological aerosols include suspensions of viruses, toxins, waste products from living things, pollen, bacteria, fungal spores and biological material fragments. Organisms can be categorized as viable and nonviable. Scientists have subdivided viable microorganisms into culturable and nonculturable to include the possibility of disease spreading by viable but nonculturable organisms. Culturable organisms are those that can grow on culture plates in a laboratory environment; however, viable organisms show activities such as ATP synthesis but are not necessarily culturable. Bioaerosol particles cover a very wide size range: viruses and biological material fragments are in the nanometer range and pollen spores are in the hundreds of microns range. Table 10.1 shows typical particle sizes of bacteria, viruses, fungal spores and pollen.[6,7]

10.4 Particle Size

10.4.1 Diameter

Particle size is usually described in terms of the diameter of the particle and provided in units of micrometers (1 μm = 10^{-6} m) or nanometers (1 nm = 10^{-9} m). Particle diameter is usually denoted by the symbol d_p and it is the most important property that characterizes the behavior of particles in a gas. Particles smaller than 0.1 μm are most influenced by diffusion and Brownian motion, whereas larger particles are most influenced by inertial and gravitational forces. The particle diameter of spherical particles is

Table 10.1 Typical particle sizes of viruses, bacteria, fungal spores and pollen.

Particle	Typical diameter range/μm
Viruses	0.02–0.3
Bacteria	0.3–10
Fungal spores	1–100
Pollen	5–>200

easy to determine; however, many commonly encountered particles are not spherical and describing irregularly shaped particles is complicated. Many techniques are employed to solve this problem, the most common being to ascribe an "equivalent diameter" to irregularly shaped particles. The term "equivalent diameter" is used to assign an equivalently sized sphere to an irregularly shaped particle based on a certain physical property. An equivalent diameter represents the diameter of a sphere that has the same value of a specific physical property as the particle under consideration. Two particles with similar equivalent diameters based on one property could have different diameters based on other properties.

The projected area diameter can be used to describe a particle viewed under a microscope; it is the diameter of a circle that has the same projected area as the irregular particle observed. Visual observation and estimation are cumbersome; therefore, image analysis software is currently used to automate this process. In addition to the projected area diameter, light scattering, aerodynamic behavior, surface area, volume, mass and particle mobility in an electric field are also used to determine equivalent diameters. In the inhalation toxicology field, where health effects are linked to the size of the inhaled particles, the particle behavior in air and the respiratory system is important. This behavior is well represented by the aerodynamic particle size, which is described later in this chapter.

Particles are also categorized based on size as micrometer, sub-micrometer, ultrafine and nanometer.[8] In general, micrometer particles are defined as particles in the 1–10 μm diameter range and sub-micrometer particles are those with diameters <1 μm. Ultrafine particles are defined as particles having diameters <0.1 μm and nanoparticles are those in the range 1–100 nm. Figure 10.1 shows the ranges of these defined particle sizes and examples of common bioaerosols and their sizes.

The particle size of aerosols of interest spans over five orders of magnitude, from 0.001 μm (1 nm) to larger than 100 μm, as shown in Figure 10.1. Particle size may also change over time depending on the particle concentration,

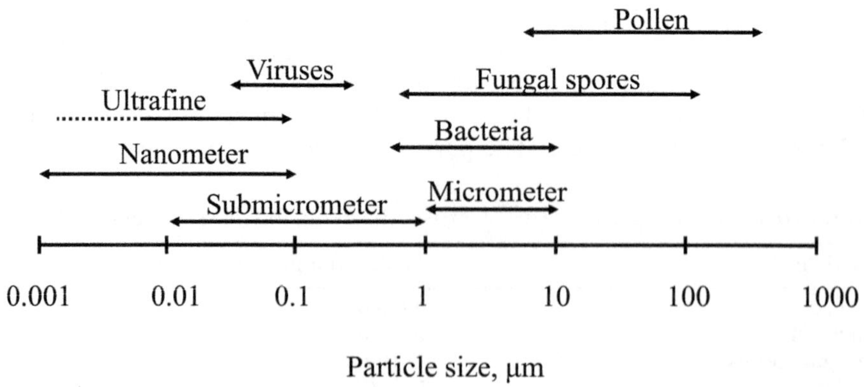

Figure 10.1 Particle size and various bioaerosol size ranges (in μm).

relative humidity and temperature of the environment. Particles can absorb water and increase in size in a high-humidity environment; they are called hygroscopic particles. These include sea salt particles, tobacco smoke and various medicinal aerosols. Similarly, water molecules are lost from particles in dry environments, resulting in a reduction in the particle size. Additionally, particle size may increase when high concentrations of particles in the environment agglomerate to form larger particles.

10.4.2 Knudsen Number (Kn)

In addition to particle diameter, the *Knudsen number* (Kn) is also used to describe particle behavior in a gas. Kn is the ratio of the molecular mean free path λ_g (of the gas) to the particle radius r_p (or particle diameter, d_p) as shown in eqn (10.1). The molecular mean free path represents the average distance traveled by gas molecules between successive collisions with other gas molecules. The mean free path of air under standard laboratory conditions (1 atm and 293 K) is 0.066 μm. Very small particles (radius <0.005 μm) have Kn >10 and experience the surrounding gas molecules as individual particles. On the other hand, particles >0.65 μm in radius have Kn <0.1 and experience the outside air as a continuous medium.

$$\text{Kn} = \frac{\lambda_g}{r_p} = \frac{2\lambda_g}{d_p} \tag{10.1}$$

10.5 Particle Size Distribution

A particle size distribution describes the distribution of a particle property such as number, surface area and mass over a range of particle sizes. A monodisperse aerosol contains particles that are mainly the same size and can be described by the size of the particle and the number of particles. These particles are rarely found in nature, but are very useful for laboratory studies and instrument calibration. In contrast, more commonly found polydisperse aerosols contain a range of particle sizes and amounts; statistical methods are required to describe this aerosol adequately. An illustration of a monodisperse and a polydisperse aerosol is shown in Figure 10.2. In general, the particle size distribution is characterized by the mean (arithmetic or geometric average of all sizes) or the mode (most frequent size). The count median diameter (CMD) divides an equal number of particles above and below this size. The geometric standard deviation (GSD) is a mathematical term that describes the spread of the distribution. Similarly, surface area median diameter (SMD) and mass median diameter (MMD) divide the surface area and mass to provide 50% above and below the median size.

Naturally occurring aerosols are usually log-normally distributed and the distribution looks normal (Gaussian or bell-shaped curve) when plotted on a log scale. Log-normal distributions are described with a mathematical

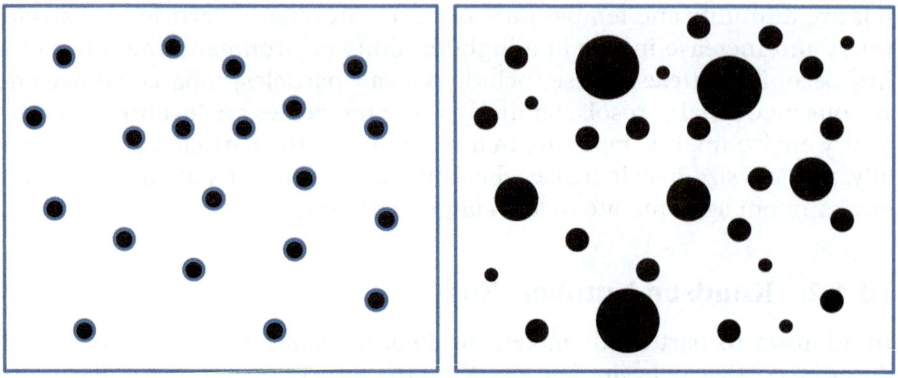

Figure 10.2 Illustration of a monodisperse and a polydisperse aerosol.

equation that provides a median and a geometric standard deviation. These can be used to construct the original distribution and also convert the distribution from number to mass or surface area and *vice versa*.

Biological particles are often collected on slides or filters, then sized and counted using a microscope. From these data, the median size of the distribution and the geometric standard deviation are determined using a methodology described here. The measured particles are grouped evenly or unevenly by size to facilitate analysis of the data. The percentage of particles in each group and the cumulative percentage of particles greater or less than the group size are then calculated. The percentage of particles less/more than the indicated sizes as a function of particle size is subsequently plotted on a log-probability graph, as shown in Figure 10.3. A linear function is fitted to the data points; a straight line can only be fitted to the data if the size distribution is log-normal. The particle size at the 50th percentile is the CMD and the slope of the curve is the GSD. A particle count distribution is obtained when particle count and size are measured with microscopic counting; this is considered a number-based distribution. In contrast, particle mass distribution is obtained with cascade impactor sampling, which provides particle mass and size; an MMD and GSD will be obtained from this distribution.

Figure 10.3 shows a CMD of 2.2 µm, which is the 50th percentile size ($d_{50\%}$). The GSD is the ratio of the $d_{84\%}$ and $d_{50\%}$ or $d_{50\%}$ and $d_{16\%}$ and is calculated to be 1.7 for this example. When the particle distribution is log-normal, the Hatch–Choate equations can be used to convert between CMD, SMD and MMD, as shown in eqn (10.2) and (10.3).

$$\text{SMD} = \text{CMD}\exp\left(2\ln^2 \text{GSD}\right) \tag{10.2}$$

$$\text{MMD} = \text{CMD}\exp\left(3\ln^2 \text{GSD}\right) \tag{10.3}$$

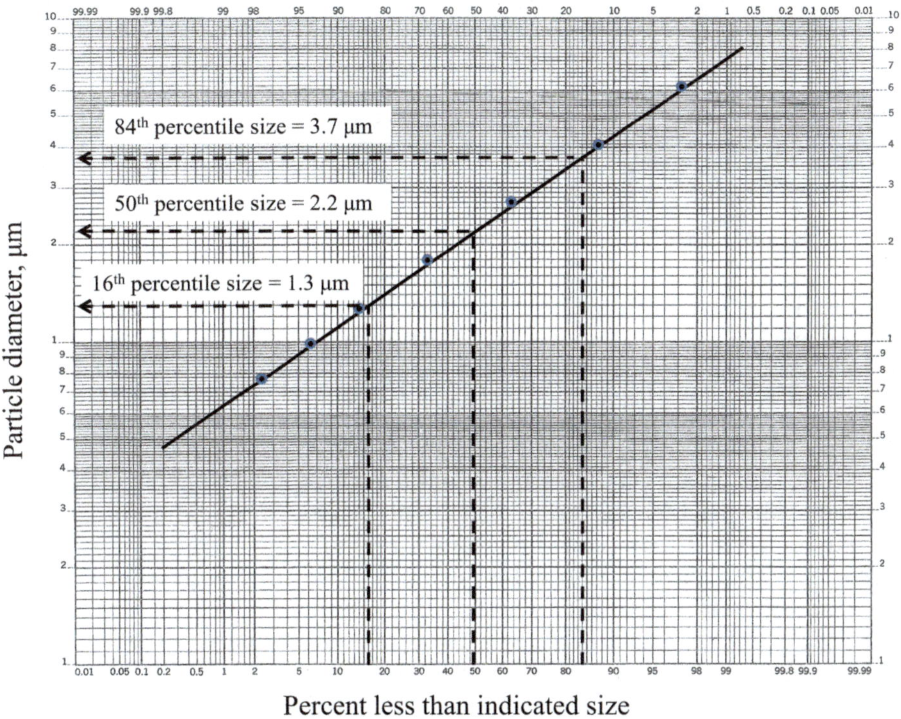

Figure 10.3 An example of log-probability distribution of collected particles.

The GSD is the same for the count, surface area and mass distribution of the same aerosol. The count distribution is preferred for infectious organisms. The surface area distribution is important for aerosols that produce allergic effects and mass distribution is more desirable for aerosols that produce toxic effects.

10.6 Particle Shape

Particles encountered in daily life have various shapes, depending largely on their origin. Calibration and liquid particles are spherical in shape, whereas dust and biological particles could have various irregular shapes. Particles are divided into three major shape classes. Class one consists of particles that have equal dimensions in all three axes. Spheres and cubes are examples of these globular particles. Class two represents particles with large dimensions on two axes and a small dimension for the third axis. Platelets are a good example of these particles. The third class of particles has two dimensions which are small and a large third dimension. Fibers are included in this class. An illustration of each of the three shape classes is shown in Figure 10.4.

Particle shape and size are critical factors in the resulting health effects, as they determine how far into the respiratory system the particles can

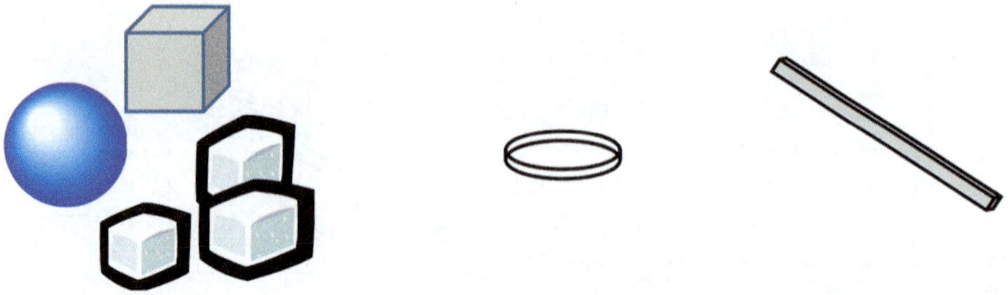

Figure 10.4 Illustration of the three shape classes.

penetrate. Fibers are very hazardous to the respiratory system, largely owing to their shape in which the length of a fiber is three or more times longer than its diameter. Fibers can reach the alveolar region more easily than spherical particles with a similar volume because they have smaller geometric diameters and therefore a smaller aerodynamic diameter. Since the fibers align themselves with the airflow along the axis, they are capable of reaching the deep lung. The long length of fibers also prevents them from being engulfed by macrophages and removed from the body, thereby prolonging any deleterious health effects associated with these fibers. Bioaerosols with fiber-like geometric characteristics will behave similarly.

Examples of bioaerosols with fiber-like geometric characteristics are vegetative bacteria and spores grouped as chains. These large chain-like structures have large surface areas and small densities and remain airborne for a very long time. Owing to their small aerodynamic size, they can reach the alveolar region and produce toxic health effects and/or infection.

10.7 Particle Density

Particle density is another important property that affects airborne particle behavior. Particle density is defined as particle mass divided by particle volume and is denoted by ρ_p. The units of density are usually expressed in terms of kg m^{-3} or g cm^{-3}. The density of a particle can be, in general, determined from the parent (bulk) material; however, porous particles with voids have significantly lower densities compared with the parent material. The density of the parent material can be found in handbooks but the porosity, or void space, of the particle needs to be considered. Smoke and fume particles are composed of chain-like structures with significant void space that reduces the overall density of the particle. Similarly, clustered biological particles may also have significant void space and lower particle density compared with single biological particles. Particle density has significant effects on particle behavior in air, especially the particle settling velocity. For example, a particle with higher density will settle faster than a low-density particle with a similar geometric diameter.

The density of a material is affected by many factors, including the preparation method and the water content of the material. For example, densities of differently prepared *Bacillus atrophaeus* (BG) spores, a simulant for *Bacillus anthraces*, and ovalbumin, a simulant for toxins, were determined at the US Army Edgewood Chemical Biological Center (ECBC) using an autopycnometer.[9] The densities of these materials were determined before and after drying them in an oven to remove the absorbed and adsorbed water. The densities of BG were found to be affected by the preparation method and the water content of the material and ranged from 1.33 to 1.78 g cm^{-3}. Similarly, the density of ovalbumin was affected by the preparation method and ranged from 1.36 to 1.41 g cm^{-3}.[9]

10.8 Particle Surface Area and Mass

All other things being equal, the health response to a toxic material is primarily dependent on its mass; however, it can also depend on its surface area and the chemical properties of the material. Particle surface area is proportional to particle diameter raised to the second power (squared). Particle mass is proportional to volume or particle diameter raised to the third power (cubed). For example, a 10 µm particle has 100 times more surface area and 1000 times more mass compared with a 1 µm particle. Considering aerosols with equal mass concentrations, an aerosol consisting of only 1 µm particles will have significantly more surface area than an aerosol consisting of only 10 µm particles. In this case, the aerosol consisting of 1 µm particles will have 1000 times more particles than the aerosol consisting of 10 µm particles. Comparatively, the higher number of 1 µm particles will provide a higher total surface area, which may adversely affect the health of humans more severely than particles with larger diameter but lower total surface area. Recent studies have also shown the importance of surface area in producing toxic effects by nanometer-sized particles or nanoparticles.[10–12] Additional studies have also indicated that ultrafine particles may be more toxic per unit mass than larger particles.[13]

10.9 Aerosol Concentration

Aerosol concentration is a measure of particle property per unit of air. It can be described in terms of number, surface area and mass. Particle number concentration measurements are more commonly used in clean rooms and in microbial and pollen quantification. The units are usually number of particles per m^3 or cm^3. Particle surface area concentration is important in toxicity and allergenicity studies and the units are $\mu m^2\ m^{-3}$ or $\mu m^2\ cm^{-3}$. The particle mass concentration is important in toxic studies and the units are $\mu g\ m^{-3}$ or $mg\ m^{-3}$. Health effects due to biological particles in the atmosphere depend on the number, surface area or mass of the particles and also on the mechanism of effect, such as chemical *versus* biological. Low concentrations of some particles may not produce any effects whereas high concentrations of these particles may cause adverse health effects. For example, an average of 8000–10 000 *Bacillus anthracis* spores are needed to produce an infection (infectious dose), whereas low concentrations of this bacterium may not cause the disease.[14] Developing a disease will also depend on other factors, such as the age and health status of the individual.

Many studies have evaluated the number and mass of total and biological particles in common indoor and outdoor environments. Although these numbers vary from season to season and geographically, outdoor sampling studies using fluorescence detection methods have reported mass concentrations of biological particles ranging from 4×10^{-3} to 123 μg m^{-3}, with a mean of 1.3 μg m^{-3}. The number concentration of the fluorescent particles ranged from 1.2×10^{-3} to 1.4 particles cm^{-3} with a mean of 0.027 particles cm^{-3}.[15] Pollen concentrations in the air are also quantified as counts and reported daily in some cities during the spring season so that people with allergies can reduce outdoor activities.

10.10 Settling Velocity

A particle that is released in air accelerates initially and in a very short time reaches its terminal settling velocity, which is a function of its particle size and density. Terminal settling velocity, also known as settling velocity, is calculated using known equations and determines how long particles are airborne, a key element in determining the hazard. Particle settling velocity is a useful parameter in many analyses of aerosol behavior. For example, it is employed to determine the time needed for aerosols to deposit on surfaces for surface sampling and decontamination experiments where seeded surfaces are used. Particle settling velocity can be determined experimentally using sedimentation cells, settling tubes, horizontal elutriators and horizontal elutriation aerosol spectrometers.

The terminal settling velocity is derived by equating the gravitational force and the air resistance (drag) force acting on a particle. Air resistance depends on the characteristics of the airflow around the particle and is determined by the Reynolds number. The following sections describe Reynolds number, air resistance and settling velocity calculations.

10.10.1 Reynolds Number (Re)

The Reynolds number (Re) is dimensionless (it has no units) and characterizes the airflow around an aerosol particle. Re must be similar to have geometrically similar flows around geometrically similar objects. The Reynolds number is the ratio of inertial forces to frictional forces acting on each element of the air and it indicates whether the airflow around a particle is laminar or turbulent. It is calculated using eqn (10.4). The factors associated with inertia are in the numerator and the factor associated with frictional force is in the denominator. The density and viscosity in eqn (10.4) are of the air and not of the particle.

$$\mathrm{Re} = \frac{\rho_g V d_p}{\eta} \tag{10.4}$$

where ρ_g is the density of gas, V is the velocity of the particle (relative to the gas), d_p is the diameter of the particle and η is the viscosity of air.

Eqn (10.4) can be simplified to eqn (10.5) when particles are in air at standard conditions and when SI units are used. In this equation, the velocity is in meters per second and diameter is in meters. A Re of <1 indicates laminar flow around a particle when viscous forces are much greater than inertial forces. Larger Re values indicate turbulent flow around a particle.

$$\mathrm{Re} = 66000 V d_p \tag{10.5}$$

10.10.2 Drag Force

Newton derived an equation to determine the resistance force experienced by particles moving in air. This force depends on the coefficient of drag (C_D), density of the gas (ρ_g), particle diameter squared and velocity squared, as shown in eqn (10.6). This equation is valid for all subsonic particle motion.

$$F_D = \frac{C_D \pi \rho_g d_p^2 V^2}{8} \tag{10.6}$$

The C_D for Re > 1000 is constant but it varies for Re < 1000 (in the transition and Stokes law regions) and this value is determined using equations or by a graphical method. The C_D for Re < 1 and for Re between 1 and 1000 are determined using eqn (10.7) and (10.8), respectively.

$$C_D = \frac{24}{\mathrm{Re}} \quad \mathrm{Re} < 1 \tag{10.7}$$

$$C_D = \frac{24}{\mathrm{Re}}\left(1 + 0.15\mathrm{Re}^{0.687}\right) \quad 1 < \mathrm{Re} < 1000 \tag{10.8}$$

In the Stokes law region where the Re is <1, inertial forces are small compared with frictional forces. The drag force depends on the particle size, velocity and air viscosity as shown in eqn (10.9).

$$F_D = 3\pi\eta V d_p \quad \mathrm{Re} < 1 \tag{10.9}$$

10.10.3 Terminal Settling Velocity

Gravity pulls particles towards the Earth with a downward force, F_G, that is equal to the mass of the particle (m_p) multiplied by the gravitational acceleration (g):

$$F_G = m_p \cdot g \tag{10.10}$$

Also, m_p can be written in terms of volume and density of the particle:

$$F_G = \frac{g \cdot \pi d_p^3 \rho_p}{6} \tag{10.11}$$

The volume of the gas or fluid displaced by the particle produces a buoyant force (F_B), which is equal to the weight of the fluid displaced by the particle. The weight of the fluid or gas is equal to the volume of the fluid multiplied by the density of the fluid and gravitational acceleration. The buoyant force is significantly smaller than the gravitational force, hence it is ignored in calculations.

When a particle is released in air, it accelerates due to the gravitational force and reaches its terminal settling velocity. At this condition, the gravitational force (F_G) is equal to the drag force (F_D) exerted on the particle:

$$F_D = F_G \tag{10.12}$$

The terminal settling velocity (V_{TS}) is determined by substituting eqn (10.9) and (10.11) into eqn (10.12) and rearranging the terms as shown in eqn (10.13). This equation determines the terminal settling velocity as a function of particle size and density. The terminal settling velocities of particles with density of 1000 kg m^{-3} are provided in Table 10.2. The values indicate that 10 μm particles settle 100 times faster than 1 μm particles.

$$V_{TS} = \frac{\rho_p d_p^2 g}{18\eta} \tag{10.13}$$

10.10.4 Nonspherical Particles

Primary particles are produced by mechanical break-up of a parent material. These types of particles have random shapes, whereas particles grown by the condensation of a liquid from the gas phase tend to be closer to spherical. Diameters of solid particles of random shapes are difficult to

Table 10.2 Terminal settling velocities of 1, 5 and 10 μm diameter particles with a density of 1000 kg m^{-3}.

Particle size/μm	Terminal settling velocity/cm s^{-1}
1	0.003
5	0.07
10	0.30

measure and classify with simple methods. Because particle size and shape vary greatly, various measurement techniques rely on different particle properties to classify particles of disparate shapes and sizes. In the case of projected area diameter, the surface area of irregularly shaped particles is measured against the area of known spherical particle sizes. Another method involves the aerodynamic diameter, in which the behavior of irregularly shaped particles in air is measured using an aerodynamic particle sizer, which effectively incorporates particle size, shape and density into its measurement and produces an equivalent size of a sphere with unit density.

Commercially available calibration particles and liquid droplets are spherical but most other particles are nonspherical in shape. Nonspherical particles often have regular geometric shapes, such as cylinders, fibers or cubes, but they can also be of other irregular shapes. A dynamic shape factor correction (χ) is applied to Stokes law to account for the effect of shape on particle motion. The drag force on an irregular particle in the Stokes law region is provided by eqn (10.14), which includes the irregular particle's equivalent volume diameter (d_e) and the dynamic shape factor (χ); d_e is the diameter of a sphere that has the same volume as the irregular particle.

$$F_D = 3\pi\eta V d_e \chi \tag{10.14}$$

The terminal settling velocity equation for an irregularly shaped particle is provided in eqn (10.15), which uses the equivalent volume diameter and dynamic shape factor to correct for the effect of shape on the settling velocity.

$$V_{TS} = \frac{\rho_p d_e^2 g}{18\eta\chi} \tag{10.15}$$

Dynamic shape factors for different particle shapes are provided in Table 10.3. As shown, spherical particles have a shape factor of 1 and lead fumes have a shape factor of 1.5–3.5.[17]

10.10.5 Cunningham and Slip Correction Factors

Stokes law assumes that the relative velocity of the gas is zero at the surface of the particle; however, this assumption does not hold true for small particles with diameters approaching the mean free path of gas (λ). The settling velocity of these particles is faster than predicted by Stokes law and the error becomes significant for particles smaller than 1 µm. For such small particles, it is important to use the Cunningham derived correction factor (C_C),[8] which is always >1 and is calculated using eqn (10.16). This equation uses λ of the gas and d_p to calculate the correction factor. This equation is valid for particles of size >0.1 µm.

$$C_C = 1 + \frac{2.52\lambda}{d_p} \tag{10.16}$$

Table 10.3 Dynamic shape factors for different particle shapes.[8,16,17]

Shape	Dynamic shape factor, χ
Sphere	1.00
Cube	1.08
Cylinder (orientation averaged with axial ratio of 2)	1.09
Straight chain with an axial ratio of 2	1.10
Compact clusters – three spheres	1.15
Compact clusters – four spheres	1.17
Glass fiber ($L/D = 5$)	1.71
Quartz	1.36
Sand	1.57
Talc	1.88
High-ash soft coal	1.95
Carbonaceous smoke	3.26–6.77
Lead fume	1.5–3.5

An empirical equation is used to calculate the slip correction factor for particles smaller than 0.1 μm, as shown in eqn (10.17). The slip correction factor increases significantly as the particle size decreases.

$$C_C = 1 + \frac{\lambda}{d_p}\left[2.34 + 1.05\exp\left(-0.39\frac{d_p}{\lambda}\right)\right] \tag{10.17}$$

C_C reduces the drag force acting on the particles, as shown in eqn (10.18):

$$F_D = \frac{3\pi\eta V d_p}{C_C} \tag{10.18}$$

The slip correction factor must be used in terminal settling velocity corrections for small particles ($d_p < 1$ μm), as shown in eqn (10.19). The slip correction factor and terminal settling velocity with and without C_C corrections for a few examples of particle sizes are provided in Table 10.4. This table shows that the C_C is high for smaller particles with a value of 22.98 for a 0.01 μm particle compared with 1.02 for a 10 μm particle. This indicates the importance of including the C_C in the V_{TS} calculations for small particles.

$$V_{TS} = \frac{\rho_p d_p^2 g C_C}{18\eta} \text{ for Re} < 1 \tag{10.19}$$

The drag force of a small nonspherical particle is determined by using the shape factor correction and slip factor correction as shown in eqn (10.20). The terminal settling velocity for the same particle is given by eqn (10.21).

$$F_D = \frac{3\pi\eta V d_e \chi}{C_C} \tag{10.20}$$

Table 10.4 Slip correction factors and terminal settling velocities for particles of various size.

Particle size, d_p/μm	Slip correction factor (C_C)	V_{TS} without C_C/m s^{-1}	V_{TS} with C_C/m s^{-1}
0.01	22.98	3.01×10^{-9}	6.92×10^{-8}
0.1	2.93	3.01×10^{-7}	8.82×10^{-7}
1.0	1.16	3.01×10^{-5}	3.48×10^{-5}
5.0	1.03	7.53×10^{-4}	7.76×10^{-4}
10.0	1.02	3.01×10^{-3}	3.06×10^{-3}

$$V_{TS} = \frac{\rho_p d_e^2 g C_C}{18\eta\chi} \tag{10.21}$$

10.11 Aerodynamic Diameter (d_a)

The aerodynamic diameter is a parameter used to classify the diameter of particles with various shapes and densities such that similar aerodynamic diameter particles behave similarly in air. The aerodynamic diameter is defined as the diameter of a unit density (1 g cm^{-3} or 1000 kg m^{-3}) spherical particle that has the same terminal settling velocity as the particle of interest. The aerodynamic diameter takes into account the size, shape and density of the particle, so this information does not have to be calculated or measured. This type of diameter is best applied to particles larger than 0.5 μm, which are less affected by Brownian motion and diffusion. The terminal settling velocity of an irregularly shaped particle with a density ρ_p is equated to that of a spherical particle with unit density ρ_0 as shown in eqn (10.22), where they both have the same settling velocity. Particles with similar aerodynamic diameters deposit in the same location in a cascade impactor and elutriator and have the same size when measured using an aerodynamic particle sizer (APS, TSI Inc., Shoreview, MN, USA).

$$V_{TS} = \frac{\rho_p d_e^2 g}{18\eta\chi} = \frac{\rho_0 d_a^2 g}{18\eta} \tag{10.22}$$

The aerodynamic diameter of a spherical particle with density other than 1 g cm^{-3} (1000 kg m^{-3}) can be determined using eqn (10.23), which is simplified from eqn (10.22).

$$d_a = d_p \sqrt{\frac{\rho_p}{\rho_0}} \tag{10.23}$$

The aerodynamic diameter of an irregularly shaped particle can be determined using the volume equivalent diameter, shape factor and density of the particle as shown in eqn (10.24). This is a simpler form of eqn (10.22).

$$d_a = d_e \sqrt{\frac{\rho_p}{\rho_0 \chi}} \tag{10.24}$$

10.12 Relaxation Time

As a particle traverses through space, it is affected by external forces or conditions. The particle's relaxation time represents the time required for its velocity to adjust to a new condition or force and it is determined using eqn (10.25). Note that the particle's relaxation time is not dependent on the nature of the external force.

$$\tau = \frac{\rho_p d_p^2 C_C}{18\eta} \tag{10.25}$$

A particle's terminal settling velocity can also be determined using its relaxation time, as shown in eqn (10.26). The relaxation times for various particle sizes are provided in Table 10.5. The relaxation time varies from 7.05×10^{-9} to 3.12×10^{-4} s for particle sizes from 0.01 to 10 μm in diameter.

$$V_{TS} = \tau g \tag{10.26}$$

10.13 Brownian Motion and Diffusion

Brownian motion and diffusion refer to the irregular motion of particles caused by the constant bombardment of gas molecules on particles. Particles move from higher to lower concentration areas due to diffusion. This is the primary transport and deposition mechanism for particles smaller than 0.5 μm in diameter and causes small particles to deposit in filters and in the lungs where transport distances are small. Brownian motion and diffusion are characterized by the particle diffusion coefficient, D, which is expressed in units of $m^2\ s^{-1}$. Smaller particles have larger diffusion coefficients, more vigorous Brownian motion and rapid transfer of mass in a concentration gradient. The diffusion coefficient is given by eqn (10.27), which uses Boltzmann's constant (k) and absolute temperature (T). Boltzmann's constant is the gas constant per molecule (gas constant, R, divided by Avogadro's number, N_A).

$$D = \frac{kTC_C}{3\pi\eta d_p} \tag{10.27}$$

The diffusion coefficient, mean thermal velocity due to Brownian motion, Brownian displacement in 1 s and gravitational settling in 1 s for various

Table 10.5 Relaxation time and time to reach the final condition of various standard density particles under standard conditions.

Particle size, d_p/μm	Relaxation time, τ/s	Time to reach the final condition/s
0.01	7.05×10^{-9}	2.12×10^{-8}
0.1	8.99×10^{-8}	2.70×10^{-7}
1.0	3.54×10^{-6}	1.06×10^{-5}
5.0	7.91×10^{-5}	2.37×10^{-4}
10	3.12×10^{-4}	9.36×10^{-4}

Table 10.6 Diffusion coefficient, mean thermal velocity, Brownian displacement in 1 s and gravitational settling in 1 s for various standard density spheres under standard conditions.

Particle size µm	Diffusion coefficient $m^2\ s^{-1}$	Mean thermal velocity $m\ s^{-1}$	Brownian displacement in 1 s m	Gravitational settling in 1 s m
0.01	5.4×10^{-8}	4.4	3.3×10^{-4}	6.9×10^{-8}
0.1	6.9×10^{-10}	0.14	3.7×10^{-5}	8.8×10^{-7}
1.0	2.7×10^{-11}	0.0044	7.4×10^{-6}	3.5×10^{-6}
10	2.4×10^{-12}	0.00014	2.2×10^{-6}	3.1×10^{-3}

standard density spheres under standard conditions are provided in Table 10.6.[8] From the table, it is clear that smaller particles are transported predominantly by diffusion and larger particles predominantly by gravitational force. A 10 µm particle travels 3.1×10^{-3} m by gravitational force in 1 s but travels only 2.2×10^{-6} m by Brownian motion. On the other hand, a 0.01 µm particle travels 6.9×10^{-8} m by gravitational force but travels a larger distance of 3.3×10^{-4} m by Brownian displacement.

10.14 Coagulation

When two particles impact with each other, they usually stick together and form one particle that is larger in size. This process is called coagulation and it is enhanced when the particle concentration is high. Coagulation happens when there is diffusion or a net difference in velocities between the particles. These differences occur due to varying diffusion coefficients, different settling velocities, different velocities due to laminar and turbulent flows, differences in electrical mobility, *etc.*

During coagulation, there is a decrease in the number of particles and an increase in particle size, which can be described by equations. The simplest case to describe is coagulation of monodisperse particles due to diffusion. The rate of change in number of particles is negative (decrease) and proportional to the number concentration of particles squared (N^2) and the coagulation coefficient, K, as given by

$$\frac{\mathrm{d}N}{\mathrm{d}t} = -KN^2 \tag{10.28}$$

K is related to the slip correction factor, which depends on particle size and can be calculated by

$$K = 3\times 10^{-16} C_\mathrm{C}\,\mathrm{m^3 s^{-1}} \quad d_\mathrm{p} > 0.1\,\mu\mathrm{m} \tag{10.29}$$

Particles tend to coagulate over time and the number concentration of the system can be obtained for a time t, if the initial number concentration, N_0, is known, as shown in eqn (10.30):

$$N(t) = \frac{N_0}{(1 + N_0 K_0 t)} \tag{10.30}$$

The size of the particles also changes over time, as shown by eqn (10.31), where the final particle size, $d_p(t)$, is determined using the initial particle size, d_{p_0}, initial particle number concentration, N_0, coagulation coefficient, K, and time, t. K depends on the particle size and must be corrected as the particle size changes. Factors such as air currents and particle charge also affect coagulation.

$$d_p(t) = d_{p_0}(1 + N_0 K t)^{\frac{1}{3}} \tag{10.31}$$

10.15 Electrical Properties

Freshly generated particles are highly charged and they lose their charge over time. This charge affects the particle behavior in electric fields and their deposition on surfaces and in the respiratory system. An example of this effect is powder coating of metallic surfaces, which disseminates highly charged metallic particles onto a metal surface, causing them to adhere to the surface before they are baked onto it.

Usually, dry dissemination generates very highly charged particles compared with wet dissemination. Particles acquire charge due to the frictional or triboelectric effect when the particles are generated. Particles become charged when they leave an aerosol generator and also when they separate from other particles. Charges acquired by particles can be calculated and are proportional to the particle diameter (d_p).[18]

Similarly, aerosolizing liquid particles also produces charged droplets due to the random motion of free ions in the liquid. The average number of charges on particles is a function of formed droplet (particle) diameter, d_p (μm), and ion concentration in the liquid (ions of one sign per cubic centimeter of liquid).

In addition to acquiring charges during generation, particles also acquire and lose charges in air. Charged particles in air interact with air ions and gain and lose charge, until an equilibrium state called Boltzmann equilibrium is eventually reached. The average number of charges, for a given particle size, at the Boltzmann equilibrium stage is given by

$$\bar{n} = 2.37 d_p^{\frac{1}{2}} \tag{10.32}$$

Laboratory-generated particles are highly charged and are often neutralized using an ^{85}Kr neutralizer, ^{210}Po or nonradioactive ion generators. Particles lose their charge by interacting with air ions following an exponential decay. The charge decay of particles is described by the half-time, which is 400 s under standard conditions in air. In general, freshly generated particles reach Boltzmann equilibrium in 4 h.[18]

Particle charging is utilized in particle sampling and sizing applications of particles less than 0.1 μm in diameter. These particles are difficult to measure optically because of their small size and they are also too small for inertial or gravitational separation and collection. Electrostatic precipitators separate particles from the airstream by electrical forces and collect them. These instruments function by charging the particles in the first section of the sampler using a high-voltage field or corona discharge and collecting the charged particles in the second stage, using an oppositely charged or grounded collection plate. Samplers that employ this mechanism require less power and less pressure drop than inertial impactors. In addition to sampling particles, electrical charge is utilized in sizing small particles. The electrical mobility analyzer, electrical aerosol analyzer and differential mobility analyzer are examples of instruments that leverage the electrical properties of aerosols to determine their size.

10.16 Adhesion and Resuspension of Particles

There are three main adhesive forces that attach a particle to a surface: van der Waals force, electrostatic force and the force arising from the surface tension of an adsorbed liquid.[8] Particles are removed (resuspended) from surfaces by external forces such as air flow, vibration and centrifugal forces. Measurements of resuspension are usually expressed as either resuspension factors or resuspension rates. The resuspension factor is the ratio of the airborne concentration to surface contamination. The resuspension rate is the fraction of the surface species removed per unit time.[19,20] Many factors affect resuspension. Parameters that play a role in resuspension are particle diameter, density, shape, relative humidity, force and duration causing the resuspension, time, adhesive forces, duration of contact between the particle and the surface and roughness of the particle and surface.

Many studies have been conducted in recent years to evaluate the resuspension of inert, biological and radiological particles in laboratories, in wind tunnels and in the field. Laboratory studies have quantified the resuspension immediately after deposition and compared the results with those of outdoor studies that evaluated resuspension of contaminated sites. Many assumptions are made in extrapolating laboratory data to the field as the environmental and surface variables are changing and uncontrollable. Models also have been developed to predict resuspension using particle size, air velocity, time, particle type, density, surface features and surface roughness as independent variables.[19] These models do not consider all the factors that can affect resuspension as they are hard to quantify to be included in models. Experimental results have shown that wet deposition produced less resuspension than dry deposition, even though the materials and other surface and particle properties are the same.[21] This indicates that other properties that are not currently considered in models need to be included in explaining the differences in resuspension. Use of an atomic force microscope to measure the adhesive forces and using this information to predict resuspension may be a better method for predicting resuspension.

10.17 Aerosol Sampling and Samplers

Air sampling is conducted to collect and analyze airborne particles for military, public health and research purposes. Commonly, the high-volume air samplers used for defense-related bioaerosol sampling are designed to collect 1–10 μm particles. These are specialized samplers made for sampling biological aerosols with a gentle collection process to preserve viability and prevent the destruction of the nucleic materials needed for identifying and quantifying the target organisms. Similarly, samplers have been designed with specific characteristics for exposure assessment and research-related purposes. These different samplers and detection systems must be thoroughly tested and their performance efficiencies determined so they can be properly matched to the application at hand, as selection of the wrong sampler can result in meaningless data.

Most aerosol samplers have several common components: an inlet, transmission tubes, pre-separator/skimmer to reject large particles, aerosol concentrating stages and a collector substrate as in an impactor. Figure 10.5 gives an exploded illustration of these components. A sampler can be as simple as a single component such as a filter or as complicated as a multistage virtual impactor aerosol concentrator. The performance of an aerosol sampler, usually denoted by the sampling efficiency, is the overall end-to-end ratio of the amount of aerosol contained in the final sample to the amount of aerosol contained in the volume of ambient air entering the system's inlet. In a multistage system, it

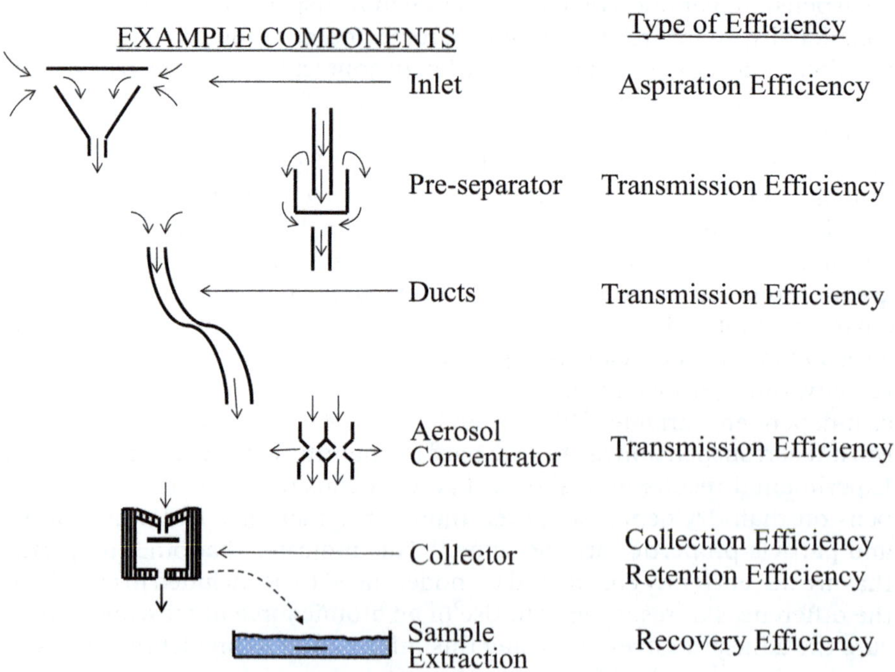

Figure 10.5 Illustration of aerosol sampler components and the related efficiencies.

is the product of the individual efficiencies of the sampler's individual components, variously aspiration, transmission and collection efficiencies.

An ideal sampler would have an efficiency of 100%, in which all the particles in the sampled air are captured and retained in the final collected sample. Unfortunately, this is never possible, as particle losses occur at all stages of a sampling system and larger particles (>10 μm) become increasingly difficult to sample. The aspiration efficiency of a sampler's inlet quantifies the efficiency with which particles are extracted from the air and transmitted through the sampler inlet to the remaining components of the system. It is dependent on particle aerodynamic size and wind speed. Transmission efficiency describes the efficiency with which particles are transported from the intake of a component to its outlet and the collection efficiency describes the efficiency with which particles are captured by the final collection component. Retention efficiency indicates how efficiently particles are retained by the sampler during a long sampling time, as in an impinger or a wetted wall cyclone, which stores the collected particles in the active collection fluid throughout the collection time, as particles in the collection fluid can escape into the air (re-aerosolization) and be ejected with the exhaust. The particles in the final collected sample are recovered for analysis; the efficiency with which they are recovered from the solution is called the recovery efficiency. None of these processes are 100% efficient; there are losses along the entire sampling chain, but in a well-designed system they can be minimized.

Sampling of bioaerosols is often combined with various analytical techniques, including culturing, polymerase chain reaction (PCR) and viability assays. Sometimes these techniques are automated directly into the system and give detection results in near real-time. Other techniques must be applied later in the laboratory on the collected sample. Survival of an organism can be measured by flow cytometry using different dyes that reveal viable *versus* nonviable organisms and by other life function measures such as adenosine triphosphate (ATP). Culturability is usually determined by plating.[22] Both are reported as fractions rather than efficiencies because they are characteristics of the aerosol in the sample, not the amount of aerosol in the sample. Aerosol samplers that do not adversely affect the viability and culturability must be properly selected when this information is needed.

Samplers and each component of a sampler are designed based on the application and the type of environment into which the sampler will be placed. Components such as an inlet may be simple with just a straight tube sampling into the air; however, other inlets are more complex and may have a rain hat, bug screens and a pre-separator. A rain hat prevents rain from entering the inlet and a pre-separator removes unwanted large particles. Similarly, the transmission region can be a simple tube connecting the inlet to the collection site or it can be outfitted with aerosol concentrators, bends, constrictions and expansions. The collection site is also designed based on the application and may use one or more of the four main collection mechanisms: impaction, interception, diffusion or electrostatic forces.

It is useful to know the sampling efficiency and the sampler characteristics before the selection of a sampler; some samplers and techniques are better suited to the particle size of interest or target concentration range than others. Many studies have been carried out to characterize commercial and military samplers for this reason. The various materials/methods and techniques used to characterize the sampling efficiency of aerosol samplers used at the ECBC and at universities are provided in detail in many publications.[23–27]

Some samplers, such as wetted-wall cyclones and large-format filters, can collect the aerosol from a high volumetric flow; others make use of aerosol concentrators followed by an aerosol collection stage in a much lower volumetric flow. Aerosol concentrators concentrate aerosol particles from a larger volume into a smaller volume of air without losing the sampled particles. Once the particles have reached the collection stage of the sampler, they can be captured with various methods, including filters, impactors, impingers and electrostatic collectors.

Filters are a well established method used in many aerosol samplers[28] and have many advantages over other methods. Most filters have a particle collection efficiency of 100% and the sampling time can be very long because there is no liquid to evaporate. The sampling method is very simple and less susceptible to error. There are, however, significant disadvantages in using a filter collection system. For example, the recovery of particles from filters may not be 100%. Further, bioparticles sampled onto filters may desiccate over time and lose their culturability.

Impaction is another collection mechanism used in many samplers.[29] Impaction is a form of particle collection onto a dry surface, wet surface, fan blades, rotating surface or an agar plate. In traditional impactors, an aerosol stream is first accelerated and then its direction is rapidly shifted by 90° so that the inertial force on a particle will cause it to move perpendicularly to the direction of the turning air stream and impact on a surface. Larger particles will impact on the surface and smaller particles will follow the airstreams and escape. Impactors can be operated in series, known as cascade impactors. With increasing acceleration, they can collect smaller and smaller particles on the downstream stages.

Impingers work on the same principle as impactors; however, the collection surface of the impinger is a liquid or a solid (*e.g.* glass) immersed in liquid.[29] Particles larger than about 1 μm are captured by the inertial mechanisms in the liquid; impingers have also been used for collecting nanoparticles.[23] Particles are collected in the same liquid volume over time (*i.e.* the collection is on a batch basis and the particles are thereby concentrated in the liquid). Some disadvantages exist with impingers, as particles can be re-aerosolized over time as the air bubbles through the liquid and evaporation of the liquid medium will affect the sample time and the final liquid volume.

Electrostatic forces are used in many samplers to collect particles.[30] Particles are charged in the first region of the sampler and collected in a subsequent region of the sampler. This is an efficient collection mechanism for small (<1 μm) particles and it is also mild enough to collect biological particles without disturbing their viability.

Virtual impactors are used to concentrate the aerosols and operate under the same principle as traditional impactors. Virtual impactors can be operated in series, where the minor flow from a first stage becomes the inflow to a second stage, which can provide much higher concentrations than a single stage alone. With a traditional impactor, particles are collected on a real surface placed transverse to the initial airflow direction; however, in a virtual impactor, the larger particles are driven into a port and passed on to the next stage. About 10% of the air stream is drawn into the port to transport the large particles away from the fractionation zone, while the remaining 90% is exhausted away from the port. The two streams are referred to as the minor (concentrated aerosol) and major flows, respectively.[8]

The physical characteristics of solid, liquid and biological particles also affect the sampling efficiency. Solid particles can bounce when they impact on internal surfaces or they can be re-entrained into the airflow after deposition. Liquid particles are permanently captured upon impact on surfaces; any bends in the transmission lines will inadvertently capture larger particles. Biological particles containing microorganisms and biological substances such as proteins have characteristics of both solid and liquid particles with respect to particle bounce characteristics. Bioparticles may also be sticky and attach to the sampler surfaces, significantly affecting recovery by elution of the sample from the collector and adversely affecting the aerosol collector's overall performance. Some bioparticles are hydrophilic whereas others are hydrophobic, and this may determine how easily they can be recovered from the collection surfaces. Newer samplers for collecting smaller particles are using a process whereby small particles are grown to micrometer-sized particles using water or other liquid vapor so that more efficient collection mechanisms such as impaction and interception can be used to sample these particles.

10.18 Conclusion

Bioaerosols are ubiquitous in the environment and can produce adverse effects on the humans, plants and animals in that environment. These effects are dependent on the physical, chemical and biological properties of the airborne particles, many of which can be readily leveraged to generate, monitor and collect them. This chapter has provided an introduction to some fundamental aerosol physics topics necessary to understand and optimize aerosol applications for the evaluation of toxicological effects of bioaerosols. This chapter provides only an introduction and detailed information on aerosol physics should be obtained from full-length aerosol physics text books.[8,31,32]

Nomenclature

d_p	Particle diameter (μm)
d_a	Aerodynamic diameter (μm)
V	Particle velocity (cm s^{-1} or m s^{-1})
V_{TS}	Particle terminal settling velocity (cm s^{-1} or m s^{-1})
m_p	Particle mass (g or kg)
ρ_p	Particle density (g cm^{-3})

ρ_0	Unity density (g cm^{-3})
F_D	Drag force on a particle (N)
τ	Relaxation time (s)
C_C	Slip correction factor (dimensionless)
D	Diffusion coefficient (cm^2 s^{-1} or m^2 s^{-1})
C_D	Particle drag coefficient (dimensionless)

References

1. F. H. Top and P. F. Wehrle, *Communicable and Infectious Diseases*, Mosby Press Company, Saint Louis, 7th edn, 1972.
2. P. R. Murray, E. J. Baron, M. A. Pfaller, F. C. Tenover and R. H. Yolken, *Manual of Clinical Microbiology*, American Society for Microbiology Press, Washington, DC., 1999.
3. C. J. Hurst, G. R. Knudsen, M. J. Mclnerney, L. D. Stetzenbach and M. V. Walter, *Manual of Environmental Microbiology*, American Society for Microbiology Press, Washington, DC., 1997.
4. S. Soubeyrand, L. Held, M. Hohle and I. Sache, Modelling the spread in space and time of an airborne plant disease, *J. Roy. Stat. Soc. C Appl. Stat.*, 2008, **57**(**3**), 253–272.
5. H. A. Burge, *Bioaerosols*, Lewis Publishers, Ann Arbor, 1995.
6. J. M. Macher, J. Douwes, B. Prezant and T. Reponen, *Aerosols Handbook: Measurement, Dosimetry and Health Effects*, in Bioaerosols, ed. L. S. Ruzer and N. H. Harley, CRC Press, New York, 2013, pp. 285–344.
7. A. Rantio-Lehtimaki, *Bioaerosols Handbook*, in Aerobiology of Pollen and Pollen Antigens, ed. C. S. Cox and C. M. Wathes, Lewis Publishers, Boca Raton, 1995, pp. 387–406.
8. W. C. Hinds, *Aerosol Technology: Properties, Behavior and Measurement of Airborne Particles*, John Wiley & Sons, Inc., New York, 1999.
9. J. Kesavan and R. Doherty, *Density measurements of materials used in aerosol studies*, Aberdeen Proving Ground, US Army Edgewood Chemical Biological Center, 2000.
10. Z. Wei, L. Chen, D. M. Thompson and L. D. Montoya, Effect of particle size on *in vitro* cytotoxicity of titania and alumina nanoparticles, *J. Exp. Nanosci.*, 2014, **9**(**6**), 625–638.
11. G. Oberdorster, E. Oberdorster and J. Oberdorster, Nanotoxicology: An emerging discipline evolving from studies of ultrafine particles, *Environ. Health Perspect.*, 2005, **113**, 823–839.
12. G. Oberdorster, J. N. Finkelstein and C. Johnston, *et al.*, Acute pulmonary effects of ultrafine particles in rats and mice, *Res. Rep. – Health Eff. Inst.*, 2000, **96**, 5–74, discussion 75–86.
13. D. M. Brown, M. R. Wilson, W. MacNee, V. Stone and K. Donaldson, Size-dependent proinflammatory effects of ultrafine polystyreneparticles: a role for surface area and oxidativestress in the enhanced activity of ultrafines, *Toxicol. Appl. Pharmacol.*, 2001, **175**, 191–199.

14. T. J. Cieslak and E. M. Eitzen, Clinical and Epidemiologic Principles of Anthrax, *Emerging Infect. Dis.*, 1999, **5**, 552–555.
15. J. A. Huffman, B. Treutlein and U. Poschl, Fluorescent biological aerosol particle concentrations and size distributions measured with an Ultraviolet Aerodynamic Particle Sizer (UV-APS) in central Europe, *Atmos. Chem. Phys.*, 2010, **10**, 3215–3233.
16. P. A. Baron and K. Willeke, *Aerosol Measurement: Principles, Techniques and Applications*, in Gas and Particle Motion, ed. P. A. Baron and K. Willeke, Wiley Interscience, New York, 2001, pp. 61–82.
17. I. Colbeck, *Introduction to aerosol science*, in Physical and chemical properties of aerosols, ed. I. Colbeck, Blackie Academic & Professional, New York, 1998, pp. 1–30.
18. R. F. Phalen, *Inhalation Studies: Foundations and Techniques*, Informa Health Care, New York, 2009.
19. G. A. Loosmore, Evaluation and development of models for resuspension of aerosols at short times after deposition, *Atmos. Environ.*, 2003, **37**, 639–647.
20. K. W. Nicholson, A review of particle resuspension, *Atmos. Environ.*, 1988, **22**, 2639–2651.
21. J. A. Layshock, B. Pearson, K. Crockett, M. J. Brown, S. V. Cuyk and W. B. Daniel, Reaerosolization of bacillus spp. in outdoor environments: A review of the experimental literature, *Biosecur. Bioterrorism*, 2012, **10**(**3**), 299–303.
22. A. M. Rule, J. Kesavan, K. J. Schwab and T. J. Buckley, Application of flow cytometry for the assessment of preservation and recovery efficiency of bioaerosol samplers spiked with Pantoea agglomerans, *Environ. Sci. Technol.*, 2007, **41**(7), 2467–2472.
23. Z. Wei, R. Rosario and L. D. Montoya, Collection efficiency of a midget impinger for nanoparticles in the range of 3–100 nm, *Atmos. Environ.*, 2010, **44**(**6**), 872–876.
24. J. Kesavan and E. Stuebing, Aerosol Sampling Efficiency Evaluation Methods at the US Army Edgewood Chemical Biological Center, in *Atmospheric and Biological Environmental Monitoring*, ed. Y. J. Kim, U. Platt, M. B. Gu and H. Iwahashi, Springer, New York, 2009.
25. J. Kesavan, D. Schepers and A. R. McFarland, Sampling and Retention Efficiencies of Batch-Type Liquid-Based Bioaerosol Samplers, *Aerosol Sci. Technol.*, 2010, **44**, 817–829.
26. J. Kesavan, J. R. Bottiger and A. R. McFarland, Bioaerosol concentrator performance: comparative tests with viable and with solid and liquid nonviable particles, *J. Appl. Microbiol.*, 2008, **104**, 285–295.
27. E. Lennox, N. Kreisberg and L. D. Montoya, Design and Characterization of a New Area Aerosol Sampler Based on Microtrap Collection Technology, *Aerosol Sci. Technol.*, 2013, **47**, 626–633.
28. M. Lippmann, *Air Sampling Instruments for Evaluation of Atmospheric Contaminants*, in Filters and Filter Holders, ed. B. S. Cohen and S. V. Hering, ACGIH, Cincinnati, Ohio, 1995.

29. S. Hering, *Air Sampling Instruments for Evaluation of Atmospheric Contaminants*, in Impactors, Cyclones and Other Inertial and Gravitational Collectors, ed. B. Cohen and S. Hering, ACGIH, Cincinnati, Ohio, 1995, pp. 279–322.
30. D. L. Swift and M. Lippmann, *Air Sampling Instruments for Evaluation of Atmospheric Contaminants*, in Electrostatic and Thermal Precipitators, ed. B. Cohen and S. Hering, ACGIH, Cincinnati, Ohio, 1995, pp. 323–336.
31. J. H. Vincent, *Aerosol Science for Industrial Hygienists*, Pergamon, New York, 1995.
32. *Aerosol Measurement: Principles, Techniques and Applications*, ed. P. Kulkarni, P. A. Baron and K. Willeke, John Wiley & Sons, New Jersey, 2011.

CHAPTER 11

Respiratory Protection Against Some Pathogens and Toxins

GENEVIEVE BONA[†a] AND SIDNEY A. KATZ*[a]

[a]Rutgers University, 450 Kingston Drive, Cherry Hill, NJ 08034, USA
*E-mail: skatz@camden.rutgers.edu

11.1 Introduction

Pathogens are disease-producing microorganisms. Among the naturally occurring pathogens are bacteria, mycoplasma, rickettsia, fungi and viruses. Toxins are poisons produced through the metabolic activities of living organisms. They are organic compounds such as proteins, polypeptides and alkaloids that come from a variety of biological sources. Pathogens and toxins are normal constituents of the environment. However, industrial processes for fermentation and synthesis are capable of producing large quantities of pathogens and toxins for environmental release by military or terrorist actions.

The human body is vulnerable to injury from harmful exposures to environmental pathogens and toxins. The atmosphere is by far the largest component of the environment, and of the three portals for intoxication of the human body, the lungs are the largest surface (100 m^2) exposed to the environment. Protecting the lungs from exposures to hazardous materials in the environment has been accomplished with a variety of air filtration devices.

[†]Present affiliation: Honeywell International, Inc., Claymont, DE 19703, USA.

Issues in Toxicology No. 25
Aerobiology: The Toxicology of Airborne Pathogens and Toxins
Edited by Harry Salem and Sidney Katz

Published by the Royal Society of Chemistry, www.rsc.org

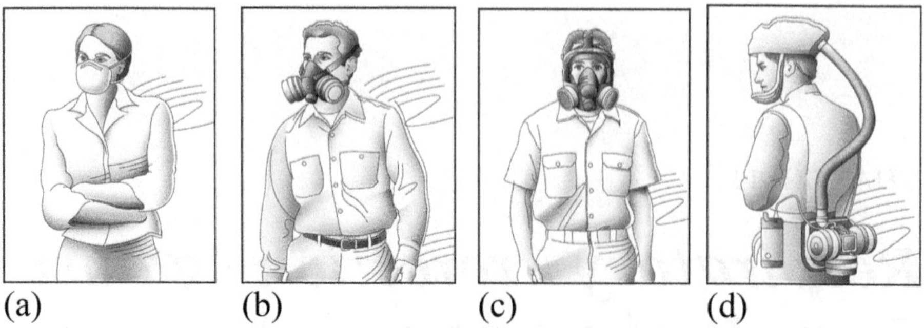

Figure 11.1 Respiratory protection devices: (a) dust or surgical mask; (b) half-face respirator; (c) full face respirator; (d) forced-air respirator.

In the United States, an estimated five million workers are required to wear respirators for compliance with the OSHA Respiratory Protection Standard. In addition, police, fire fighters and other first responders use such devices when conditions require respiratory protection. Respirators protect the user in one of two basic ways: by removing contaminants from the air by filtration or adsorption, or by supplying respirable air from a remote source. The former is achieved with surgical masks, dust masks, gas masks and forced-air purifying respirators; the latter makes use of self-contained breathing apparatuses (SCBA). Some typical examples of the former devices are depicted in Figure 11.1.

11.2 Pathogens

Characteristics for some of the environmental pathogens are listed in Table 11.1. Most are larger than 0.1 μm and few, if any, are smaller than 0.01 μm. Consequently, the devices for protecting the respiratory system must efficiently filter airborne particulates of such dimensions. With respect to contagion, Roy *et al.*[1] suggested, "Naturally produced infectious bioaerosols generated from an infected host possess a size distribution that can range from 0.1 to 20.0 μm." The aerodynamic behavior of airborne particulate matter is dependent upon its size, shape and mass. For this reason, bioaerosols are frequently characterized by aerodynamic equivalent diameter (see Section 11.2.1).

11.2.1 The MMAD

The mass median aerodynamic diameter (MMAD) is used along with its geometric standard deviation to describe statistically the particle size distribution of any aerosol on the basis of the mass, the size and the shape of the particles in question. By mass, half of the particles will be smaller than the MMAD and half will be larger. The aerodynamic diameter of an irregular particle is normalized to that of a unit density sphere with the same settling velocity, *i.e.* $d_e = d_a(\rho_p/\rho_0\chi)^{1/2}$, where d_e is the equivalent aerodynamic

Table 11.1 Bacteria, fungi and viruses, sizes and significance. Source: Army Field Manual No. 3–9, Navy Publication No. P-467, Air Force Manual No. 355-7, Potential Military Chemical/Biological Agents and Compounds, Headquarters, Department of the Army, Department of the Navy, Department of the Air Force, Washington, DC, 12 December 1990, PCN 320 008457 00.

Organism	Microbial group	Rod length/μm	Rod or coccus diameter/μm	Source	Significance
Absidia corymbifera	Fungi		3.8	Environmental	Zygomycosis
Acetobacter melanogenus	Bacteria	1.0–2.0	0.4–0.8		Strong beer/vinegar bacterium
Acinetobacter	Bacteria		1.3	Environmental	Opportunistic infections
Acremonium spp.	Fungi		2.5	Environmental	Extrinsic allergic aveons
Actinomyces israelii	Bacteria		1.0	Humans	Antinomycosis
Adenovirus	Virus		0.08	Humans	Colds
Alcaligenes viscolactis	Bacteria	0.8–2.6	0.6–1.0		Causes ropiness in milk
Alkaligenes	Bacteria		0.75	Humans	Opportunistic infections
Alternaria alternata	Fungi		14.4	Environmental	Mycotoxicosis
Arenavirus	Virus		0.18	Rodents	Hemorrhagic fever
Aspergillus spp.	Fungi		3.5	Environmental	Aspergillosis, volatile organic compound
Aureobasidium pullulans	Fungi		5	Environmental	Chromomycosis
Bacillus anthracis	Bacteria	3.0–10.0	1.0–1.3 (average 1.1)	Environmental	Causes anthrax in mammals
Bacillus stearothermophilus	Bacteria	2.0–5.0	0.6–1.0		Biological indicator for steam sterilization
Bacillus subtilis	Bacteria	2.0–3.0	0.7–0.8		Biological indicator for ethylene oxide sterilization
Blastomyces dermatitidis	Fungi		14	Environmental	Blastomycosis
Bordetella pertussis	Bacteria		0.25	Humans	Whooping cough
Botrytis cinera	Fungi		7	Environmental	Extrinsic allergic aveons
Cardiobacterium	Bacteria		0.63	Humans	Opportunistic infections
Chaetomium globosum	Fungi		5.5	Environmental	Chromomycosis, volatile organic compound
Chlamydia psittaci	Bacteria		0.3	Birds	Psittacosis

(*continued*)

Table 11.1 (*continued*)

Organism	Microbial group	Rod length/μm	Rod or coccus diameter/μm	Source	Significance
Chlamydia pneumoniae	Virus		0.3	Humans	Pneumonia
Cladosporium spp.	Fungi		9	Environmental	Chromblastomycosis
Clostridium botulinum (B)	Bacteria	3.0–8.0	0.5–0.8		Produces exotoxin causing botulism
Clostridium perfringens	Bacteria	4.0–8.0	1.0–1.5		Produces toxin causing food poisoning
Clostridium tetani	Bacteria	4.0–8.0	0.4–0.6		Produces exotoxin causing tetanus
Coccidioides immitis	Fungi		4	Environmental	Coccidiodomycosis
Coronavirus	Virus		0.11	Humans	Colds
Corynebacterium diphtheriae	Bacteria		1.0	Humans	Diphtheria
Coxiella burnetii	Bacteria		0.5	Cattle, sheep	Q fever
Coxsackievirus	Virus		0.027	Humans	Colds
Cryptococcus neoformans	Fungi		5.5	Environmental	Cryptococcosis
Diplococcus pneumoniae	Bacteria		0.5–1.25		Causes lobar pneumonia
Echovirus	Virus		0.028	Humans	Colds
Emericella nidulans	Fungi		3.3	Environmental	Mycotoxicosis, volatile organic compound
Epicoccum nigrum	Fungi		20	Environmental	Extrinsic allergic aveons
Erwinia aroideae	Bacteria	2.0–3.0	0.5		Causes soft rot in vegetables
Escherichia coli (*E. coli*)	Bacteria	1.0–3.0	0.5		Indicator of fecal contamination in water
Eurotium spp.	Fungi		5.8	Environmental	Extrinsic allergic aveons
Exophiala jeanselmei	Fungi		2	Environmental	Chromomycosis
Francisella tularensis	Bacteria		0.2	Wild animals	Tularemia
Geomyces pannorum	Fungi		3	Environmental	Extrinsic allergic aveons
Haemophilus influenzae	Bacteria	0.5–2.0	0.2–0.3		Causes influenza and acute respiratory infections
Haemophilus influenzae	Bacteria		0.43	Humans	Meningitis, pneumonia

Haemophilus parainfluenzae	Bacteria		1	Humans	Opportunistic infections
Hantavirus	Virus		0.07	Rodents	Hantavirus
helminthosporium	Fungi		12.5	Environmental	Extrinsic allergic aveons
Histoplasma capsulatum	Fungi		3	Environmental	Histoplasmosis
Influenza	Virus		0.1	Humans, birds	Flu
Klebsiella pneumoniae	Bacteria	5	0.4–0.5	Environmental	Opportunistic infections, causes pneumonia and other respiratory inflammation
Lactobacillus delbrueckii	Bacteria	2.0–9.0	0.5–0.8		Causes souring of grain mashes
Legionella pneumophila	Bacteria		0.6	Environmental	Pontiac fever
Micromonospora faeni	Actinomycetes		1	Agricultural	Farmers lung, hypersensitivity, pneumonitis
Micropolyspora faeni	Actinomycetes		0.69	Agricultural	Farmers lung, hypersensitivity pneumonitis
Moraxella catarrhalis	Bacteria		1.3	Humans	Opportunistic infections
Moraxella lacunata	Bacteria		1	Humans	Opportunistic infections
Morbillivirus	Virus		0.12	Humans	Measles (rubeola)
Mucor plumbeus	Fungi		7.5	Environmental	Mucormycosis
Mycobacterium avium	Bacteria		1.2	Environmental	Cavitary pulmonary disorder
Mycobacterium intracellulare	Bacteria		1.2	Environmental	Cavitary pulmonary disorder
Mycobacterium kansasii	Bacteria		0.86	Unknown	Cavitary pulmonary disorder
Mycobacterium tuberculosis	Bacteria	1.0–4.0	0.2–0.5 (average 0.86)	Humans	Hard swelling of body tissues, tuberculosis
Mycoplasma pneumoniae	Bacteria		0.25	Humans	Pneumonia
Mycoplasma pneumoniae (PPLO)	Bacteria		0.3–0.5		Smallest known free-living organism
Neisseria meningitidis	Bacteria		0.8	Humans	Meningitis

(continued)

Table 11.1 (*continued*)

Organism	Microbial group	Rod length/μm	Rod or coccus diameter/μm	Source	Significance
Nocardia brasiliensis	Actinomycetes		1.5	Environmental	Pulmonary mycetoma
Nocardia asteroides	Actinomycetes		1.1	Environmental	Nocardiosis
Paecilomyces variotii	Fungi		3	Environmental	Mucormycosis
Paracoccidioides brasiliensis	Fungi		23	Environmental	Paracoccidioidomycosis
Parainfluenza	Virus		0.22	Humans	Flu
Paramyxovirus	Virus		0.23	Humans	Mumps
Parvovirus B19	Virus		0.022	Humans	Filth disease, anemia
Pediococcus acidilactci	Bacteria		0.6–1.0		Causes mash spoilage in brewing
Pediococcus cerevisiae	Bacteria		1.0–1.3		Causes deterioration in beer
Penicillium spp.	Fungi		3.3	Environmental	Mycotoxicosis, volatile organic compound
Phialophora spp.	Fungi		1.5	Environmental	Chromomycosis
Phoma spp.	Fungi		3.3	Environmental	Mucormycosis
Pneumocystis carinii	Bacteria		2	Environmental	Pneumocystosis
Poxvirus – Vaccinia	Virus		0.23	Agricultural	Cowpox
Pseudomonas aeruginosa	Bacteria		0.57	Environmental	Opportunistic infections
Pseudomonas mallei	Bacteria		0.77	Environmental	Opportunistic infections
Pseudomonas pseudomallei	Bacteria		0.57	Environmental	Opportunistic infections
Pseudomonas diminuta	Bacteria	1.0	0.3		Test organism for retention on 0.2 μm membranes
Rhinorvirus	Virus		0.023	Humans	Colds
Rhizopus stolonifer	Fungi		8	Environmental	Zygomycosis
Rhodoturula spp.	Fungi		14	Environmental	Extrinsic allergic aveons
Salmonella enteritidis	Bacteria	2.0–3.0	0.6–0.7		Causes food poisoning
Salmonella enteritidis	Bacteria	2.0–3.0	0.6–0.7		Causes food poisoning
Salmonella hirschfeldii	Bacteria	1.0–2.5	0.3–0.5		Causes enteric fever
Salmonella typhimurium	Bacteria		1.0–1.5 (average 0.5)		Causes food poisoning in humans
Salmonella typhosa	Bacteria	2.0–3.0	0.6–0.7		Causes typhoid fever

Sarcina maxima	Bacteria		4.0–4.5		Isolated from fermenting malt mash
Scopulariopsis spp.	Fungi		6	Environmental	Onychomycosis
Serratia marcescens	Bacteria	0.5–1.0	0.5		Test organism for retention on 0.45 μm membranes
Serratia marcescens	Bacteria		1.3	Environmental	Opportunistic infections
Shigella dysenteriae	Bacteria	1.0–3.0	0.4–0.6		Causes dysentery in humans
Sporothrix schenckii	Fungi		6.5	Environmental	Sporotrichosis
Stachybotrys spp.	Fungi		5.7	Environmental	Stachybotryotoxicosis
Staphylococcus aureus	Bacteria		0.8–1.0	Humans	Causes pus-forming infections, opportunistic infections
Streptococcus lactis	Bacteria		0.5–1.0		Contaminant in milk
Streptococcus pneumoniae	Bacteria		0.9	Humans	Pneumonia, otitis media
Streptococcus pyogenes	Bacteria		0.6–1.0 (average 0.9)	Humans	Causes pus-forming infections, scarlet fever, pharyngitis
Thermoactinomyces sacchari	Actinomycetes		0.86	Agricultural	Bagassosis
Thermoactinomyces vulgaris	Actinomycetes		1	Agricultural	Farmers lung, hypersensitivity, pneumonitis
Thermomonspora viridis	Actinomycetes		0.6	Agricultural	Farmers lung, hypersensitivity, pneumonitis
Togavirus	Virus		0.063	Humans	Rubella (German measles)
Trichoderma spp.	Fungi		4.1	Environmental	Mycotoxicosis, volatile organic compound
Ulociadium spp.	Fungi		15	Environmental	Extrinsic allergic aveons
Varicella zoster	Virus		0.3	Humans	Chickenpox
Wallemia sebi	Fungi		3	Environmental	Extrinsic allergic aveons
Yersinia pestis	Virus		0.75	Humans	Pneumonic plague

diameter, d_a is the aerodynamic diameter of the particle in question, ρ_p is the density of the particle in question, ρ_0 is 1.000 g cm^{-3} and χ is the dynamic shape factor of the particle in question. First[2] wrote, "For example, a 1.0 mm sphere of UO_2, density 11 g cm^{-3}, has an aerodynamic equivalent diameter of 3.3 mm."

11.3 Toxins

Characteristics of the biological toxins are summarized in Table 11.2. Although the organisms from which they are derived are filterable, these compounds are of molecular dimensions. For example, ricin is a two-chain polypeptide (dimer) with a molecular mass of between 60 and 65 kDa. Tetanus toxin is also a dimer having a molecular mass of ~150 kDa. Several sub-types of botulinum toxin are known. Like the tetanus toxin, they too are dimers with molecular masses of ~150 kDa. The biological toxins are of low volatility; however, they become airborne when adsorbed on the surfaces of dusts or other forms of finely divided particulate matter. As suggested by the CDC,[3] DHS[4] and NRT,[5] HEPA (high-efficiency particulate air) filters provide adequate protection against the inhalation of aerosolized forms of biological toxins. Level C protective ensemble and a full facepiece air-purifying respirator with P100 or N100 filters or Level C protective ensemble and a full facepiece forced-air-purifying respirator equipped with high-efficiency filters are recommended to be worn along with disposable hooded coveralls, shoe coverings and nitrile or vinyl gloves. The characteristics of filters designated N, R and P are given in Table 11.3.

11.4 Air Filtration

In the industrial sector, the focus of air filtration is most often on dusts and smokes containing inorganic materials such cadmium or lead and on organic vapors and fumes such as benzene or carbon tetrachloride. Individual protection devices range from fabric or fiber masks covering the nose and mouth to more sophisticated gas masks containing charcoal filters.

NIOSH and the FDA have approved a variety of dust and surgical masks to protect industrial and medical workers from inhaling toxic chemicals and pathogenic organisms. The filter media are efficient in trapping airborne particulate matter, but the fit of the mask on the face often compromises the protection factor. Half-face and full-face respirators offer higher degrees of protection with superior fit and options for removing both particulate and gaseous contaminants. Some of such devices are shown in Figure 11.1a and b. These are typical of the NIOSH-approved P100 particle-filtering facepiece respirator. This half-face version provides 99.97% filter efficiency against oil and non-oil particulate matter, certain dusts, fumes, mists and radionuclides and asbestos-containing dusts and mists. The properties of some of the filters certified by NIOSH are given in Table 11.3.

Table 11.2 Characteristics of some toxins.

Toxin	$LD_{50}{}^{a}/\mu g\ kg^{-1}$	Molar mass	Source
Abrin[43,44]	0.04	65 000	Plant (rosary pea)
Aconitine[43]	100	647	Plant (monkshood)
Aerolysin[44]	7.0	54 000	Bacterium
Anatoxin-A(s)[43]	50	500	Blue-green alga
Batrachotoxin[43]	2	539	Arrow-poison frog
Botulinum toxins[43,44]	0.001	150 000	Bacterium
Bungarotoxins[44]	8	8000	Elapid snake venom
Caeruleotoxin[45]	40	13 000	Elapid snake venom
Cereolysin[46]	40–80	23 000	Bacterium
Ciguatoxin[43,44]	0.4	1000	Marine dinoflagellate
Clostridium perfringens toxins[43,44]	0.1–5	35 000–40 000	Bacterium
Conotoxin	5	1500	Cone snail
Diphtheria toxin[43,44]	0.1	62 000	Bacterium
Listeriolysin[44]	3–12	55 800	Bacterium
Maitotoxin[43,44]	0.1	3400	Marine dinoflagellate
Microcystin[43]	50	994	Blue-green alga
Modeccin[44]	1.3	56 000	Plant
Notexin[44]	5	13 500	Snake venom
Palytoxin[43]	0.15	2700	Marine soft coral
Pertussis toxin[44]	18	10 500	Bacterium
Pneumolysin[47]	1.5	5300	Bacterium
Pseudomonas aeruginosa exotoxin[48]	14	71 500	Bacterium
Ricin[43,44]	3	64 000	Plant (castor bean)
Saxitoxin[43]	10^{b}	299	Marine dinoflagellate
Shiga toxin[43,44]	0.002	55 000	Bacterium
Shigella dysenteriae neurotoxin[44,49]	0.25	72 000	Bacterium
Staphylococcus aureus exotoxin[43,50]	27^{c}	22 000	Bacterium
Streptolysin O[44,51]	8	60 500	Bacterium
T-2 toxin[43,44]	1210	466	Fungal mycotoxin
Taipoxin[43,44]	5.0	46 000	Elapid snake venom
Tetanus toxins[43,44]	0.002	150 000	Bacterium
Tetrodotoxin[43,44]	8	319	Puffer fish
Textilotoxin[43]	0.6	80 000	Elapid snake venom
Tityustoxin[43]	9	8000	Scorpion

[a]i.v./i.p.
[b]Inhalation:[43] 2.0.
[c]Inhalation:[43,49] $ED_{50} \sim$ pg.

The C2A1 canister for the M45 Army field mask contains both a particulate filter and a chemical filter, thereby offering protection against both biological and chemical warfare agents. The C2A1 canister shown in Figure 11.2(a) is used with the M45 protective mask shown in Figure 11.2(b). After ambient air enters through the orifice on the left side of the canister, it passes

Table 11.3 Properties of NIOSH-certified filters.[a]

Class of filter	Efficiency/%	Test agent	Test maximum loading/mg	Type of contaminant	Service time
N series		NaCl	200	Solid and water-based particulates (*i.e.* non-oil aerosols)	Nonspecific
N100	99.97				
N99	99				
N95	95				
R series		DOP oil	200	Any	One work shift
R100	99.97				
R99	99				
R95	95				
P series		DOP oil	Stabilized efficiency	Any	Nonspecific
P100	99.97				
P99	99				
P95	95				

[a]Source: *NIOSH Guide to the Selection and Use of Particulate Respirators*, DHHS (NIOSH) Publication No. 96–101, Centers for Disease Control and Prevention, Atlanta, GA, January 1996, http://www.cdc.gov/niosh/docs/96-101, accessed 6 January 2016.

(a) (b)

Figure 11.2 (a) C2A1 canister and (b) M45 gas mask. Source: *Technical Manual Operator Instructions: Mask, Chemical – Biological: Land Warrior, M45, TM 3-4240-348-10*, Department of the Army, 21 January 2000.

first through the pleated white filter where particulates are removed, then through the layer of TEDA–ASZM charcoal for the absorption of chemical warfare agents, then through a second filter to remove charcoal dust and finally exits the canister through the orifice on the right side. TEDA is an acronym for triethylenediamine and ASZM signifies that the charcoal has been impregnated with compounds of copper, silver, zinc and molybdenum. As a particulate filter, it is tested to be at least 99.99% efficient against an ~0.3 μm MMAD aerosol challenge with dioctyl phthalate (DOP).

In addition, the military has developed group protection systems for mobile and stationary use. The M98 200 cfm (cubic feet per minute) CBRN (chemical, biological, radiological, nuclear) filter set removes toxic gases and particulates to provide purified air for a variety of applications. These filter sets are used on military H200–GPFU vehicle collective protection systems. Toxic gas removal is achieved with carbon type 12 × 30 ASZM–TEDA and the removal of airborne particulate matter is 99.97% for 0.3 μm particle size. The M28, M20A1 and M521 collective protection systems make use of these HEPA filters to provide clean room environments for mission operations in the presence of CBRN hazards. The M51 shelter is a trailer-mounted system that features an overpressure and environmental control system. The shelter is predominately used by battalion aid stations and other medical units. It can also be used as a temporary rest and relief shelter.

The military technology for air filtration has been adopted in the civilian sector. Filters are of the HEPA and ULPA (ultra low particle air) type with efficiencies ranging from 95% for 0.3 μm to 99.99999+% for 0.12 μm particles, respectively. Filters are utilized where disk drives are assembled, in clean room operations where medical device are fabricated and in the nuclear, semiconductor, food, pharmaceutical and other industries that require contamination control. Special HEPA filters are manufactured for hospital applications such as vector filters and ceiling filters for surgical and patient environments. HEPA filters are found in consumer products such as vacuum cleaners and devices for control of asthma and allergy triggers. First[2] described the historical development and operational characteristics of HEPA filters and their military, industrial, biomedical and consumer applications.

11.4.1 Mechanisms of Air Filtration

First,[2] Lavoie *et al.*[6] and others[7–9] have described models for the mechanisms of air filtration. The most predominant mechanisms contributing to the efficiency of the filter in capturing particles are inertial impaction, interception and diffusion.

11.4.1.1 Inertial Impaction

Particles having higher momentum relative to that of the gas carrying them possess sufficient inertia to resist following the gas stream around the fibers of the filter. Such particles deviate from the gas stream and impact on the fibers. This effect becomes greater with increasing aerodynamic equivalent diameter of the particles and with increasing velocity of the gas approaching the filter. Inertial impaction is the favored mechanism when the particles are large and the velocity of the gas stream is high.

11.4.1.2 Interception

Particles with geometric diameters close to those of the filter fibers that approach the fiber to within one particle radius adhere to the fiber by interception. The diameters of filter fibers limit the size of the particles that are filterable. The particle size should not be much smaller than the diameter of the filter fiber. The velocity of the gas has little influence on the interception mechanism.

11.4.1.3 Diffusion

Very small particles, smaller than 0.1 μm, have little inertia. They demonstrate Brownian motion, and when these random movements result in contact of the particles with the fibers of the filter, adhesion removes the particles from the gas stream. A concentration gradient is established between the surface of the filter fibers and the body of the gas stream. Diffusion of the particles to the surface of the filter fibers is enhanced under the influence of this concentration gradient. The diffusion mechanism is favored by small particle size. The rate of diffusion is increased as the magnitude of the concentration gradient increases. Hence the diffusion mechanism is favored by low gas velocity.

As shown in Figure 11.3, the filter collection efficiency varies with particle size. For very small particles, <0.1 μm in diameter, the filter is very efficient. The primary filtration mechanism for the removal of these very small particles from the gas stream is diffusion. For particles between 0.1 and 0.4 μm in diameter, filtration becomes less efficient because the particles are too large for significant diffusion effects and too small for interception effects. Particles larger than 0.4 μm in diameter are efficiently collected by interception and inertial impaction mechanisms. It is for this reason that the C2A1 canister for the M45 Army field mask and other military air filtration devices are tested with a 0.3 μm MMAD aerosol challenge with DOP, the region of minimum efficiency.

11.5 Some Comparisons and Applications

The efficiencies of the air filtration devices have been compared and their applications have been evaluated under a variety of test and field conditions.

11.5.1 Comparisons

Guyton *et al.*[10] described a method for evaluating the respiratory protection provided by representative types of contagion masks against an aerosol of *Bacillus subtilis* var. *niger* spores. From this comparison, they concluded that

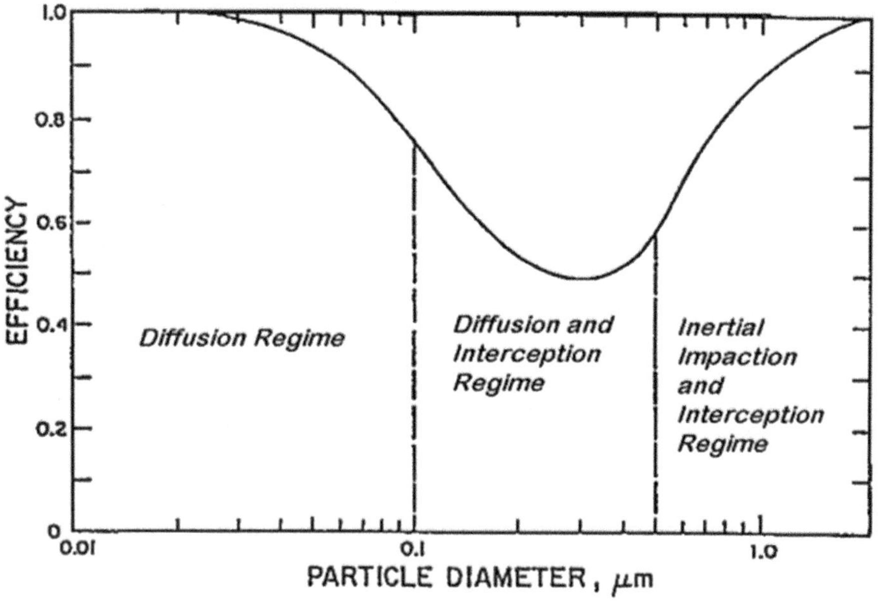

Figure 11.3 Air filtration mechanisms. Source: NIOSH Science Blog, *N95 Respirators and Surgical Masks*, Centers for Disease Control and Prevention, blogs.cdc.gov/niosh-science-blog/2009/10/14/n95, accessed 8 October 2014.

the contagion masks evaluated offer poor respiratory protection against airborne microorganisms in the particle diameter range 1.0–5.0 μm. Commercially available dust respirators and related masks were ~2.5–5 times more efficient than contagion masks and an even higher degree of protection could be obtained by the use of full face masks.

Subsequently, Guyton and Decker[11] reported on the effectiveness of five recently (at that time) developed contagion masks in filtering airborne particles (1–5 μm diameter). They found that one mask, available in four sizes, was 99% efficient. This mask could be reused after sterilization. The other four masks were available in only one size and were intended to be used one time only. Two of these four disposable types were more than 90% efficient, but the variability of their respective test results was much greater than that for the reusable mask. The two remaining disposable types were less than 80% efficient. The first two of these disposable contagion mask types were worn by hospital personnel for periods of up to 8 h to determine the effect of such prolonged use on aerosol filtration efficiency. No significant decrease in filtration efficiency was noted. They suggested that physicians, nurses and other hospital personnel who wear masks would benefit from the increased individual respiratory protection afforded by improved contagion masks; concurrently, the incidence of hospital patient airborne infections would be greatly reduced.

McCullough *et al.*[12] compared the efficiencies of a variety of NIOSH-approved respirator filters (*i.e.* dust/mist filters and HEPA filters) and surgical masks by challenging them with three aerosolized bacteria of different shapes: *Mycobacterium abscessus* (a rod), *Staphylococcus epidermidis* (a sphere) and *Bacillus subtilis* (a rod). Tests were conducted at two flow rates (45 and 85 L min^{-1}) and two humidity levels (30 and 70%). Aerosols were measured with a total particle, direct-reading spectrometer and a viable particle cascade impactor. Aerosol penetration was determined by evaluating the filter or mask up- and downstream; the fit of the respirator or surgical mask was not evaluated. Bioaerosol penetration measured with the two aerosol sampling instruments was found to correlate. Additionally, the bioaerosol test parameters were evaluated with respect to their effect on penetration. Increasing flow rate resulted in increased penetration of all organisms whereas an increase in relative humidity did not exert a consistent effect on all organisms. As expected, of the respirators approved by NIOSH, the filter of the dust/mist respirators had the lowest efficiency and the HEPA filters had the highest. Surgical masks were the least efficient of all filters tested; these are not certified by NIOSH. In this study, the bioaerosol penetration was compared with that of a polystyrene latex sphere aerosol. Penetration of the test aerosols was predicted on the basis of particle aerodynamic diameter and the extent of penetration was expected to decrease in the order polystyrene latex sphere > *Mycobacterium abscessus* > *Staphylococcus epidermidis* = *Bacillus subtilis*. The polystyrene latex sphere aerosol was the most penetrating, as predicted. However, the results showed that the rod-shaped *Bacillus subtilis* was more penetrating than the spherical *Staphylococcus epidermidis*. Therefore, the aerodynamic diameter may not be the best parameter for predicting aerosol penetration of these filters by nonspherical particles.

Rengasamy *et al.*[13] stressed the importance of research on respiratory protection against biological agents in addressing major concerns such as occupational safety and terrorist attack. The respiratory protection against biological warfare agents such as anthrax in military and civilian situations was also described. Their work reviewed the literature on respiratory protection against bioaerosols and identified research gaps. The complexities of respiratory protection were considered with respect to a number of factors: the efficiency of respirator filter material, the fitting of the facepiece and the maintenance, storage and reuse of the respirators. Also, studies on nonpathogenic microorganisms with physical characteristics similar to those of *Mycobacterium tuberculosis* were used to analyze microbial penetration through respirators. These comparative studies showed that high-efficiency particulate air and N95 filters provided a higher level of protection than dust/mist and dust/mist/fume filters. Variables such as flow rate and relative humidity appeared to alter the level of penetration of microorganisms through respirator filters. The relationships between microbial penetration through respirator filters and the aerodynamic diameter, length and other physical characteristics of microorganisms remained controversial. Whether bioaerosol particle reaerosolization, which is the reintroduction of previously airborne particles that have settled onto a surface back into the atmosphere, should be a concern was unclear even though one study demonstrated significant reaerosolization of 1–5 μm particles loaded onto respirator filters.

This work also mentioned that maintenance, storage and decontamination of respirators are important factors to be considered when reusing them.

Because there are differences between the NIOSH test protocols and the ENs (European Norms) for evaluating air-purifying particulate respirators (filtering facepiece respirators), Rengasamy *et al.*[14] conducted a side-by-side comparison of the relative performances of NIOSH-approved and EN-certified CE (Conformité Europëene)-marked filtering facepiece respirators. NIOSH required minima of 95 and 99.97% efficiencies for the N95 and P100 filtering facepiece respirators, respectively, whereas the EN required 94 and 99% efficiencies for class P2 and P3 filtering facepiece respirators, respectively. To understand better the filtration performance of NIOSH-approved and CE-marked filtering facepiece respirators, initial penetration levels of N95, P100, P2 and P3 respirators were measured and compared using a series of polydisperse and monodisperse aerosol test methods. Initial penetration levels of polydisperse NaCl aerosols (MMAD of 238 nm) were measured using a method similar to the NIOSH respirator certification test method. Monodisperse aerosol penetrations were measured using silver particles for the 4–30 nm range and NaCl particles for the 20–400 nm range. Two models for each of the N95, P100, P2 and P3 filtering facepiece respirator types were selected and five samples from each type were tested against charge-neutralized aerosol particles at a flow rate of 85 L min^{-1}. The penetration results from the test with 238 nm MMAD polystyrene aerosol were <1% for N95 and P2 models and <0.03% for P100 and P3 models. The penetration levels of this monodisperse aerosol showed that the most penetrating particle size was in the 30–60 nm range for all models of filtering facepiece respirators tested in the study. Penetrations at the most penetrating particle sizes were <4.28, <2.22, <0.164 and <0.009% for the N95, P2, P3 and P100 filtering facepiece respirator models, respectively. The most penetrating particle sizes obtained for all four filtering facepiece respirator types suggested particle capture by an electrostatic mechanism. Liquid 2-propanol treatment of filtering facepiece respirators for electrostatic charge removal shifted the most penetrating particle sizes to 200–300 nm. This treatment also dramatically increased both polydisperse and monodisperse aerosol penetrations of all four filtering facepiece respirator types, indicating that all of these respirator types share filtration characteristics of electret filters. Furthermore, electrostatic charge removal from all four filtering facepiece respirator types increased the penetration levels of 400–1000 nm range particles. The particle penetration data obtained in this study showed that each model of NIOSH-approved N95 and P100 and CE-marked P2 and P3 respirators that were used provided expected levels of laboratory filtration performance against nanoparticles.

Earlier, Jasper *et al.*[15] studied the performance of electret filters for respirators when exposed to ethylbenzene for charge removal. They reported that in their study, the percentage of DOP aerosol penetrating the electret filter media was not affected by its exposure to ethylbenzene vapor, whereas exposure to ethylbenzene liquid caused a 40% decrease in the performance. They attributed this decrease in performance to changes in the static electric field caused by the presence of liquid ethylbenzene on the surfaces of the filter fibers. The performance of the mechanical fiber media was unaffected after charge removal by exposure to ethylbenzene in either the liquid or gaseous phases.

Subsequently, Rengasamy *et al.*[16] employed polydisperse DOP aerosols for testing filter penetration with loading of R- and P-series particulate filters for NIOSH certification. Penetration for filters must not exceed NIOSH-allowed maximum levels during the entire loading of 200 mg of DOP to indicate no filter degradation. To understand better the degradation of respirators with electrostatic filter media, two models of N95 and P100 filtering facepiece respirators were loaded with polydisperse DOP aerosols up to 200 mg as employed for NIOSH certification testing with simultaneous measurement of filter penetration. In parallel experiments, both N95 and P100 filtering facepiece respirators were loaded with 10–200 mg of DOP and challenged with polydisperse NaCl aerosol employed for NIOSH certification testing and monodisperse NaCl aerosol and filter penetration was also measured. The results showed that filter penetration for both N95 models increased with increasing amount of DOP loading and exceeded NIOSH-allowed maximum penetration (5%) with both DOP and NaCl aerosols, indicating filter degradation. Monodisperse NaCl aerosols (20–400 nm) gradually increased the penetration and shifted the most penetrating particle size from ~40 nm to larger sizes. In the case of P100 filtering facepiece respirators, DOP aerosol penetration was below 0.03% for up to a 200 mg DOP loading, as required for NIOSH certification. Interestingly, one of the two P100 filtering facepiece respirator models loaded with 10–50 mg of DOP showed >0.03% penetration with polydisperse, in addition to monodisperse, for NaCl aerosol testing. Overall, the results obtained in the study indicate that some P100 models loaded with DOP at lower concentrations may show higher penetration with other aerosols such as NaCl. To understand better the filter degradation of DOP-loaded respirators, further studies are needed.

Lawrence *et al.*[17] compared the elastomeric N95 respirators and N95 filtering facepiece respirators provided for American workers to guard against potential inhalation hazards. This study compared the performances of these types of respiratory protection during a simulated workplace test that measured both filter penetration and faceseal leakage. A panel of 25 test subjects with varying face sizes tested 15 models of elastomeric N95 respirators, 15 models of N95 filtering facepiece respirators and six models of surgical masks. Simulated workplace testing was conducted using a TSI PORTACOUNT Plus Model 8020 respirator fit tester and consisted of a series of seven exercises. Six simulated workplace tests were performed with redonning of the respirator/mask occurring between each test. The results of these tests produced a simulated workplace protection factor. The geometric mean and the fifth percentile values of the simulated workplace protection factors were computed by category of respiratory protection using the six overall simulated workplace protection factor values. The levels of protection provided by each of the three respiratory protection types were compared. The geometric mean and fifth percentile simulated workplace protection factor values without fit testing were used for the comparison, as surgical masks were not intended to be fit tested. The geometric mean values were 36 for elastomeric N95 respirators, 21 for N95 filtering facepiece respirators and 3 for surgical masks. An analysis of variance demonstrated a statistically significant difference between all three. Elastomeric N95 respirators had the highest fifth percentile simulated workplace protection factor of 7. N95 filtering facepiece respirators and surgical masks had fifth percentile simulated workplace protection factors of 3 and 1, respectively. A Fisher exact test revealed that the fifth percentile simulated workplace protection factors for all three types of respiratory protection were statistically different. In addition, both the qualitative Bitrex and saccharin fit tests[18] and the quantitative N95 Companion fit test[19] were performed on the N95 filtering and elastomeric facepiece respirators. It was found that passing a fit test generally improves the protection afforded the wearer. Passing the Bitrex fit test resulted in fifth percentile simulated workplace protection factors of 11.1 and 7.9 for elastomeric and filtering facepiece respirators, respectively. After passing the saccharin tests, the elastomeric respirators provided a fifth percentile of 11.7 and the filtering facepiece respirators provided a fifth percentile of 11.0. The fifth percentiles after passing the N95 Companion fit test were 13.0 for the elastomeric respirators and 20.5 for the filtering facepiece respirators. The data supported fit testing as an essential element of a complete respiratory protection program.

Cho *et al.*[20] compared workplace protection factors for five different contaminants [endotoxin, fungal spores, (1 → 3)-β-D-glucan, total particle mass and total particle number] provided by an N95 elastomeric respirator and an N95 filtering facepiece respirator. Farm workers ($n = 25$) wore the elastomeric respirator and filtering facepiece respirator while performing activities at eight locations representing horse farms, pig barns and grain handling facilities. For the determination of workplace protection factors, particles were collected on filters simultaneously inside and outside the respirator during the first and last 15 min of 60 min intervals. One field blank per subject was included in the experiment. A reporting limit was established for each contaminant based on geometric means of the field blanks as the lowest possible measurable values. Depending on the contaminant type, 38–48% of data points were below the reporting limit. Therefore, a censored regression model was used to estimate a censored workplace protection factor. The censored workplace protection factors for the two types of respirators were not significantly different. In contrast, significant differences were found in the censored workplace protection factors for different types of contaminants. The geometric means for censored workplace protection factors for the two types of respirators combined were 154, 29, 18, 19 and 176 for endotoxin, fungal spore count, (1 → 3)-β-D-glucan, total particle mass and total particle number, respectively. The censored workplace protection factor was strongly associated with concentrations measured outside the respirator for endotoxin, fungal spores and total particle mass except for total particle number.

However, when only data points with outside concentrations higher than 176 times the reporting limit were included, the censored workplace protection factor increased and the association between the outside concentrations and the censored workplace protection factors became weaker. The results indicate that difference in censored workplace protection factors observed between different contaminants may be attributed to differences in the sensitivity of analytical methods to detect low inside concentrations, rather than the nature of particles (biological or nonbiological).

The degree of worker protection provided by N95 filtering facepiece respirators appears to be dependent upon both the filter efficiency and the leakage through the face to facepiece seal.[17,19] Rengasamy and Eimer[21] considered the combination of particle penetration through the filter media and particle leakage through the faceseal and any component interfaces as total inward leakage. To understand better the significance of particle penetration through filter media and through faceseal leakage, they measured total inward leakage using 20–800 nm NaCl particles at 8–40 L breathing minute intervals for four models of N95 full-face respirators in a manikin test system. The results of these measurements showed that the most penetrating particle size was ~45 nm for all four models of respirators. Filter penetration for 45 nm particles was significantly ($p < 0.05$) higher than that for 400 nm particles. The total inward leakage of the 45 nm particles increased with leak size and breathing minute volume and was consistently higher than the total inward leakage of the 400 nm particles. This finding suggests that (1) faceseal leakage is responsible for the presence of test aerosol inside the full-face respirator regardless of particle size and (2) filter penetration is responsible for the greater total inward leakage for the smaller particles. The full-face respirators with relatively small filter penetration values showed lower total inward leakage at the smaller leak sizes than did those with higher filter penetrations, indicating the dependence of total inward leakage on filter penetration. When the electrostatic charge was removed, the full-face respirators showed a shift in the most penetrating particle size to 150 nm with the same test aerosols (mode size ~75 nm) at different hole sizes and breathing minute volumes, confirming the interaction between filter penetration and faceseal leakage processes. The shift in the most penetrating particle size from 45 to 150 nm for the filters after charge removal indicated that mechanical filters may perform better against nanoparticles than electrostatic filters rated for the same filter efficiency. The results suggest that among the different sized particles that enter inside the N95 respirators, relatively high concentrations of the most penetrating particle size particles in the breathing zone of respirators can be expected in workplaces with high concentrations of nanoparticles. Overall, the data obtained in the study suggested that good-fitting respirators with lower filter penetration values would provide better protection against nanoparticles. Subsequently, Rengasamy *et al.*[22] proposed that these data indicated that good-fitting filtering facepiece respirators might offer higher protection against submicron-sized bioaerosols. The "faceseal leakage-to-filter" penetration ratios determined by He *et al.*[23] suggested that faceseal leakage was the primary penetration pathway.

Davidson *et al.*[24] evaluated the filtration efficiencies of 11 models of N95 respirators and surgical masks using aerosolized *B. anthracis* endospores and polystyrene latex particles. The geometric mean for the relative efficiencies of all respirators and masks challenged with the endospores ranged from 34 to 65% whereas that for the respirators and masks challenged with the polystyrene latex particles ranged from 35 to 64%. The geometric means for the filtration efficiencies of the filter media from all respirators and masks were >99% when challenged with polystyrene latex particles larger than 0.1 μm in diameter. However, the geometric mean filtration efficiencies for the surgical mask filter media ranged from 70 to 83% with particles having diameters >0.1 μm and from 74 to 93% using 1.3 μm polystyrene latex particles. The results of these observations further highlighted the importance of faceseal leakage in the respiratory protection provided by N95 respirators. Harnish *et al.*[25] reported a similar study in which five N95 filtering facepiece respirators were challenged with viable N1H1 influenza aerosols. Observing that the efficiencies of the respirators were >95%, they concluded that inhalation exposure to airborne influenza virus could be reduced when the respirators were properly fitted. A similar conclusion was reached by Gardner *et al.*,[26] who investigated the filtration efficiencies of N95 and P100 filtering facepiece respirators and filtering cartridges against viable MS2 virus aerosolized from liquid suspension. Mean penetrations were typically less than 2 and 0.03% for the respirators and the filtering cartridges, respectively.

11.5.2 Applications

Lai *et al.*[27] reported the results of experiments assessing the levels of protection against contagion provided by surgical face masks. The experiments were based on simulated expiratory emissions and inhalations of a polydisperse NaCl aerosol (35 nm with a geometric mean of 1.18 nm) and two electronically controlled manikins. When mask fit and when simulated breathing pattern and distances between the expiring and inhaling manikins were varied, performances protection were found to vary from 33 to 100%. The distance between the expiring and inhaling manikins was the most influential parameter affecting facemask protection. Lindsley *et al.*[28] used a coughing patient simulator and a breathing worker simulator to investigate the exposure of healthcare workers to cough aerosol droplets and to examine the efficiency of face shields in reducing such exposure. At a distance of 46 cm, the use of face shields reduced the inhalation exposure of the simulated worker to 8.5 μm volume median diameter influenza aerosol by 96% in the period immediately following the simulated cough. With 3.4 μm volume median diameter influenza aerosol, exposure was reduced by 68% under the same experimental conditions. By 30 min after the simulated cough, exposure was reduced only 23% because the larger particles had settled from the aerosol. Increasing the distance between the coughing patient simulator and the breathing worker simulator to 183 cm from 46 cm reduced the inhalation exposure to 8.5 μm volume median diameter influenza aerosol by 92% in the period immediately following the simulated cough. These observations demonstrate that the use of face shields can reduce the short-term exposure of health-care workers to large infectious aerosol particles, but protection against inhalation of smaller particles that remain aerosolized for longer times after the cough was not significantly diminished. Face shields may be a useful adjunct to full-face respirators in protecting healthcare workers treating patients with respiratory infections.

Project BREATHE (Better Respiratory Equipment using Advanced Technologies for Healthcare Employees), a multi-disciplinary team with expertise in pandemic and emergency preparedness, infectious disease medicine and epidemiology, respirator and personal protective equipment policy and regulation, occupational and environmental medicine, respirator and materials science, infection control, respirator physiology and physics and bio-security, recommended improvements for respiratory protective equipment used by healthcare workers. This set of recommendations constituted an idealized view of the features included in the next generation of respirators for healthcare workers. Each of recommendations was included in one of four categories of desirable characteristics: (1) the respirators should perform their intended functions safely and effectively, (2) the respirators should support, not interfere with, occupational activities, (3) the respirators should be comfortable and tolerable for the duration of wear and (4) the respiratory protective programs should comply with Federal standards and guidelines, state regulations and local policies. These recommendations may be regarded as (1) an action and research agenda for the Federal government, (2) a guide for the US healthcare sector that identifies activities which might yield strong returns on their resource investment and (3) a research and development roadmap for the next generation of respirators used by the US healthcare workforce. Reflected in this report is a position held by the team that clinical assessment tools, such as clinical trials, are preferred over methods performed solely in experimental laboratories. However, many instances are favored where clinical assessments are not practical, in which case the use of laboratory tools that have been validated against clinical outcomes is adopted. In cases where neither is available, the team made suggestions about the types of assessment methods that should be considered for development and validation. The team favored the development of a new respirator class called Biological N95 or B95, which connotes protection against biological particulates.[29,30]

Cecala *et al.*[31] reported that many different types of surface mining equipment use enclosed cabs to protect equipment operators from health and safety hazards. Overburden removal and mining processes were extremely dusty and caused excessive dust exposure. They investigated this issue as a cooperative research effort between the National Institute for Occupational Safety and Health (NIOSH), U.S. Silica Co., Clean Air Filter Co. and Red Dot

Corp. in an effort to lower respirable dust levels in an enclosed cab on an older surface drill at a silica sand operation. Throughout this research effort, a number of modifications were incorporated into the drill's filtration and pressurization system, and also in other areas, to improve its design and performance. An average cab efficiency of 93.4% was determined with gravimetric sampling instruments on comparing the outside with the inside cab dust levels in the final design. Although this study considered just one operation, the goal was to identify cost-effective improvements that could be implemented on all types of enclosed cabs to lower respirable dust concentrations. Two critical components for an effective enclosed cab system are having a properly designed, installed and maintained filtration and pressurization system, along with a method for maintaining structural cab integrity, which allows the cab to be positively pressurized. Another important component is maintaining cab cleanliness. Although this research was originally directed towards the mining industry, it is also applicable to agricultural equipment for protection against bioaerosols.

The applicability of cab enclosures to protect workers at composting facilities was investigated by Schlosser *et al.*[32] The focus of their investigation was aimed at estimating the mean protection efficiency of the vehicle cab environment against bioaerosols with higher precision. In-cab measurement results were also analyzed to ascertain if these protection systems reduced workers' exposures to tolerable levels. Five front-end loaders, one mobile mixer and two agricultural tractors pulling windrow turners were investigated. Four vehicles were fitted with a pressurization and high-efficiency particulate air filtration system and the other four were equipped with only pleated paper filters without pressurization. Bacteria, fungi and endotoxins were measured in 72 pairs of air samples collected simultaneously inside the cab and on the outside of the cab. A front-end loader fitted with a pressurization system and high-efficiency particulate air filtration system exhibited a mean protection efficiency of between 99.47 and 99.91% depending on the biological agent when the cab was clean. It is likely that the lower protection efficiency demonstrated in other vehicles was caused by penetration through the only moderately efficient filters, by the absence of pressurization, by leakage in the filter sealing system and by the resuspension of particles, which accumulated in dirty cabs. The mean protection efficiency with regard to bacteria and endotoxins ranged between 92.64 and 98.61% and between 92.68 and 98.43%, respectively. The mean protection efficiency was the lowest when confronted with fungal spores, from 59.76 to 94.71%. The probability that in-cab exposure to fungi exceeded the benchmark value for short-term respiratory effects suggested that front-end loaders and mobile mixers in composting facilities should be fitted with a pressurization and high-efficiency particulate air filtration system, regardless of the facility being indoors or outdoors. Regarding the tractors, exposure inside the cabs was not significantly reduced. However, in this study, there was a $<0.01\%$ risk of exceeding the benchmark value associated with fungi-related short-term respiratory effects during a 1 h per day windrow turning operation. Pressurization in conjunction with a HEPA filtration system can provide safe working conditions inside loaders and mobile mixer with respect to airborne bacteria, fungi and endotoxins in composting facilities. However, regular thorough cleaning of the vehicle cab and workers' overalls and shoes and mitigation of leakage in the filter sealing system were necessary to achieve high levels of protection efficiency.

Lee *et al.*[33] determined the workplace protection factors for dust and bioaerosols in agricultural environments. This field study was performed with a subject wearing an N95 filtering facepiece respirator while performing animal feeding, grain harvesting and unloading and routine inspection of facilities. The geometric means of the workplace protection factors increased with increasing particle size ranging from 21 for 0.7–1 μm particles to 270 for 5–10 μm particles ($p < 0.001$). The workplace protection factor for total cultivable fungi (geometric mean = 35) was significantly greater than for total cultivable bacteria (geometric mean = 9, $p = 0.01$). Among the different microorganism groups, the workplace protection factors of *Cladosporium*, cultivable fungi and total fungi were significantly correlated with the workplace protection factors for particles of the same sizes. Compared with the workplace protection factors for dust particles, the workplace protection factors for bioaerosols were found more frequently below 10, which is a recommended assigned protection factor for N95 filtering facepiece respirators. More than 50% of the workplace protection factors for microorganisms (mean aerodynamic diameter <5 μm) were less than the proposed assigned protection factor of 10. Even lower workplace protection factors were calculated after correcting for dead space and lung deposition. Thus, the assigned protection factor of 10 for N95 filtering facepiece respirators seems inadequate against microorganisms (mean aerodynamic size <5 μm). These results provide useful pilot data to establish guidelines for respiratory protection against airborne dust and microorganisms on agricultural farms.

Cho *et al.*[34] compared size-selective workplace protection factors for an N95 elastomeric respirator and for an N95 filtering facepiece respirator in agricultural environments. Twenty-five healthy workers between 20 and 30 years of age employed at eight different farms (two horse farms, three pig barns and three grain handling sites) volunteered to participate in this study. Subjects wore the respirators while performing their daily activities. Aerosol concentrations in an optical particle size range of 0.7–10 μm were determined simultaneously inside and outside the respirator during the first and last 15 min of 60 min intervals. For every subject, size-selective workplace protection factors were calculated in 1 min intervals and averaged over 30 min. For the elastomeric respirator, geometric mean workplace protection factors were 172, 321, 1013, 2097 and 2784 for particle with diameters of 0.7–1.0, 1.0–2.0, 2.0–3.0, 3.0–5.0 and 5.0–10.0 μm, respectively. The corresponding values for the filtering facepiece respirators were 67, 124, 312, 909 and 2089. The fifth percentiles for the elastomeric respirator and filtering facepiece respirator were higher than the assigned protection factor of 10 and varied from 28 to 250 and from 16 to 223, respectively. The results showed that the N95 elastomeric respirator and filtering facepiece respirator tested in this study provided an expected level of protection for workers on agricultural farms against particles ranging from 0.7 to 10 μm in diameter. Workplace protection factors for the elastomeric respirator were higher than those for the filtering facepiece respirator for all particle size ranges. Workplace protection factors for both respirator types increased with increasing particle size.

Kujundzic *et al.*[35] assessed the effectiveness of a new generation of high-volume, ceiling-mounted HEPA–ultraviolet filters for their ability to remove or inactivate bacterial aerosol. In an environmentally controlled, full-scale laboratory chamber (87 m^3) and an indoor therapy pool building, the mitigation ability of air filters was assessed by comparing concentrations of total bacteria, cultivable bacteria and airborne endotoxin with and without the air filters operating under otherwise similar conditions. Controlled chamber tests with pure cultures of aerosolized *Mycobacterium parafortuitum* cells showed that the HEPA–ultraviolet filters unit tested provided an equivalent to 11 air exchanges per hour. Using this equivalent air exchange rate as a design basis, three HEPA–ultraviolet filters were installed in an indoor therapy pool building for bioaerosol mitigation and their effectiveness was studied over a 2 year period. The HEPA–ultraviolet filters reduced the concentrations of cultivable bacteria by 69 and 80% during monitoring periods executed in the respective years. Also, the HEPA–ultraviolet filters reduced the concentrations of total bacteria by 12 and 76%, respectively. During the same monitoring period, airborne endotoxin concentrations were not affected by the HEPA–ultraviolet filter operations.

Vokurka *et al.*[36] investigated the availability of HEPA-filtered rooms on the incidence of pneumonia and mortality in post-hematopoietic stem cell transplant patients. Their investigation involved monitoring 689 patients for 100 days following the procedure. Pneumonia developed in 14 of 400 patients undergoing autologous hematopoietic stem cell transplantation and was the cause of death in two of these 14 patients. There was no significant difference in mortality between the patients in the HEPA-filtered rooms and those in the rooms without HEPA filtration. In allogeneic transplantation, pneumonia developed in 24 out of 289 patients and it was the cause of death in 11 of these 24 patients. Again, there was no significant difference in mortality between the HEPA-filtered and non-HEPA-filtered groups. The incidence of pneumonia in the autologous transplantation setting was low and more pneumonia was observed in the allogeneic hematopoietic stem cell transplantation group, especially among those stem cell transplanted patients receiving corticosteroid treatment in HEPA-filtered rooms. The conclusions drawn from this study are that autologous hematopoietic stem cell transplantation may safely be performed without HEPA filtration and that HEPA filtration might be preferable in patients undergoing allogeneic transplantation.

In the consumer sector, HEPA filters are incorporated into whole-house air purification HVAC systems, forced hot air home heaters, window air conditioners and bedside air cleaners for asthmatics, ductless kitchen fans and vacuum cleaners.

Sublett *et al.*[37] reviewed the literature for 17 studies on the efficiencies of whole-house air filtration systems and room air cleaners in the homes of people with allergic respiratory disease for information about reducing its progression. There was sufficient evidence in this review to conclude that air filtration reduced indoor levels of ambient particles that could trigger the disease process. Portable room air cleaners with HEPA filters, especially those that filtered the breathing zone air during sleep, appeared to be beneficial.

Karottki *et al.*[38] examined the potential beneficial effects of indoor air filtration in the homes of elderly subjects. Their randomized, double-blind, crossover intervention study was conducted with 48 nonsmoking subjects aged 51–81 years residing in 27 different homes for consecutive 2 week periods with or without the inclusion of a high-efficiency particle air filter in recirculating custom-built units in their living rooms and bedrooms. At days 0, 2, 7 and 14 during each exposure scenario, blood pressure and microvascular and lung function were measured; and blood samples were collected for hematological, inflammation, monocyte surface and lung cell damage markers. The particle filters reduced the median concentration of particles <2.5 μm in diameter ($PM_{2.5}$) from approximately 8 to 4 μg m^{-3} and the particle number concentration from 7669 to 5352 particles cm^{-3}. No statistically significant effects of filtration were observed on microvascular and lung function or the biomarkers of systemic inflammation among all subjects or in the subgroups taking (n = 11) or not taking (n = 37) vasoactive drugs. However, the filtration efficacy was variable. Microvascular function increased significantly within 2 days even though the actual $PM_{2.5}$ decreased in the bedroom, especially among 25 subjects who were not taking any drugs during this time. The conclusions drawn from these observations were that substantial exposure contrasts existed in the bedrooms and these were not confounded by the medications required for improved microvascular function or by air filtration. No other beneficial effect was found in this elderly population.

Brown *et al.*[39] conducted a modeling analysis of the effectiveness of filter-based interventions in the home to reduce airborne asthma and allergy triggers. In this work, the effectiveness of the filter for key allergy and asthma triggers based on applicable particle sizes for cat dander allergen, indoor and outdoor sources of particles <2.5 μm in diameter ($PM_{2.5}$) and airborne influenza and rhinovirus was assessed. The results showed that filters with removal efficiencies of >70% for cat dander particles, fine particulate matter ($PM_{2.5}$) and respiratory virus can lower the concentrations of those asthma triggers and allergens in indoor air of the home by >50%. Very high removal efficiency filters, such as those rated at 16 on the nationally recognized Minimum Efficiency Removal Value rating system, tend to be only marginally more effective than Minimum Efficiency Removal Value 12- or 13-rated filters. The results of this analysis indicated that the use of an air filter performing at a Minimum Efficiency Removal Value of 12 or higher in home ventilation systems can effectively reduce indoor levels of these common asthma and allergy triggers. These reductions in airborne allergens in turn may help reduce allergy and asthma symptoms, especially if employed in conjunction with other environmental management measures recommended for allergy and asthma patients.

Wright and Phipatanakul[40] reviewed the literature on environmental interventions for the treatment of asthma and allergies. Several airborne triggers for asthmatic and allergic episodes were identified: house dust mites, Fel d1 and Can f1 pet allergens, Bla g1 and 2 from *Blattella germanica*, Per a1 allergens from *Periplaneta americana*, Mus m1 and 2 mouse allergens, molds (*Alternaria*, *Aspergillus* and *Cladosporium*), tobacco smoke and endotoxin. Based on the evidence available, the most effective control measures targeting multiple allergens include education, thorough house cleaning, use of HEPA filters, integrated pest management and continued maintenance of these procedures.

11.6 Conclusion

Research at government agencies, academic institutions, commercial laboratories and industrial facilities has resulted in the development of an array of devices for protecting the respiratory system against the inhalation of numerous pathogens and toxins. These devices have found use in promoting public and occupational health in addition to protecting war fighters and first responders from the toxic effects of inhaling biological agents released in combat, by terrorist activities or during medical emergencies. Although airborne transmission is not the major mode for contagion, the specter of an Ebola pandemic has heightened concerns for the protection of healthcare workers.[41] Such concerns may serve as a stimulant for the development of the B95 facepiece respirator (see Project BREATHE in Section 11.5.2). Similarly, measures to reduce environmental exposure to airborne allergens and irritants have been suggested for addressing the disproportionately high incidences of asthma morbidity and mortality among minority children in urban centers. The use of HEPA filters was suggested for decreasing the concentrations of asthma triggers in urban air.[42] Continued research and development will meet these and yet to be identified challenges for protecting the respiratory system against airborne pathogens and toxins.

References

1. C. J. Roy, D. S. Reed and J. A. Hutt, Aerobiology and Inhalation Exposure to Biological Select Agents and Toxins, *Vet. Pathol.*, 2010, **47**(**5**), 779–789, DOI: 10.1177/0300985810378650, accessed 11 August 2014.
2. M. W. First, HEPA Filters, *J. Am. Biol. Saf. Assoc.*, 1998, **3**(**1**), 33–42.
3. CDC, Centers for Disease Control and Prevention, *Interim Recommendations for the Selection and Use of Protective Clothing and Respirators Against Biological Agents*, 2001, http://ww.cdc.gov/anthrax/, accessed 17 August 2014.
4. DHS, Department of Homeland Security, *Guidance for Protecting Responders' Health During the First Week Following a Wide-Area Aerosol Anthrax Attack*, 2012, http://www.dhs.gov/, accessed 17 August 2014.

5. NRT, National Response Center, *NRT Quick Reference Guide: Bacillus anthracis (Anthrax)*, 2008, http://ww.nrt.org/production/nrt/A-566ESAnthraxQRG, accessed 17August 2014.
6. J. Lavoie, Y. Cloutier, J. Lara and G. Marchand, *Guide on Respiratory Protection against Bioaerosols: Recommendations on Its Selection and Use*, Technical Guide RG – 501, Institut de recherché Robert – Sauvé en santé et an sécurité du travai, (IRSST), Québec, 2007.
7. IAEA, International Atomic Energy Agency, *Comparison of High Efficiency Particulate Filter Testing Methods*, AIEA – TECDOC – 355, Wien, 1985.
8. NIOSH, National Institute for Occupational Safety and Health, *Guidance for Filtration and Air Cleaning Systems to Protect Building Environments from Chemical, Biological or Radiological Attacks*, DHHS (NIOSH) publication no. 2003 – 136, Centers for Disease Control and Prevention, April, 2003.
9. TSI, *Mechanisms of Filtration for High Efficiency Fibrous Filters*, Application Note ITI – 041, TSI, Inc., Shoreview, MN, 2012.
10. H. G. Guyton, L. M. Buchanan and F. T. Lense, Evaluation of Respiratory Protection of Contagion Masks, *Appl. Microbiol.*, 1956, **4**, 141–143.
11. H. G. Guyton and H. M. Decker, Respiratory Protection Provided by Five New Contagion Masks, *Appl. Microbiol.*, 1963, **11**, 66–68.
12. N. V. McCullough, L. M. Brosseau and D. Vesley, Collection of Three Bacterial Aerosols by Respirator and Surgical Mask Filters under Varying Conditions of Flow and Relative Humidity, *Ann. Occup. Hyg.*, 1997, **41**(**6**), 677–690.
13. S. Rengasamy, Z. Zhuang and R. Berryann, Respiratory Protection Against Bioaerosols: Literature Review and Research Needs, *Am. J. Infect. Control*, 2004, **32**(**6**), 345–354.
14. S. Rengasamy, B. C. Eimer and R. E. Shaffer, Comparison of Nanoparticle Filtration Performance of NIOSH-Approved and CE-Marked Particulate Filtering Facepiece Respirators, *Ann. Occup. Hyg.*, 2009, **53**(**2**), 117–128, DOI: 10.1093/annhyg/men086. accessed 20 August 2014.
15. W. J. Jasper, A. Mohan, J. Hinestroza and R. Barker, Degradation Process in Corona-Charged Electret-Media with Exposure to Ethyl Benzene, *J. Eng. Fibers Fabr.*, 2007, **2**(**4**), 1–6. http://www.jeffjournal.org. accessed 26 August 2014.
16. S. Rengasamy, A. Miller, E. Vo and B. C. Eimer, Filter Performance Degradation of Electrostatic N95 and P100 Filtering Facepiece Respirators by Dioctyl Phthalate Aerosol Loading, *J. Eng. Fibers Fabr.*, 2013, **8**(**3**), 62–69.
17. R. B. Lawrence, M. G. Duling, C. A. Calvert and C. C. Coffey, Comparison of Performance of Three Different Types of Respiratory Protection Devices, *J. Occup. Health Environ. Hyg.*, 2006, **3**(**9**), 465–474.
18. NIOSH, National Institute for Occupational Safety and Health, *Particulate Respirator Qualitative Fit Test Utilizing Saccharine or Bitrex Solutions*, Procedure No. RCT – APR – STP – 0067, National Protective Technology Laboratory, Pittsburgh, PA, 31 August 2005.

19. S. Rengasamy, B. C. Eimer and R. E. Shaffer, Comparison of the Performance of the N95 – Companion: Effects of Filter Penetration and Comparison with Other Aerosol Instruments, *J. Occup. Environ. Hyg.*, 2012, **9**(7), 417–426, DOI: 1080/15459624.2012.685838, accessed 26 August 2014.
20. K. J. Cho, T. Reponen, R. McKay, A. Dwivedi, A. Adhikari, U. Singh, R. Shukla, S. Jones, G. Jones and S. A. Grinshpun, Comparison of Workplace Protection Factors for Different Biological Contaminants, *J. Occup. Environ. Hyg.*, 2011, **8**(7), 417–425, DOI: 10.1080/15459624.2011.585094, accessed 20 August 2014.
21. S. Rengasamy and B. C. Eimer, Nanoparticle Penetration through Filter Media and Leakage through Face Seal Interface of N95 Filtering Facepiece Respirators, *Ann. Occup. Hyg.*, 2012, **56**(**5**), 568–580, DOI: 10.1093/annhyg/mer122. Epub 2012, Jan 31, accessed 08 October 2014.
22. S. Rengasamy, B. C. Eimer and J. Szalaida, A Quantitative Assessment of the Total Inward Leakage of NaCl Aerosol Representing Submicron – Size Bioaerosol Through N95 Filtering Facepiece Respirators and Surgical Masks, *J. Occup. Environ. Hyg.*, 2014, **11**(**6**), 388–396, DOI: 1080/15459624.2013.866715, accessed 11 October 2014.
23. X. He, T. Reponen, R. McKay and S. A. Grinshpun, How Does Breathing Frequency Affect the Performance of an N95 Filtering Facepiece Respirator and Surgical Mask Against Surrogates of Viral Particles?, *J. Occup. Environ. Hyg.*, 2014, **11**(3), 178–185, DOI: 10.1080/15459624.2013.848037, accesses 13 October 2014.
24. C. S. Davidson, C. F. Green, S. G. Gibbs, K. K. Schmid, A. L. Panlilo, P. A. Jensen and P. V. Scarpino, Performance Evaluation of Selected N95 Respirators and Surgical Masks when Challenged with Aerosolized Endospores and Inert Particles, *J. Occup. Environ. Hyg.*, 2013, **10**(**9**), 461–467, DOI: 10.1080/15452624.2013.818243, accessed 11 October 2013.
25. D. A. Harnish, B. K. Heimbuch, M. Husband, A. E. Lumley, K. Kinney, R. E. Shaffer and J. D. Wander, Challenge of N95 Filtering Facepiece Respirators with Viable N1H1 Influenza Aerosols, *Infect. Control Hosp. Epidemiol.*, 2013, **3**(**5**), 494–499, DOI: 10.1086/670225, accessed 11 October 2014.
26. P. D. Gardner, J. P. Eshbaugh, S. D. Harpest, A. W. Richardson and K. C. Hofacre, Viable Virus Efficiency of N95 and P100 Respirator Filters at Constant Cyclic Flow, *J. Occup. Environ. Hyg.*, 2013, **10**(**10**), 564–572, DOI: 10.1080/15459624.2013.818228, accessed 11 October 2014.
27. A. C. K. Lai, C. K. M. Poon and A. C. T. Cheung, Effectiveness of Facemasks to Reduce Exposure Hazards for Airborne Infections Among General Populations, *J. R. Soc., Interface*, 2012, **9**, 938–948, DOI: 10.1098/rsif.2011.0537, accesses 18 August 2014.
28. W. G. Lindsley, J. D. Noti, F. M. Blachere, J. V. Szalajda and D. H. Beezhold, Efficiency of face Shields Against Cough Aerosol Droplets From a Cough Simulator, *J. Occup. Environ. Hyg.*, 2014, **11**(**8**), 509–518, DOI: 10.1080/15459624.2013.877951, accessed 13 October 2014.

29. L. J. Radonovich, A. Balg, R. E. Shaffer, R. Roberge, A. Levinson, D. E. Doerr and V. Davey, *Better Respiratory Equipment Using Advanced Technologies for Healthcare Employees*, A Report of an Interagency Working Group of the U.S. Federal Government, Office of Public Health and Environmental Hazards in the Veterans Health Administration in the U.S. Department of Veterans Affairs, Washington, 2013.
30. M. E. Gosch, R. E. Shaffer, A. E. Eagan, R. J. Roberge, V. J. Vevey and L. J. Radonovich, B95: A New Respirator for Health Care Personnel, *Am. J. Infect. Control*, 2013, **41**(**12**), 1224–1230, DOI: 10.1016/ajic.2013.03.293. Epub 2013 May 29, accessed 13 October 2014.
31. A. B. Cecala, J. A. Organiscak, J. A. Zimmer, W. A. Heitbrink, E. S. Moyer, M. Schmitz, E. Ahrenholtz, C. C. Coppock and E. H. Andrews, Reducing Enclosed Cab Drill Operator's Respirable Dust Exposure with Effective Filtration and Pressurization Techniques, *J. Occup. Environ. Hyg.*, 2005, **2**(**1**), 54–63.
32. O. Schlosser, A. Huyard, D. Rybacki and Z. Do Quang, Protection of the Vehicle Cab Environment Against Bacteria, Fungi and Endotoxins in Composting Facilities, *Waste Manage.*, 2012, **32**(**6**), 1106–11156, DOI: 10.1016/j.wasman.2012.01.013. Epub 2012 Feb 11, accessed 20 August 2014.
33. S. A. Lee, A. Adhikari, S. A. Grinshpun, R. McKay, R. Shukla, H. L. Zeigler and T. Reponen, Respiratory Protection Provided by N95 Filtering Facepiece Respirators Against Airborne Dust and Microorganisms in Agricultural Farms, *J. Occup. Environ. Hyg.*, 2005, **2**(**11**), 577–585.
34. K. J. Cho, S. Jones, G. Jones, R. McKay, S. A. Grinshpun, A. Dwivedi, R. Shukla, U. Singh and T. Reponen, Effect of Particle Size on Respiratory Protection Provided by Two Types of N95 Respirators Used in Agricultural Settings, *J. Occup. Environ. Hyg.*, 2010, **7**(**11**), 622–627, DOI: 10.1080/15459624.2010.513910, accessed 13 October 2014.
35. E. Kujundzic, D. A. Zander, M. Hernandez, L. T. Angenent, D. Henderson and S. L. Miller, Effects of Ceiling-Mounted HEPA-UV Air Filters on Airborne Bacteria Concentrations in an Indoor Therapy Pool Building, *J. Air Waste Manage. Assoc.*, 2006, **55**, 210–218.
36. S. Vokurka, E. Bystrická, T. Svoboda, I. K. Škoda Gorican, M. Sever, E. Mazur, A. Kopinska, V. Pavlicová, O. Mocanu, A. Tanase, R. Ghelase, M. Zítková, M. Labudíková, L. Raida, D. Hrabánková-Navrátilová and J. Bocková, The Availability of HEPA-Filtered Rooms and the Incidence of Pneumonia in Patients after Hematopoietic Stem Cell Transplantation (HSCT): Results from a Prospective, Multicentre, Eastern European Study, *J. Clin. Nurs.*, 2014, **23**(**11–12**), 1648–1652, DOI: 10.1111/jocn.12286. Epub 2013 Dec 20, accessed 21 August 2014.
37. J. L. Sublett, J. Seltzer, R. Burkhead, P. B. Williams, H. J. Wedner and W. Phipatakul, Air Filters and Air Cleaners: Rostrum by the American Academy of Allergy, Asthma & Immunology Indoor Allergen Committee, *J. Allergy Clin. Immunol.*, 2010, **125**(**1**), 32–38, DOI: 1016/j.jaci.2009.08.036, accessed 21 October 2014.

38. D. G. Karottki, M. Spilak, M. Frederiksen, L. Gunnarsen, E. V. Brauner, B. Kolarik, Z. J. Andersen, T. Sigsgaard, L. Barregard, B. Strandberg, G. Sallsten, P. Møller and S. Loft, An Indoor Air Filtration Study in Homes of Elderly Cardiovascular and Respiratory Effects of Exposure to Particulate Matter, *Environ. Health*, 2013, **28**(**12**), 116, DOI: 10.1186/1476-069X-12-116, accessed 20 August 2014.
39. K. W. Brown, T. Minegishi, J. G. Allen, J. F. McCarthy, J. D. Spengler and D. L. MacIntosh, Reducing patients' exposures to asthma and allergy triggers in their homes: an evaluation of effectiveness of grades of forced air ventilation filters, *J. Asthma*, 2014, **51**(**6**), 585–594, DOI: 10.3109/02770903.2014.895011. Epub 2014 Mar 12.
40. L. S. Wright and W. Phipatanakul, Environmental Remediation in the Treatment of Allergy and Asthma: Latest Updates, *Curr. Allergy Asthma Rep.*, 2014, **14**(**3**), 419, DOI: 1007/s11882-014-0419-7, accessed 21 October 2014.
41. B. McKay, M. Bustillo and M. Beck, Ebola Case Puts Focus on Safeguards; Dallas Nurse Contacts Disease After Caring for Liberian Patient, Raising Concerns Over Health Procedures, *Wall St. J.*, 2014, **CCLXIV**(**88**), A1–A6, Monday 13 October.
42. W. Kanchongkittiphon, J. M. Gaffin and W. Phipatanakul, The Indoor Environment and Inner-City Childhood Asthma, *Asian J. Allergy Immunol.*, 2014, **32**(**2**), 103–110.
43. D. R. Franz, *Defense Against Toxin Weapons*, MCMR – UIZ – A, U.S. Army Medical Research Institute of Infectious Diseases, Fort Detrick, MD, 1997.
44. EHS UFL, *Environmental Health and Safety*, University of Florida, Gainsville, 2014, http://ww.ehs.ufl.edu/programs/bio/toxins/toxin-table/, accessed 31 August 2014.
45. B. G. Fry, Structure – Function Properties of Venom Components from Australian Elapids, *Toxicon*, 1999, **37**, 11–32.
46. M. S. Gilmore, A. L. Cruz-Rodz, M. Leimeister-Wächter, J. Kreft and W. Goebel, A Bacillus Cereus Cytolytic Detrerminant, Cereolsin AB, which Compromises the Phospholipase C and Sphingomyelinase Genes: Nucleotide Sequence and Genetic Linkage, *J. Bacteriol.*, 1989, **171**(**2**), 744–753.
47. K. Kanclerski and R. Möllby, Production and Purification of Streptococcus Pneumoniae Hemolysin (Pneumolysin), *J. Clin. Microbiol.*, 1987, **25**(**2**), 222–225.
48. M. L. Vasil, D. Kabat and B. H. Iglewski, Structure – Activity Relationships of an Exotoxin of Pseudomonas Aeruginosa, *Infect. Immun.*, 1977, **16**(**1**), 353–361.
49. J. McIver, G. F. Grady and G. T. Keusch, Production and Characterization of Exotoxin of Shigella Dysenteria, *J. Infect. Dis.*, 1975, **131**(**5**), 559–566.
50. P. M. Schlievert, K. N. Shands, B. B. Dan, G. P. Schmidt and R. D. Nishimura, Identification and Characterization of an Exotoxin from Staphylococcus Aureus Associated with Toxic – Shock Syndrome, *J. Infect. Dis.*, 1981, **143**(**4**), 509–516.
51. D. E. Van Epps and B. R. Andersen, Streptolysin O: Sedimentation Coefficient and Molecular Weight Determinations, *J. Bacteriol.*, 1969, **100**(**1**), 527.

CHAPTER 12

An Improved Model of Human Response to Bioaerosol Exposure

GENE E. McCLELLAN*[a], KYLE K. MILLAGE[a], AND BAHMAN ASGHARIAN[b]

[a]Applied Research Associates, Inc., 801 N. Quincy St. STE 700, Arlington, VA 22203, USA; [b]Applied Research Associates, Inc., 8537 Six Forks Rd. STE 600, Raleigh, NC 27615, USA
*E-mail: gmcclellan@ara.com; kmillage@ara.com; basgharian@ara.com

12.1 Introduction

This chapter describes the Deposition and Response in the Respiratory Tract (DARRT) model developed to improve assessments of airborne biological hazards by including the influence of aerosol particle size in assessing the health effects of bioaerosols. The DARRT model accounts for (1) the physical processes during inhalation that determine where bioaerosol particles are deposited in the respiratory tract and (2) the influence of deposition location on the nature of the resulting infection and illness.

Figure 12.1 presents curves calculated with the Multiple-Path Particle Dosimetry[1] (MPPD) software application showing the fraction of particles in ambient air that will be deposited in the three main regions of the respiratory system for nasal breathing in still air. The complex shapes of the curves

Issues in Toxicology No. 25
Aerobiology: The Toxicology of Airborne Pathogens and Toxins
Edited by Harry Salem and Sidney Katz

Published by the Royal Society of Chemistry, www.rsc.org

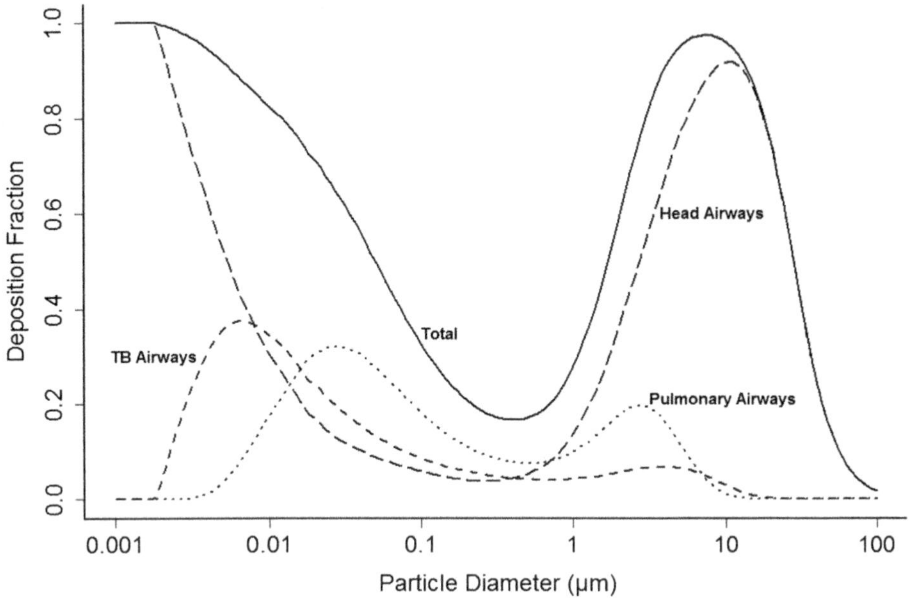

Figure 12.1 Particle deposition in the respiratory tract as a function of particle aerodynamic diameter; calculated with MPPD.[1] TB refers to the tracheobronchial airways.

represent interplay between the main processes involved in respiratory mechanics: impaction, diffusion and gravitational settling. Particles with aerodynamic diameters above ~50 µm have little chance of entering the nostrils during inspiration and those that do will be deposited entirely in the head (extrathoracic) airways by inertial impaction. In the 10–20 µm diameter range, ambient particles in inhaled air have an 80–90% probability of entering the nostrils, with most being deposited in the head airways, a few percent reaching the tracheobronchial (TB) airways and essentially none reaching the pulmonary (P) airways. Below 10 µm diameter, the impaction rate lessens and particles begin to reach the P airways with fewer depositing in the head airways. Between about 2 and 5 µm diameter, ambient air particles have nearly a 100% chance of entering the nostrils during inspiration, with about 15–20% reaching the P airways and depositing there. A few percent will deposit in the TB airways during transit to and from the P airways and 40–70% will deposit in the head airways, partly during inspiration and partly during expiration. In this 2–5 µm range, 10–40% of particles entering the nostrils will not be deposited at all, but are exhaled. Interestingly, in the 0.3–0.5 µm range, nearly 80% of inhaled particles avoid deposition and are exhaled. The 20% deposited in this limited size range are distributed throughout the respiratory tract with a small bias towards the P region. Contrary to intuition, particles below 0.01 µm diameter (the nanoparticle range) tend not to reach the pulmonary airways. For these very small particles, diffusion rates increase dramatically so that nearly all particles diffuse to the walls of the head and TB airways during inhalation and are deposited before reaching the P airways. Note that in addition to the peak in pulmonary deposition between 2 and 5 μm, there is another, larger peak in pulmonary deposition between 0.01 and 0.1 μm. This range is not of significance for bacterial organisms, which are generally larger than 0.1 μm, but could be important for viral agents, depending on source, composition and mechanism of dissemination of the bioaerosol.

Historically, computational models of biological agent aerosol hazards typically ignored large particles (>5 or sometimes >10 μm) because large particles were not considered "respirable" (*i.e.* they had a low probability of reaching the pulmonary airways of the lung). However, as shown by Figure 12.1, larger particles will deposit in the extrathoracic area, where they may cause upper respiratory infections that for some agents may be as significant as pneumonic infections. The new DARRT model improves casualty assessment by accounting for these large-particle hazards, which may be especially important in urban environments where many people near a release may be exposed to particles larger than 5 μm and where these larger particles may remain suspended in the air for longer times owing to turbulence effects. Accounting for particle size effects will provide better estimates of the types of illnesses and the number of patients seeking medical care over time from such a release and, thereby, improve medical planning.

The development process for DARRT included leveraging existing models of the human respiratory tract coupled with the particle size distribution (PSD) of atmospheric aerosol hazards to estimate the rate and location of agent deposition in the body as a function of particle size. We investigated several existing models of inhalation dosimetry and chose MPPD as the most appropriate for DARRT. Section 12.2 describes the different models and our analysis of their applicability in the bioaerosol context. The section also includes a detailed discussion of MPPD.

In addition, we developed medical models to quantify the dependence of agent health effects on deposition site and on aerosol particle characteristics. The study was focused on applying data taken from human inhalation exposure experiments and animal studies to construct mathematical models that predict the effect of particle size on the course of development of an infection or intoxication. These models can then be used to improve the fidelity of existing models for the estimation of casualties resulting from biological warfare or bioterrorism incidents. Strictly, we should distinguish between infection and illness for infective organisms. Some infective organisms can cause infection and be transmitted from a carrier to others without causing illness in the carrier. In this chapter, we assume that infection is accompanied by illness since that is the case for essentially all organisms considered in the context of biological warfare or bioterrorism.

We describe biological response models for *Francisella tularensis* (the bacterial organism causing tularemia) and ricin toxin in this chapter. We implemented these models with the respiratory deposition model for the dry particulate form of the agents and integrated these models into the Hazard Prediction and Assessment Capability (HPAC) of the US Defense

Threat Reduction Agency (DTRA). The previous biological agent dry particulate models in HPAC would account for particle size during transport in the atmosphere but not during exposure of personnel. The previous human health effects models assumed all the particles were about 5 μm in diameter (the "respirable" range) and caused pulmonary effects. Therefore, the HPAC biological agent material property files ("material files") used by bioagent source models such as CB Weapon included only a single particle size bin with bounds of 4.9–5.1 μm. Although CB Weapon would allow the user to define a PSD, in fact all agent mass was put into the single available size bin and transported as 5 μm particles. Since the new DARRT health effects model will account for PSDs, the tularemia and ricin material files have been modified to allow the PSD defined by the user in CB Weapon to be accurately assessed during both the atmospheric transport and personnel dosimetry calculations. In other words, our new model not only improves the calculation of the biological response to the agent, but also indirectly improves the actual transport and dispersion of the agent in the atmosphere by allowing an accurate representation of the PSD specified by the user.

The following two sub-sections of this introduction present results calculated with DARRT for the dry agent models for *F. tularensis*, the organism causing tularemia, and for ricin toxin. The results are compared with those from previous models that did not account for particle size effects. The remainder of this chapter describes the development of the DARRT models.

12.1.1 Dry Agent Model for Tularemia

The dry-particulate tularemia model in DARRT accounts for the regional deposition of particulates in the respiratory tract. The model calculates the dose deposited in the airways of the extrathoracic (ET) region (referred to earlier as the head airways), the tracheobronchial (TB) region and the pulmonary (P) region. Based on the presented dosage and the particle size, the model converts the inhaled dose to the number of particles and, based on the agent concentration (organisms per unit mass of agent), calculates how many of the particles actual contain an agent organism. The probability of ET or upper respiratory infection is calculated based on the number of agent-containing particles (ACPs) deposited in the ET region and the probability of P or pulmonary infection is based on the number of ACPs deposited in the P region. Output contour plots are available to indicate the total probability of infection (upper respiratory and pulmonary). Also, the casualty table individually identifies the population with upper respiratory infection, pulmonary infection and the total infected population. Those predicted to acquire both upper respiratory and pulmonary infections are listed in the pulmonary category and not in the upper respiratory category because the pulmonary infection usually develops first.

For a 5 μm particle size, the DARRT model casualty estimates for tularemia are similar to previous estimates, as expected. However, there are substantial differences for other particle sizes. Figures 12.2 and 12.3 show hazard areas

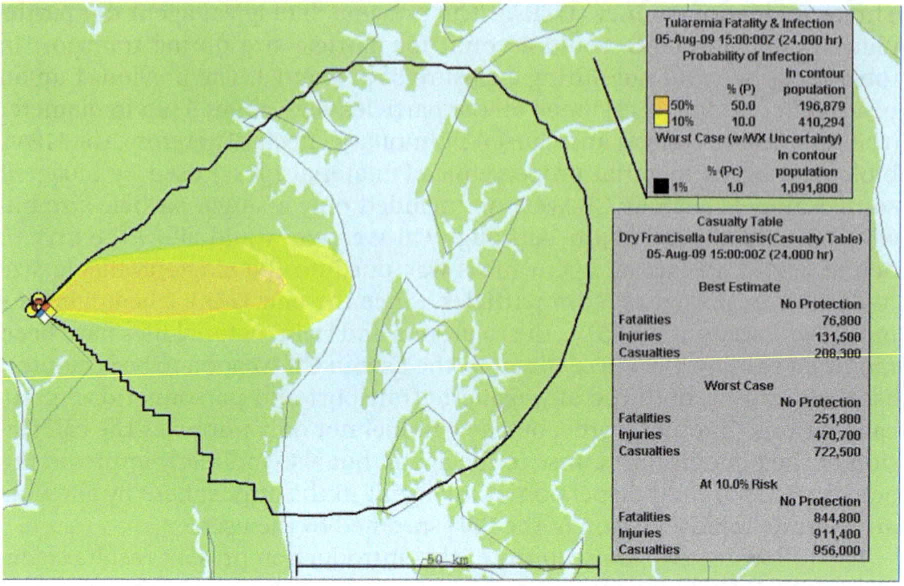

Figure 12.2 Previous model for tularemia: 5 μm particles using the old dry-agent material file (one size bin at 5 μm).

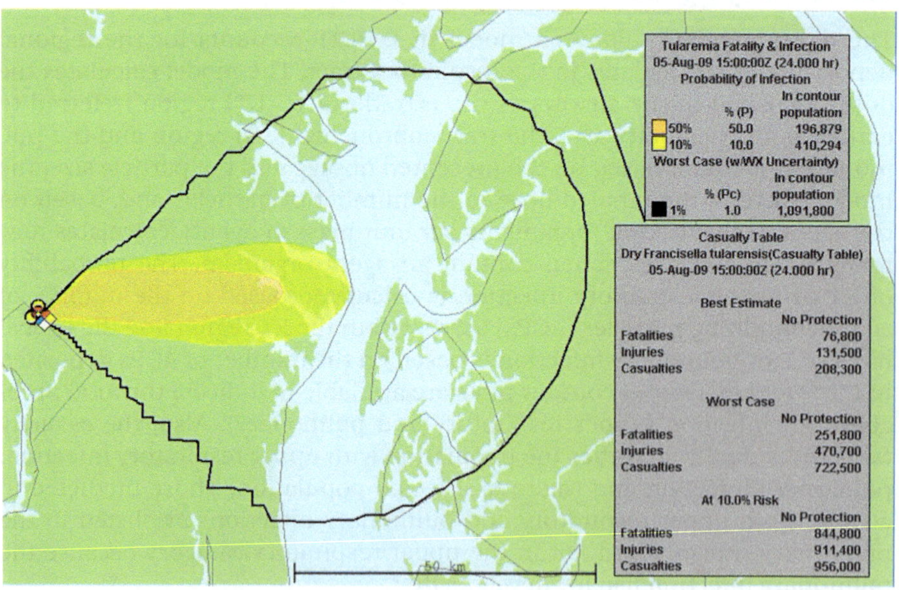

Figure 12.3 Previous model for tularemia: 80 μm particles using the old dry-agent material file (one size bin at 5 μm). Note that the results are the same as in Figure 12.2.

and casualty tables for 5 and 80 μm particles, respectively, using the previous model for tularemia and the associated material files in HPAC. Note that the plumes appear the same and the casualty table data are identical, which is the result of having only one size bin at 5 μm in which to put the agent mass. Figures 12.4 and 12.5 show hazard areas and casualty tables for the same two particle sizes, using the DARRT model and material files with improved size bins able to accommodate adequately particles ranging from 0.1 to 100 μm in diameter. The results are significantly different for the 80 μm particles, creating a much smaller hazard area and fewer casualties.

Note that the casualty tables in Figures 12.4 and 12.5 list the number of people suffering pulmonary and extrathoracic (upper respiratory) illness separately. The previous tables in Figures 12.2 and 12.3 did not distinguish these health effects, tacitly assuming pneumonic illness. Section 12.4 provides a detailed description of the tularemia model providing this improved casualty description.

12.1.2 Ricin

Like the dry tularemia material data file, the previous ricin material file included only a single particle size bin at 5 μm. In addition, there were no detailed injury or mortality models; the human response calculation provided lethal dosages, *i.e.* lethal concentration–time integrals (LCt_x), corresponding to $x = 10, 20, 50$ and 90% mortality based solely on presented dosage in mg min m^{-3} regardless of particle size. As a result, much like the tularemia

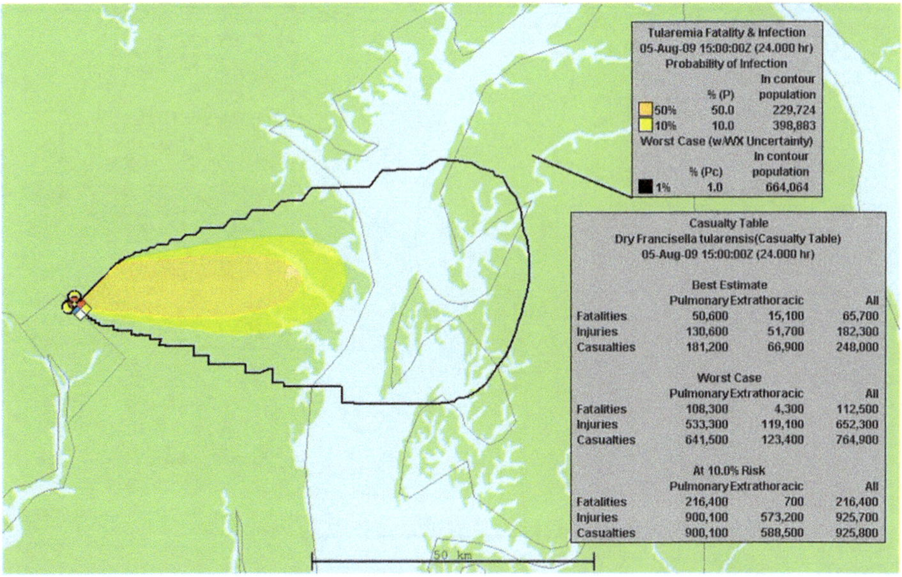

Figure 12.4 DARRT model for tularemia: 5 μm particles using the new dry-agent material file.

calculations, the lethality levels for large particles were overestimated. The new dry particulate ricin model accounts for PSD and the regional deposition of particulate in the respiratory tract. Based on our analysis of the literature, the probability of mortality calculation is primarily determined by the mass of ricin that deposits in the pulmonary region; deposition in the ET and TB regions from aerosolized ricin does not typically result in mortality. Since ricin is a toxin, the body mass of the animal is also a factor, so the probability of mortality is based on the mass of ricin deposited in the pulmonary region divided by the mass of the animal (for humans, usually taken as 70 kg for a "standard man") in units of $\mu g\ kg^{-1}$.

The ricin model does not include a probability of injury model, a task that was outside the scope of our effort. An injury model would perhaps include ET effects, in addition to pulmonary effects since both deposition locations can result in transfer of the toxin to the bloodstream. The new ricin model includes plotting options for the probability of mortality and a casualty table to calculate the population affected. A comparison of Figures 12.6 and 12.7 shows the differences between 5 and 10 μm particle sizes for the probability of mortality using the DARRT model. The 10 μm particles cause significantly less mortality. Since the probability of mortality depends on the amount of material that reaches and deposits in the lung, large particles do not pose a significant mortality threat. Figure 12.1 shows that above about 15 μm diameter, particles do not reach the pulmonary region. Section 12.4 presents a detailed description of the ricin model.

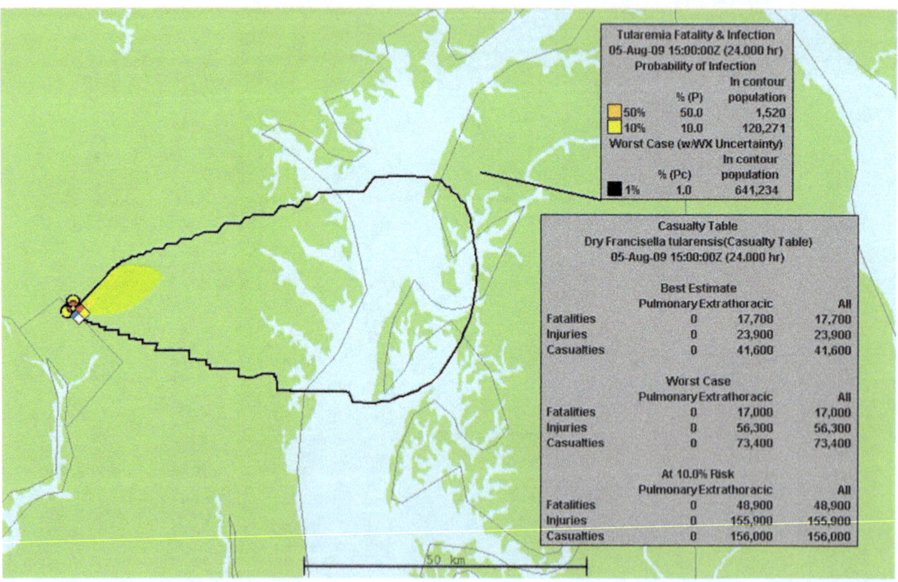

Figure 12.5 DARRT model for tularemia: 80 μm particles using the new dry-agent material file. Casualty numbers are reduced by aerosol settling to the ground and reduced inhalability.

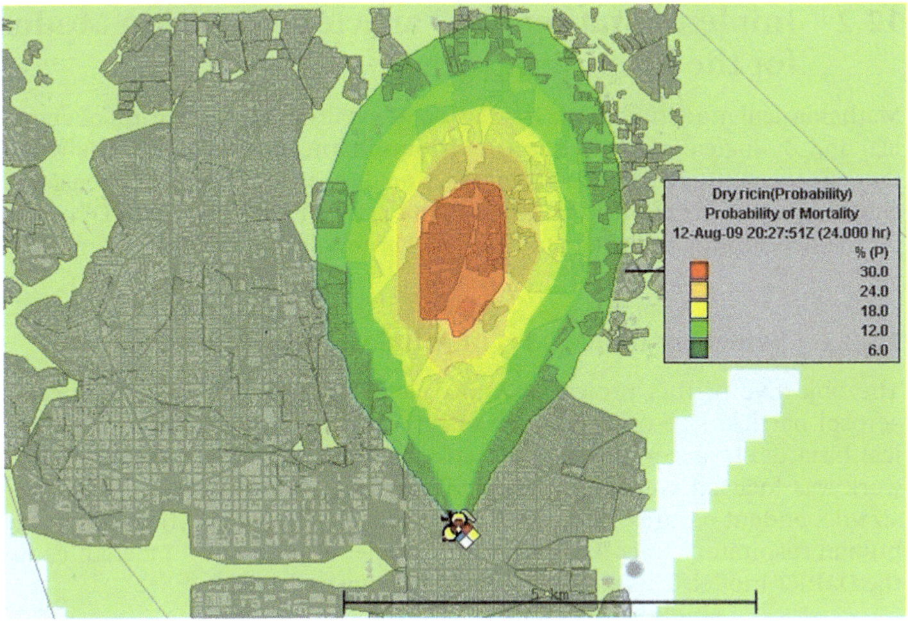

Figure 12.6 5 μm particles using the new dry ricin material file and DARRT model.

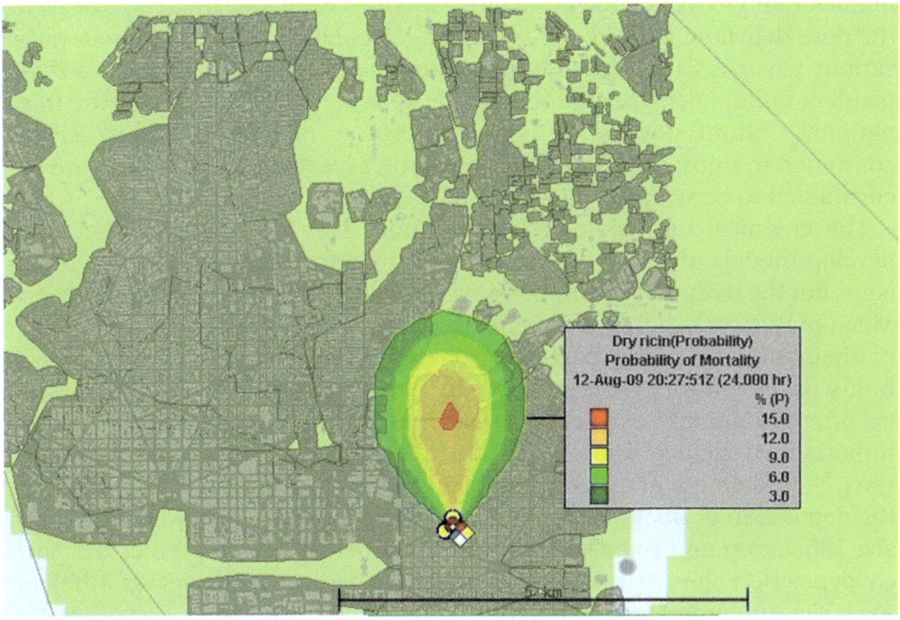

Figure 12.7 10 μm particles using the new dry ricin material file and DARRT model.

12.2 Implementation of a Particle Deposition Model for the Respiratory Tract

Mathematical modeling of the inhalation, deposition and retention of chemical and biological aerosols forms the foundation for estimating the resultant effects on the human body. This section describes the technical basis for choosing to implement the Multiple-Path Particle Dosimetry (MPPD) model for the development of the medical model of inhalation hazards.

12.2.1 Selection of a Model of Inhalation Mechanics

The objective of this project was to develop models for the influence of aerosol particle size on the health effects of inhaled chemical and biological hazards. In accordance with the first deliverable of the project, initial work was focused on examining existing models that describe the particle size-dependent nature of the inhalation and deposition of particulates in the human respiratory tract and selecting a model of inhalation mechanics for the DARRT model. This sub-section reviews our findings.

12.2.1.1 Background

The radiological community has extensively studied the biological effect of inhaled radioactive particles and has developed extensive models to describe the dose delivered to the body as a result of inhaling radioactive material of various particle sizes. The first international guidelines for particle size-dependent radiological inhalation hazards were published in 1959 by the International Commission on Radiological Protection (ICRP).[2] The ICRP has continued to improve the models and some aspects of their models are still considered to be state-of-the-art.

The chemical and biological community has more recently begun to develop models of inhalation hazards of toxic substances and microorganisms, but the nature of the modeling varies widely in both purpose and scope. Much of the work in this area is focused on animal experiments rather than mathematical modeling, particularly in the area of vapor inhalation of toxic industrial chemicals (TICs) and toxic industrial materials (TIMs). These studies often do not address particle size dependence, a factor that is critically important in the determination of the deposition pattern in the respiratory tract.[3,4] Modeling and experiments are also being performed to understand the deposition of fibers, which pose unique deposition hazards,[5] and extensive inhalation and retention work is being performed to predict the particle deposition sites to assess the delivery of therapeutic drugs *via* inhalers.[6] Extensive epidemiological modeling work is related, particularly with regard to environmental and pollution hazards,[7] but is not explicitly applicable to this project.

12.2.1.2 Investigative Discussions

We visited several research centers involved in inhalation research to investigate existing respiratory mechanics models. Each center has specific areas of expertise and, although there were some overlapping research efforts, the general area of research was much broader than the scope of this project. Although by no means an exhaustive survey, the following centers were visited and their input was solicited to determine who was currently involved in mathematical modeling of inhalation.

1. The Hamner Institutes for Health Sciences (The Hamner, formerly CIIT, the Chemical Industry Institute of Toxicology), Research Triangle Park, NC, USA. CIIT has been a leader in developing mathematical and animal models to describe particulate inhalation.[8,9] Researchers at CIIT have developed highly complex computational fluid dynamics (CFD) models to simulate inhalation flow patterns in addition to simpler, fast-running models to describe inhalation patterns, such as the MPPD software.
2. Environmental and Occupational Health Sciences Institute (EOHSI), Rutgers University, Piscataway, NJ, USA. EOHSI is involved in a variety of source–dispersion–dose modeling, and also detailed systems biology modeling. EOSHI reported that they use the MPPD model in some of their research.
3. Lovelace Respiratory Research Institute (LRRI), Albuquerque, NM, USA. The LRRI is a leader in animal studies and carries out research using human lung casts to investigate deposition pattern experimentally s. At the time of our visit, they did not have an active mathematical modeling program. They confirmed the usefulness and positive view of the inhalation community regarding MPPD.

ICRP Publication 66[10] describes a mathematical model of inhalation and deposition focused on radiological protection. It would be time consuming to develop a new software implementation of this model; however, a few commercial software programs already exist:

1. LUDEP (LUng Dose Evaluation Program).[11] LUDEP Version 1.0 (1993) was published by the National Radiation Protection Board (NRPB) in the United Kingdom to implement a late draft of the proposed lung model. Version 1.1 (1994) incorporated all of the recommendations of ICRP Publication 66 and Version 2.0 (1996) included a "bioassay" module and the facility to evaluate ingrowth of progeny radionuclides in a radioactive series. NRPB is now the Radiation Protection Division of the UK Health Protection Agency.
2. Dose and Risk Calculation (DCAL) software.[12,13] DCAL is a comprehensive software system for the calculation of tissue dose and subsequent health risk from intakes of radionuclides or exposure to radionuclides present in environmental media.

Although generally applicable, the focus of these codes is on radioactive material and the resultant internal energy deposition; extracting the inhalation and deposition portions of the code from the remainder would likely prove difficult. In addition, although many aspects of the ICRP 66 inhalation model are widely used, the foundation of the model is based solely on a semiempirical relationship between deposition and particle size. The use of the ICRP model is justified within the range of the measurements used to build the model. However, use of the model in a way that requires extrapolation beyond the data supporting the semiempirical relationship may produce unreliable results.

The MPPD model[14–16] is a widely used, fast-running, GUI-driven, Java-based set of algorithms that can calculate the deposition and retention of both mono- and polydispersed particulates and aerosol droplets in the respiratory tracts of several species, including at the time of this writing humans (children and adults), rhesus monkeys, two strains of rats, two strains of mice, guinea pigs and sheep. Version 1.0 of the MPPD code was first introduced in 2002 and version 2.0 was released in 2006. The human respiratory model includes both single-path, symmetrical calculations and also several multi-path variations of limited asymmetric, asymmetric and stochastic models. The limited asymmetric model uses a five-lobed model with subsequent symmetric airways. An age-dependent set of lung morphologies is also available. The model can calculate deposition in three regions, extrathoracic (ET), tracheobronchial (TB) and pulmonary (P) or by specific airway generation. Figure 12.8 illustrates the anatomy and functionality of these three

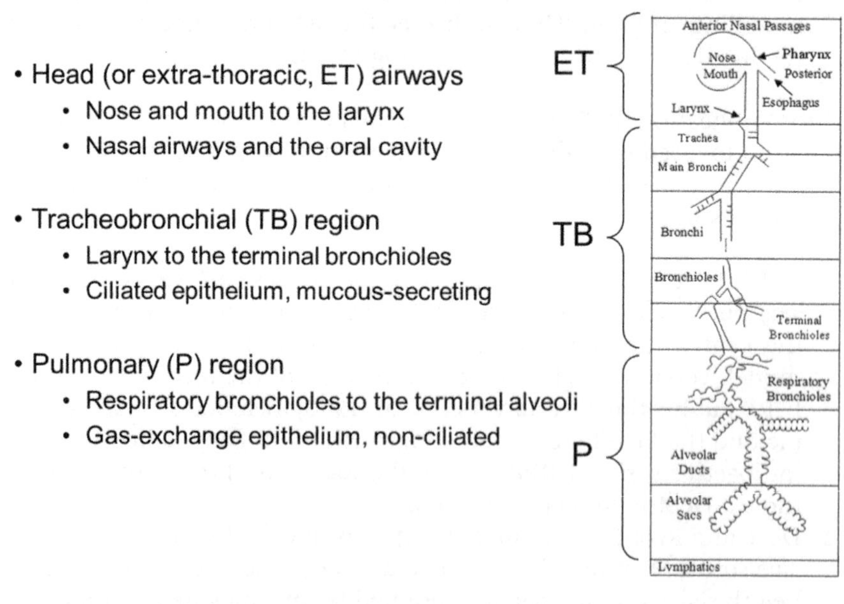

Figure 12.8 The respiratory tract may be divided into three anatomical–functional regions.[14]

regions. Deposition according to these three regions is generally adequate to correlate with human response data currently available.

12.2.1.3 Decision to Use MPPD

We decided to proceed with the MPPD model as the respiratory dosimetry component of the DARRT model based on our review of the state-of-the-art in modeling inhalation mechanics and dosimetry, the positive recommendations regarding MPPD given by EOSHI and LRRI and the willingness of The Hamner Institutes for Health Sciences to provide source code through a subcontract to Applied Research Associates (ARA). A subcontract was issued and executed, providing ARA with a functional version of MPPD adapted for the goals of this project. Subsequently, The Hamner Institutes for Health Sciences made a business decision to reduce its involvement in modeling respiratory mechanics and released its MPPD copyright to one of the authors (B.A.). The next section of this chapter provides more details on respiratory mechanics and the MPPD dosimetry model.

12.2.2 Dosimetry Assessment of Inhaled Materials in the Lungs: MPPD Modeling

Particles suspended in the atmosphere, whether generated naturally or for commercial use or even for malicious intent (such as biological or chemical warfare), may create adverse health effects when inhaled. Once the materials enter the respiratory tract, they may deposit on the airway surfaces, depending on their physical and chemical properties. In addition to inflicting injury at the site of deposition (for hazardous materials), deposited particles may be translocated throughout the lung and removed from airway surfaces by various clearance mechanisms that are operative at different locations in the respiratory tract. The retained portion (dose) in the lung may pose short- and long-term health effects. Therefore, estimations of the initial and retained dose throughout the lung are needed to correlate biological response from exposure to airborne agents.

Two different approaches are adopted in calculating deposition and retention of particles in the respiratory tract, depending on the available information on airway geometry and the desired level of details for the dose. The first approach is site-specific modeling, for which the precise airway geometry for the region of interest and also information on airflow and compound transport through the region at inlets and outlets are known. This approach is most useful at the entry port to the respiratory tract (*i.e.* nasal and oral airways) for which airway passages are reconstructed from scanned images of airway cross-sections and combined with breathing information to form anatomically and physiologically realistic models. Since the complete form of the transport equations with the full account of dominant physical mechanisms are solved, detailed information regarding the airflow and compound losses to the airway walls is obtained, but at the expense of a heavy computational toll.

The second approach is whole-lung modeling in which airways are described dimensionally, but precise geometric features (such as curvature and smoothness of airway walls) are neglected. The entire respiratory tract geometry can be constructed from information on airway length, diameter and orientation and gravity angles. Averaged transport equations that include dominant mechanisms for the flow and particle transport and deposition, when applied to idealized lung geometries, will greatly reduce computations. However, deposition predictions may only be made per airway or region of the lung. In this modeling approach, higher order effects such as local variation in airflow and fluctuations in particle trajectory due to non-smooth geometry can be neglected in favor of higher order assessment of particle losses in the lung. Approximations in predictions are justified considering uncertainties in lung and breathing parameters and inter-subject variability in lung geometry across a population. The whole-lung modeling approach is ideal for regional calculations of deposition in the entire lung, whereas site-specific modeling is handicapped by the enormity of the lung airways and lack of knowledge on airflow mechanics in the lung. By making simplifying assumptions, computations can be carried out in a relatively short time when detailed, site-specific deposition information is either unavailable or unreliable.

12.2.2.1 Whole-Lung Modeling/Regional and Total Dosimetry Calculations

Lung airway dimensions span multiple scales, ranging from centimeters to micrometers. Currently, the resolution of imaged scans is at the 10 μm level; therefore, reconstruction of the entire respiratory tract based on imaged scans is not feasible with today's technology. Lung reconstruction from scanned images can be made only for the first few airway generations of the lung conducting tree. Even if the entire geometry of the respiratory tract were available from scanned images, limited information on airflow mechanics in the lung and particularly deep lung would hamper a thorough investigation of the airflow field and particle deposition. Detailed fluid mechanics computations of the airflow in the lung would require information on local airway compliance and boundary conditions at the exit of the most distal airways. Outlet boundary conditions that encompass lung elastic properties (compliance) must be constructed based on pressure information in the pleural cavity. Pressure in the pleural cavity is gravity dependent and varies vertically. As a result, lung expansion is nonlinear or simply each lobe of the lung expands at a different rate. In addition, pleural pressure varies with the time into the breathing cycle and breathing frequency, thus adding further challenges to detailed computations of lung ventilation. Hence fluid dynamics analysis (site-specific modeling) in the deep lung is severely limited by the lack of detailed information on lung geometry and time- and volume-dependent pleural pressure.

A whole-lung deposition model contains three major components. First, a complete lung geometry extending from the nasal and oral airways to the most distal alveolar airways (sacs) must be constructed. Second, the airflow through this geometry must be determined. Inhaled air will preferentially carry the suspended particles to various airways of the lung depending on the mechanics of breathing. Finally, a mathematical description of particle transport and deposition is needed based on the mass balance on the inhaled material. Typically, a modified version of the convective–diffusion equation is employed for particle deposition calculations.

To facilitate computations of losses to the lung from inhaling airborne materials, simplifications regarding lung geometry and airflow mechanics are needed. The lung geometry is assumed to be a collection of cylindrically shaped tubes arranged as a dichotomous branching network. Airway dimensions and orientations are found from morphometric measurements. Additional volumes are added to respiratory bronchial airways in the acinar region to account for the alveolar space.

The most popular lung geometry for humans is the typical-path lung structure obtained from available airway measurements in the lung.[17] The lung geometry is assumed to consist of 23 airway generations, of which the first 16 comprise the tracheobronchial region (conducting airways) and the last seven form the alveolar region. Using lung airway measurements,[18] Yeh and Schum[19] constructed a typical path, but also developed a five-lobe lung model with each lobe being symmetric but different in dimensions. Koblinger and Hofmann[20] further expanded the human geometric model by using measurements of Raabe *et al.*[18] to create stochastic lung geometries that were asymmetric and closely resembled the anatomical structure of the lung. However, airway shapes were still cylindrical and deviated from reality. In rats, Anjilvel and Asgharian[1] created an asymmetric lung geometry based on published measurements.[18]

Lung ventilation is calculated based on uniform expansion and contraction of airways for slow breathing and by the lumped parameter method or RLC network analogy for fast breathing. In addition, the airflow velocity profile in an airway is assumed to be uniform with a magnitude equal to the average parabolic velocity. Using the above geometries and airflow distribution model, mathematical models of particle transport in the model geometry were developed and solved to calculate particle deposition fraction in the entire respiratory tract.[1,9]

A number of deposition models are available in the literature for the calculations of particle deposition. Early deposition models were compartmental and limited in predictive capability. More realistic deposition models assumed the lung to expand and contract uniformly and resembled a trumpet in shape when symmetric lung geometry was assumed.[21,22] These models have been confirmed for regional and lobar fractions. The human lung has an asymmetric branching structure that affects both airflow and particle deposition. Asgharian *et al.*[9] used the multiple-path analogy in 10 stochastically generated lungs of human adults to calculate site-specific and regional deposition of particles. The stochastic lungs were based on morphometric measurements[18] and were developed by Koblinger and Hofmann.[20] Significant dose variations among the lung structures were found, which indicated inter-subject variability in the population.

Deposition models based on asymmetric lung geometries[9] offer an advantage over previous models in that they account for lung physiology and morphometric variability within each generation. These two parameters are shown to be most critical in accurate microdosimetry predictions of particles in the lung.[9,23,24] To make the deposition model available to the public and to allow other investigators in related fields such as toxicology and risk assessment to use the model, a software package with a graphical user interface was developed based on the models to calculate particle deposition and retention in human adults and children and rats.[8] The software was used to reduce the uncertainty factor due to dosimetric adjustment in animal-to-human dose extrapolation. It was also used by the US Environmental Protection Agency (EPA) in the toxicology assessment[12] of particulate material (PM).

12.2.2.2 MPPD Software Development

In collaboration with scientists from the Dutch Ministry of Health, Asgharian and colleagues at CIIT developed a software package, Multi-Path Particle Deposition (MPPDep), based on the deposition models to aid computations and simulations required by toxicologists, risk assessors and the drug delivery community.[8,23,25] The computer algorithms in the software calculated the deposition of aerosols in the respiratory tract of several species for particles ranging from ultrafine (0.01 μm) to coarse (20 μm) sizes. The graphical user interface (GUI) feature of the software enabled easy and quick calculations of deposition. The MPPDep software allowed various inputs such as selection of lung geometry, lung parameters such as functional residual capacity, particle characteristics (including size distribution) and breathing types (oral, nasal, oronasal, tracheal) and parameters (tidal volume, breathing frequency). A set of default values for the above parameters is used if the parameters were not specified. The software provided graphical and textual outputs. Simulation results were provided for total, regional and lobar deposition and as a function of airway generation number. The option of bar plots of total, regional (head, TB, P) and lobar deposition was provided. Also included with the software was an extensive Help Menu that described model details, software features and operation procedure. Several tutorials and a quick start guide helped the user to get launched quickly. An upgraded version of the software (MPPD) was released to the public in 2002. It included additional asymmetric lung geometries for humans that permitted more realistic calculations of deposition and also models of particle clearance in various regions of the lung. The software also included predictions of particle deposition in children.[26] Recently, a new version of MPPD has been made available through an ARA web site.[27] New features in this version include a deposition model specifically for nanoparticles, nonuniform lung ventilation to include the effect of body orientation (posture) on deposition and the availability of 30 stochastic lung geometries for more realistic assessment of variations of dose in the population. In addition, various improvements were made to the input and output of MPPD. Figure 12.9 shows sample screenshots from MPPD.

12.2.3 Inhalability

Inhalability was one of the parameters investigated. Inhalability is a measure of the ratio of the concentration of material just after inhalation, *i.e.* just after entering the nostrils or mouth, to the concentration in the ambient air. Deposition takes place subsequently. Some inhaled material does not deposit in the respiratory system and is simply exhaled, but that effect is not considered in inhalability. Inhalability I refers strictly to the intake process and is defined as a function of particle size d_{ae}:

$$I(d_{ae}) = \frac{C_{inspired}(d_{ae})}{C_{ambient}(d_{ae})} \quad (12.1)$$

where $C_{inspired}(d_{ae})$ is the concentration of particles of a given size present in inhaled air and $C_{ambient}(d_{ae})$ is the concentration of particles of the same size in the ambient breathing zone. The particle size d_{ae} is the aerodynamic diameter.

The literature was reviewed, the data from inhalability measurements were examined and the current mathematical models of inhalability were investigated. A mathematical model[28] that provided the best description of the data across a variety of conditions while minimizing mathematical inconsistencies was developed and published on the basis if this review. Figure 12.10 shows a family of inhalability curves for various ambient wind speeds. Note that the value approaches 0% for 100 μm particles at a wind speed of 0 m s^{-1} corresponding to still air as assumed in MPPD for the curves shown in Figure 12.1.

12.2.4 MPPD as Implemented in the DARRT Model

The MPPD code has several different lung models and many parameter settings to modify the lung performance. The code has a symmetric model that is commonly used because of its simplicity and speed, but the human respiratory system is not symmetric. The code also includes a five-lobed asymmetric model, which is a better representation, having two lobes in the left and three lobes in the right lung. Finally, because lobe structure varies from person to person, the code also includes 10 stochastically developed lung models. The airway branching angles and the length and diameter of segments differ and even the number of generations can differ slightly from person to person. It was decided to represent the average human respiratory system for the present purposes by calculating the deposition fractions with each of 10 stochastic models and averaging the results for each particle size. In addition, the standard deviation for each deposition value was reported.

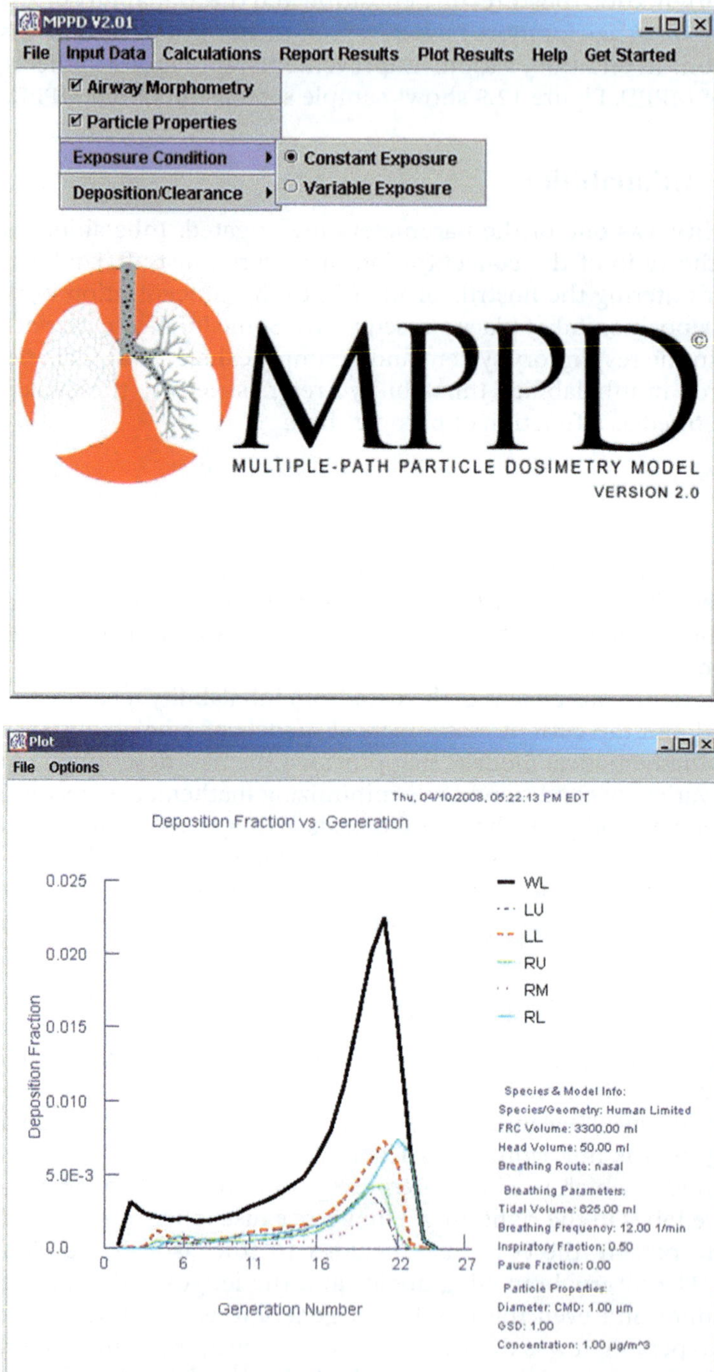

Figure 12.9 MPPD screenshots.

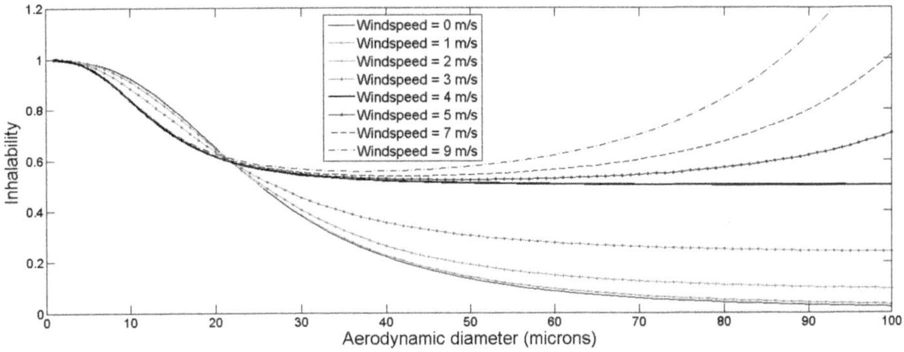

Figure 12.10 Recommended set of inhalability curves for a normal augmenter at a low breathing rate.

Another input parameter that changes the regional deposition is the breathing rate. The human response models that are implemented in HPAC allow the user to specify breathing rate with suggested values ranging from 15 L min^{-1} (lpm) to 75 lpm. Clearly, the airflow rates are significantly different over this breathing rate range. In addition, with a normal augmenter (a person who normally breaths through their nose until the breathing rate increases to the point that they need to augment with oral breathing), the transition from 100% nasal breathing to a combination of oral and nasal breathing occurs at around 35 lpm.[29] The nose is a very good particulate filter; so much of the material that would be trapped in the ET region at low breathing rates (100% nasal breathing regime) may actually reach deeper into the respiratory tract at higher breathing rates because airflow is being supplemented with mouth breathing, bypassing the nasal filter.

The MPPD was implemented in the DARRT model using the stochastic lung models. To avoid running multiple cases and averaging the results in an operational code, a look-up table of regional deposition for the desired range of parameters was built and a linear interpolation scheme for breathing rate and particle size was implemented. The 10 stochastic lung models, including the largest and smallest number of TB airways for the analysis, were used. All 10 lungs had the same lung volume. Each lung model was run at eight different breathing rates, ranging from 15 to 75 lpm, and for 22 particle sizes. Particle sizes were based on the bin midpoints that were intended to be used for the dry particulate materials; however, this choice does not constrain future choice of bin sizes. The regional deposition fractions for the ET, TB and P regions were collected and the average deposition and associated standard deviation in each respiratory tract region for tabulation were calculated for each breathing rate and particle size.

Figure 12.11 shows the average fractional deposition and standard deviations as a function of particle size for a breathing rate of 15 lpm and Figure 12.12 shows the same for a breathing rate of 40 lpm. Note that there are no

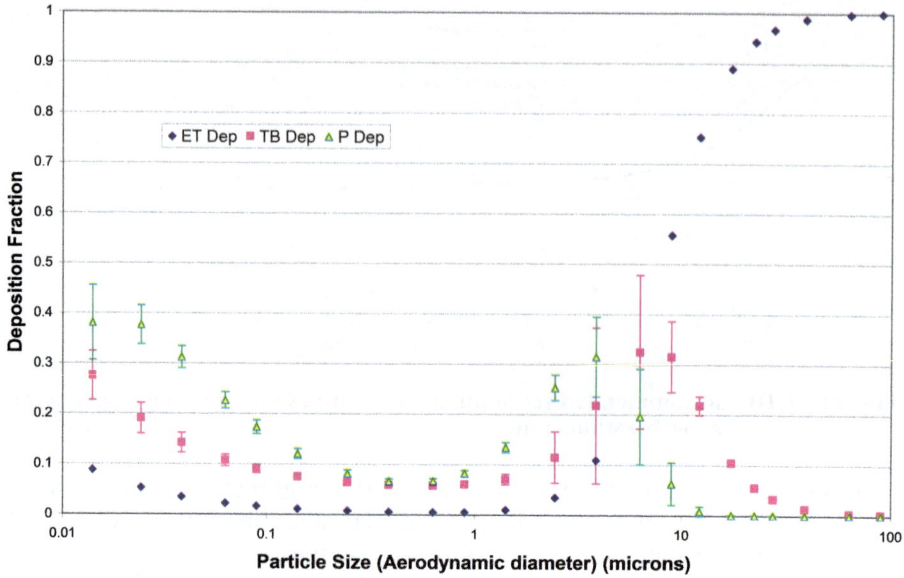

Figure 12.11 Fractional deposition and standard deviations of inhaled particles for a 15 lpm breathing rate.

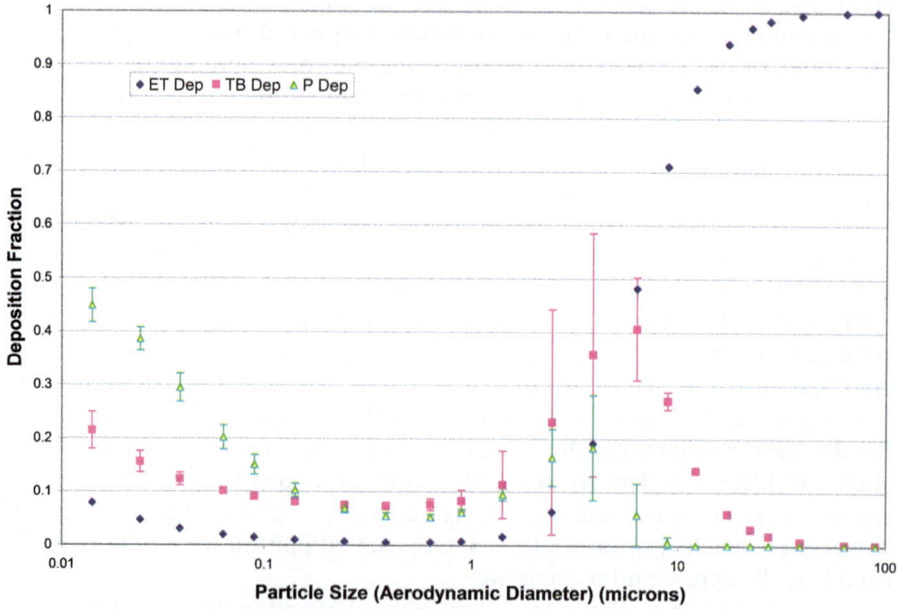

Figure 12.12 Fractional deposition and standard deviations of inhaled particles for a 40 lpm breathing rate.

standard deviations associated with the ET region. The stochastic portion of the MPPD model is associated only with the TB and P regions, hence there is no variability incorporated in the ET deposition. These curves may be compared with the older work of Snipes.[30] One difference is that the curves of Snipes include inhalability, leading to the apparent 50% deposition fraction for 100 μm particles. In DARRT, the calculation of inhalability is separated from the deposition calculation and it is not included in the curves plotted in Figures 12.11 and 12.12. These figures show correctly that 100% of inhaled 100 μm particles are deposited by impaction and settling.

Another notable feature at both breathing rates is the large standard deviation (variability) associated with TB and pulmonary deposition in the 1–5 μm particle size range. In this size range, impaction is a primary mechanism for deposition. The variability in deposition is due to changes in the number and size of pathways in the TB region for the 10 stochastic models and how that affects impaction. With fewer and smaller pathways through the TB region, airflow rate and, hence, impaction are increased, causing increased deposition in the TB region. In this case, the filtering effect of increased deposition in the TB region reduces deposition in the P region. More and larger pathways through the TB region reduce deposition there and increase it in the P region, so variability in deposition for the two regions is negatively correlated. This variability fairly is attributed to the person-to-person variability of the human population. Research is being conducted in the toxicology community on how this variability influences the interpretation of data from animal models of asthma and the susceptibility to and treatment of asthma in humans.[31,32] For the present purposes, it was noted that the pulmonary retained dose for different persons exposed to the same 5 μm aerosol environment is expected to vary by a factor of at least three. The data plotted in Figures 12.11 and 12.12 do not include the inhalability factor. For modeling, the inhalability factor is calculated and applied to the ambient air concentration independently.

Implementing the MPPD stochastic models as a look-up table utilizes the best estimate of the regional deposition for the average human respiratory system. Also, the standard deviation associated with the deposition fraction can be determined. Further development of the inhalation model should include the deposition uncertainty with the dosage uncertainty to improve the uncertainty of overall calculations of probability of effect.

12.2.5 Agent-Containing Particle Calculation

The physiological response to biological agents depends on a number of factors; as demonstrated, the particle size and resultant deposition location is one factor. Another factor for some agents is that the probability of infection correlates best with the number of ACPs deposited in a given region of the respiratory tract rather than the total number of infective organisms deposited. For these agents, adding more organisms to a given particle does not significantly increase its probability of causing infection. In other words, the number of organisms deposited is less important than the total number of ACPs deposited. For other agents, the probability of infection is a function of both the size (deposition location) and the total number of organisms deposited. The data indicate that *F. tularensis* is of the former type[33] and spores of *Bacillus anthracis* (causing anthrax) are of the latter type.[34] Bacteria are fairly large, on the order of 1 μm or more, so small particles prepared from a medium of both carrier and organisms may not in fact contain an organism. Recall that agent releases are usually characterized by a PSD, typically a log-normal distribution, so the aerosol can contain small particles that may be composed only of carrier. An approximate means of predicting the probability that a particle of a given size is actually an ACP has been developed employing the organism concentration of the bulk material before aerosolization, that is, the number of organisms per unit mass (or volume), and a Poisson distribution to model the random distribution of organisms among aerosol particles. The next sub-section briefly describes the calculation and how it is used in the DARRT model.

12.2.5.1 Discussion

Experimental aerobiology studies use various methods to determine the number of organisms in a particle, including simple optical scanning and counting. More often, data on the number of organisms in an aerosol are simply reported as the bulk concentration, either in organisms per unit mass or per unit volume. For a large particle, the bulk concentration value times the particle volume gives a reasonable approximation of the number of organisms in the particle. However, as the particle size decreases, the situation in which only one organism or less is expected is the approach employed here. At this point, whether or not the particle contains an organism becomes a random process. Since particles can only contain integral numbers of viable organisms, Poisson statistics may be applied to estimate whether or not a particle contains one or more organisms.

To test this approach, the average number of organisms per unit volume as a bulk property was compared with experimental data on organism count per particle. Reports of *F. tularensis* aerobiology studies performed in the early 1970s included data regarding the average number of organisms per particle for particles of various sizes. Table 12.1 shows data for four different aerosol generation conditions from a report by Day and Berendt.[33]

Day and Berendt also quoted a bulk concentration of 4×10^{10} cells (organisms) per milliliter for their preparation.[33] Assuming a spherical particle shape, the volume of each particle can be calculated, and by multiplying by the bulk concentration, the expected (average) number of organisms per particle can be obtained. Figure 12.13 shows the measured median number of cells per particle *versus* particle size for the four aerosols in Table 12.1. For comparison, the dashed line shows the predicted average based on simple volumetric calculation. As can be seen in the graph, simply scaling the bulk cell concentration value by the particle volume gives a reasonable

Table 12.1 Aerosol characteristics from Day and Berendt[33]

Particle size range/μm	Geometric mean size/μm	Particle frequency/%	Average no. of cells/particle	ONMD[a]/μm	Geometric standard dev.
Atomizer aerosol 1				2.10	2.92
1.0–1.4	1.2	53.8	0.047		
1.4–1.9	1.6	24.9	0.097		
1.9–2.7	2.3	13.5	0.205		
2.7–3.8	3.2	6.4	0.47		
3.8–5.4	4.5	1.4	1.1		
Spinning top aerosol 1				7.5	1.24
3.8–5.4	4.5	4.8	5.1		
5.4–7.6	6.4	85.4	12.2		
7.6–10.8	9.1	9.8	26.7		
Spinning top aerosol 2				12.5	1.22
7.6–10.8	9.1	7.2	35.6		
10.8–12.5	11.6	83.8	78.4		
12.5–17.6	14.8	9	151.3		
Spinning top aerosol 3				24.0	1.33
12.5–17.6	14.8	3.6	241.5		
17.6–24.9	20.9	83.6	490.0		
24.9–35.0	29.5	12.8	895.0		

[a]Organism number median diameter.

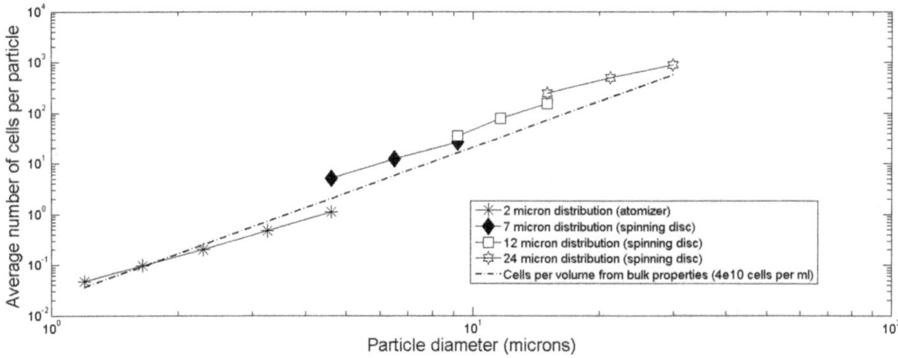

Figure 12.13 Experimentally measured[33] median number of cells per particle compares well with an estimate (dashed line) based on bulk concentration.

approximation of the average number of cells per particle over four orders of magnitude, from less than 0.1 to about 1000.

As noted above, particles can only have integral numbers of cells, but one can see from Figure 12.13 that small particles may have, on average, much less than one cell per particle. At this point, a method for calculating the probability that a particle of a given size does or does not contain at least one cell, *i.e.* whether or not it is an ACP, must be developed.

12.2.5.2 ACP Model

It was assumed that the number of cells per particle is a random variable following a Poisson distribution and that the bulk material concentration provides a reasonable estimate of the average number of cells per unit particle volume. For a Poisson distribution, the probability P that a particle of diameter d_{ae} contains zero agent cells (*i.e.* is not an agent-containing particle) is

$$P(0,d_{ae}) = e^{-C_{ave}(d_{ae})} \tag{12.2}$$

where $C_{ave}(d_{ae})$ is the average (expected) number of cells in a particle of diameter d_{ae} based on the volume of the particle and the bulk concentration of the medium. Conversely, the probability that a particle of diameter d_{ae} contains one or more cells (*i.e.* is an ACP) is

$$P(\geq 1,d_{ae}) = 1 - P(0,d_{ae}) \tag{12.3}$$

This relationship was included in the DARRT model as part of the overall estimation of the number of ACPs, of a given size, that are deposited in a region of the respiratory tract. This relationship also provides a parameter to judge the potential biological effectiveness that one might attribute to a given biological threat. For example, the *F. tularensis* slurry mixture used in the experiments described above contained a cell concentration of 4×10^{10} cells mL^{-1}. The material used to generate the aerosol for the Operation Whitecoat studies had a cell concentration of 3×10^{11} cells mL^{-1}.[35] ("Operation Whitecoat" was the name given to a series of experiments involving human volunteer test subjects;[35,38–40] see below.) For a particle of given size, the probability of being an ACP increases with increase in bulk concentration. Figure 12.14 shows an example using the two concentrations just described.

As can be seen in Figure 12.14, for the higher concentration material, a particle of 3 μm diameter has nearly a 100% probability of containing a cell. However, at the lower concentration, the same size particle has less than a 50% probability of containing a cell. This calculation is performed as part of the estimate of the number of ACPs deposited in a specific region of the respiratory tract and should always be factored in when evaluating potential threats.

12.2.5.3 ACP Model Summary

The bulk material property of concentration of infectious organisms (cells) is used to estimate the average number of organisms in a given volume. The parameter is an accurate estimate for relatively large volumes, but as the

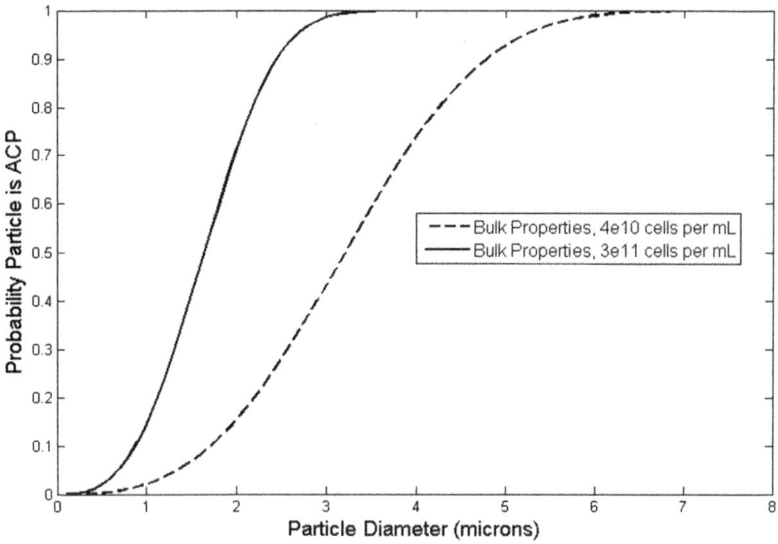

Figure 12.14 The agent bulk cell concentration influences the probability that a dispersed particle of a given size is an ACP.

volume of interest becomes smaller, the bulk property is less useful. In particular, as the volume associated with a particle becomes small, the bulk property characteristic must be replaced with the expected number of cells per particle. For small particles, it is necessary to calculate the probability that a particle of a given size does or does not contain an agent cell. Poisson statistics are assumed and the bulk cell concentration is used to estimate the probability that a particle of a given size contains at least one cell, *i.e.* is an ACP.

12.3 Development of a Human Response Model Accounting for Aerosol Size – Prototype Agent *Francisella tularensis*

An extensive literature review was conducted around 2008 to extract experimental data from the literature concerning the relationship between the particle size of inhaled bioaerosols and the development of respiratory illness. Many of the studies reviewed were concerned with respiratory illness caused by *F. tularensis*, listed by the Centers for Disease Control and Prevention (CDC) as a Class A bioterrorism agent. The nature of illness caused by *F. tularensis* is strongly related to the port of entry of the organism into the body.[36,37] The primary motivation for developing the DARRT model was the observation discussed below that in nonhuman primates, even for a route of entry restricted to inhalation, the nature of the ensuing illness depends on the deposition site within the respiratory tract, which in turn depends on the aerosol particle size.

The effects of aerosol exposure to *F. tularensis* were extensively studied for both human and animal subjects in the US offensive biological warfare program before the program was dismantled in 1969. "Operation Whitecoat" was the name given to a series of experiments involving human volunteer test subjects[35,38–40] that took place during the period 1954–1973. Most of the 2300 volunteers were conscientious objectors who agreed as an alternative to combat duty to be exposed to live agent aerosol clouds in chamber tests and in field tests simulating battlefield deployment of biological agents,[41] including *F. tularensis*. During the height of the program, a 10^6 L aerosol test chamber was constructed at Fort Detrick, MD, to study agent characteristics under controlled conditions. The 40 ft high, gas-tight globe was nicknamed the "Eight Ball." Enclosed in a 60 × 60 ft laboratory building, the structure had control panels and exposure chambers at the base of the test sphere, and also catwalks that allowed scientists to collect samples from portals. Both animal and human test subjects were exposed to biological aerosols generated in the chamber. Human subjects ascended to equator level to be exposed at specially designed exposure cubicles. Animals were exposed at other portals. The human exposures were conducted prospectively in the 1960s as tests of vaccine efficacy. Several decades later, in retrospective studies, this unique set of experimental data provided a valuable resource from which mathematical models could be constructed to inform biological defense better and to aid in understanding the impact of particle size on the disease development process. Availability of both human and animal data facilitated the modeling process.

12.3.1 Development of a Human Response Model

Particles from an aerosol hazard that can be inhaled and deposited in different parts of the respiratory tract can cause location-specific disease manifestations. For example, biological particles containing *F. tularensis* that deposit in the pulmonary region can cause a pulmonary infection resulting in pneumonitis or, once the infection progresses, can transfer to the bloodstream and cause a subsequent systemic infection. Likewise, biological particles that deposit in the extrathoracic region can cause localized infections in the cervical lymph nodes. Hence it is important to estimate the probability of infection initiating in a given location as a function of the deposition in that region. The probability of infection can be a function either of the number of ACPs deposited or of the number of organisms, whichever independent variable is more predictive as indicated by experimental data.

Naturally occurring disease, accidents and experimental studies with aerosol challenge from *F. tularensis* indicate at least three likely initial infection sites, the pulmonary (P), extrathoracic (ET) and ocular regions.[33] Since this study is focused on the respiratory tract, only the P and ET regions will be considered. The infection probability functions utilize data from human exposures wherever possible. When necessary, the human exposure information is supplemented with animal data.

Prior to the current study, probability of infection calculations were based on the inhaled dose, that is, the estimated number of organisms breathed during the exposure. *F. tularensis* from the controlled exposures of the Operation Whitecoat studies have been used to estimate infection rates for human subjects.[42,43] The data from these studies serve as the foundation of the dose–response curve for pulmonary infection. Based on the estimated particle size of the aerosol, the inhaled dose data are modified to account for the likely fraction that will deposit in the pulmonary region. Further, the concentration of organisms per unit volume in the material to be aerosolized is used to estimate the number of aerosol particles that include one or more organisms (that is, are ACPs). Data described in the next sub-section indicate that the probability of infection for *F. tularensis* is better described by the number of ACPs that are deposited rather than the number of organisms deposited. (Note that for *F. tularensis*, the terms "organism" and "cell" are used interchangeably in the discussion.)

Tularemia can be caused by very few organisms within the respiratory tract.[42,43] In the Operation Whitecoat studies, 28 human volunteers received doses of fewer than 52 organisms of Type A *F. tularensis* challenge, but most were still infected (22/28). The data associated with these 28 subjects are used to estimate the probability of pulmonary infection. References indicate that the aerosol challenge to these individuals was less than 1 μm in diameter.[35] The literature search for preparing this chapter did not find controlled human exposures with large particle sizes that would deposit primarily in the ET region of the respiratory tract; therefore, modeling of ET infections must be based on experiments with nonhuman primates.

12.3.1.1 ET Infection Calculation

Day and Berendt[33] conducted a tularemia infection study in the early 1970s using *Macaca mulatta* (rhesus monkeys). Four different aerosol PSDs were used with median diameters of approximately 2, 7, 12 and 24 μm. Table 12.2

Table 12.2 Results[33] for rhesus monkeys exposed to *Pasteurella tularensis*.

Particle mean size	2.1 and 7.5 μm	12.5 and 24.0 μm
No. of animals exposed	48	45
% infected and ill	100%	84%
Primary infection site	Lower respiratory system, pneumonitis	Upper respiratory tract, nasal pharyngeal area, cervical and mandibular lymph nodes; eyes (15%)
Onset of illness	2–3 days	6–10 days
% fatal (of those ill)	69%	63%
Time to death	4–8 days	8–21 days
LD_{50} (organisms)	14 and 378	872 and 4447
LD_{50} (ACPs)	14 and 28	11 and 8

summarizes results of the Day and Berendt experiments, grouping the data for the two smaller and two larger sized particles according to differences in observed pathology. Note that since the 1970s, nomenclature has changed so that the bacterium *Pasteurella tularensis* is now referred to as *Francisella tularensis*.

As shown in Table 12.2, the smaller particle sizes, 2 and 7 μm, resulted in 100% pulmonary infection with 69% mortality and a time-to-death of 4–8 days. The larger particle sizes, 12 and 24 μm, resulted in 84% infected animals with significant levels of infection of the cervical lymph nodes and the nasal pharyngeal area and onset times ranging from 6 to 10 days; the initial infection site appeared to be the ET area. Of the infected animals, 63% died with a time-to-death of 8–21 days. The next to last row of Table 12.2 shows the median lethal dose (LD_{50}) values reported by Day and Berendt[33] for the four particle sizes, ranging from 14 organisms for the smallest particles (2.1 μm) to 4447 for the largest particles (24 μm), a variation of more than two orders of magnitude. The last row of the table shows the LD_{50} values expressed in terms of ACPs. Here, the variation is only about a factor of three, showing that the probability of infection for *F. tularensis* is more nearly a universal function of the number of ACPs inhaled rather than the number of organisms.

The data from this study can be used to provide an estimate of probability of infection for particles deposited in the ET region. Table 12.3 summarizes the raw data for the exposures in the Day and Berendt paper.[33]

Note that for most particle sizes and doses, the infection response was 100%. Only the lower two doses of 24 μm particles and the lowest dose at 12 μm showed animals with no infection. In practice, only the 24 μm particle size data provide a range of doses and incidence of infection that can be used to

Table 12.3 Rhesus monkey infection and death data[33]

ONMD[a]/μm	Inhaled dose (cells)	No. exposed	No. infected	No. dead
2.1	5	6	6	1
	11	6	6	3
	32	6	6	4
	65	6	6	6
7.5	240	6	6	3
	720	6	6	4
	2208	6	6	6
	4416	6	6	6
12.5	556	6	5	3
	1141	6	6	5
	2745	7	7	6
	29 863	3	3	3
24.0	146	6	2	0
	873	6	4	1
	2315	6	6	2
	11 616	5	5	4

[a]Organism number median diameter.

estimate a dose–response function for infection. The working hypothesis is that the probability of an infection for *F. tularensis* depends on the number of ACPs deposited in a given region regardless of the size of the particle or the number of organisms. In other words, the total number of organisms deposited is of secondary importance compared with how many ACPs are deposited in the region.

The characteristics of the 24 μm PSD are shown as "Spinning top aerosol 3" in Table 12.1. The distribution is described by three size ranges, with the center bin (17.6–24.9 μm) accounting for 83.6% of all the particles in the distribution, in other words, a nearly monodispersed aerosol. Since the distribution is assumed to be represented by a log-normal distribution, the geometric mean of the bin endpoints as shown in Table 12.1 is the best approximation of the typical particle size within each range. The data also include the average number of cells in each particle. These data allow for the conversion from total dose in organisms to the number of particles. Note that given the large average number of cells per particle, all particles can be assumed to have one or more cells.

The final input for determining the probability of infection is an estimate of the fraction of particles of a given size deposited in a specific respiratory region, which is calculated using the MPPD model. At the time of this study, MPPD did not specifically model the rhesus monkey respiratory system (it has since been added), but the rhesus monkey is a reasonable animal surrogate model for human respiratory mechanics.[44] MPPD accounts for a number of variables that can influence the inhalation and deposition of particles. For example, the model includes a calculation of a particle's inhalability, breathing rate and the breathing mode (nasal or oral). In the example of the rhesus monkey study, the inhalability fraction was set to 1.0. The animals were anesthetized, so it was assumed that they were prone and masked, and this environment would probably allow for all the particles in the breathing zone to be inhaled. Further, since the animals were anesthetized, the breathing rate was equivalent to a sleeping breathing rate of 7.5 lpm. MPPD results for the deposition fractions for the ET region, using these parameters, are given in Table 12.4. A detailed discussion of the respiratory mechanics model is presented in the previous section.

The ET dose D^{ET} for estimating the probability of ET infection can now be created using the equation

$$D^{\mathrm{ET}} = D\sum_{i=1}^{3}\frac{f_i d_i^{\mathrm{ET}}}{\bar{C}_i} \tag{12.4}$$

Table 12.4 Extrathoracic deposition fraction from MPPD.

Geometric mean particle size/μm	ET deposition fraction, d^{ET}
14.8	0.9891
20.9	0.9974
29.52	0.9994

where

D^{ET} = number of ACPs deposited in the ET region
D = total inhaled dose as number of organisms
f_i = fraction of aerosol in each size range
d_i^{ET} = fractional deposition in the ET region for each particle size range
$\bar{C}_i$ = average number of cells per particle for a specific size range
i = index for each of three particle size ranges.

The application of eqn (12.4) to the dose levels associated with the 24 μm particle size data from Table 12.3 results in the values of D^{ET} given in Table 12.5.

Note that the lowest inhaled dose resulted in an expected ET deposition of less than one particle. Obviously fractional particle deposition is not physical, so this value shows the stochastic nature of the dose received by an individual in a population. The data for numbers infected/exposed in Table 12.3 can be fitted to a cumulative log-normal distribution using a nonlinear least-squares fit. This fit was performed using the Curve Fit Routine in MATLAB. Eqn (12.5) shows the log-normal dose–response function followed by the values of the fitted parameters:

$$P^{ET}\left(D^{ET}\right) = 0.5\left[1 + \operatorname{erf}\frac{\left(\ln D^{ET}\right) - \mu)}{\sigma\sqrt{2}}\right] \tag{12.5}$$

where $\mu = -0.719$ is the mean of the natural logarithm of the dose (also the natural logarithm of the median infective dose or ID_{50}) and $\sigma = 1.53$ is the standard deviation of the natural logarithm of the dose.

The statistical analysis of the Curve Fit Routine in MATLAB included a sum of the squares due to error (SSE) = 0.01651, an R^2 = 0.946 and a root mean squared error (RMSE) = 0.09085. Figure 12.15 shows a plot of the dose–response function for ET infection and points representing the response data from Table 12.5 according to Day and Berendt.[33]

The median infective dose (ID_{50}) for this dose–response function is an average of 0.49 ACPs deposited in the ET region per individual and the ID_{90} is 3.5 particles deposited. This result shows that the likelihood of infection resulting from a single ACP deposited is near unity for these experimental conditions.

Table 12.5 ET deposition of ACPs for the 24 μm size group shown in Table 12.3.

Inhaled dose (cells)	Expected no. of ACPs deposited in ET region	Infected/total[33]
146	0.29	2/6
873	1.74	4/6
2315	4.61	5/5
11 616	23.14	6/6

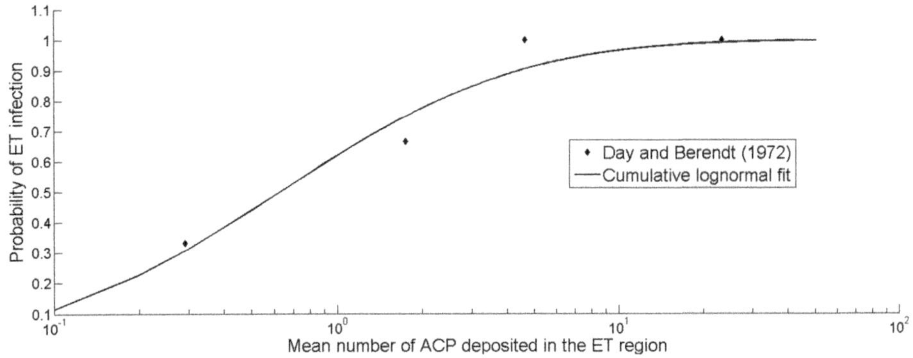

Figure 12.15 Probability of ET infection.

12.3.1.2 Pulmonary Infection Calculation

The probability of infection was calculated using the Operation Whitecoat data[42,43] based on the inhaled dose. The dose–response probability curve for infection can be correlated with the number of particles deposited in the pulmonary region by estimating the fraction of the inhaled dose that will result in deposition of ACPs in the pulmonary region.

According to one of the original papers[35] for these data, the aerosolization method used in the set of experiments generated an average particle size of 0.7 µm diameter and involved inhaled doses of 52 organisms or fewer. Figure 12.14 shows that particles in this size range have only a small chance of containing an organism. Furthermore, this particle size is roughly the size of an individual organism. Consequently, it must be assumed that for this experiment, no particles contained more than one organism. In this case, we may assume that organism dose and ACP dose are the same. This small particle size is not very efficiently retained in the pulmonary region; MPPD estimates a deposition fraction of only 0.067 at a breathing rate of 15 lpm. The dose–response curve derived[42] from the Operation Whitecoat data using a cumulative log-normal distribution has a mean natural logarithm of the inhaled dose $\mu^{\mathrm{IN}} = 2.28$ with a standard deviation $\sigma^{\mathrm{IN}} = 1.19$. After the dose–response curve has been scaled by the pulmonary deposition factor of 0.067, the geometric standard deviation σ^{P} will be the same as σ^{IN}, but the mean of the natural logarithm of deposited dose μ^{P} will change as follows:

$$\mu^{\mathrm{P}} = \ln\left(0.067\,\mathrm{e}^{\mu^{\mathrm{IN}}}\right) = -0.423 \tag{12.6}$$

As an alternative, the same nonlinear least-squares curve-fit routine used for the ET calculations was used to fit the Operation Whitecoat data with inhaled dose scaled to pulmonary deposition. As shown in Figure 12.16, this new fit compares very well with the scaled inhaled dose fit.

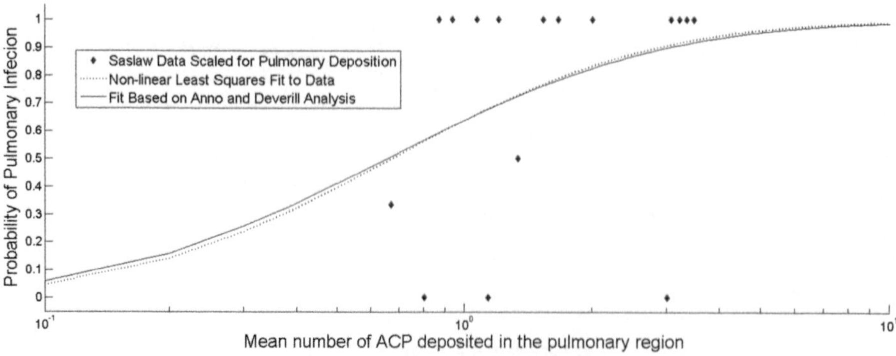

Figure 12.16 Probability of pulmonary infection from Operation Whitecoat data scaled to deposited dose in the P region.

12.3.2 Summary

Estimates of the probability of pulmonary or ET infection can be calculated from available data. The Operation Whitecoat studies were used to calculate a probability of infection and the original function can be scaled to estimate the probability of pulmonary infection. The resultant probability function is

$$P^{\mathrm{P}}\left(D^{\mathrm{P}}\right)=0.5\left[1+\operatorname{erf}\frac{\left(\ln D^{\mathrm{P}}\right)-\mu^{\mathrm{P}}}{\sigma^{\mathrm{P}}\sqrt{2}}\right] \tag{12.7}$$

where $\mu^{\mathrm{P}} = -0.423$ and $\sigma^{\mathrm{P}} = 1.19$.

The probability of ET infection can be estimated from rhesus monkey studies conducted by Day and Berendt.[33] Eqn (12.8) shows the ET probability function:

$$P^{\mathrm{ET}}\left(D^{\mathrm{ET}}\right)=0.5\left[1+\operatorname{erf}\frac{\left(\ln D^{\mathrm{ET}}\right)-\mu^{\mathrm{ET}}}{\sigma^{\mathrm{ET}}\sqrt{2}}\right] \tag{12.8}$$

where $\mu^{\mathrm{ET}} = -0.719$ and $\sigma^{\mathrm{ET}} = 1.53$.

12.4 Development of a Model for Particle Size-Dependent Health Effects of Ricin Exposure

In addition to the prototype biological agent for tularemia, we developed a biological toxin model for inhaled dry ricin. The toxic effects of ricin vary according to the route of administration, but the clinical symptoms frequently are related to a severe inflammatory response and multiorgan failure. Numerous studies, including a study[45] with BALB/c mice, have shown that inhaled ricin appears to cause significant damage only if it reaches the pulmonary region. Therefore, a reasonable working assumption is that the probability of mortality by the inhalation route is based on the inhaled dose delivered to the lung. Since ricin is a toxin, the dose should be scaled to the body mass and the resultant dose is calculated as (mass of deposited toxin)/(body mass) expressed in μg kg^{-1}. This section describes the calculations involved in estimating a probability of mortality function for inhaled ricin.

12.4.1 Background

The study[45] with BALB/c mice showed that 1 μm aerosolized ricin particles caused mortality, whereas 5 or 12 μm particles did not. Few of the larger particles can reach the pulmonary region of the mouse, but the toxin showed up in other organs due to extrathoracic deposition and clearance. Further, pathology of rodents and nonhuman primates shows aerosolized ricin resulting in significant lesions and edema in the alveolar and terminal bronchial region.[46,47] Therefore, one can hypothesize that after an inhalation exposure, the probability of significant effect or mortality is a function of the mass of ricin that is deposited in the pulmonary region of the respiratory tract.

Griffiths *et al.*[46] performed a series of exposures to rats using a ricin aerosol with a PSD of 0.9 μm mass-median aerodynamic diameter (MMAD) and a σ of 1.6 – a fairly narrow distribution. The study used six groups of six rats each (one group had only five) and used dosages ranging from 1.51 to 119.3 mg min m^{-3}. Table 12.6 provides a summary of the data.

The goal is to make a meaningful estimate of the lung dosimetry associated with this study, in other words, to use the data available and make additional assumptions as necessary to estimate the mass of ricin deposited in the lung and correlate it with the mortality rates from the experiment. Some assumptions regarding their data must be made in order to use the MPPD Rat model to calculate the lung dosimetry associated with the study.

The report on the rat study[46] noted that the calculation of inhaled dose is an estimate based on the time-integrated concentration (*Ct*). Unfortunately,

Table 12.6 Response to ricin exposures in a rat study[46]

Exposure duration/min	Aerosol concentration/ μg L^{-1}	Exposure *Ct*/ mg min m^{-3}	Inhaled dose/ μg kg^{-1}	No. exposed	No. dead
40	2.98	119.3	85.77	5	5
4	2.98	11.93	7.47	6	6
2	2.98	5.96	3.83	6	4
3	1.48	4.54	3.55	6	1
2	1.48	3.01	2.02	6	0
2	0.57	1.51	1.05	6	0

the report does not provide guidance on the data used to calculate the dose, *i.e.* the ventilation rate (VR) and body mass of the animals (BM). At its simplest, the inhaled dose calculation, D_I, is

$$D_{\mathrm{I}}\left(\mu\mathrm{g\,kg}^{-1}\right)=\frac{Ct\cdot \mathrm{VR}}{\mathrm{BM}} \tag{12.9}$$

MPPD is applied to the calculation of the fraction of the inhaled dosage that is deposited in the rat lung. MPPD uses data from Mauderly[48] for the rat respiratory dynamics data and sets the default values of 102 bpm for breathing rate and 2.1 mL for tidal volume. Based on the PSD parameters noted above and using the default ventilation rate parameters, MPPD calculates a pulmonary lung deposition fraction of 0.070.

12.4.2 Calculations

A nonlinear least-squares fit to a cumulative log-normal distribution of the mortality rate can be calculated as a function of the presented dosage provided by Griffiths *et al.*[46] The fit was calculated using the MATLAB CurveFit Toolkit. Based on the fit, shown in the upper half of Figure 12.17, the ICt_{50} is 5.5 mg min m^{-3}, which corresponds to a μ of 1.707 and a σ of 0.1946.

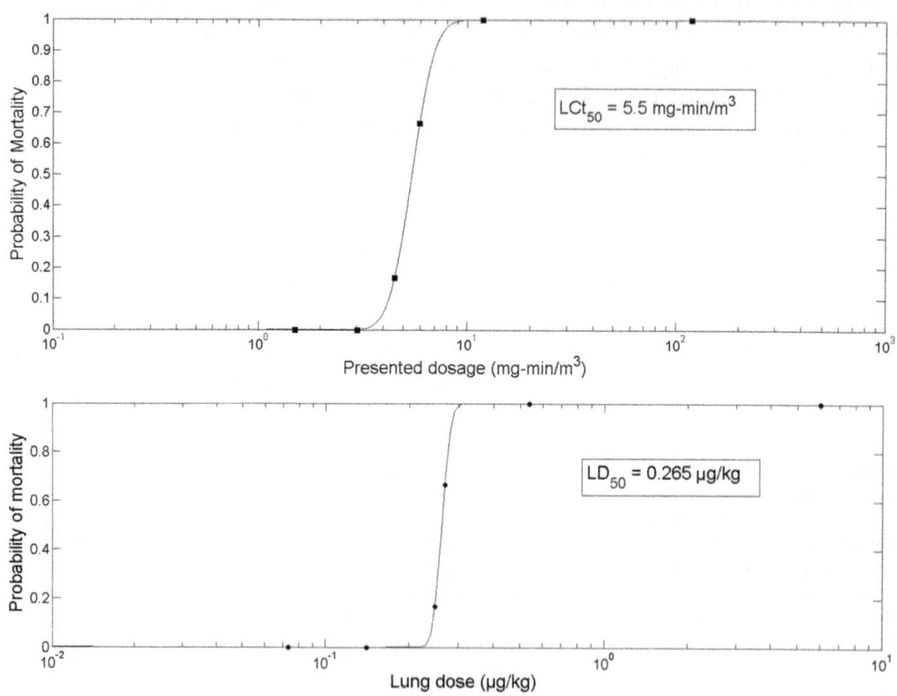

Figure 12.17 Curve fits to dosage and lung dose based on Griffiths *et al.*'s rat study.[46] Lung dose is normalized to total body weight.

The presented dosage is converted to deposited lung dose, D_L. This conversion is based on

$$D_L\left(\mu g\,kg^{-1}\right) = \frac{Ct \cdot VR}{BM} f_P \tag{12.10}$$

In other words, the fraction of the inhaled dose that is deposited in the lung will be calculated. As before, a cumulative log-normal response function is fitted to the mortality rate as a function of lung dose. The fit results in an LD_{50} of 0.265 μg kg^{-1}, which corresponds to a μ of −1.34 and a σ_g of 0.0543. The curve is shown in the lower half of Figure 12.17.

Note the steepness of the curve compared with the fit to presented dosage. The data from Table 12.6 show that the two highest dosages (and dose) result in 100% mortality, whereas the two lowest dosages result in no mortality. Hence the two intermediate data points are key drivers in the shape of the curve. For the dosage calculation, the 1/6 mortality data point is associated with a dosage of 4.54 mg min m^{-3}, whereas the 4/6 mortality is associated with 5.96 mg min m^{-3}, a difference of nearly 25%. However, the inhaled dose values associated with those data points have only a difference of about 7%. In other words, compared with dosage, a much smaller increase in dose will result in a significantly higher mortality rate. This result is consistent with laboratory studies, which have shown that in mice and dogs there are narrow ranges between the highest ricin dose at which all animals survive and the lowest dose at which all animals die.[49] Of course, when the data are analyzed in terms of dosage without regard for the varying weight of individual animals, the weight variations among individuals result in a shallower dose–response curve (smaller slope), as in the upper half of Figure 12.17.

How does the calculated LD_{50} lung dose in the rat compare with human estimates? Multiple sources[50–52] indicate an LCt_{50} of 30 mg min m^{-3}. MPPD can be used to provide lung dosimetry for a comparison if some assumptions are made. Unfortunately, none of the references for the 30 mg min m^{-3} have PSD data. Consequently, the same particle size as was used in the rat study, 0.9 μm MMAD, and a σ of 1.6 were assumed. Also assumed was a ventilation rate of 0.015 m^3 min^{-1} and a body mass of 70 kg (Standard Man). MPPD estimates a pulmonary deposition fraction of 0.104. Therefore, the calculated human LD_{50} lung dose is 0.665 μg kg^{-1} or about 2.5 times higher than what was calculated for the rat model using Griffiths *et al.*'s data. Given the uncertainty associated with the PSD of the human data and the fact that the calculations were performed using completely different starting points and different species, the comparison is adequate.

12.4.3 Summary for Ricin Model

Multiple studies have indicated that aerosolized ricin toxicity is of primary concern when the aerosol is deposited in the pulmonary region. Deposition in the extrathoracic or tracheobronchial regions does not appear to result in significant mortality risk. As a result, it is reasonable to postulate that the best parameter to model the probability of mortality from aerosolized ricin is the mass of ricin deposited in the pulmonary region of the lung. One must also take the animal's body mass into account. What remains is a parameter of deposited pulmonary dose measured in mass of ricin deposited in the lung per unit mass of animal, typically μg kg^{-1}. MPPD can be used to provide a dosimetry model to translate experimental dosage and inhaled dose data into lung deposition.

Based on the experimental data from the rat study[46] and modeling from MPPD, an LD_{50} lung dose of 0.265 μg kg^{-1} can be calculated. Also, MPPD can be used to translate the currently used 30 mg min m^{-3} LCt_{50} value for humans that is given in the literature into a lung dose. The translation results in an LD_{50} lung dose of 0.665, a factor of 2.5 higher than from the rat model. Therefore, the use of a lung dose value of 0.264 μg kg^{-1} as a basis for calculating the probability of mortality from aerosolized ricin is recommended. The variability in body mass across a normal human population will result in a dose–response curve with a significantly shallower slope than that from the rat study, which normalized out body mass.

12.5 Parameters for the Inhalation Mechanics of Aerosol Hazards

In this section, several parameters that are relevant to modeling particle size effects for aerosol inhalation hazards are defined. These definitions provide a context for current and future work for both biological and chemical agents.

12.5.1 Parameter Definitions

Generally, the following parameter definitions focus on the agent fill of the dissemination device, whether sophisticated or improvised, and are they aimed at supporting calculations of the disseminated aerosol downwind of the device. In a hazard prediction code, some of the properties may be specified or derived from data in a material properties file and others should be under direct user control *via* the graphical user interface (GUI) of the hazard source model.

- V = volume of agent filling a container (m^3)
  The container may be for storage or transport or it may be the tank of a dissemination device. If the container is only partially filled, then V refers only to the portion of the container that actually contains agent.
- M_W = wet bulk mass (kg)
  M_W refers to the total mass of wet agent loaded into some container, where the agent may be an emulsion, solution, suspension or slurry.

- M_D = dry bulk mass (kg)
  M_D refers to the total mass of dry agent loaded into some container. This parameter is also applicable to the mass of agent remaining when a wet agent dries out after being fully disseminated as an aerosol.
- M_x = active mass (kg)
  M_x refers to the mass of that portion of the agent which is biologically active in producing the toxic or pathogenic effect of the agent. For biological agents, the active mass is generally assumed to be part of the dry bulk mass rather than part of the volatile component of the wet bulk mass.
- N = number of viable infectious organisms in V
  N must be referenced to a specific time, usually just before dissemination of agent. Note that viability may have decreased during storage and transport in the weapon system and viability may decrease further during dissemination and transport in the atmosphere. Furthermore, viability of infectious organisms may depend on the method by which it is assayed, particularly on whether by *in vitro* culture growth or by *in vivo* infectivity in an animal model.
- P = purity (a dimensionless ratio)
  Purity is the mass fraction of agent that is biologically active in producing the toxic or pathogenic effect of the agent.
- C = concentration (number per unit mass)
  Concentration is the number of viable infectious organisms per unit mass of agent.
- ρ_W = wet bulk density (kg m^{-3})
  Wet bulk density, generally near the density of water, is equal to M_W/V.
- ρ_D = dry bulk density (kg m^{-3})
  Dry bulk density is equal to M_D/V. For particulate matter, the dry bulk density is generally less than the density of individual particles. This definition applies only when V is the fill volume for the dry agent.
- P_W = wet bulk purity (dimensionless ratio)
  Wet bulk purity is defined as the ratio M_x/M_W.
- P_D = dry bulk purity (dimensionless ratio)
  Dry bulk purity is defined as the ratio M_x/M_D.
- C_W = wet bulk concentration (no. kg^{-1})
  Wet bulk concentration is defined as the ratio N/M_W.
- C_D = dry bulk concentration (no. kg^{-1})
  Dry bulk concentration is defined as the ratio N/M_D.

12.5.2 Definitions Applied to Dissemination of Dry Preparations

Figure 12.18 illustrates relationships among M_x, M_D and N for dry preparations of biological agents, which may be either infective organisms or toxins. If the agent is a biological toxin, then the active mass should be measured in mass units. If the agent is an infective organism, then the quantity may be further specified according to the number of organisms by dividing the active mass M_x by the mass of a single organism.

12.5.3 Definitions Applied to Dissemination of Wet Preparations

Figure 12.19 illustrates the relationships among M_x, M_W and N for wet preparations of biological agents, which might be either toxins or infective organisms.

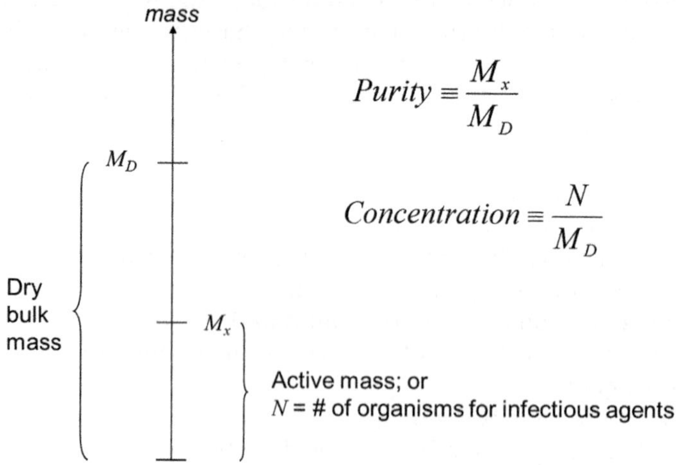

Figure 12.18 Purity and concentration for dry biological agents.

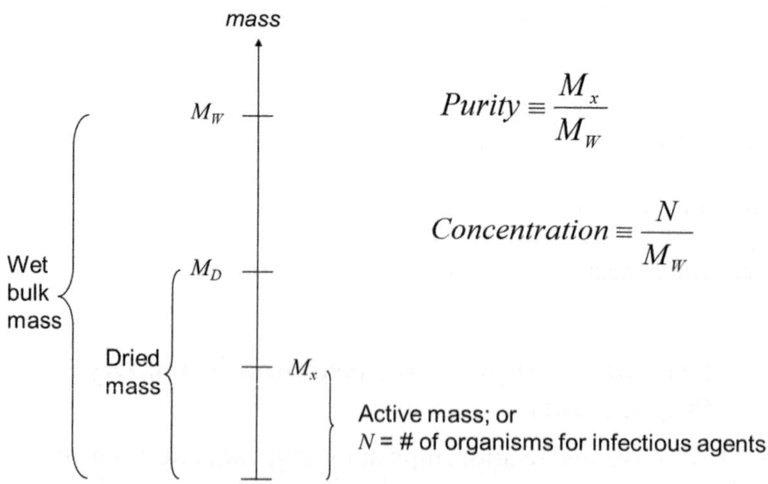

Figure 12.19 Purity and concentration for wet biological agents.

From these definitions, we see that the ratio of the purity of the wet agent to that of the dry agent is the same as the ratio of the concentrations for the wet and dry agent, so that the following relation should be valid:

$$P_W = \left(\frac{C_W}{C_D}\right) P_D \tag{12.11}$$

In practice, it is unusual for all four of these quantities to be reported for the same agent batch, so the relationship is only approximate when the quantities are taken from different reports. Moreover, concentration for wet agents is frequently reported in number of viable organisms per unit volume, so the density of the wet agent must be either known or assumed in order to derive the concentration in terms of unit mass.

12.5.4 Applications to Inhalation Modeling

The calculations of primary concern regarding the inhalation model are inhalability, hygroscopicity and the amount of active agent per particle.

12.5.4.1 *Inhalability*

The inhalation model first calculates inhalability for aerosol hazards. This calculation requires

1. the aerodynamic diameter distribution of the ambient aerosol particles at the location of the head; and
2. the ambient wind speed.

These two requirements are the same for both wet and dry bioagents. The diameter required is the actual physical diameter at the time of inspiration, regardless of the composition of the particle at that time. That is, the diameter must include both carrier material and nonviable agent in addition to viable agent.

12.5.4.2 *Hygroscopicity*

Aerosol particles may either gain or lose water content by condensation or evaporation, respectively, depending on the ambient water vapor concentration (determined by temperature and humidity) and the hygroscopicity of the material making up the particle. Condensation and evaporation are strongly affected by particle size because of the influence of surface tension, which is in turn influenced by curvature of the particle surface. For hygroscopic particles, there is an equilibrium size that depends on ambient humidity. The equilibration time is fast enough that the size of particles entering the respiratory tract will essentially follow the local humidity during inspiration.

Furthermore, the air entering the respiratory tract rapidly reaches body temperature and 100% humidity as it passes through the ET region and into the TB region. Therefore, the relevant size for calculation of deposition in the TB and ET regions, especially, is this new equilibrium diameter accounting for humidity, as opposed to the diameter in the ambient atmosphere. Hence this diameter depends on whether the aerosol particles are hygroscopic.

These processes would ideally be modeled at the same level of fidelity in the respiratory tract as they are modeled during transport in the atmosphere and *vice versa*. Therefore, hazard simulation codes should make available to the inhalation model the required parametric data for these calculations.

12.5.4.3 Active Agent

The inhalation model requires sufficient data from the atmospheric transport model to calculate the amount (mass) of active agent in each inhaled particle, whether the active agent is an infective organism or a toxin of biological origin. If the agent is an infective organism, it is necessary that the inhalation model be able to calculate the expected number of viable organisms corresponding to the mass of active agent in each particle.

In the context of infective agents, we equate "active" with "viable," *i.e.* able to cause infection. The loss of viability during atmospheric transport should be calculated by the atmospheric transport code but the particle size must still include the nonviable toxin or organisms.

For some infectious agents, such as anthrax spores, the infective organisms in each deposited particle tend to disperse on deposition so that each organism contributes equally to the probability of infection. For other agents, such as *F. tularensis*, the probability of infection correlates strongly with the number of agent-containing particles deposited, as if the organisms in a single particle stay together and suffer a common fate as they encounter the immune system.

12.6 Discussion

In this project, Applied Research Associates (ARA) developed the Deposition and Response in the Respiratory Tract (DARRT) model; an improved medical model of particle size effects for inhalation hazards from chemical and biological (CB) agents. This project was initiated to improve modeling of the variation in human response due to differences in inhaled particle sizes, in particular, for coarse particle sizes that may be present in the atmosphere near a dissemination source and may remain suspended long enough in an urban environment to expose large numbers of people. In addition, plume dispersion calculations with the HPAC consequence assessment tool show that, contrary to common wisdom, particles as large as 20–40 μm in diameter can be a significant hazard for tens of kilometers downwind of a release. DARRT predicts the probability of infection or injury from exposure to bioaerosol hazards with aerodynamic diameters between 0.01 and 100 μm and describes the resulting medical impact. Many models have assumed that only 1–5 μm "respirable" particles capable of reaching the pulmonary region of the respiratory system are of concern. However, inclusion of coarse (>5 μm) particle deposition was also found to be important. Coarse particle deposition in the nose, mouth and throat can pose a substantial health risk, particularly for infectious agents.

The DARRT model uses the Multiple-Path Particle Dosimetry (MPPD) model to estimate the deposited (retained) doses of inhaled aerosols by region of the respiratory tract. Several modifications have been made to MPPD to improve its applicability to PSDs of interest to the CB defense community and provide a solid foundation for improving associated human response models. An improved model of inhalability, a measure of the fraction of the ambient concentration of particles that actually enter the head through the nose or mouth during inspiration, was developed and implemented with MPPD. In addition, the particle size range was extended to include particle sizes from 10 nm to 100 μm and an improved clearance model for the extrathoracic region was implemented. The MPPD model was also modified to accept input based on particle size bins to be consistent with data transfer from transport and dispersion models such as the SCIPUFF (Second-Order Closure Integrated PUFF) model in HPAC. The DARRT model requires material property data for the bioaerosol of interest, such as material density and organism concentration. In the current software implementation of the DARRT model for HPAC and the Joint Effects Model (JEM), this information is read from the material file. With the new DARRT model, the user can now also control some inhalation variables such as breathing rate.

Models of human response to biological agents previously implemented in HPAC and JEM all assume that the presented dose (the time-integrated, ambient agent concentration multiplied by the breathing rate) is completely effective in producing disease or injury. Consistent with this assumption, the material property files for dry agents previously specified only a single particle size bin of 4.9–5.1 μm. As noted above, this simplifying assumption was based on the understanding that most effects, particularly for biological agents, are associated with pulmonary deposition of "respirable" particles of the agent. Although some of the source modules in HPAC and JEM, such as CBWPN, allow the user to set a PSD in the form of a log-normal distribution for the released biological agent, the transport and dispersion calculation performed by SCIPUFF had only the single particle size bin provided by the material file at its disposal and so put all the material in that bin, resulting in a monodispersed aerosol of diameter 5 μm. As part of the DARRT implementation, size bins were added to the material files for the dry particulate form of ricin and *F. tularensis* to allow realistic PSDs. As a result, not only does the implementation of DARRT improve the human response modeling of ricin and tularemia, but additionally improves the transport and dispersion modeling of these two agents.

The current implementation of the DARRT model for bioaerosols includes human response models for *F. tularensis* and ricin; data for additional agents have been reviewed and models can be developed in the future. The DARRT human response model predicts the differences in disease manifestation for tularemia depending on the particle deposition location. Nonhuman primate data indicate that the probability of infection for tularemia is a function of the number of agent-containing particles deposited in the extrathoracic (ET) or pulmonary (P) regions. ET deposition results in an upper respiratory infection, whereas P deposition results in a pneumonic infection.

As noted, the probability of tularemia infection is a function of the number of deposited ACPs. During this project, we developed a method for assessing the probability that a particle of a given size is actually an ACP. Particles can consist of carrier material in addition to *F. tularensis* organisms. Larger particles will often contain multiple organisms and the smaller particles can contain no organisms. As the particle size decreases towards that of an organism (~1 μm for *F. tularensis* bacteria), some particles may contain no cells and simply consist of carrier material, and cannot cause an infection. Calculating the probability of a particle of a given size being an ACP is essential when evaluating the potential hazard of threats developed at varying levels of sophistication.

The DARRT model of human response to ricin, a toxin of biological origin, is based on the mass of ricin deposited in the pulmonary region and correlates the human response with pulmonary deposition of the particles. Whereas ricin can be lethal by other routes of exposure such as injection, aerosolized ricin has been shown to be lethal only by pulmonary deposition.

The DARRT model provides a much needed improvement in understanding the physical location of the deposition of bioaerosol particulates in the respiratory tract and correlates the biological effects with the deposition location.

12.7 Recommendations

During this study, several areas of science and technology work that will further improve the ability to model the effects of chemical and biological hazards were identified. Much of the data used in the modeling come from animal experiments. High-quality data for human exposures such as that for the tularemia studies from Operation Whitecoat are the exception rather than the rule. As a result, the data analysis will greatly benefit from further development of animal modeling and techniques for extrapolation of experimental results from animals to humans. Developing models of both respiratory mechanics and also biological response mechanisms for animals will greatly improve our understanding of how the animal effects translate to human effects.

Additional work should be carried out to understand the impact of inhalation hazards on specific sub-populations such as children and the elderly, and also to understand gender differences. This will improve our ability not only to model effects on general populations, a significant concern for current trends in military support for civilian emergency response, but also to model the effects of the changing demographic makeup of modern military forces.

During the development of the DARRT model, data that could be used to model the effect of particle size on inhalation effects of several other agents that may be of interest, including *Coxiella burnetti* (causative agent of Q fever) and *Yersinia pestis* (causative agent of plague), were discovered. The fidelity of existing casualty prediction models will be significantly improved by utilizing these data.

Acknowledgements

This work was sponsored by the US Defense Threat Reduction Agency under contract DTRA1-03-D-0014-0025. The authors wish to acknowledge support and encouragement from Mrs Stephanie Hamilton, Mr Chuck Fromer and Dr Chris Kiley. In addition, the authors wish to acknowledge contributions to this work made by colleagues at Applied Research Associates: Mr Josh Bergman, Dr Darren Oldson and Dr Sharon Watson.

References

1. S. Anjilvel and B. Asgharian, A multiple-path model of particle deposition in the rat lung, *Fundam. Appl. Toxicol.*, 1995, **28**, 41.
2. K. Z. Morgan, W. Binks, A. M. Brues, W. H. Langham, L. D. Marinelli, W. G. Marley, M. K. Nakaidzumi, G. J. Neary, M. N. Pobedinski, E. E. Pochin and C. G. Stewart, *ICRP, Recommendations of the International Commission on Radiological Protection: Report of Committee II on Permissible Dose for Internal Radiation*, 1959, ICRP Publication 2.
3. J. M. Benson, B. M. Tibbets and K. D. Thrall, Uptake, Tissue Distribution, and Fate of Inhaled Carbon Tetrachloride: Comparison of Rat, Mouse, and Hamster, *Inhalation Toxicol.*, 2001, **13**, 207.
4. J. M. Benson, B. M. Tibbetts and E. B. Barr, The Uptake, Distribution, Metabolism, and Excretion of Methyl Teriary-Butyl Ether Inhaled Alone and in Combination with Gasoline Vapor, *J. Toxicol. Environ. Health, Part A*, 2003, **66**, 1029.
5. W. C. Su and Y. S. Cheng, Deposition of Fiber in the Human Nasal Airway, *Aerosol Sci. Technol.*, 2005, **39**, 888.
6. T. D. Sweeney and J. D. Brain, Pulmonary Deposition: Determinants and Measurement Techniques, *Toxicol. Pathol.*, 1991, **19(4)**, 384.
7. P. G. Georgopolous, S. W. Wang and V. M. Vyas, A Source-to-Dose Assessment of Population Exposures to fine PM and ozone in Philadelphia, PA, During a Summer 1999 Episode, *J. Exposure Anal. Environ. Epidemiol.*, 2005, **15**, 439.

8. B. Asgharian, F. J. Miller and R. P. Subramanian, Dosimetry software to predict particle deposition in humans and rats, *CIIT Act.*, 1999, **19**(**3**), 1.
9. B. Asgharian, W. Hofmann and R. Bergmann, Particle deposition in a multiple-path model of the human lung, *Aerosol Sci. Technol.*, 2001, **34**, 332.
10. ICRP, Human Respiratory Tract Model for Radiological Protection. ICRP Publication 66, *Ann. ICRP*, 1994, **24**(**1–3**).
11. A. Birchall, M. R. Bailey and A. C. James, LUDEP: A Lung Dose Evaluation Program, *Radiat. Prot. Dosim.*, 1991, **38**(**1–3**), 167.
12. U.S. EPA, Dosimetry of Particulate Matter, in *Air Quality Criteria for Particulate Matter*, Report No. EPA/600/P-99/002bF, Office of Research and Development, Research Triangle Park, NC, 2004, ch. 6, vol. II.
13. U.S. EPA web site, http://www.epa.gov/radiation/assessment/dcal.html, accessed February, 2015.
14. B. Asgharian, W. Hofmann and F. J. Miller, Toxicology of the Lung, in *Dosimetry of particles in humans: from children to adults*, ed. D.E. Gardner, CRC Press, Boca Raton, Florida, 4th edn, 2006, pp. 151–194.
15. B. Asgharian, O. T. Price and W. Hofmann, Prediction of Particle Deposition in the Human Lung Using Realistic Models of Lung Ventilation, *J. Aerosol Sci.*, 2006, **7**, 1209.
16. B. Asgharian and O. T. Price, Airflow Distribution in the Human Lung and its Influence on Particle Deposition, *Inhalation Toxicol.*, 2006, **8**, 795.
17. E. R. Weibel, *Morphometry of the Human Lung*, Springer Verlag, Berlin, 1963.
18. O. G. Raabe, H. C. Yeh, G. M. Schum and R. F. Phalen, *Tracheobronchial geometry: human, dog, rat, hamster*, Lovelace Foundation, Albuquerque, New Mexico, 1976, Report no. LF-53.
19. H. C. Yeh and G. M. Schum, Models of human lung airways and their application to inhaled particle deposition, *Bull. Math. Biol.*, 1980, **42**, 461.
20. L. Koblinger and W. Hofmann, Analysis of human lung morphometric data for stochastic aerosol deposition calculations, *Phys. Med. Biol.*, 1985, **30**, 541–556.
21. P. W. Scherer, L. H. Shendalman and N. M. Greene, Simultaneous diffusion and convection in single breath lung washout, *Bull. Math. Biophys.*, 1972, **34**, 393.
22. C. P. Yu, Exact analysis of aerosol deposition during steady state breathing, *Powder Technol.*, 1978, **21**, 55.
23. R. P. Subramaniam, B. Asgharian, J. I. Freijer, F. J. Miller and S. Anjilvel, Analysis of lobar differences in particle deposition in the human lung, *Inhalation Toxicol.*, 2003, **15**, 1.
24. B. Asgharian, M. G. Ménache and F. J. Miller, Modeling Age-Related Particle Deposition in Humans, *J. Aerosol Med.*, 2004, **17**, 213.
25. F. R. Cassee, J. I. Freijer, R. Subramaniam, B. Asgharian, F. J. Miller, L. Van Bree and P. J. A. Rombout, *Development of a model for human and rat airway particle deposition: implications for risk assessment*, in National Institute of Public Health and the Environment (RIVM), Bilthoven, The Netherlands, 1999, Report no. 650010018.

26. R. de Winter-Sorkina and F. R. Cassee, *From concentration to dose: factors influencing airborne particulate matter deposition in humans and rats*, in National Institute of Public Health and the Environment (RIVM), Bilthoven, The Netherlands, 2002, Report no. 650010031.
27. Applied Research Associates, Inc., MPPD web site, http://www.ara.com/products/mppd.htm, accessed March 2015.
28. K. K. Millage, J. Bergman and B. Asgharian, A review of inhalability fraction models: discussion and recommendations, *Inhalation Toxicol.*, 2010, **22**(**2**), 159.
29. V. Niinimaa, P. Cole, S. Mintz and R. J. Shephard, Oronasal Distribution of Respiratory Airflow, *Respir. Physiol.*, 1981, **43**, 69.
30. M. B. Snipes, Internal Radiation Dosimetry, Health Physics Society Summer School, in *Kinetics of inhaled radionuclides*, ed. O. G. Raabe, Medical Physics Publishing, Madison WI, 1994.
31. O. R. Moss and M. J. Oldham, Dosimetry counts: Molecular hypersensitivity may not drive pulmonary hyperresponsiveness, *J. Aerosol Med.*, 2006, **19**(**4**), 555.
32. O. R. Moss, Toxicology of the Nose and Upper Airways, in *Particle Dosimetry in the Nose and Upper Airways of Humans*, ed. J.B. Morris and J. Dennis, Informa Healthcare, New York, NY, Target Organ Toxicology Series 30, 2010, ch. 8, p. 480.
33. W. C. Day and R. F. Berendt, Experimental Tularemia in *Macaca mulatta*: Relationship of Aerosol Particle Size to the Infectivity of Airborne Pasteurella tularensis, *Infect. Immun.*, 1972, **5**(**1**), 77.
34. H. A. Druett, D. W. Henderson, I. Packman and S. Peacock, Studies on Respiratory Infection: The Influence of Particulate Size on Respiratory Infection with Anthrax Spores, *J. Hyg.*, 1953, **31**, 359.
35. S. Saslaw, H. T. Eigelsbach, J. A. Prior, H. E. Wilson and S. Carhart, Tularemia Vaccine Study: II – Respiratory Challenge, *Arch. Intern. Med.*, 1961, **107**, 134.
36. D. T. Dennis, T. V. Inglesby, D. A. Henderson, J. G. Bartlett, M. S. Ascher, E. Eitzen, A. D. Fine, A. M. Friedlander, J. Hauer, M. Layton, S. R. Lillibridge, J. E. McDade, M. T. Osterholm, T. O'Toole, G. Parker, T. M. Perl, P. K. Russell and K. Tonat, AMA Consensus Statement, Tularemia as a Biological Weapon, Medical and Public Health Management, *JAMA, J. Am. Med. Assoc.*, 2001, **285**(**21**), 2763.
37. M. J. Hepburn, A. M. Friedlander and Z. F. Dembek, Medical Aspects of Biological Warfare, in *Tularemia*, ed. Z. F. Dembek, Borden Institute, Walter Reed Army Medical Center, Washington, DC, 2007, ch. 8, pp. 167–184.
38. R. B. Hornick and H. T. Eigelsbach, Aerogenic Immunization of Man with Live Tularemia Vaccine, *Bacteriol. Rev.*, 1966, **30**, 532.
39. W. D. Sawyer, H. G. Dangerfield and A. L. Hogge, Antibiotic Prophylaxis and Therapy of Airborne Tularemia, *Bacteriol. Rev.*, 1966, **30**, 542.
40. R. S. Pekarek, K. A. Bostian and P. J. Bartelloni, The Effects of *Francisella tularensis* Infection on Iron Metabolism in Man, *Am. J. Med. Sci.*, 1969, **258**(**1**), 14.

41. P. R. Pittman, S. L. Norris, K. M. Coonan and K. T. McKee, An assessment of health status among medical research volunteers who served in the Project Whitecoat program at Fort Detrick, Maryland, *Mil. Med.*, 2005, **170**(**3**), 183.
42. G. H. Anno, S. K. Sanemitsu, G. E. McClellan, M. A. Dore and A. P. Deverill, *Consequence Analytic Tools for NBC Operations: Volume 1 – Biological Agent Effects and Degraded Personnel Performance for Tularemia, Staphylococcal Enterotoxin B (SEB) and Q-Fever*, Defense Special Weapons Agency, Alexandria, VA, 1998, DSWA-TR-97-61-V1.
43. R. M. Jones, M. Nicas, A. Hubbard and M. D. Sylvester, The Infectious Dose of *Francisella tularensis* (Tularemia), *Appl. Biosaf.*, 2011, **10**(**4**), 227.
44. A. Hislop, S. Howard and D. V. Fairweather, Morphometric studies on the structural development of the lung in *Macaca fascicularis* during fetal and postnatal life, *J. Anat.*, 1984, **1**, 95–112.
45. C. J. Roy, M. Hale, J. M. Hartings and L. Pitt, Impact of Inhalation Exposure Modality and Particle Size on the Respiratory Deposition of Ricin in BALB/c Mice, *Inhalation Toxicol.*, 2002, **15**, 619.
46. G. D. Griffiths, C. D. Lindsay, P. Rice and D. G. Upshall, The Toxicology of Ricin and Abrin Toxins – Studies on Immunisation Against Abrin Toxicity, *Med. Def. Biosci. Rev., Proc.*, 1993, 1407.
47. E. K. Leffel, J. M. Hartings, M. L. M. Pitt and E. Stevens, Comparison of Deposition Patterns for Small and Large Particle Aerosolized Toxins and Resulting Disease in Guinea Pigs and African Green Monkeys, in *Def. Eff. Chem. Hazards: Toxicol. Diagn. Med. Countermeas*, 2007, Meeting Proceedings RTO-MP-HFM-149, 11–1.
48. J. L. Mauderly, J. E. Tesarek and L. J. Sifford, Respiratory measurements of unsedated small laboratory mammals using nonrebreathing valves, *Lab. Anim. Sci.*, 1979, **29**, 323.
49. O. Fodstad, J. V. Johannessen, L. Schjerven and A. Pihl, Toxicity of abrin and ricin in mice and dogs, *J. Toxicol. Environ. Health*, 1979, **5**, 1073.
50. S. Franke, *Textbook of Military Chemistry*, Military Publisher of the German Democratic Republic, Berlin, 1977, vol. 1, p. 711.
51. C. E. Whalley, *Toxins of Biological Origin*, Army Armament Munitions Chemical Command, Aberdeen Proving Ground, MD, 1990, CRDEC-SP-021, pp. 38–40.
52. J. P. Robinson and M. Leitenberg, *The Problem of Chemical and Biological Warfare Volume 1: The Rise of CB Weapons*, Almqvist & Wiksell, Stockholm, 1971.

CHAPTER 13

Aerosol Exposure to Pathogenic Bacteria and Virus Particles: Standard Operating Procedure

DOUGLAS S. REED[a], DRITAN XHILLARI[b], ARLENE L. WEISS[c], AND RUDOLPH J. JAEGER*[b,c]

[a]Center for Vaccine Research, University of Pittsburgh, Pittsburgh, PA 15261, USA; [b]CH Technologies (USA) Inc., Westwood, NJ 07675, USA; [c]Environmental Medicine Inc., Westwood, NJ 07675, USA
*E-mail: jaegerr@chtechusa.com

13.1 Introduction

Bioaerosol research or, more correctly, Aerobiology, is described in Wikipedia as "**Aerobiology** (from Greek ἀήρ, *aēr*, "air"; βίος, *bios*, "life"; and -λογία, *-logia*) is a branch of biology that studies organic particles, such as bacteria, fungal spores, very small insects, pollen grains and viruses ..." An early text covering this area was written by Dimmick and Akers.[1]

Bioaerosols consist of airborne particles of various sizes. Viruses are the smallest infective particles, about 0.2–0.3 μm in size, whereas bacterial and fungal spores cover a range from ~0.3 to ~100 μm. Pollen, algae, protozoa and dander can range in size up to hundreds of micrometers in diameter.[2] Bioaerosols have specific characteristics that influence their aerial properties and their potential to be collected and sampled. They may be (a) single spores,

Issues in Toxicology No. 25
Aerobiology: The Toxicology of Airborne Pathogens and Toxins
Edited by Harry Salem and Sidney Katz

Published by the Royal Society of Chemistry, www.rsc.org

pollen grains, bacterial cells or viruses, (b) aggregates of several spores or cells or other biological material, (c) products or fragments of spores and cells or (d) biological material carried by other nonbiological particles. When the bioaerosols exist as aggregates, their migration and deposition depend on the overall size of the ensemble through the exposure system.[2]

Some aerobiology research cited above has been associated with the potential human exposure to naturally occurring sources of pollen and the consequential allergic reactions that are due to these airborne antigens. However, more recently, the role of airborne dissemination of bacteria and viruses has become a concern for homeland security and biodefense.[3,4]

Our goal in this chapter is to discuss and outline methods that can be used to produce standardized infections and, in this way, to provide the reader with an example of a Standard Operating Procedure (SOP) as a basis for those investigators who desire to perform animal studies with infectious agents. Examples from the literature using similar equipment to that described here are provided.

Animal studies with infectious agents are typically carried out with either a nose-only inhalation exposure system (NOIES), a whole-body exposure system (WBIES) or a head-only inhalation exposure (typically employed for non-human primates) system (HOIES).

Figure 13.1 shows examples of the latter two chamber types. The choice of chamber used is typically determined by the species being exposed and by concerns regarding contamination of fur (WBIES) or stress induced as a result of animal restraint in the tubes (NOIES).[5]

For the purposes of this chapter, we will focus on the NOIES; however, the general principles for aerosol exposure and infectious agent generation are the same between the three types of exposure chambers. The differences lie in animal handling, animal status during exposure (*i.e.* anesthesia) and air volumes and flow that may be used under such conditions. Discussions of these principles can be found in the literature.[6–9]

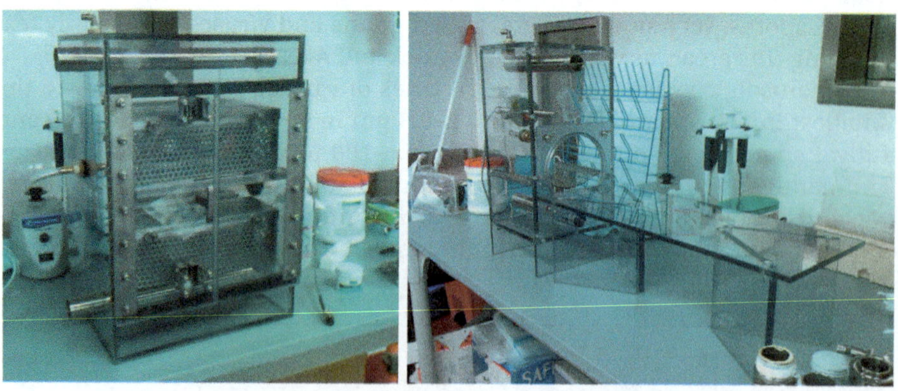

Figure 13.1 Typical whole-body (left) and head-only (right) exposure chambers used in the exposure of animals to pathogenic agents.

13.2 Exposure System Considerations

Prior to conducting exposures to bioaerosols, the Institutional Biosafety Officer or the person with that responsibilty should be involved in discussing how and where aerosol exposures will be conducted. They must participate in the development of institution-specific protocols for conducting such aerosol exposures and their subsequent decontamination. The entire exposure apparatus should be placed inside a suitable biosafety cabinet, one specifically constructed or sized to contain the unit to be employed. A class III biological safety cabinet (BSC) or glove-box, such as that shown in Figure 13.2, is preferred.

A hard-ducted, non-recirculating (to the room) class II BSC can also be used (Figure 13.3). This is typically employed for less virulent or more readily treatable microorganisms, such as nonpathogenic strains of *Mycobacterium tuberculosis*.

For maximum safety, the cabinet should be located within a properly designed and operated biosafety facility at the appropriate level of containment for the pathogen being aerosolized. Personnel should be wearing appropriate personal protective equipment. Access to the room in which aerosol exposures are conducted should be restricted while infectious agents are being used.

Almost without exception, the air balance between adjoining areas and the place where exposures are conducted should be from higher pressure to lower presure so that air flow is directed to the exposure area and then out.

Within the exposure chamber itself, ideally, the air balance during exposures in an exposure system consists of equal volumes of air going into and

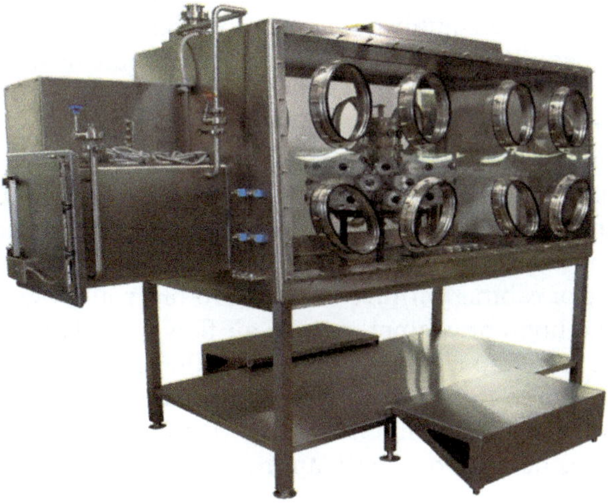

Figure 13.2 Class III biological safety cabinet (a.k.a. glove-box) containing a nose-only exposure system. Not shown are the HEPA-filtered supply and exhaust, decontamination ports, gloves and common accessories such as a dunk tank or beta port for removal of samples.

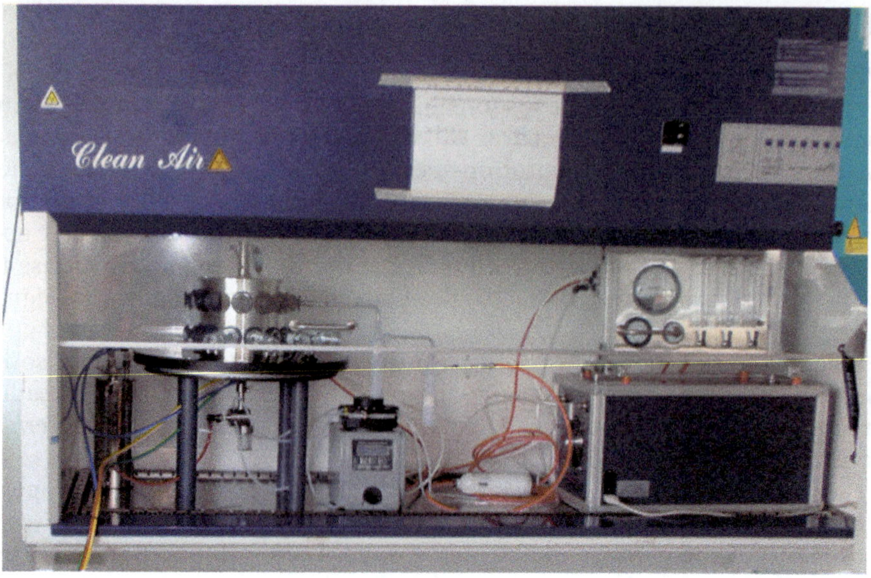

Figure 13.3 Nose-only system in a Class II BSC.

out of the system simultaneously (*e.g.* if a volume of 8 L min^{-1} (lpm) is going into the system, then 8 lpm should be exhausted). Typically, the volume used is from half to one times the total volume of the chamber (per minute), so that there is one complete air change in the chamber every 2 min or less. Air flow may also be set by the breathing rate of the animals being exposed times the number of animals in the system and considerations of time needed to reach 95 or 99% or the equilibrium concentration. Generally, for well-mixed chambers, this equilibrium will be reached more quickly at higher air flow rates.

The total air input includes both the air from the aerosol generator and any wet or dry dilution air. Total exhaust includes that taken from the air stream by aerosol sampling devices. Such exposure systems using minimal or non-pathogenic organisms may be operated in a suitable ventilated cabinet or laboratory hood at atmospheric pressure or they may be slightly positive relative to the hood or cabinet within the room. However, air flow should be from the room to the hood or cabinet *and* the air flow in and around the cabinet face should be tested to ensure that no air spillage occurs. Some principles for testing hood and safety cabinet function can be found elsewhere.[10–12]

Exposure systems can be run under positive pressure; however, it is recommended that this be done only with appropriate safety precautions (preferably in a class III BSC).

When conducting positive-pressure aerosol exposures, the restraint tubes may be either open or closed. If open tubes are used, these will allow leakage past the head and body of the animal. In such cases, and if done in class I containments systems, careful control of hood face velocity and associated air flow is critical to prevent the release or to allow the re-entry of aerosol into the room. Therefore, only nontoxic or nonpathogenic aerosol or gas exposure should be done in this way.

Alternatively, and much preferred for use (if not obligatory) with pathogenic agents, the exposure system is sealed and operated at negative pressure, approximately −0.5 to −1.0 inches of water column (inH_2O) compared with the surrounding glove-box or the room. The typical input air flow rate is similar to that in the positive-pressure system but the chamber exhaust flow rate is set slightly larger than the total air input flow rate. This flow difference results in the desired slight negative pressure inside the exposure chamber. Then, makeup air, necessary to balance the input with output flow, is provided by a calibrated leak (a high-resistance opening), equipped with an appropriate filter, at the outlet of the chamber. The use of a calibrated leak was originally advocated by Henderson in the 1950s at Porton Down in the United Kingdom. Its use continues to this day in the Henderson Apparatus.[13]

A 24-port exposure tower (shown in Figure 3.3) is fed aerosol through the use of a vertical bioaerosol nebulizing generator (BANG) (single-jet version), as shown in Figure 13.4 (left). Similarly, a vertical outlet collison nebulizer (either one-, three- or six-jet) is displayed in Figure 13.4 (right). The latter allows

Figure 13.4 Aerosol generators typically used in exposures. On the left is the bioaerosol nebulizing generator (BANG) connected to an exposure tower and on the right is a three-jet collison nebulizer connected to a Jaeger–NYU exposure tower.

for 2, 6 or 12 lpm of multi-pass aerosol at 1–1.5 bar pressure. The operation of the collison nebulizer, in one-, three- and six-jet versions, with H1N1 virus was described by Ibrahim *et al.*[14]

Not shown, but available for use in cases where shear force is damaging to the microorganism, the sparging liquid aerosol generator (SLAG) may be employed to produce aerosls with shear forces or impaction. Operation of this device was first described by Mainelis *et al.*[15] and subsequently by Simon *et al.*[16]

An alternative and newly developed generation system that allows selective single- *versus* multi-pass operation is the Blaustein atomizing (BLAM) generator (Figure 13.5). The design allows for thin liquid films to be sprayed by a jet or air created by a ruby nozzle. The distance between the exit from the generator can be varied and the device used in the horizontal (single-pass atomizer or SPA mode) or the vertical wet impaction that is illustrated here. Operational characteristics and a comparison with other bioaerosl sprayers devices of the SPA version of the BLAM were reported by Zhen *et al.*[17]

The operating principle is based on a thin liquid film that is sheared one time only by one, four or eight radially arrayed air jets. The air flow is generated by a positive-pressure critical flow orifice centered above the one, four or eight expansion nozzles. The liquid layer is between two confining disks and is fed from above (single-pass mode) or below (multi-pass mode). The device may be operated in vertical mode with soft impaction into a liquid layer or

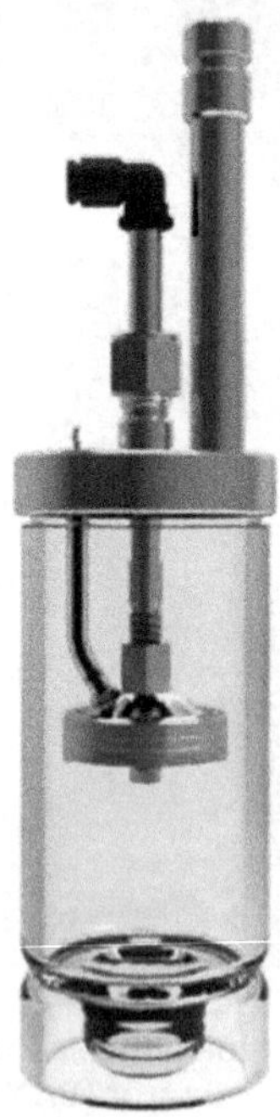

Figure 13.5 BLAM generator shown in single-pass mode.

horizontal as an atomizer without secondary droplet formation on a surface (as occurs in the collison nebulizer). The particle output is less in such a case but viability is greater in the absence of impaction.

Each of these generation devices has applicability in this setting. The objective is to develop a respirable aerosol with droplets in the 1–3 μm range. For inhalation systems of the type shown here (vertical upwards), the aerosol generator unit that is chosen should have a vertical discharge that attaches directly to the bottom of the exposure system. Figures 13.4 and 13.5 show this type of connection. This configuration allows any condensation or impactive desposit that occurs within the central plenum to drain into the generator reservoir by gravity flow. In contrast, top–down systems may result in infective liquids pooling at the bottom of the inhalation tower. Such an event complicates decontamination.

Depending upon the pathogen, the animal being exposed and the desired presented infectious aerosol dose, one property of the chosen aerosol generator may be how the generation process impacts the viability of the microorganisms in the test aerosol being generated. Depending on their design, nebulizers or atomizers can operate in one-pass mode (the liquid test article goes only once through the jet atomization process) or multi-pass mode (the liquid is recirculated several times in the jets). One-pass devices may preserve better the viability of microorganisms in the aerosol. In a multi-pass aerosol generator, the pathogen suspension within the dose jar or reservoir is repeatedly drawn by aspiration through the aerosol nozzle. In a nebulizer, the primary aerosol impacts the generator wall to yield secondary impactive particles. The resulting primary and secondary droplets pass through an elbow (horizontal collison) or a droplet trap (vertical collison). This process results in an aerosol that is stripped of large particles. The droplet filter/impactor or elbow that is built into the outlet refines the output and makes it smaller.

By contrast, in a single-pass generator, the liquid test article is fed into the aerosol nozzle by an external liquid metering device, *e.g.* a syringe or peristaltic pump, giving the operator total control on the amount of liquid that enters the aerosolization process. In the case of an atomizer where no secondary impaction occurs, the droplets are further refined by the 90° elbow or by the droplet trap. In any case, the organism is subjected to a single passage through the jet. Previous work showed enhanced viability from such processes.[15,17]

In Figure 13.6, a centrifugal tangential aerosol generator (CenTAG) is illustrated. Based on previous work with spinning top pneumatic generators, this motor-driven design is more stable and allows for the generation of large-particle monodisperse aerosols with deposition characteristics such that droplets are desposited into the upper airways. Thomas and co-workers[18–20] have shown that such particles produce different patterns of infectious disease compared with the same organism when presented by alveolar deposition.

The operation and application of the device in exposures of nonhuman primates was described by Bohannon *et al.*[21]

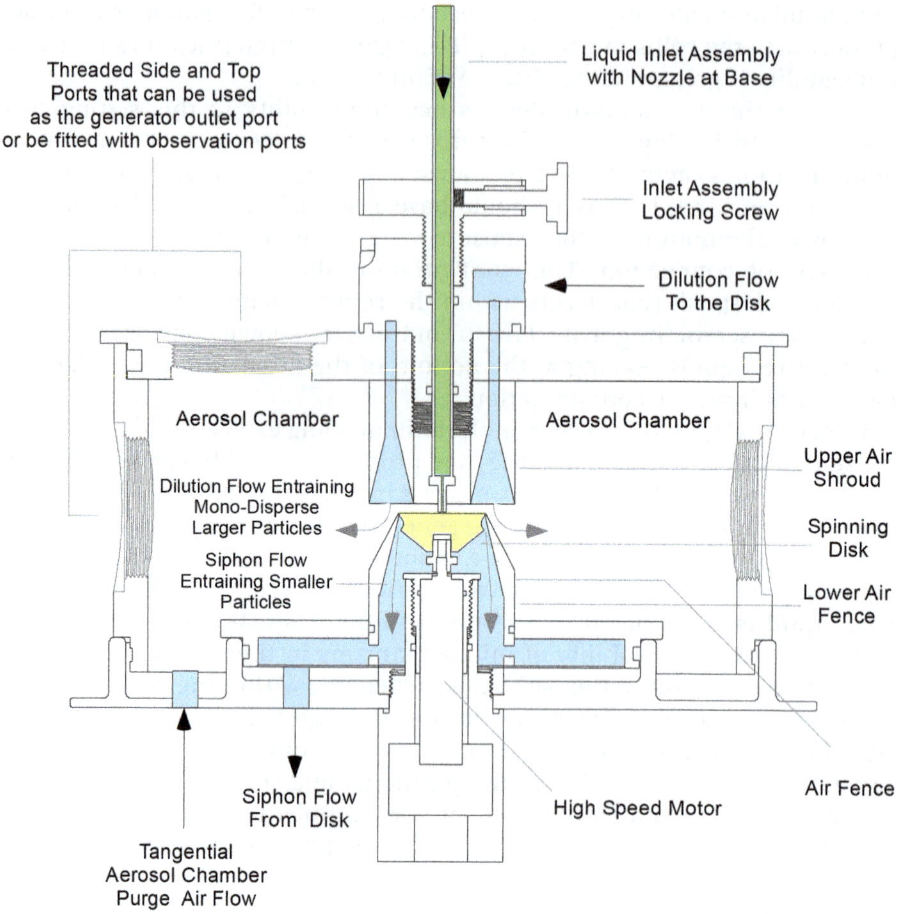

Figure 13.6 Schematic of CenTAG.

13.3 Exposure

The generated aerosol leaving the nebulizer or atomizer meets a secondary air stream (either wet, dry or blended of mixed humidity air). This dilution step allows for further drying of the aerosol. Depending on the flow rates of one *versus* the other and the mixing vessel, a reduction in aerosol concentration occurs such that the aerosol is ready to be inhaled by the test animal. In the case of the Jaeger–NYU system, the mixed aerosol then enters the system exposure unit at its base. After entering the inner plenum and being delivered outwards, each animal inhales the delivered aerosol *via* directed flow from the central plenum. Exhaled gases and the noninhaled aerosol exit the individual nose cones *via* the outer exhaust plenum shown in Figure 13.7. The exhaust air is conducted upward and out through a HEPA filter that is connected to the vacuum line. In this way, the aerosol is confined and more easily decontaminated by filtration.

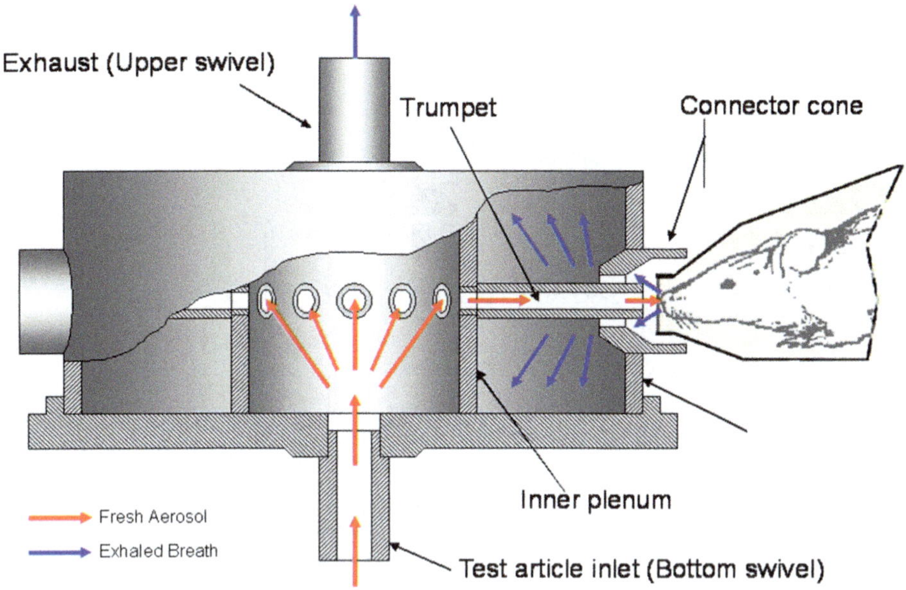

Figure 13.7 Schematic of the directed-flow nose-only exposure chamber.

It is thought that such negative-pressure systems with sealed tubes are safer than positive-pressure systems. This can be argued based on the systems used, but in the sealed system, the reduction of pressure, even with the calibrated leak, allows greater confinement of the aerosol stream.

To determine aerosol concentration, an in-line sampler or impinger may be attached to the outflow line (prior to the HEPA), allowing collection of an aliquot of the pathogen for subsequent analysis. Alternatively, the collection of the pathogen can be carried out in one of the exposure ports (Figure 13.8).

Sampling at an exposure port should provide a better estimate of the concentration of live microorganisms that are presented to the animal's breathing zone. Typically, liquid-filled impingers are used for aerosol sampling. The optimal flow rate of the impinger should be provided by the vendor. The investigator should select an impinger flow rate compatible with the desired air flow at each animal nose port. For example, an AGI-30 typically operates at 6 lpm, so if used at the outlet of the chamber, the total exhaust flow needs to exceed 6 lpm slightly so as to maintain the system at negative pressure. Typically, 10 mL of liquid collection medium are used in the impinger. Depending on the collection medium used, antifoaming agents may be needed (*e.g.* Antifoam A, glycerol). The optimal concentration should be determined empirically. Fetal calf serum and albumin should be minimized in the collection medium as they cause considerable foaming.

Prior to conducting aerosol exposures with animals, it is recommended that aerosol exposures be conducted without animals but with the pathogen of interest to determine the behavior of the pathogen in the aerosol and

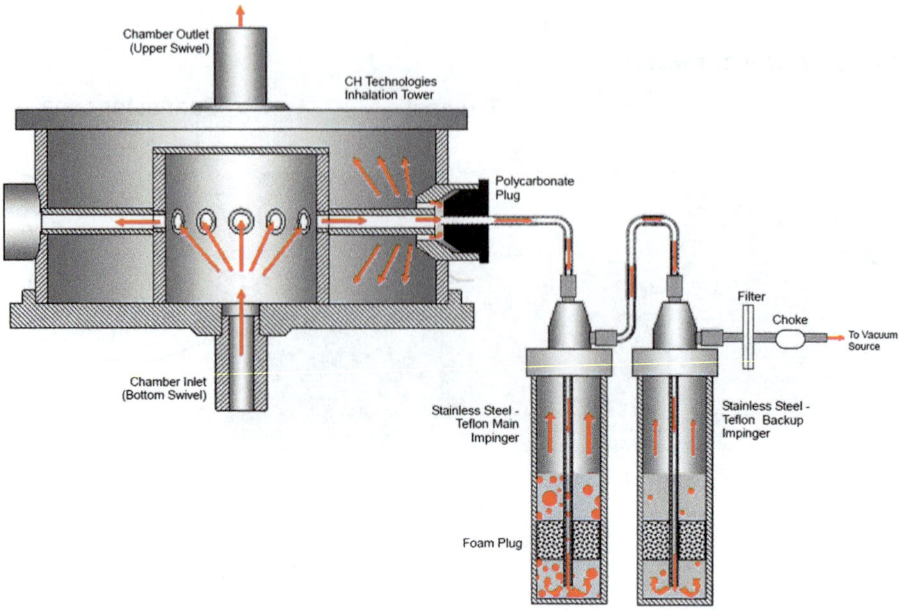

Figure 13.8 Schematic of pathogen sampling from an exposure port.

evaluate system performance (negative control). This will allow the determination of an appropriate concentration to achieve the desired presented (or inhaled) dose in animal studies.

13.4 Outline of System Start-up

- Confirm that the flow meters, aerosol generators and aerosol sampling devices to be used have been calibrated with suitable instruments. Aerosol generators and sampling devices should be sterilized (typically by autoclaving).
- In negative-pressure systems, the first task is to turn on the vacuum pump and initiate vacuum-controlled outflow. This causes the pressure inside the exposure chamber to become negative. Remove and reinsert port plugs to demonstrate establishment of negative pressure. Use a pressure limiter or vacuum regulator so that vacuum is *not* excessive.
- Verify and record pressure readings within the exposure chamber relative to the pressure within the hood or glove-box. This can be done by opening one port in the system and observing the pressure gauge. Recheck all stoppers, plugs, tubing connections, *etc.*, and confirm that there are no obvious leaks. Test various locations by moving the connections and observing the pressure gauge.
- Operate the various devices in the absence of animals or infectious agents. Confirm all flow rates and the pressure generated within the chamber.

- Turn on the air compressor or in-room air supply and activate air input to the chamber. Note that the chamber becomes less negative as air is added to the system. If a manual control unit is being used, balance the chamber pressure by adjusting accordingly air input and chamber exhaust. The recommended negative pressure level is −0.5 to −1.0 inH_2O but slightly higher negative pressures such as −0.5 inHg can also be used. In the case of a 24-port mouse system, the dilution air (a.k.a. breathing air) is set to 4.5 lpm [90% of outflow rate if a 5 lpm flow restrictor (choke) is used]. Under appropriate safety conditions, the system may be run under positive pressure relative to the cabinet.
- Record the chamber pressure relative to the hood. You may wish to choose between wet and dry air sources for diluting air to adjust the relative humidity in the chamber. Wet and dry thermometers or a temperature/humidity probe can be used to monitor environmental parameters in the exposure chamber. Note that the aerosol generator used and the flow rate of air through the chamber may also affect relative humidity levels. Humidity levels and temperature in the external environment may also impact these parameters.
- In the animal quarters or in a holding room adjacent to the chamber area, insert each animal into a restraint tube. Personnel loading animals into the tubes should be trained prior to aerosol use to insure that animals are properly loaded. Move the animals from the holding room to the exposure facility, preferably inside a sealed container but which provides sufficient air for the animals for the duration of transport. In some facilities, a portable class III BSC (a.k.a. mobile transport container) is used, which connects to the class III BSC where aerosol exposures will be conducted.
- The animals are moved into the exposure cabinet typically *via* an air lock, pass-through or rapid transfer port. Carefully insert the test animals into ports, one after another, removing plugs or stoppers and then setting them aside as each animal is inserted. The stoppers should be easily accessible for later use. Unused ports should be plugged.
- Confirm that connections are good and that flow meters are operating properly. Confirm and record the flow rate. Check all animal positions to be certain that none have turned around or are retracted within the tube.

13.5 System Operation with Pathogens

- Prepare a pathogen suspension in advance (typically 10 mL per aerosol exposure or "run") and place in the nebulizer cup or jar. Alternatively, a pump-fed delivery system may be used depending on the type of aerosol generator being used (one-pass *versus* multi-pass). The particle density of the suspension may vary and should be chosen based on the desired infectivity, *e.g.* desired number of colony-forming units (cfu) mL^{-1}.

- Confirm or adjust the flow of dilution air as required (or shut off if not used) and set the aerosol flow to the desired flow rate needed for the aerosol generator being used. For example, with a three-jet collison nebulizer attached to a two-tier/ring 32-port chamber, the aerosol flow rate is 7.5 lpm with dilution air at 0.5 lpm for a total of 8 lpm. This results in a port flow of 250 mL min^{-1} per animal.
- Start aerosol generation by turning on the air flow to the aerosol generator. If you are operating a negative-pressure system with manual controls, quickly increase the system exhaust until the chamber negative-pressure balance is restored. If the system is operated automatically, make sure that the system responds quickly to the added flow of the aerosol generator.
- Aerosol exposures are typically between 10 and 30 min in duration. The exact duration will depend on the concentration of the organism, the behavior of the pathogen in the aerosol (as determined in earlier test aerosol runs without animals, see the previous section), the respiratory minute volume of the animal and the desired inhaled dose.[1]
- When the exposure time is reached, end exposure by shutting down the nebulizer air flow. Adjust the chamber exhaust flow to compensate so that the total air into the system is the same as during exposure and the chamber pressure remains negative (sealed systems).
- Allow a period, typically 5–10 min, of clean air operation to remove the residual aerosol from the system. This is typically called the air "wash."
- Remove the impingers from the system, reseal the ports and pour the contents into a 15 mL conical tube. Detach and pour the contents of the aerosol generator into a 15 mL conical tube. Put both the impinger and generator collection tubes into Ziploc bags and store on ice if not removed immediately from the cabinet.
- Remove the infected animals, one at a time, from the exposure space. Prepare to reseal the ports in succession. As soon an animal tube has been removed, plug each exposure port in quick succession to avoid any potential exposure of the infected surfaces of the chamber to cabinet air. Return the animals to their holding room using the transport container and approved methods used to bring the animals to the exposure space.
- Remove each animal from its restraint tube carefully (typically within a class II biological safety cabinet located in the animal holding room).
- If needed, the tubes can be reloaded and additional animals exposed using the same procedures asoutlined above. The generator is loaded with fresh medium containing the pathogen at the desired concentration and new impingers containing fresh collection medium are attached to the system. If different presented doses are desired, it is recommended to perform the lower dose first and to increase the dose with each "run."
- Once the study is completed, put 10 mL of 70% ethanol or 10% sodium hypochlorite (bleach) into the aerosol generator and run the system for

5–10 min. The purpose of this step is to clean, disinfect and sterilize the internal components (*e.g.* tubing, chamber) of the exposure system.

- If required, confirm the effectiveness of the decontamination steps.
- The animal restraint tubes are placed in a container with a lid and disinfected in a suitable solution. After a set time, *e.g.* 10 min or more, the restraint tubes are washed in a suitable solution such as hot water and detergent. The aerosol generator is disinfected externally in a suitable solution. Depending on the aerosol generator, it can be autoclaved and then cleaned.
- Exposure chamber tiers are sprayed with 70% ethanol or a suitable disinfectant as determined by the biosafety officer. The entire chamber is disassembled. If sodium hypochlorite is used, the layers should be rinsed with distilled water (or 70% ethanol) and the system air dried.
- Note that aluminum-plated or anodized parts may corrode when left wet or when kept in contact with alkaline solutions. Physical damage of aluminum parts can lead to leaks and/or corrosion. Therefore, the integrity of system elements must be confirmed before each experiment.
- As noted above, the exposure unit should be rinsed with clear water followed by a distilled water rinse and then allowed to air dry. Note that a separate SOP is used for steps involving an autoclave.
- Depending on the pathogen used, decontamination of the cabinet and exposure system using formaldehye, vaporous hydrogen peroxide or other decontaminating agents may be necessary. This should be discussed with the biosafety officer. Prior to initiation of aerosol exposures, the method of decontamination should be validated. It is recommended that during validation and for all aerosol exposures, biological indicators be used. The most commonly used are spore strips containing *Geobacillus stearothermophilus.*
- *In lieu* of the inhaled/presented dose, the deposited pathogen dose can be determined by sacrificing one or more animals immediately after exposure; 2 h is a standard sacrifice time. Viruses or bacteria may begin to replicate so it should be as soon as practicable after the exposure. Remove the lungs and homogenize and plate on appropriate agar medium.

13.6 Conclusion

The objective of this chapter is to inform the reader of the origin and application of methods suitable for use with pathogens or highly toxic agents that are delivered as inhalable aerosols. The information should be considered general in nature but specific in intent, namely to allow the user the ability to generate more viable bioaerosols for delivery to the breathing zone of the test subject. Case-by-case determinations of particle size, particle number and infectivity requirements may alter the methods that are applied in individual cases.

References

1. R. L. Dimmick and A. B. Akers, *An Introduction to Experimental Biology*, John Wiley & Sons Ltd, Chichester, 1969.
2. T. Reponen, K. Willeke and S. Grinshpun, Biological Particle Sampling, in *Aerosol Measurement. Principles, Techniques, and Applications*, ed. P. A. Baron and K. Willeke, Wiley-Interscience, 2nd edn, 2001, pp. 751–777.
3. C. J. Roy, D. S. Reed and J. A. Hutt, Aerobiology and Inhalation Exposure to Biological Select Agents and Toxins, *Vet. Pathol.*, 2010, **47**, 779–789.
4. C. J. Roy and M. L. M. Pitt, *Biodefense Research Methodology and Animal Models*, CRC Press, 2012.
5. D. R. Reed, L. M. Bethel, D. S. Powell, A. L. Caroline and A. L. Hartman, Differences in aerosolization of Rift Valley fever virus resulting from choice of inhalation exposure chamber: impliacations for animal challenge studies, *Pathog. Dis.*, 2015, **2014**, 227–233.
6. H. Salem, Principles of Inhalation Toxicology, in *Inhalation Toxicology Research Methods Applications and Evaluation*, ed H. Salem, Marcel Dekker Inc., New York NY 1987, pp. 7–9.
7. R. F. Phalen, *Inhalation studies: foundations and techniques*, CRC Press, Inc., Boca Raton, 1984.
8. R. F. Phalen, *Inhalation Studies*, Informa Healthcare, USA Inc., New York, 2nd edn, 2009.
9. R. O. Mcclellan and R. F. Henderson, *Concepts in inhalation toxicology*, Taylor & Francis, Washington, 2nd edn, 1995.
10. K. Huttunen, A. Hyvarinen, A. Nevalainen, H. Komulainen and M. Hirvonen, Production of Proinflammatory Mediators by Indoor Air Bacteria and Fungal spores in Mouse and Human Cell Lines. *Environ. Health Perspect.*, 2003, **111**, 85–92.
11. L. DiBerardinis, M. First and E. Party, *et al.*, Report of the Howard Hughes Medical Institute's Workshop on the Performance of Laboratory Chemical Hoods, *AIHA*, 2003, **64**, 228–237.
12. M. First, Laboratory Chemical Hoods: A Historical Perspective, *AIHA*, 2003, **64**, 251–259.
13. D. W. Henderson, An apparatus for the study of airborne infection, *J. Hyg.*, 1952, **50**, 53–68.
14. E. Ibrahim, D. Harnish, K. Kinney, B. Heimbuch and J. Wander, An experimental investigation of the performance of a Collison Nebulizer generating H1N1 influenza aerosols, *Biotechnol. Biotechnol. Equip.*, 2015, **29(6)**, 1142–1148.
15. G. Mainelis, D. Berry and H. R. An, *et al.*, Design and performance of a single-pass bubbling bioaerosol generator, *Atmos. Environ.*, 2005, **39**, 3521–3533.
16. X. Simon, P. Duquenne, V. Koehler, C. Piernot, C. Coulais and M. Faure, Aerosolisation of Escherichia coli and associated endotoxin using an improved bubbling bioaerosol generator, *J. Aerosol Sci.*, 2011, **42**, 517–531.

17. H. Zhen, T. Han, D. E. Fennell and G. Mainelis, A Systematic comparison of four bioaerosol generators: affect on culturability and cell membrane integrity when aerosolizing Escherichia coli bacteria, *J. Aerosol Sci.*, 2014, **70**, 67–79.
18. R. Thomas, C. Davies and A. Nunez, *et al.*, Influence of particle size on the pathology and efficacy of vaccination in a murine model of inhalation anthrax, *J. Med. Microbiol.*, 2010, **59**, 1415–1427.
19. R. J. Thomas, D. Webber and W. Sellors, *et al.*, Characterization and Deposition of Respirable Large- and Small-Particle Bioaerosols, *Appl. Environ. Microbiol.*, 2008, **74**, 6437–6443.
20. R. J. Thomas, D. Webber and A. Collinge, *et al.*, Different Pathologies but Equal Levels of Responsiveness to the Recombinant F1 and V Antigen Vaccine and Ciproflaxican in a Murine Model of Plague Caused by Small- and Large-Particle Aerosols, *Infect. Immun.*, 2009, **77**, 1315–1323.
21. J. K. Bohannon, M. G. Lackemeyer and J. H. Kuhn, *et al.*, Generation and characterization of a large-particle aerosols using a center flow tangential aerosol generator with a non-human-primate, head-only aerosol chamber, *Inhalation Toxicol.*, 2015, **2015**, 247–253.

CHAPTER 14

Programming an Agent-Based Model for Disease Dynamics with Multiple Sources of Infection[†]

THOMAS INGERSOLL*[a] AND K. Y. WILLIAMS[a]

[a]Department of Defense, Defense Threat Reduction Agency, Bldg: E2800 2nd Floor, Aberdeen Proving Ground, MD 21010, USA
*E-mail: tomingersoll@yahoo.com

14.1 Introduction

Acute infectious disease modeling is used to predict the number of people who can be affected by a communicable disease that has been introduced into a community. Epidemiologists have applied such models to diseases such as anthrax released into communities, smallpox infections, Ebola outbreaks and seasonal infections involving influenza. To characterize the dynamics of an acute infectious disease in a host community, the host population has been divided into categories based on the infectious state of the

[†]The views and opinions expressed in this chapter are those of the author and should not be construed as an official Department of the Army position, policy, or decision, unless so designated by other official documentation. Citation of trade names in this chapter does not constitute an official Department of the Army endorsement or approval of the use of such commercial items.

Issues in Toxicology No. 25
Aerobiology: The Toxicology of Airborne Pathogens and Toxins
Edited by Harry Salem and Sidney Katz

Published by the Royal Society of Chemistry, www.rsc.org

individuals within the community. A set of categories useful for describing the state of a host includes the susceptible (S), infective (I) and resistant (or removed: R) states.

14.1.1 The Compartmental Susceptible, Infective, Resistant (SIR) Models

SIR models[1,2] are ways to describe how diseases are spread through a population. Most SIR models assume that the spread of disease occurs through close contact between infected and healthy individuals within a community, yet it is known that many diseases can be acquired through contact with contaminated environments. Within the SIR model, three categories of infectious state are taken into account: susceptibility (S), infective (I) and resistant/removed (R). Population sizes within each category are dynamic and change over time. Susceptible (S) individuals who are not infected in the population and are not at risk of being infected can change to being infective when they come into adequate contact with an infected individual for disease transmission. Infective (I) individuals are infected and capable of transmitting the disease to other members of the population, increasing the number of infected individuals. When infected individuals recover, they move to the resistant/removed (R) category within the model. Removed/resistant (R) members of the population are no longer able to contract the virus for one of three reasons: the infected individual has developed immunity to the disease, the infected individual has been placed in isolation or the infected individual has died as a result of the disease. With any of these cases the individual is only counted once even if they move from isolation to death. Simple models assume that members are mixing well within the population and that it is possible that any two individuals will encounter each other in a given period of time. Simplifying assumptions can also include population closure, that is, the population is not growing or declining over the course of the model, except mortality due to disease. Model extensions that relax these assumptions can be included.

Models containing these categories are called SIR models.[1,3,4] Susceptible individuals are not yet infected, but are available to become hosts to the disease. Infective individuals have acquired the disease and may pass it to others. Resistant individuals have acquired immunity so they can no longer become hosts or transmit the disease to others.

14.1.2 The Compartmental Susceptible, Exposed, Infective, Resistant (SEIR) Models

An elaboration of the SIR model adds an exposed class (SEIR[5]), where infection undergoes a latent period, where infected hosts are unable to transmit the infection to susceptible hosts. The individuals within the susceptible class can be exposed to the infection, hence they become exposed, but during the latent period they are not yet infective. These individuals are not infective and cannot transmit the infection to other members of the community until an adequate level of infection develops. The introduction of the exposed category makes the model more complex, but increases realism for many diseases.

14.1.3 The Compartmental Susceptible, Infective, Susceptible (SIS) Models

In a simplification of the SIR model, SIS models include no resistant category.[1] Within the SIS model, infective individuals return to susceptibility upon clearance of the disease. SIS provides a model for contaminated environment, where no mechanism exists for adaptive immunity.

SIR, SEIR and SIS models are diagramed as series of boxes, or compartments, indicating the count of individuals (stocks) within each category (Figure 14.1). Movements between categories, often called flows, are represented by arrows. These compartmental models summarize the movement between categories, assist in the development of mathematical expressions describing these movements and clarify assumptions inherent in model structure.[5]

Single infective source SIR models have a single row of compartments representing the host population as in Figure 14.1, but generalist pathogens can have multiple host species, or multiple resources for pathogen population growth or persistence. These provide multiple sources of contact for uninfected hosts with pathogens, such as from infected hosts, or contaminated environmental patches. For example, epidemiology of cholera and other water-borne diseases may be characterized by models with multiple infective sources, as these diseases may also be passed directly from host to host.[6] When alternative sources exhibit different disease dynamics, these should be represented as separate rows in the box diagram (Figure 14.2). Rows need not contain the same complement of compartments for each infection source. For example, infection may exhibit SEIR dynamics in a vertebrate host, while exhibiting SIS dynamics in contaminated environments surfaces or aerosols.

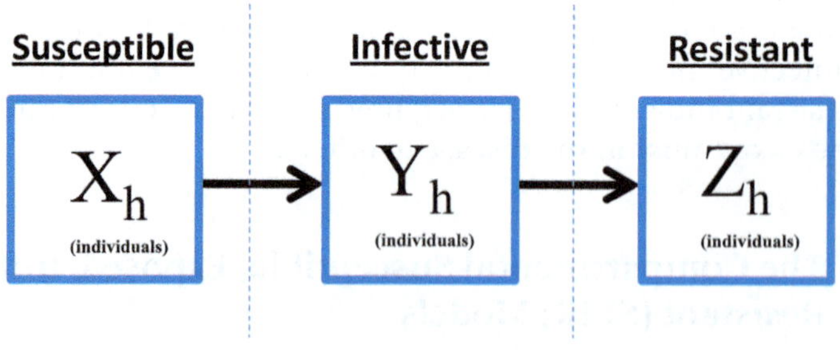

Figure 14.1 Compartmental SIR model and governing assumptions for dynamics of an infectious disease. Hosts are initially susceptible, then become infective following exposure to disease and finally become resistant to reinfection following recovery.

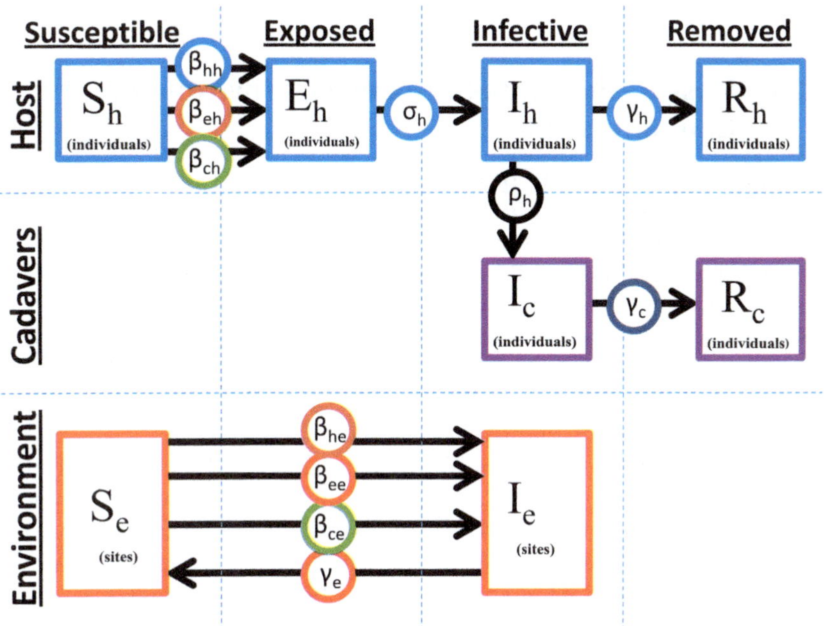

Figure 14.2 Compartmental SEIR model and governing assumptions for dynamics of an infectious disease with environmental exposure. With each time step, susceptible hosts (S_h) transform to exposed hosts (E_h) with contact probability β_{hh} on exposure to infective hosts (I_h),with contact probability β_{eh} conditional on exposure to contaminated environmental patches (I_e) and with contact probability β_{ch} conditional on exposure to infective cadavers. Assuming density-dependent transmission, susceptible host transformation probabilities (per individual) are $\beta_{hh}(I_h/N_h)$ for host to host transmission, $\beta_{eh}(I_e/N_e)$ for environment-to-host transmission and $\beta_{ch}[I_c/(I_c + N_h)]$ for cadaver-to-host transmission, where N_h and N_e are total host population and number of environmental patches. Exposed hosts transform to infective hosts at probability σ_h. Infective hosts die due to disease and are transformed to infective cadavers (I_c) with probability ρ_h, or are removed from the system (R_h) at probability γ_h upon development of immunologic resistance to infection. Infective cadavers are removed at probability γ_c upon clearing infection. With each time step, uncontaminated environmental patches (S_e) transform to infective environment with probability β_{he} conditional on exposure to infective hosts and with probability β_{ee} conditional on transport of contaminated material between patches. Note that β_{ee} will be zero for many, but not all, systems. Assuming density-dependent transmission, unconditional susceptible environmental transformation rates are $\beta_{he}(I_h/N_h)$ for host-to-environment transmission and $\beta_{ee}(I_e/N_e)$ for environment-to-environment transmission. Infective patches are returned to susceptibility at probability γ_e upon clearing infection.

Cross-contact with alternative sources can be included in the estimate of infection probability.

14.1.4 Host-to-Host Transmission and Environmental Contamination

A box-diagram describing a compartmental model for infectious disease that can be acquired from both infected hosts and contaminated environments is given in Figure 14.2. Sources of infection that differ by possible infection state are separated into rows, while infection state is given by columns. Contact along rows (within a single source) is controlled by primary contact probabilities such as β_{hh}, the probability of susceptible host contact with disease through an infected host. Contact between rows is controlled by interactive contact probabilities such as β_{eh}, the probability of host contact with disease through a contaminated environmental patch.

Infectious cadavers are typical for disease that can be contacted through contaminated environments. Cadavers are included as a separate row, since these include fewer states than living hosts and different states than contaminated environment. As examples, such infectious cadavers can be observed for diseases such as anthrax, smallpox and Ebolavirus disease.

14.1.5 Density Dependence and Contact Probabilities

Deterministic disease models include a contact rate, usually given as β, which is the rate at which susceptible individuals come into contact adequate for disease transmission with other individuals. Because rates can vary within stochastic models, the contact probability, β, is defined as the probability of contact adequate for disease transmission. By definition, a probability has a value bounded by 0 and 1. The contact probability is generalized to include cross-contact between alternative sources by adding subscripts, so that β_{aa} is the probability of disease-transmitting contact between conspecifics of species a and β_{ab} is contact between species a and b, adequate for transmission to b.

For well-mixed, motile populations, exposure of susceptible individuals to infection can be a function of the proportion of infective hosts within the population $I_{\mathrm{h}}/N_{\mathrm{h}}$, where I_{h} is the count of infective hosts and N_{h} is the count of hosts in all categories. This is called density-dependent transmission and is useful for describing many human diseases.[5] Because the density term $I_{\mathrm{h}}/N_{\mathrm{h}}$ is bounded by 1 and 0, it is a probability, the probability that an individual chosen at random is in an infective state.

The probability that a susceptible individual becomes exposed is the probability that adequate contact occurs and that contact is with an infective individual. Therefore, the probability of transitioning to the exposed state is the product of the contact probability and the density term, $\beta_{\mathrm{hh}}(I_{\mathrm{h}}/N_{\mathrm{h}})$,for exposure between hosts h. Note that this product remains bounded by 0 and 1, and therefore is also a probability.

14.1.6 Double Counting and Multiple Routes of Exposure

A probability space including all possible outcomes for contact with infectious disease (Figure 14.3, large white circle) is assumed. By the fundamental properties of probability, the probability associated with this space must be equal to 1, since it encompasses all outcomes.[7]

Two host species, red hosts and blue hosts, are also assumed. Hence there are two possible sources of infection. Suppose contact of red with red has a probability of 0.6 and contact of red with blue has a probability of 0.5. To estimate the combined probability of contact of red with all infective sources, the simple sum of the two single-source probabilities (infected by red or infected by blue) cannot be assumed because the net probability would then be 1.1, a probability exceeding that of all possible outcomes. This impossibility results from double counting the purple region when summing across the probabilistic "or." Instead, all cases where contact is with red and not blue (red region), or contact is with blue and not red (blue region), or contact is with red and blue (purple region) must be accounted for.

The product of probabilities is used to designate "and," the complement of probabilities to designate "not" and the sum of probabilities to designate "or," which sum can never exceed 1.

Contact is with red and not blue (red region): $\beta_{\text{red}}(1 - \beta_{\text{blue}})$, or contact is with blue and not red (blue region): $\beta_{\text{blue}}(1 - \beta_{\text{red}})$, or contact is with red and blue (purple region): $\beta_{\text{blue}}\,\beta_{\text{red}}$.

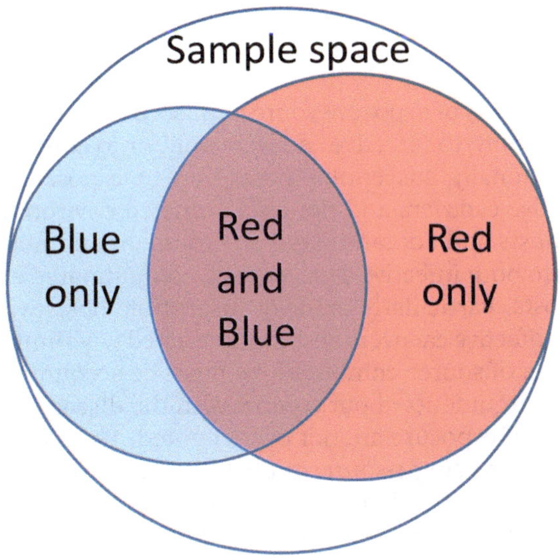

Figure 14.3 Venn diagram representing two possible sources of infection. The large white circle is the probability space including all possible outcomes for contact with an infectious disease. Contact can be with red not blue (red region), blue not red (blue region), or both red and blue (purple region). Simply summing the area of the red and blue circle results in counting the purple region twice.

The net probability of contact is the probability of contact with red and not blue, or contact is with blue and not red, or contact is with red and blue; therefore, we sum for each "or":

$$[\beta_{\text{red}}(1-\beta_{\text{blue}})] + [\beta_{\text{blue}}(1-\beta_{\text{red}})] + \beta_{\text{red}}\,\beta_{\text{blue}}$$

This formula correctly prevents the probability of contact with red and blue from being counted twice, so is equivalent to:

$$\beta_{\text{red}} + \beta_{\text{blue}} - \beta_{\text{red}}\,\beta_{\text{blue}}$$

The formulation can be expanded to include three (or more) possible routes of contact:

$$\begin{aligned}&\left[\beta_{\text{red}}(1-\beta_{\text{blue}})(1-\beta_{\text{yellow}})\right]+\left[\beta_{\text{blue}}(1-\beta_{\text{red}})(1-\beta_{\text{yellow}})\right]\\&+\left[\beta_{\text{yellow}}(1-\beta_{\text{red}})(1-\beta_{\text{blue}})\right]+\left[\beta_{\text{red}}\beta_{\text{blue}}(1-\beta_{\text{yellow}})\right]\\&+\left[\beta_{\text{red}}\beta_{\text{yellow}}(1-\beta_{\text{blue}})\right]+\left[\beta_{\text{blue}}\beta_{\text{yellow}}(1-\beta_{\text{red}})\right]+\beta_{\text{red}}\beta_{\text{blue}}\beta_{\text{yellow}}\end{aligned}$$

14.1.7 Net Transformation Probability

For the model given in Figure 14.2, there are three primary routes for exposure of susceptible hosts to disease. Susceptible hosts may be exposed to infective hosts, infective cadavers, contaminated environmental patches, or any combination of these. Therefore, there are

$$\binom{3}{1}+\binom{3}{2}+\binom{3}{3}=7$$

possible combinations of exposure sources. Susceptible hosts may be exposed to infective hosts only, to infective cadavers only, or to contaminated environment only. Additionally, susceptible hosts may be exposed to both infective hosts and infective cadavers and not contaminated environment, exposed to both infective hosts and contaminated environment and not infective cadavers, or exposed to both infective cadavers and contaminated environment and not infective hosts. Particularly unlucky susceptible hosts will be exposed to infective hosts, infective cadavers and contaminated environments. The "only" and "not" aspects of source combinations must be accounted-for when making probabilistic arguments about contact with the disease so that hosts with multiple sources of exposure are not over-counted. This is done by including the compliment of the single-source probability in the "not" cases.

Assuming density-dependent transmission, per-time-step contact probabilities for each source combination are as follows:

Infective hosts only: $\beta_{\text{hh}}\left(\dfrac{I_{\text{h}}}{N_{\text{h}}}\right)\left(1-\beta_{\text{ch}}\left(\dfrac{I_{\text{c}}}{I_{\text{c}}+N_{\text{h}}}\right)\right)\left(1-\beta_{\text{eh}}\left(\dfrac{I_{\text{e}}}{N_{\text{e}}}\right)\right)$

Infective cadavers only: $\beta_{\text{ch}}\left(\dfrac{I_{\text{c}}}{I_{\text{c}}+N_{\text{h}}}\right)\left(1-\beta_{\text{hh}}\left(\dfrac{I_{\text{h}}}{N_{\text{h}}}\right)\right)\left(1-\beta_{\text{eh}}\left(\dfrac{I_{\text{e}}}{N_{\text{e}}}\right)\right)$

Contaminatedenvironmentonly: $\beta_{\text{eh}}\left(\frac{I_{\text{e}}}{N_{\text{e}}}\right)\left(1-\beta_{\text{ch}}\left(\frac{I_{\text{c}}}{I_{\text{c}}+N_{\text{h}}}\right)\right)\left(1-\beta_{\text{hh}}\left(\frac{I_{\text{h}}}{N_{\text{h}}}\right)\right)$

Infective hosts and cadavers: $\beta_{\text{hh}}\left(\frac{I_{\text{h}}}{N_{\text{h}}}\right)\beta_{\text{ch}}\left(\frac{I_{\text{c}}}{I_{\text{c}}+N_{\text{h}}}\right)\left(1-\beta_{\text{eh}}\left(\frac{I_{\text{e}}}{N_{\text{e}}}\right)\right)$

Infective hosts and contaminated environment:

$$\beta_{\text{hh}}\left(\frac{I_{\text{h}}}{N_{\text{h}}}\right)\beta_{\text{eh}}\left(\frac{I_{\text{e}}}{N_{\text{e}}}\right)\left(1-\beta_{\text{ch}}\left(\frac{I_{\text{c}}}{I_{\text{c}}+N_{\text{h}}}\right)\right)$$

Infective cadavers and contaminated environment:

$$\left(1-\beta_{\text{hh}}\left(\frac{I_{\text{h}}}{N_{\text{h}}}\right)\right)\beta_{\text{ch}}\left(\frac{I_{\text{c}}}{I_{\text{c}}+N_{\text{h}}}\right)\beta_{\text{eh}}\left(\frac{I_{\text{e}}}{N_{\text{e}}}\right)$$

Infective hosts and cadavers and contaminated environments:

$$\beta_{\text{hh}}\left(\frac{I_{\text{h}}}{N_{\text{h}}}\right)\beta_{\text{ch}}\left(\frac{I_{\text{c}}}{I_{\text{c}}+N_{\text{h}}}\right)\beta_{\text{eh}}\left(\frac{I_{\text{e}}}{N_{\text{e}}}\right)$$

The net contact probability for susceptible hosts is the sum of all:

$$\begin{aligned}
&\left[\beta_{\text{hh}}\left(\frac{I_{\text{h}}}{N_{\text{h}}}\right)\left(1-\beta_{\text{ch}}\left(\frac{I_{\text{c}}}{I_{\text{c}}+N_{\text{h}}}\right)\right)\left(1-\beta_{\text{eh}}\left(\frac{I_{\text{e}}}{N_{\text{e}}}\right)\right)\right]\\
&+\left[\left(1-\beta_{\text{hh}}\left(\frac{I_{\text{h}}}{N_{\text{h}}}\right)\right)\beta_{\text{ch}}\left(\frac{I_{\text{c}}}{I_{\text{c}}+N_{\text{h}}}\right)\left(1-\beta_{\text{eh}}\left(\frac{I_{\text{e}}}{N_{\text{e}}}\right)\right)\right]\\
&+\left[\left(1-\beta_{\text{hh}}\left(\frac{I_{\text{h}}}{N_{\text{h}}}\right)\right)\left(1-\beta_{\text{ch}}\left(\frac{I_{\text{c}}}{I_{\text{c}}+N_{\text{h}}}\right)\right)\beta_{\text{eh}}\left(\frac{I_{\text{e}}}{N_{\text{e}}}\right)\right]\\
&+\left[\beta_{\text{hh}}\left(\frac{I_{\text{h}}}{N_{\text{h}}}\right)\beta_{\text{ch}}\left(\frac{I_{\text{c}}}{I_{\text{c}}+N_{\text{h}}}\right)\left(1-\beta_{\text{eh}}\left(\frac{I_{\text{e}}}{N_{\text{e}}}\right)\right)\right]\\
&+\left[\beta_{\text{hh}}\left(\frac{I_{\text{h}}}{N_{\text{h}}}\right)\beta_{\text{eh}}\left(\frac{I_{\text{e}}}{N_{\text{e}}}\right)\left(1-\beta_{\text{ch}}\left(\frac{I_{\text{c}}}{I_{\text{c}}+N_{\text{h}}}\right)\right)\right]\\
&+\left[\left(1-\beta_{\text{hh}}\left(\frac{I_{\text{h}}}{N_{\text{h}}}\right)\right)\beta_{\text{ch}}\left(\frac{I_{\text{c}}}{I_{\text{c}}+N_{\text{h}}}\right)\beta_{\text{eh}}\left(\frac{I_{\text{e}}}{N_{\text{e}}}\right)\right]\\
&+\left[\beta_{\text{hh}}\left(\frac{I_{\text{h}}}{N_{\text{h}}}\right)\beta_{\text{ch}}\left(\frac{I_{\text{c}}}{I_{\text{c}}+N_{\text{h}}}\right)\beta_{\text{eh}}\left(\frac{I_{\text{e}}}{N_{\text{e}}}\right)\right]
\end{aligned}$$

A similar argument can be made for the transformation probability for susceptible environment, except that the net transformation will include return of infective patches to the susceptible stock as infectivity clears (infective hosts are removed through development of resistance and so do not return to the susceptible state). If it is assumed that cadavers do not spread contamination between patches (*i.e.* cadavers remain on the last patch contaminated while living, or are removed without exposure to new patches: $\beta_{ce} = 0$), the net transformation probability is

$$\left[\beta_{he}\left(\frac{I_h}{N_h}\right)\left(1-\beta_{ee}\left(\frac{I_e}{N_e}\right)\right)\right]+\left[\left(1-\beta_{he}\left(\frac{I_h}{N_h}\right)\right)\beta_{ee}\left(\frac{I_e}{N_e}\right)\right]+\left[\beta_{he}\left(\frac{I_h}{N_h}\right)\beta_{eh}\left(\frac{I_e}{N_e}\right)\right]$$

With each time step, stocks of susceptible hosts and susceptible environment transform to exposed hosts and infective environment with the given probabilities. Demographic stochasticity is included by using the transformation probability. This may be done by picking a random binomial number at each time step for each susceptible host or patch and using the transformation probability as the binomial probability. The equivalent (and expedient) programming method of drawing a number at random for each susceptible host or patch from a uniform distribution between 0 and 1 is chosen. If the random number is less than or equal to the transformation probability, the host or patch transforms to the exposed condition.

Other transformations, the mortality, recovery and removal probabilities, are simpler than contact probabilities, although they may be made time dependent by adding a time-in-class counter and making the argument for the probability a function of the counter's value: $\gamma_e = F(t)$, where $F(t)$ is a function of time-in-class.

14.2 Modeling Building with R

The R programming language offers a variety of statistical and graphical techniques that are highly extensible.[8] Statisticians use R as a means for programming and imaging in statistics because of its ease of use and versatility with linear and nonlinear modeling, classical statistical tests, time-series analysis, classification and clustering and its ability to incorporate additional packages into the program. The Comprehensive R Archive Network (CRAN) repository makes it possible for users to download, develop and incorporate the various packages into their research studies. Within the R program that is supplied in this chapter, various parts of the program must be declared before the programming language will be able to work with any information that is being supplied. For simplicity, we have reduced the program to 50 lines of R code that can be run on any platform with R installed. The models presented here assume that the scientist has a working knowledge of and familiarity with R programming. These models and codes can be easily modified for diseases with dynamics that differ from our examples.

14.2.1 Building a Model with an Example Data Set

While it might be hoped to have real data on the number and condition of hosts and environmental patches, for demonstration purposes a host population and environment will be simulated. A function can be written in R that will create a simulated population with initial values for stocks in SEIR classes, for a host and environmental patches. The simulated population can then be used to explore disease dynamics by applying a second function controlling transformation from initial states. A population will then be built containing the following stocks: Sh the susceptible hosts, Eh the exposed hosts, Ih the infective hosts, Rh the resistant hosts, Ic the infective cadavers, Rc the non-infective cadavers, Se the susceptible environment and Ie the infective environment. Units of hosts are in individuals, while units of environment are in patches. Values are then selected for each of these stocks to use as arguments to our R function, which then builds a matrix of categorical values representing the state of each individual and patch.

Enter the following function into the R console:

```
ABInitials <-function(Sh0,Eh0,Ih0,Ic0,Rc0,Rh0,Se0,Ie0){

   ## create a matrix to hold the state values
   DATA0 <-matrix(NA,sum(Sh0 + Eh0 + Ih0 + Ic0 + Rc0 + Rh0 + Se0 + Ie0),2)

   ## create a matrix to hold the time within state
   Counters0 <-matrix(0,length(DATA0[,1]),3)

   ## allocate hosts
   if(Sh0 > 0)
   for (i in 1:Sh0){
      DATA0[i,1] <-"h"
      DATA0[i,2] <-"S"
      }
   if(Eh0 > 0)
   for (i in (Sh0 + 1):(Sh0 + Eh0)){
      DATA0[i,1] <-"h"
      DATA0[i,2] <-"E"
      }
   if(Ih0 > 0)
   for (i in (Sh0 + Eh0 + 1):(Sh0 + Eh0 + Ih0)){
      DATA0[i,1] <-"h"
      DATA0[i,2] <-"I"
      }
   if(Ic0 > 0)
   for (i in (Sh0 + Eh0 + Ih0 + 1):(Sh0 + Eh0 + Ih0 + Ic0)){
      DATA0[i,1] <-"c"
      DATA0[i,2] <-"I"
      }
```

```
if(Rc0 > 0)
for (i in (Sh0 + Eh0 + Ih0 + Ic0 + 1):(Sh0 + Eh0 + Ih0 + Ic0 + Rc0)){
    DATA0[i,1] < -"c"
    DATA0[i,2] < -"R"
    }
if(Rh0 > 0)
for (i in (Sh0 + Eh0 + Ih0 + Ic0 + Rc0 + 1):(Sh0 + Eh0 + Ih0 + Ic0 + Rc0 +
Rh0)){
    DATA0[i,1] < -"h"
    DATA0[i,2] < -"R"
    }

## allocate environment
if(Se0 > 0)
for (i in (Sh0 + Eh0 + Ih0 + Ic0 + Rc0 + Rh0 + 1):(Sh0 + Eh0 + Ih0 + Ic0 + Rc0
+ Rh0 + Se0)){
    DATA0[i,1] < -"e"
    DATA0[i,2] < -"S"
    }
if(Ie0 > 0)
for (i in (Sh0 + Eh0 + Ih0 + Ic0 + Rc0 + Rh0 + Se0 + 1):(Sh0 + Eh0 + Ih0 + Ic0
+ Rc0 + Rh0 + Se0 + Ie0)){
    DATA0[i,1] < -"e"
    DATA0[i,2] < -"I"
    }

## count the stocks
Sh = length(DATA[,1][DATA[,1] == "h" & DATA[,2] == "S"])
Eh = length(DATA[,1][DATA[,1] == "h" & DATA[,2] == "E"])
Ih = length(DATA[,1][DATA[,1] == "h" & DATA[,2] == "I"])
Ic = length(DATA[,1][DATA[,1] == "c" & DATA[,2] == "I"])
Rc = length(DATA[,1][DATA[,1] == "c" & DATA[,2] == "R"])
Rh = length(DATA[,1][DATA[,1] == "h" & DATA[,2] == "R"])
Se = length(DATA[,1][DATA[,1] == "e" & DATA[,2] == "S"])
Ie = length(DATA[,1][DATA[,1] == "e" & DATA[,2] == "I"])

COUNTS < -c(Sh,Eh,Ih,Ic,Rc,Rh,Se,Ie)

## make values available to the global environment
DATA0 ≪ -DATA0
DATA ≪ -DATA0
Counters ≪ -Counters0
COUNTS ≪ -COUNTS
Counters0 ≪ -Counters0
Counters ≪ -Counters0
}
```

The function ABInitials creates a matrix, DATA0, containing initial values for the host population and environmental patches. Each individual is assigned two letters representing a category. The first letter, h, c or e, represents host, cadaver or environmental patch, respectively. The second letter, S, E, I or R, represents the infective state of the individual.

The number of individuals in each class is determined by arguments to the function call ABInitials (Sh0,Eh0,Ih0,Ic0,Rc0,Rh0,Se0,Ie0), where Sh0 is the initial count of susceptible hosts, Eh0 exposed hosts, Ih0 infective hosts, Ic0 infective cadavers, Rc0 removed cadavers, Rh0 resistant hosts, Se0 susceptible environmental patches, Ie0 contaminated environment. For example, entering ABInitials(90,10,0,0,0,0,100,0) (followed by return) creates a matrix with 90 susceptible hosts, 10 exposed hosts and 100 susceptible environmental patches.

Using a series of for-loops and the counts given by the function arguments to allocate the correct categorical codes for each host or patch, h, c or e, into column 1 for type and S, E, I or R for infection state into column 2, in this way either real data representing the number of individuals in each category or simulated data may be used to fill the matrix of initial values.

Entering DATA0 displays the matrix. Entering COUNTS returns a vector of counts in each class.

A second matrix, Counters, records the number of time steps an individual host has remained in the exposed, infective or infective cadaver classes. Counters also record time in infective class for environmental patches. This is useful for diseases where infectivity increases with time or where the probability of transforming from the infective category is time dependent. For the matrix of initial values, all values in Counters are zero (enter Counters to see).

```
## example
ABInitials(90,10,0,0,0,0,100,0)

## view the initial values
COUNTS
Counters
```

14.2.2 Perform the Transition

Type the following function into the R console:

```
Transition <-function(DATA,BETAhh,BETAeh,BETAhe,BETAee,SIGMAh,
GAMMAh,GAMMAe,GAMMAc,RHOh,BETAch){

DATAT <-DATA
CountersT <-Counters

## count the stocks
Sh = length(DATA[,1][DATA[,1] == “h” & DATA[,2] == “S”])
Eh = length(DATA[,1][DATA[,1] == “h” & DATA[,2] == “E”])
Ih = length(DATA[,1][DATA[,1] == “h” & DATA[,2] == “I”])
```

```
Ic = length(DATA[,1][DATA[,1] == "c" & DATA[,2] == "I"])
Rc = length(DATA[,1][DATA[,1] == "c" & DATA[,2] == "R"])
Rh = length(DATA[,1][DATA[,1] == "h" & DATA[,2] == "R"])
Se = length(DATA[,1][DATA[,1] == "e" & DATA[,2] == "S"])
Ie = length(DATA[,1][DATA[,1] == "e" & DATA[,2] == "I"])

## the contact probabilities
## to host
ifelse(Ih > 0, HH < −(BETAhh * (Ih/(Ih + Sh))), HH < −0)
ifelse(Ie > 0, EH < −(BETAeh * (Ie/(Ie + Se))), EH < −0)
ifelse(Ic > 0, CH < −(BETAch * (Ic/(Ic ++ Ih + Sh))), CH < −0)
## to environment
ifelse(Ih > 0, HE < −(BETAhe * (Ih/(Ih + Sh))), HE < −0)
ifelse(Ie > 0, EE < −(BETAee * (Ie/(Ie + Se))), EE < −0)

## the transitions
for (i in 1:length(DATA[,1])){

ifelse(DATA[i,1] == "h" && DATA[i,2] == "S",
x < −runif(1,0,1),x < −1)
if(x<
### infected by environment not hosts or cadavers
((EH * (1 − HH) * (1 − CH))+
### infected by hosts not environment or cadavers
(HH * (1 − EH) * (1 − CH))+
### infected by cadavers not environment or hosts
(CH * (1 − HH) * (1 − EH))+
### infected by environment and hosts not cadavers
(EH * HH * (1 − CH))+
### infected by hosts and cadavers not environment
(HH * CH * (1 − EH))+
### infected by environment and cadavers not hosts
(EH * (1 − HH) * CH)+
### infected by environment and cadavers and hosts
(EH * HH * CH))
)
DATAT[i,2] < −"E"
}

for (i in 1:length(DATA[,1])){
    ifelse(DATA[i,1] == "h" && DATA[i,2] == "E",
    CountersT[i,1] < −Counters[i,1] + 1,
    CountersT[i,1] < −0)
    ifelse(DATA[i,1] == "h" && DATA[i,2] == "E",
    x < −runif(1,0,1), x < −1)
    if(x < SIGMAh)
    DATAT[i,2] < −"I"
    }
```

```
for (i in 1:length(DATA[,1])){
    ifelse(DATA[i,1] == "h" && DATA[i,2] == "I",
    CountersT[i,2] <-Counters[i,2] + 1,
    CountersT[i,2] <-0)
    ifelse(DATA[i,1] == "h" && DATA[i,2] == "I",
    x <-runif(1,0,1), x <-1)
    if(x < RHOh)
    DATAT[i,2] <-"I"
    if(x < RHOh)
    DATAT[i,1] <-"c"
    if(x > RHOh && x < RHOh + GAMMAh)
    DATAT[i,2] <-"R"
    }

for (i in 1:length(DATA[,1])){
    ifelse(DATA[i,1] == "c" && DATA[i,2] == "I",
    CountersT[i,3] <-Counters[i,3] + 1,
    CountersT[i,3] <-0)
    ifelse(DATA[i,1] == "c" && DATA[i,2] == "I",
    x <-runif(1,0,1), x <-1)
    if(x < GAMMAc)
    DATAT[i,2] <-"R"
    }

for (i in 1:length(DATA[,1])){
    ifelse(DATA[i,1] == "e" && DATA[i,2] == "S",
    x <-runif(1,0,1), x <-1)
    if(x<
    ### infected by environment not hosts
    ((EE * (1 - HE))+
    ### infected by hosts not environment
    (HE * (1 - EE))+
    ### infected by hosts and environment
    (HE * EE)))
DATAT[i,2] <-"I"
}

for (i in 1:length(DATA[,1])){
    ifelse(DATA[i,1] == "e" && DATA[i,2] == "I",
    CountersT[i,2] <-Counters[i,2] + 1,
    CountersT[i,2] <-0)
    ifelse(DATA[i,1] == "e" && DATA[i,2] == "I",
    x <-runif(1,0,1), x <-1)
    if(x < GAMMAe)
    DATAT[i,2] <-"S"
    }
```

```
    DATA ≪ −DATAT
    ShT = length(DATAT[,1][DATAT[,1] == "h" & DATAT[,2] == "S"])
    EhT = length(DATAT[,1][DATAT[,1] == "h" & DATAT[,2] == "E"])
    IhT = length(DATAT[,1][DATAT[,1] == "h" & DATAT[,2] == "I"])
    IcT = length(DATAT[,1][DATAT[,1] == "c" & DATAT[,2] == "I"])
    RcT = length(DATAT[,1][DATAT[,1] == "c" & DATAT[,2] == "R"])
    RhT = length(DATAT[,1][DATAT[,1] == "h" & DATAT[,2] == "R"])
    SeT = length(DATAT[,1][DATAT[,1] == "e" & DATAT[,2] == "S"])
    IeT = length(DATAT[,1][DATA[,1] == "e" & DATAT[,2] == "I"])

    COUNTST < −c(ShT,EhT,IhT,IcT,RcT,RhT,SeT,IeT)
    COUNTST ≪ − COUNTST
    Counters ≪ − CountersT
}
```

The function Transition takes the matrix of initial values, DATA0, and transfers individuals between infection states according to probabilities entered as arguments to the function. These probabilities are BETAhh the host-to-host contact probability, BETAeh the environment-to-host contact probability, BETAhe the host-to-environment contact probability, BETAee the environment-to-environment contact probability, BETAch the infective cadaver-to-host contact probability, SIGMAh the host transition from latency probability, GAMMAh the host recovery probability, GAMMAe the environmental recovery probability, RHOh the host mortality from infection probability and GAMMAc the cadaver removal (loss of infectivity) probability. Probabilities given by function arguments can be fixed values or can call heterogeneous values from other functions, including functions using time-in-class given by Counters.

A series of ifelse conditional statements are used to determine each individual's category, then the correct transition probability for that category is applied. To introduce demographic stochasticity, a random uniform number between 0 and 1 is drawn for each individual. If this number is less than the given transformation probability, the individual transforms to the next infection state, else remains in the current condition.

If the individual is in a potentially time-sensitive infection state, such as exposed or infected, a value of 1 is added to the appropriate column of Counters with each time step, so that Counters exhibits the sum of time steps in category. Upon transitioning to a new infection state, the individual's value in Counters returns to 0.

Renaming is used to keep the matrices of original values separate from new matrices constructed when state transformations take place. DATA0 and Counters0 are first renamed DATA and Counters. DATA, COUNTS and Counters are renamed DATAT, COUNTST and CountersT after each transformation and can be displayed by entering DATAT, COUNTST, or CountersT, respectively.

14.2.2.1 Example 1

```
## display initial values
COUNTS
Counters
## allocate the initial values to matrices DATA and Counters
DATA < −DATA0
Counters < −Counters0

## perform the first transition
Transition(DATA,0.5,0.5,0.5,0,0.2,0.2,0.5,0.2,0.2,0.2)

## values at t = 1
COUNTST
Counters

## perform a second transition
Transition(DATA,0.5,0.5,0.5,0,0.2,0.2,0.5,0.2,0.2,0.2)

## values at t = 2
COUNTST
Counters
```

14.2.2.2 Example 2: Performing 100 Time Steps and Recording Values

A script can be used to run function Transition any number of time-steps. The DATA and Counters are first reset to initial values:

```
DATA < −DATA0
Counters < −Counters0
```

Then, a matrix to hold the initial values for COUNTS followed by COUNTS at each of 100 transitions will be created:

```
T = 101
Epi < −matrix(NA,T,length(COUNTS) + 1)
Time < −matrix(NA,T,1)
   ShT = length(DATA[,1][DATA[,1] == "h" & DATA[,2] == "S"])
   EhT = length(DATA[,1][DATA[,1] == "h" & DATA[,2] == "E"])
   IhT = length(DATA[,1][DATA[,1] == "h" & DATA[,2] == "I"])
   IcT = length(DATA[,1][DATA[,1] == "c" & DATA[,2] == "I"])
   RcT = length(DATA[,1][DATA[,1] == "c" & DATA[,2] == "R"])
   RhT = length(DATA[,1][DATA[,1] == "h" & DATA[,2] == "R"])
   SeT = length(DATA[,1][DATA[,1] == "e" & DATA[,2] == "S"])
   IeT = length(DATA[,1][DATA[,1] == "e" & DATA[,2] == "I"])

   COUNTS < −c(ShT,EhT,IhT,IcT,RcT,RhT,SeT,IeT)
```

COUNTS

A time step = 1 is allocated to the first column first row, followed by the initial values in the rest of the first row:

```
Epi[1,] < −c(1,COUNTS)
```

Then, a for-loop is used to allocate subsequent time steps:

```
for (h in 2:T){
transition(DATA,0.5,0.5,0.5,0,0.2,0.2,0.5,0.2,0.2,0.2)
Epi[h,] < −c(h, COUNTST)
}
```

Display the values:

```
Epi
Counters
```

14.2.2.2.1 Plotting the Results. Although R has several built-in functions for standard plots, building custom plots with extra spaces for labels, legends, and custom axes is preferred.

First, a graphics from with custom margins and multiple plotting regions is established:

```
par(mar = c(1.4, 1.4, 0.6, 0.6))

layout(
rbind(
c(4,1,2),
c(6,3,5)),
widths = c(0.7,5,3),
heights = c(5,0.7))
```

Next, the first graphics to plot the trajectories for each infection category is established:

```
plot(Epi[,1],Epi[,2],type = "n",
ylim = c(0,100),
xlim = c(0,100),
axes = FALSE, ann = FALSE,
)
```

Now lines are added, following the trajectories from subsequent transformations of COUNTS:

```
points(Epi[,1]-1,Epi[,2], type = "l", lty = "solid", lwd = 2, col = "blue")
points(Epi[,1]-1,Epi[,3], type = "l", lty = "solid", lwd = 2, col = "purple")
points(Epi[,1]-1,Epi[,4], type = "l", lty = "solid", lwd = 2, col = "red")
```

```
points(Epi[,1]-1,Epi[,5], type = "l", lty = "solid", lwd = 2, col = "orange")
points(Epi[,1]-1,Epi[,6], type = "l", lty = "solid", lwd = 2, col = "yellow")
points(Epi[,1]-1,Epi[,7], type = "l", lty = "solid", lwd = 2, col = "light blue")
points(Epi[,1]-1,Epi[,8], type = "l", lty = "solid", lwd = 2, col = "green")
points(Epi[,1]-1,Epi[,9], type = "l", lty = "solid", lwd = 2, col = "brown")
```

The custom axes:

```
axis(2, at = seq(0,100,50),labels = c("0","50","100"),las = 1, pos = 0,
cex.axis = 1.2)
axis(1,at = c(0, 50, 100),
labels = c("0","50","100"),pos = 0,cex.axis = 1.2)
```

A dummy variable for the legend and label plotting regions is created:

```
k = seq(0,1,0.1)
l = seq(0,1,0.1)
```

Plot the legend:

```
plot(k,l,type = "n",
axes = FALSE, ann = FALSE,
)

legend(0,0.6, legend = c("Sh","Eh","Ih","Ic","Rc","Rh","Se","Ie"),
col = c("blue","purple","red","orange","yellow","light blue","green","brown"),
lty = c("solid","solid","solid","solid"), lwd = c(2,2,2,2),
title = "Stocks", bty = "n",cex = 1.1)
```

Reset the margins and plot the labels:

```
par(mar = c(0, 0, 0, 0))

k = seq(0,1,0.1)
l = seq(0,1,0.1)

plot(k,l,type = "n",
axes = FALSE, ann = FALSE,
)
text(0.5,0.5,"Time",cex = 2)

plot(k,l,type = "n",
axes = FALSE, ann = FALSE,
)
text(0.3,0.5,"Count",srt = 90,cex = 2)

## Two unused plotting regions remain (in the corners):

plot(k,l,type = "n",
axes = FALSE, ann = FALSE,
   )
```

```
plot(k,l,type = "n",
axes = FALSE, ann = FALSE,
)
```

14.2.2.3 Example 3: Multiple Interactions and Confidence Levels

Incorporation of stochasticity creates conditions where outcomes vary from trial to trial. To estimate expected values therefore requires averaging across many trial iterations. Confidence intervals can be estimated likewise, by using standard deviations from many trials. A script with nested for-loops to repeat trials of 100 time steps across 10 000 iterations is used, then means and confidence intervals for stocks in each infection category at each time step are estimated.

```
## First, a matrix to hold the data is created:

Reps = 10000
T = 100
Epis <-matrix(NA,T * Reps,length(COUNTS) + 1)
Time <-matrix(NA,T,1)

## Next, a for-loop for the 10 000 iterations:

for(r in 0:(Reps-1)){

## set DATA back to initial values
DATA <-DATA0

Sh = length(DATA[,1][DATA[,1] == "h" & DATA[,2] == "S"])
Eh = length(DATA[,1][DATA[,1] == "h" & DATA[,2] == "E"])
Ih = length(DATA[,1][DATA[,1] == "h" & DATA[,2] == "I"])
Rh = length(DATA[,1][DATA[,1] == "h" & DATA[,2] == "R"])
Mh = length(DATA[,1][DATA[,1] == "h" & DATA[,2] == "M"])
Xh = length(DATA[,1][DATA[,1] == "h" & DATA[,2] == "X"])
Se = length(DATA[,1][DATA[,1] == "e" & DATA[,2] == "S"])
Ie = length(DATA[,1][DATA[,1] == "e" & DATA[,2] == "I"])
COUNTS <-c(Sh,Eh,Ih,Rh,Mh,Xh,Se,Ie)

Epis[(r * T) + 1,] <-c(1,COUNTS)

## Within the loop for iterations, nest a loop for time steps:

for (h in 2:T){
transition(DATA,BETAhh,BETAeh,BETAhe,BETAee,SIGMAh,
GAMMAh,GAMMAe,GAMMAc,RHOh,BETAch)
Epis[(r * T) + h,] <-c(h, COUNTST)
}
}
```

```
Epis

## estimate the per time-step means:
Epimeans <-matrix(NA,T,length(Epis[1,]))

for(j in 1:T){
  Epimeans[j,1] <-j
  for(k in 2:length(Epis[1,])){
        Epimeans[j,k] <-mean(Epis[Epis[,1]==j,k])
}
}

## estimate the per time step standard deviations:

Episds <-matrix(NA,T,length(Epis[1,]))

for(j in 1:T){
     Episds[j,1] <-j
     for(k in 2:length(Epis[1,])){
        Episds[j,k] <-sd(Epis[Epis[,1]==j,k])
}
}

## display the results
Epimeans
Episds
```

14.2.2.3.1 Plot with Expected Values and Confidence Intervals.

```
## Create a plotting frame:

par(mar = c(1.4, 1.4, 0.6, 0.6))

layout(
rbind(
c(4,1,2),
c(6,3,5)),
widths = c(0.7,5,3),
heights = c(5,0.7))

plot(Epimeans[,1],Epimeans[,2],type = "n",
ylim = c(0,100),
xlim = c(0,100),
axes = FALSE, ann = FALSE,
)

## Add lines for expected values:

points(Epimeans[,1], Epimeans[,2], type = "l", lty = "solid", lwd = 2, col = "blue")
points(Epimeans[,1], Epimeans[,3], type = "l", lty = "solid", lwd = 2, col = "purple")
```

```
points(Epimeans[,1], Epimeans[,4], type = “l”, lty = “solid”, lwd = 2, col = “red”)
points(Epimeans[,1], Epimeans[,5], type = “l”, lty = “solid”, lwd = 2, col = “orange”)
points(Epimeans[,1], Epimeans[,6], type = “l”, lty = “solid”, lwd = 2, col = “yellow”)
points(Epimeans[,1], Epimeans[,7], type = “l”, lty = “solid”, lwd = 2, col = “light blue”)
points(Epimeans[,1], Epimeans[,8], type = “l”, lty = “solid”, lwd = 2, col = “green”)
points(Epimeans[,1], Epimeans[,9], type = “l”, lty = “solid”, lwd = 2, col = “brown”)

## Assuming normally distributed deviations, plot the confidence intervals:
points(Epimeans[,1], Epimeans[,2] + (1.96 * Episds[,2]), type = “l”, lty = “dotted”, lwd = 1, col = “blue”)
points(Epimeans[,1], Epimeans[,3] + (1.96 * Episds[,3]), type = “l”, lty = “dotted”, lwd = 1, col = “purple”)
points(Epimeans[,1], Epimeans[,4] + (1.96 * Episds[,4]), type = “l”, lty = “dotted”, lwd = 1, col = “red”)
points(Epimeans[,1], Epimeans[,5] + (1.96 * Episds[,5]), type = “l”, lty = “dotted”, lwd = 1, col = “orange”)
points(Epimeans[,1], Epimeans[,6] + (1.96 * Episds[,6]), type = “l”, lty = “dotted”, lwd = 1, col = “yellow”)
points(Epimeans[,1], Epimeans[,7] + (1.96 * Episds[,7]), type = “l”, lty = “dotted”, lwd = 1, col = “light blue”)
points(Epimeans[,1], Epimeans[,8] + (1.96 * Episds[,8]), type = “l”, lty = “dotted”, lwd = 1, col = “green”)
points(Epimeans[,1], Epimeans[,9] + (1.96 * Episds[,9]), type = “l”, lty = “dotted”, lwd = 1, col = “brown”)

points(Epimeans[,1], Epimeans[,2]-(1.96 * Episds[,2]), type = “l”, lty = “dotted”, lwd = 1, col = “blue”)
points(Epimeans[,1], Epimeans[,3]-(1.96 * Episds[,3]), type = “l”, lty = “dotted”, lwd = 1, col = “purple”)
points(Epimeans[,1], Epimeans[,4]-(1.96 * Episds[,4]), type = “l”, lty = “dotted”, lwd = 1, col = “red”)
points(Epimeans[,1], Epimeans[,5]-(1.96 * Episds[,5]), type = “l”, lty = “dotted”, lwd = 1, col = “orange”)
points(Epimeans[,1], Epimeans[,6]-(1.96 * Episds[,6]), type = “l”, lty = “dotted”, lwd = 1, col = “yellow”)
points(Epimeans[,1], Epimeans[,7]-(1.96 * Episds[,7]), type = “l”, lty = “dotted”, lwd = 1, col = “light blue”)
points(Epimeans[,1], Epimeans[,8]-(1.96 * Episds[,8]), type = “l”, lty = “dotted”, lwd = 1, col = “green”)
points(Epimeans[,1], Epimeans[,9]-(1.96 * Episds[,9]), type = “l”, lty = “dotted”, lwd = 1, col = “brown”)
```

```
## Add the axes

axis(2, at = seq(0,100,50),labels = c("0","50","100"),las = 1, pos = 0,
cex.axis = 1.2)
axis(1,at = c(0, 50, 100),
labels = c("0","50","100"),pos = 0,cex.axis = 1.2)

## create a dummy variable for plotting the other regions

k = seq(0,1,0.1)
l = seq(0,1,0.1)

plot(k,l,type = "n",
axes = FALSE, ann = FALSE,
)

legend(0,0.9, legend = c("Sh","Eh","Ih","Ic","Rc","Rh","Se","Ie"),
col = c("blue","purple","red","orange","yellow","light blue","green","brown"),
lty = c("solid","solid","solid","solid","solid","solid","solid","solid","solid"),
lwd = c(2,2,2,2,2,2,2,2,2),
title = "Expected values", bty = "n",cex = 1.1)

legend(0,0.4, legend = c("Sh","Eh","Ih","Ic","Rc","Rh","Se","Ie"),
col = c("blue","purple","red","orange","yellow","light blue","green","brown"),
lty = c("dotted","dotted","dotted","dotted","dotted","dotted","dotted",
"dotted","dotted"),
lwd = c(1,1,1,1,1,1,1,1,1),
title = "95% CI", bty = "n",cex = 1.1)

## Add the labels:

par(mar = c(0, 0, 0, 0))

k = seq(0,1,0.1)
l = seq(0,1,0.1)

plot(k,l,type = "n",
axes = FALSE, ann = FALSE,
)
text(0.5,0.5,"Time",cex = 2)

plot(k,l,type = "n",
axes = FALSE, ann = FALSE,
)
text(0.3,0.5,"Count",srt = 90,cex = 2)

plot(k,l,type = "n",
axes = FALSE, ann = FALSE,
)
```

```
plot(k,l,type = "n",
axes = FALSE, ann = FALSE,
)
```

14.3 Results

Results of a single trial of one iteration are presented in Figure 14.4. Most contact probabilities are set to intermediate values (see Figure 14.4 caption). Beginning with a small number of exposed hosts, exposure occurs rapidly until all hosts are exposed by around time step 25, with around half of all environmental patches contaminated. Infected hosts are depleted by time step 50, with about 50% mortality, the rest recovering to the resistant class. With the absence of infected hosts, contaminated environment clears by around time step 50. The effect of stochasticity is apparent in the irregular

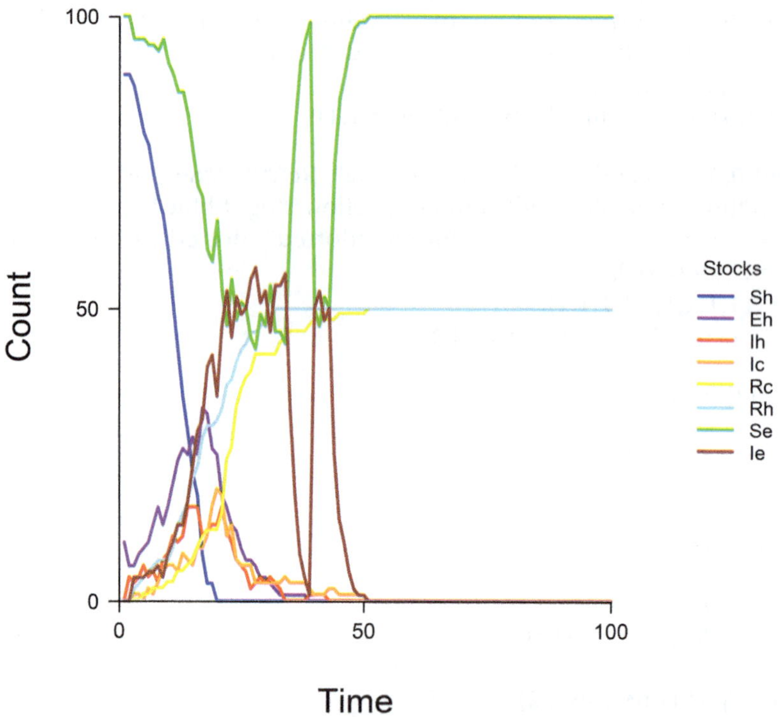

Figure 14.4 Single iteration of a stochastic infectious disease model with environmental exposure across 100 time steps. Stocks are as follows: Sh (susceptible hosts), Eh (exposed hosts), Ih (infectious hosts), Ic (infectious cadavers), Rc (removed cadavers), Rh (removed hosts), Se (uncontaminated patches), Ie (contaminated patches). Transformation probabilities are as follows: $\beta_{hh} = 0.5$, $\beta_{eh} = 0.5$, $\beta_{he} = 0.5$, $\beta_{ee} = 0$, $\sigma_h = 0.2$, $\gamma_h = 0.2$, $\gamma_e = 0.5$, $\gamma_c = 0.2$, $\rho_h = 0.2$ and $\beta_{ch} = 0.2$. No births or deaths other than those caused by disease occur.

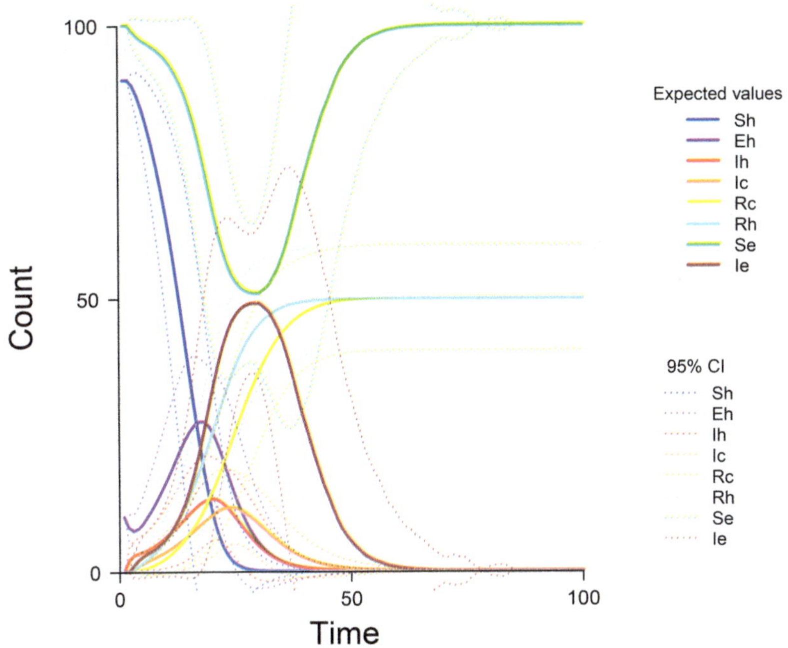

Figure 14.5 Expected values and confidence intervals for dynamics of an infectious disease with environmental exposure across 100 time steps and 10 000 iterations. Stocks are as follows: Sh (susceptible hosts), Eh (exposed hosts), Ih (infectious hosts), Ic (infectious cadavers), Rc (removed cadavers), Rh (removed hosts), Se (uncontaminated patches), Ie (contaminated patches). Transformation probabilities are as follows: $\beta_{hh} = 0.5$, $\beta_{eh} = 0.5$, $\beta_{he} = 0.5$, $\beta_{ee} = 0$, $\sigma_h = 0.2$, $\gamma_h = 0.2$, $\gamma_e = 0.5$, $\gamma_c = 0.2$, $\rho_h = 0.2$ and $\beta_{ch} = 0.2$. No births or deaths other than those caused by disease occur. Confidence intervals determined by ±1.96 standard deviations. Local disease extinction occurs at around time step 50, with about half of all hosts surviving.

oscillations of most stocks, prior to the final equilibrium. Demographic stochasticity causes variation between trials (not shown), so some user results may differ somewhat from those shown in Figure 14.4.

Averaging the results of many iterations results in smoothing of all lines (Figure 14.5). Confidence intervals may also be estimated when many iterations are performed (dotted lines).

14.4 Discussion

14.4.1 Utility of Models

The agent-based model is used to predict trajectories in time, of individuals within each infection state. As such, it can be used to plan for the response to epidemics, allocate resources and predict the response to various management decisions such as vaccine or isolation strategies.

Although the model is given in discrete time, decreasing the time interval, with appropriate rate adjustments, creates a model that more closely approximates continuous time. However, this has a cost in computing time. A compromise is needed to balance fineness of the time scale, computational requirements and available computational resources.

As given, the current model represents small, evenly mixed populations such as those found in small communities. This model may be expanded to a network or metapopulation model by writing an additional code that allows uneven mixing between populations. Such models require extensive computing resources and are beyond the scope of this chapter.

14.4.2 Heterogeneous Transformation Probabilities

Including heterogeneous transformation probabilities allows observation and process stochasticity to be incorporated and can allow probabilities to be functions of data. Instead of fixed values for arguments to the function Transition, some arguments can be calls to other functions. For example: Transition(DATA, 0.5, 0.5, 0.5, 0, Sigma(CountersT[*i*,1]), 0.2, 0.5, 0.2, 0.2, 0.2) applies the function Sigma to the counter value for the individual's time in the exposed class, a method for making transition from latency time dependent.

Probabilistic correction for multiple sources of exposure adds flexibility over single-source models. For example, the multiple-source model to include multiple hosts, accommodating the variety of emerging zoonotic diseases could easily be modified. Use of R as the modeling platform allows a wide range of R library functions to be called as model arguments, further linking dynamic models to data.

14.5 R Program Code

```
ABInitials < −function(Sh0,Eh0,Ih0,Ic0,Rc0,Rh0,Se0,Ie0){

  DATA0 < −matrix(NA,sum(Sh0 + Eh0 + Ih0 + Ic0 + Rc0 + Rh0 + Se0 + Ie0),2)

  Counters0 < −matrix(0,length(DATA0[,1]),3)

  if(Sh0 > 0)
  for (i in 1:Sh0){
     DATA0[i,1] < −"h"
     DATA0[i,2] < −"S"
     }
  if(Eh0 > 0)
  for (i in (Sh0 + 1):(Sh0 + Eh0)){
     DATA0[i,1] < −"h"
     DATA0[i,2] < −"E"
     }
  if(Ih0 > 0)
```

```
for (i in (Sh0 + Eh0 + 1):(Sh0 + Eh0 + Ih0)){
    DATA0[i,1] < −"h"
    DATA0[i,2] < −"I"
    }
if(Ic0 > 0)
for (i in (Sh0 + Eh0 + Ih0 + 1):(Sh0 + Eh0 + Ih0 + Ic0)){
    DATA0[i,1] < −"c"
    DATA0[i,2] < −"I"
    }
if(Rc0 > 0)
for (i in (Sh0 + Eh0 + Ih0 + Ic0 + 1):(Sh0 + Eh0 + Ih0 + Ic0 + Rc0)){
    DATA0[i,1] < −"c"
    DATA0[i,2] < −"R"
    }
if(Rh0 > 0)
for (i in (Sh0 + Eh0 + Ih0 + Ic0 + Rc0 + 1):(Sh0 + Eh0 + Ih0 + Ic0 + Rc0 + Rh0)){
    DATA0[i,1] < −"h"
    DATA0[i,2] < −"R"
    }
if(Se0 > 0)
for (i in (Sh0 + Eh0 + Ih0 + Ic0 + Rc0 + Rh0 + 1):(Sh0 + Eh0 + Ih0 + Ic0 + Rc0
+ Rh0 + Se0)){
    DATA0[i,1] < −"e"
    DATA0[i,2] < −"S"
    }
if(Ie0 > 0)
for (i in (Sh0 + Eh0 + Ih0 + Ic0 + Rc0 + Rh0 + Se0 + 1):(Sh0 + Eh0 + Ih0 + Ic0
+ Rc0 + Rh0 + Se0 + Ie0)){
    DATA0[i,1] < −"e"
    DATA0[i,2] < −"I"
    }

Sh = length(DATA[,1][DATA[,1] == "h" & DATA[,2] == "S"])
Eh = length(DATA[,1][DATA[,1] == "h" & DATA[,2] == "E"])
Ih = length(DATA[,1][DATA[,1] == "h" & DATA[,2] == "I"])
Ic = length(DATA[,1][DATA[,1] == "c" & DATA[,2] == "I"])
Rc = length(DATA[,1][DATA[,1] == "c" & DATA[,2] == "R"])
Rh = length(DATA[,1][DATA[,1] == "h" & DATA[,2] == "R"])
Se = length(DATA[,1][DATA[,1] == "e" & DATA[,2] == "S"])
Ie = length(DATA[,1][DATA[,1] == "e" & DATA[,2] == "I"])

COUNTS < −c(Sh,Eh,Ih,Ic,Rc,Rh,Se,Ie)

DATA0 ≪ −DATA0
DATA ≪ −DATA0
Counters ≪ −Counters0
COUNTS ≪ −COUNTS
```

```
Counters0 ≪ −Counters0
Counters ≪ −Counters0
}
```

References

1. L. Allen, Some Discrete-Time SI, SIR, and SIS Models, *Math. Biosci.*, 1994, **124**, 83–105.
2. R. M. Anderson and R. M. May, *Infectious Diseases of Humans: Dynamics and Control*, Oxford University Press, Oxford, 1991.
3. W. O. Kermack and A. G. McKendrick, A Contribution to the Mathematical Theory of Epidemics, *Proc. R. Soc. London, Ser. A*, 1927, **115**, 700–721.
4. K. Dietz, Epidemics and rumors: A survey,*J. R. Stat. Soc., A*, 1967, **130**, 505–528.
5. M. J. Keeling and P. Rohani, *Modeling Infectious Diseases in Humans and Animals*, Princeton University Press, Princeton, New Jersey, 2008, p. 368.
6. J. H. Tien and J. D. Earn, Multiple Transmission Pathways and Disease Dynamics in a Waterborne Pathogen Model, *Bull. Math. Biol.*, 2009, **72**, 1506–1533.
7. C. J. Stone, *A Course in Probability and Statistics*, Wadsworth Publishing, Belmont, California, 1996, p. 837.
8. R Core Group, *R: A language and environment for statistical computing. R Foundation for Statistical Computing*, Vienna, Austria, 2014, URL http://www.R-project.org/.

Subject Index